MacRoberts on Scottish Construction Contracts

Third Edition

WILEY Blackwell

MacRoberts on Scottish Construction Contracts

Third Edition

MacRoberts
Solicitors

Foreword by
The Rt Hon Lord Hope of Craighead

WILEY Blackwell

This edition first published 2015

Registered office
John Wiley & Sons, Ltd, The Atrium, Southern Gate, Chichester, West Sussex, PO19 8SQ, United Kingdom.

Editorial offices:
9600 Garsington Road, Oxford, OX4 2DQ, United Kingdom.
The Atrium, Southern Gate, Chichester, West Sussex, PO19 8SQ, United Kingdom.

For details of our global editorial offices, for customer services and for information about how to apply for permission to reuse the copyright material in this book please see our website at www.wiley.com/wiley-blackwell.

Library of Congress Cataloging-in-Publication Data

MacRoberts on Scottish building contracts
MacRoberts on Scottish construction contracts / MacRoberts Solicitors;
foreword by The Rt Hon Lord Hope of Craighead, KT.–Third edition.
pages cm
Includes bibliographical references and index.
ISBN 978-1-118-27345-6 (hardback)
1. Construction contracts – Scotland. I. MacRoberts (Firm), author. II. Title.
KDC501.M33 2014
343.41107′8624 – dc23
2014016562

A catalogue record for this book is available from the British Library.

Wiley also publishes its books in a variety of electronic formats. Some content that appears in print may not be available in electronic books.

Cover images: iStockphoto © blackjake, iStockphoto © Martin McCarthy
Cover design by Jeffrey Goh

Typeset in 10/12pt MinionPro by Laserwords Private Limited, Chennai, India

MIX
Paper from responsible sources
FSC
www.fsc.org
FSC® C013604

To David Henderson – without whom this book would never have seen the light of day. We are forever in his debt.

Contents

Foreword

The Rt Hon Lord Hope of Craighead KT

A textbook of this kind, which seeks to serve the day-to-day needs of those who enter into and have to administer significant contractual relationships, must keep itself up to date if it is to do its work properly. This is no easy task, especially in a field such as that occupied by contracts entered into in the Scottish construction industry. There are many factors at work which promote changes in law and practice in this area. The standard forms are subject to constant revision to react to the demands of the marketplace. There are changes in the legislation, as it seeks to promote best practice in the industry, to react to the requirements of modern competition law and to combat the increasing menace of bribery. There is also a steady stream of case law, as the limits of existing rules and principles are constantly being tested to resolve the disputes that come before the courts. It was for this reason that I said in my Foreword to the first edition that I hoped that it might be possible for this book, like the forms, to be kept up to date by the issuing of revised editions at appropriate intervals. That indeed is what has happened and, as I welcome this third edition, I have not been disappointed.

As one would expect, the editors have been careful to take account of all these changes. But this has not just been a mechanical exercise. Several important editorial initiatives have been taken to keep pace with changes in practice. At the time of the first edition the standard forms that were most commonly used in the industry in Scotland were those published by the Scottish Building Contracts Committee (the SBCC). As was to be expected, the focus in that edition was on those standard forms. But the New Engineering (NEC3) forms of contract have grown in popularity throughout the United Kingdom, especially in the public sector. So it made sense for them to be given equal consideration in this new edition together with the SBCC forms. Also the focus in previous editions was on the With Quantities version of the Scottish Building Contract which was the version that was most frequently used at that time. This edition has broadened its outlook by including an analysis of important differences between that version and the Design and Build form of contract. This is to be welcomed, as design and build is being increasingly used as the preferred method of procurement in major projects. Chapter 1 also has an extended commentary on professional consultants, who have an increasing role in projects of that kind.

Chapter 17 on arbitration has been rewritten to provide a detailed commentary on the Arbitration (Scotland) Act 2010, and two new chapters have been added. One (Chapter 21) deals with the increasingly important matter of competition law in the context of the construction industry, to which the Office of Fair Trading is now paying close attention. It is plain from its enforcement activities that contractors and those engaged in the supply chain need to be aware of the rules and to conduct their own activities accordingly. The other (Chapter 22) deals with the far-reaching measures introduced by the Bribery Act 2010, which are of concern to everyone engaged in the construction industry both domestically and internationally. As the editors point out, the construction industry is one of the sectors which is likely to be the focus of investigating authorities because of the complex framework of contracts that it routinely uses and the cross-border and international nature of its activities. The penalties for those found guilty of offences are severe. This chapter is essential reading for those in positions of responsibility throughout the industry.

There is more than a hint of unfinished business. At the end of the chapter on procurement we are told that this area of the law is currently undergoing significant transformation as new Public Procurement Directives were adopted by the European Parliament in January 2014 and a Procurement Reform (Scotland) Bill, which will enable Scottish Ministers to make regulations as to the assessment of bidder suitability to tender for public contracts, has been introduced into the Scottish Parliament. The Courts Reform (Scotland) Bill, which will set out the framework for the implementation of the programme of reform of the Scottish Civil Justice system recommended in Lord Gill's Report, is currently passing through the Parliament also. So there are important changes on the way with which this edition cannot deal. Then there is the problem that has been created by differences in approach between the English and the Scottish Courts regarding the question whether a contractor is entitled to an extension of time where there are concurrent causes of delay. The editors express a clear preference for Lord Carloway's impressive dissenting judgment in the Inner House in the *City Inn* case, which is in harmony with the English approach. It would not be in the least surprising if the UK Supreme Court, by which the issue must surely now be resolved as soon as possible, were to agree with them.

The editors are, of course, entitled to a well-earned rest from their labours. Users of the book will no doubt hope that a fourth edition will be forthcoming before too long, so that it will continue to serve so well the needs of busy practitioners. For the time, however, they will be grateful for the practical approach that it takes to the many issues with which it deals, the broader coverage that it gives to the forms of contract and for the clear and accessible way that its guidance is presented.

David Hope

Preface

In the Preface to the second edition of this book published in 2007, we remarked upon the significant changes in the landscape of construction contracts since the first edition published eight years earlier. Predictably, exactly the same can be said about this third edition. It is, however, not only the law of construction contracts which continues to spin down Tennyson's ringing grooves of change, but also the practice of the users of such contracts. While the previous editions focussed on the SBCC With Quantities and Design and Build standard forms of building contract, the third edition gives equal prominence to the NEC3 Engineering and Construction Contract, an editorial decision made somewhat inevitable by the increasing use in Scotland of that contract. Perhaps not so inevitable was the decision to change the title of the book from *MacRoberts on Scottish Building Contracts* to *MacRoberts on Scottish Construction Contracts*, but we think that the latter reflects what has become the more commonly used terminology.

We are indebted to Lord Hope of Craighead for providing, as he did for the first edition, the Foreword. In that Foreword, he mentions a number of the other changes which have demanded substantial revisions to the relevant chapters, most notably to take account of the long-awaited (and welcome) Arbitration (Scotland) Act 2010 and the significant amendments to Part II of the Housing Grants, Construction and Regeneration Act 1996. Completely new chapters dealing with competition law and the Bribery Act illustrate the increasing impact on construction law of the expanding regulatory environment.

It is also interesting to note that while (as pointed out in the last edition) there remains a close relationship between the law of Scotland and that of England, the courts of the respective jurisdictions do not always reach the same result on the same point, as demonstrated by the schism between the 'English approach' and the 'Scottish approach' following the decision of the Scottish court in *City Inn v Shepherd Construction* (see Chapter 6).

In addition to those of the editors, individual contributions have been provided by our colleagues: David Arnott, Richard Barrie, Neil Kelly, Duncan Osler, Robin Fallas, Sarah Pengelly, Madeleine Young, David Wilson, Alison Horner, Julie Hamilton, Jennifer McKay, Ainsley MacLaren, Colette McGinley, John Reid, David Flint and Valerie Surgenor, and our now former colleagues Alexandra Lavery and

Gavin Thomson. Alan McAdams, yet again, has provided invaluable support to many, if not all, of the contributors, and Jennifer Burns, Kate Moffett and Magdalena Urbanowska played an invaluable role in preparing tables and checking drafts.

Finally, our thanks to Paul Sayer and all of his colleagues at Wiley for their patience and support.

We have endeavoured to state the law as at 1 July 2014.

David Henderson
Craig Turnbull
Shona Frame

MacRoberts LLP
60 York Street
Glasgow

Chapter 1
Construction Contracts in General

1.1 *Introduction*

Numerous books have been written on the subject of construction contracts. However, many of those are of a specialist nature and most are written from the perspective of English law. The aim of this book is to provide a practical guide to construction contracts governed by the law of Scotland.

There is no doubt that the construction industry not only in Scotland, but throughout the UK, is currently being presented with continuing challenges of an almost unprecedented nature. In the first quarter of 2013, the construction industry suffered a contraction of 2.5% compared with the same period in the previous year, despite a growth in overall Scottish gross domestic product (GDP) of 1.2%. While some major public sector infrastructure projects have lessened the impact of the economic downturn on the Scottish construction sector, the lack of external funding has been a major factor in the sharp reduction in the number of new private commercial development projects.

However, the very size of the construction industry means that it is of huge significance to the economy. According to figures contained in the report by Construction Scotland, *Building for the Future: The Scottish Construction Industry's Strategy, 2013–2016*, the construction industry generates £27.4 billion (GDP) to the Scottish economy every year, contributing 10% of Scotland's total economic output, and 170,000 people work in construction in Scotland, around 10% of the total Scottish jobs.

1.2 *Definition of a construction contract*

1.2.1 General

The definition of a construction contract is not straightforward. The construction industry encompasses building and engineering projects which differ enormously in nature, size and complexity. The terms 'building contract' and 'construction contract' are often used interchangeably. The term 'construction contract' was given a statutory meaning for the first time by section 104 of the Housing Grants, Construction

MacRoberts on Scottish Construction Contracts, Third Edition. MacRoberts.

and Regeneration Act 1996 ('the 1996 Act') and the breadth of the definition illustrates the wide-ranging nature of construction contracts. The 1996 Act has been amended by the Local Democracy, Economic Development and Construction Act 2009 in relation to contracts entered into in England and Wales from 1 October 2011 and in Scotland from 1 November 2011. One of the key amendments was the repeal of the requirement that a construction contract for the purposes of the 1996 Act must be in writing. The following sections summarize the principal requirements of the statutory definition of 'construction contract'. However, it is important to bear in mind that this statutory definition is relevant only for the purposes of the 1996 Act and for determining whether the provisions of Part II the 1996 Act, such as adjudication and payment requirements, will apply to a particular contract. The fact that a contract is not a 'construction contract' for the purposes of the 1996 Act does not mean that it may not be a construction contract in the commonly understood sense of the term, and indeed some contracts falling with the statutory definition, such as for architectural services, would not be described as a construction contract in everyday terms. An English court has recently held that a collateral warranty may, depending on its terms, be a construction contract for the purposes of the 1996 Act and thus subject to the statutory adjudication provisions, see *Parkwood Leisure Ltd* v. *Laing O'Rourke Wales and West Ltd* (2013) and the discussion of this case in Section 13.4. The 1996 Act has no relevance in determining whether a contract exists in the first place since this will be determined by a matter of law and evidence; putting it at its simplest, whether there was sufficient consensus as to the essential terms and sufficient proof of such consensus.

It should also be noted that the Scottish Government's legislative programme for 2013–2014 includes the Conclusion of Contracts, etc. Bill, which is intended to allow contracts to be concluded by email and to provide a simpler process for formal execution (signing) of contracts.

1.2.2 Construction contracts under the 1996 Act

Part II of the 1996 Act applies to 'construction contracts', being agreements in relation to 'construction operations'. These terms are defined respectively by sections 104 and 105.

Section 104 provides as follows:

1. *In this Part a 'construction contract' means an agreement with a person for any of the following –*
   (a) *the carrying out of construction operations;*
   (b) *arranging for the carrying out of construction operations by others, whether under sub-contract to him or otherwise;*
   (c) *providing his own labour, or the labour of others, for the carrying out of construction operations.*
2. *References in this Part to a construction contract include an agreement –*
   (a) *to do architectural, design or surveying work, or*

(b) *to provide advice on building, engineering, interior or exterior decoration or on the laying-out of landscape, in relation to construction operations.*

3. *References in this Part to a construction contract do not include a contract of employment (within the meaning of the Employment Rights Act 1996).*

4. *The Secretary of State may by order add to, amend or repeal any of the provisions of subsection (1), (2) or (3) as to the agreements which are construction contracts for the purposes of this Part or are to be taken or not to be taken as included in references to such contracts ...*

It will be noted that the 1996 Act applies to matters beyond the carrying out of building works. It applies to architectural, design and surveying works and to advising on building, engineering, interior or exterior decoration or on the laying-out of landscape in relation to construction operations.

The definition of 'construction operations' is central to Part II. This term is defined by section 105(1) which provides as follows:

1. *In this Part 'construction operations' means, subject as follows, operations of any of the following descriptions –*

   (a) *construction, alteration, repair, maintenance, extension, demolition or dismantling of buildings, or structures forming, or to form, part of the land (whether permanent or not);*
   (b) *construction, alteration, repair, maintenance, extension, demolition or dismantling of any works forming, or to form, part of the land, including (without prejudice to the foregoing) walls, roadworks, power-lines, electronic communications apparatus, aircraft runways, docks and harbours, railways, inland waterways, pipe-lines, reservoirs, water-mains, wells, sewers, industrial plant and installations for purposes of land drainage, coast protection or defence;*
   (c) *installation in any building or structure of fittings forming part of the land, including (without prejudice to the foregoing) systems of heating, lighting, air-conditioning, ventilation, power supply, drainage, sanitation, water supply or fire protection, or security or communications systems;*
   (d) *external or internal cleaning of buildings and structures, so far as carried out in the course of their construction, alteration, repair, extension or restoration;*
   (e) *operations which form an integral part of, or are preparatory to, or are for rendering complete, such operations as are previously described in this subsection, including site clearance, earthmoving, excavation, tunnelling and boring, laying of foundations, erection, maintenance or dismantling of scaffolding, site restoration, landscaping and the provision of roadways and other access works;*
   (f) *painting or decorating the internal or external surfaces of any building or structure.*

Section 105(2) details a number of operations that are not construction operations for the purposes of Part II. The exceptions relate to oil and gas and mining, both underground and opencast; certain specified operations on a site where the primary purpose is nuclear processing, power generation, water or effluent treatment or the production, transmission, processing or bulk storage (other than warehousing) of chemicals, pharmaceuticals, oil, gas, steel or food and drink; the manufacture or delivery to site of components or equipment where the contract does not also provide for their installation; and the making, installation and repair of wholly artistic works. The scope of excluded operations insofar as applicable to works at power stations was considered in the case of *North Midland Construction plc* v. *AE & E Lentjes UK Ltd* (2009).

1.2.3 Excluded contracts

Section 106 provides that contracts with residential occupiers are excluded from the operation of Part II of the 1996 Act, as is any other description of construction contract excluded by order of the Secretary of State (or in Scotland, the Scottish Ministers).

A construction contract with a residential occupier is one which principally relates to operations on a dwelling which one of the parties to the contract occupies, or intends to occupy, as his residence.

The term 'dwelling' means a dwelling-house or a flat; and for s.106(2) 'dwelling-house' does not include a building containing a flat and 'flat' means separate and self-contained premises constructed or adapted for use for residential purposes and forming part of a building from some other part of which the premises are divided horizontally. In itself, section 106 is self-explanatory. However, it should be noted that a residential occupier cannot be a limited company for the purposes of this section, see *Absolute Rentals Ltd* v. *Gencor Enterprises Ltd* (2001).

The Construction Contracts (Scotland) Exclusion Order 1998 ('the 1998 Order') came into force with Part II of the 1996 Act on 1 May 1998. This excluded from the scope of Part II of the 1996 Act project agreements under the Private Finance Initiative (PFI) (provided certain criteria were met) but not sub-contracts such as the construction contract and facilities management or operation and maintenance contract. One of the amendments to the 1996 Act referred to in section 1.2.1 above was the introduction of section 110(1A) which prohibits payment provisions in construction contracts which are conditional on the performance of obligations under another contract or a decision by any person as to whether obligations under another contract have been performed. To avoid this prohibition applying to first-tier sub-contracts under PFI projects and thus prohibiting what are commonly known as 'equivalent project relief' clauses in such sub-contracts, the Construction Contracts (Scotland) Exclusion Order 2011 excludes from the operation of section 110(1A) a construction contract if it is a contract pursuant to which a party to a relevant contract (i.e. excluded by the 1998 Order) has subcontracted obligations under that contract to carry out construction operations.

Certain development agreements are also excluded from the operation of Part II under article 6 of the 1998 Order. A contract is a development agreement if it includes provision for the grant or disposal of a relevant interest in the land on which take place the principal construction operations to which the contract relates.

A relevant interest in land means either ownership or a tenant's interest under a lease for a period which is to expire not earlier than 12 months after the completion of the construction operations under the contract.

In *Captiva Estates Ltd v. Rybarn Ltd (In Administration)* (2006), Captiva entered into a contract with Rybarn to construct 28 flats. Captiva owned the land on which the development was to take place. The contract provided that, as consideration for the works, Captiva would pay Rybarn and would also grant to Rybarn an option to purchase leases in respect of 7 of the 28 flats. The question arose as to whether the contract was a development agreement within the meaning of the English equivalent to the 1998 Order. The court held that the definition of a 'development agreement' in the 1998 Order is wide and the contract was caught by it.

1.2.4 Agreements in writing

Section 107 of the 1996 Act as originally enacted provided that Part II only applied to agreements in writing or evidenced in writing or recorded by one of the parties or a third party who has been duly authorized to do so. However, this was repealed with effect from 1 October 2011 in England and 1 November 2011 in Scotland, which means that oral contracts may now be subject to Part II of the 1996 Act. This does not, however, avoid the practical difficulty of proving the terms, or even the existence, of an oral contract in the first place.

1.3 Parties involved in a construction project

The parties involved in a building project can vary considerably depending on the nature and complexity of the project. At one end of the scale, a private individual may engage a joiner, electrician or builder to carry out work to his home. In such an instance, the employment of anyone other than the tradesman or builder may not be necessary. At the other end of the scale, major projects, such as the construction of public buildings, motorways, hotels or power stations, can involve a considerable number of parties from different professional and non-professional disciplines. It is therefore crucial to identify, particularly in a large project, the parties involved in that project, the terms of their respective appointments, the scope of each individual's involvement, and their roles within the project. The following parties are commonly involved in building projects:

1.3.1 Employer and consultants

The term 'employer' is used throughout this book as meaning the party for whose benefit the building works are being carried out. This is the term generally used in the standard form building contracts and associated documentation. Other terms such as 'the owner', 'the client' or 'the authority' are also sometimes used. The employer usually assembles the advisory team, though there is no obligation to do so (see obligations of employers in Chapter 4) and smaller projects often do not require the involvement of anyone other than the employer and the contractor. The nature of the team varies depending upon the nature of the project and the choice of procurement method. In the case of a large project, the team may consist of an architect, civil and structural engineer, mechanical and electrical engineer, quantity surveyor, construction design and management (CDM) co-ordinator, one or more specialist consultants depending on the nature of the project, such as environmental consultant, acoustic consultant, etc., and possibly a project manager, a clerk of works and a BIM information manager (see Section 1.7). The team will also vary depending on the procurement method chosen, e.g. design and build, and may also be transferred at a later stage, at least in part, to the contractor in the case of novation (see Section 12.7). The terms of appointment of each member of the team are very important and must ensure that each member's obligations are clearly defined. Where this is not done, difficulties can arise with unnecessary overlap of work or, more importantly, in crucial issues failing to be addressed by any of the members of the team due to lack of clarity as to allocation of responsibility. An example of the type of problem that can arise is to be found in the case of *Chesham Properties Ltd* v. *Bucknall Austin Project Management Services Ltd and Others* (1997).

To avoid this kind of situation arising, an employer may appoint a single entity as an integrated multi-discipline professional team. This can take the form of a single multi-discipline consultancy practice taking overall responsibility for a number of consultancy services and carrying these out in-house, or a single discipline practice, normally the architect, engaging directly with the employer for a range of services and then sub-contracting these to other firms, such as civil and structural engineers and mechanical and electrical engineers. This provides the employer with 'one-stop' responsibility on the part of the architect, since in such circumstances (unless the appointment expressly states otherwise), the architect will be responsible for the performance of the services, including any design, by those to whom he has delegated such performance under a sub-contract. In contrast, the architect will have no responsibility (subject again to the terms of the appointment) for the work of other consultants whom the employer has appointed directly.

It may not always be the architect who assumes this single point of responsibility, as it is not uncommon for a project manager to enter into an appointment with the employer, which includes not only project management and CDM co-ordinator services, but also the full range of design services and, in turn, to sub-contract the design services to others. This is indeed the model used for the project manager-led integrated design team services which may be procured under the Government Procurement Services 'Buying Solutions' framework (see Section 1.4.9). In a design and build scenario, the design sub-consultants' appointments may in due course be novated to a contractor, with the project manager's appointment remaining with the employer.

In relation to novation of design appointments, see Section 12.7. It should also be borne in mind that though the most common practice in design and build contracts is for the employer's design team to be novated to the contractor, it is not unusual for the contractor to form its own design team by engaging sub-consultants directly. In practice, the more advanced the design at the time the contractor is appointed, the greater the likelihood of novation being appropriate.

The roles of the most commonly used consultants in construction projects are briefly described below, including the forms of consultant appointment produced by the relevant professional body. However, in many cases (and in almost all cases where the appointment is subject to the public procurement regime), the form of appointment will be selected by the employer. That is often a bespoke form of appointment rather than the professional body's standard form, though the use of the NEC3 Professional Services Contract has gained considerable popularity.

1.3.2 Architect

In a traditional building contract, it is the architect who usually has overall responsibility for the project from its conception to its conclusion. An architect is the agent of their client (in most cases, the employer, but under a design and build contract following novation, the contractor) and the general law of Scotland in relation to agency applies to their actions. The scope of their actual authority depends upon the terms of the agreement, comprising the appointment by their client.

Chartered architects in Scotland may be members of the Royal Incorporation of Architects in Scotland (RIAS) and/or the Royal Institute of British Architects (RIBA). The RIAS is a charitable organization founded in 1916 as the professional body for all chartered architects in Scotland. It is independent of the RIBA though it consults with the RIBA on UK-wide professional issues. The RIAS produces a suite of five Appointment documents, most recently revised in November 2011: SCA 2000; Sub Consultant Form of Appointment; ASP 2005 (Small Projects); Design and Build Appointment DBE/2000 (where the client is the employer); and Design and Build Appointment DBC/2000 (where the client is the contractor). The RIBA also publishes its own suite of contract documents for the appointment of architects, consultants and sub-consultants, the most recent edition of which is known as the RIBA Agreement 2010 (2012 Revision).

In a conventional building project, the architect will normally work in conjunction with the structural engineer, the former being engaged to produce at the initial stages the plans and elevations and an outline design, whereas the latter will be responsible for the design of the sub-structure and load-bearing elements of the building and for producing structural calculations and drawings. In May 2013, the RIBA launched a fairly radical change to its 'Plan of Work' which set out a recognized model for the building design and construction process, split into a number of stages, identified as A–L. The RIBA Plan of Work 2013 (endorsed by the RIAS) replaces these stages A–L with a new list of eight stages (along with eight 'task bars'). The stated intention is to align these stages with the unified industry stages agreed with the Construction Industry Council (CIC). The wider aim of the new Plan of Work is to cover the various procurement routes, to emphasize the project team as a whole, including client, contractors and designers, and to integrate building information modelling (BIM) into the work process.

The eight stages contained in the RIBA Plan of Work 2013 are:

Stage 0 – Strategic Definition
Stage 1 – Preparation and Brief
Stage 2 – Concept Design
Stage 3 – Developed Design
Stage 4 – Technical Design
Stage 5 – Construction
Stage 6 – Handover and Close Out
Stage 7 – In Use.

The architect may also, if so appointed by the employer, act as certifier under the building contract and will normally fulfil that role as contractor administrator under the SBC (but not the SBC/DB). In exercising the role of certifier, the architect will have a duty of care to the employer (see below) but will not normally owe any duty of care to the contractor (*Pacific Associates* v. *Baxter* (1980)). For a full description of the role and responsibilities of the certifier, see Section 7.6.

The architect's duty to the employer in the performance of his obligations in relation to both design activities and other services under the appointment, as with any other professional consultant or adviser, is to use reasonable skill and care. Where the professional holds himself out as having a special skill, such as in the case of an architect, the standard of care is that of the ordinary skilled man exercising and professing to have such a skill (*Bolam* v. *Friern Hospital Management Committee* (1957)). In many cases this duty of care will be expressly set out in the appointment, but in the absence of such an express term, it will nonetheless be implied. That same standard will apply to the architect's duty of care to the employer where he is carrying out the function of certifier under the building contract on behalf of the employer. On the other hand, a stricter duty than that of exercising reasonable skill and care may, by agreement, be imposed on the architect by an express term to that effect. This will generally be resisted by the architect as his professional indemnity insurance will normally cover only acts of 'negligence', i.e. breaches of the common law duty of reasonable skill and care and not a breach of a higher contractual duty. It is also possible that a stricter 'fitness for purpose' duty may be implied as matter of fact on the appointment by the particular circumstances and what is demonstrated to be the common intention of the parties (*Greaves & Co (Contractors) Ltd* v. *Baynham Meikle & Partners* (1975)).

1.3.3 Quantity surveyor

A quantity surveyor may be engaged by the employer to discharge specific functions. These tend to be of a financial nature and can include, for example, acting as cost consultant in preparing cost estimates, preparing bills of quantities, valuing work done for the purposes of both interim and final certificates, ascertaining direct loss and expense under the provisions of the building contract, and preparing the final account. Like architects, chartered surveyors are members of a professional organization, in this case the Royal Institution of Chartered Surveyors (RICS), which was founded in 1861.

It is now a world-wide organization, with RICS Scotland being one of the four UK regional divisions. The RICS produces a suite of documents for consultancy appointments. This consists of the core appointment (either the standard form of consultant's appointment or the short form of consultant's appointment, in each case with a separate version for use in Scotland) along with a standard schedule of services relevant to the appropriate discipline, i.e. building surveyor services, CDM co-ordinator services, employer's agent services, project manager services, project monitor services, and quantity surveyor services.

The various schedules of service mentioned above illustrate that the role of the chartered surveyor in a project may not be limited to the traditional, albeit still essential, role of quantity surveyor, but has expanded into other areas such as project management.

The traditional principal role of the quantity surveyor is to value the work carried out by the contractor, and not to inspect that work for quality (*Sutcliffe* v. *Chippendale & Edmondson* (1982)). This was clarified in the case of *Dhamija* v. *Sunningdale Joineries and Others* (2010) in which Coulson J held that a quantity surveyor is concerned with quantities, not the quality of the work, and that it was for the architect to advise the quantity surveyor, and not for the quantity surveyor to see for himself, any defective work which should be excluded from the valuation.

1.3.4 Engineer

Historically, contracts for the construction of infrastructure such as roads, tunnels, railways or bridges, where the design process is led by the civil engineer rather than the architect, have generally been known as engineering contracts (as opposed to contracts for 'buildings'). In a traditional engineering contract, the engineer normally undertakes design responsibility and carries out a similar administrative role to that of the architect under a building contract, such as the certification of payments and of completion. The relevant UK professional organization for civil engineers is the Institution of Civil Engineers (ICE). There are also business associations for the consultancy and engineering industry such as the Association for Consultancy and Engineering (ACE), which in turn is represented on the European Federation of Engineering Consultancy Associations (EFCA) and the International Federation of Consulting Engineers, or Fédération Internationale des Ingénieurs-Conseils (FIDIC).

The Civil Engineering Contractors Association (CECA) is a representative association for the UK civil engineering contractors. The CECA in Scotland represents over 100 civil engineering contractors, ranging in size from small rural contractors to multinationals.

While civil engineers are concerned primarily with structures and the physical environment, mechanical engineers (almost invariably conjoined with electrical engineers and often referred to in that combined role as building services consultants) are responsible for the mechanical and electrical systems in a building. The professional institution in the UK for mechanical engineers is the Institution of Mechanical Engineers (IMechE), while the equivalent body for electrical engineers is the Institution of Engineering and Technology (IET) which was formed in 2006 by the merger of the

Institution of Electrical Engineers (IEE) and Institution of Incorporated Engineers (IIE). The Chartered Association of Building Services Engineers (CIBSE) is the business organization that promotes building services engineers. This was formerly the Institution of Heating and Ventilating Engineers and was granted its Royal Charter in 1976.

In a typical building contract (as opposed to an engineering contract), the specialist consultant engineering input will be provided to the employer by the structural engineer and the mechanical and electrical (M&E) (or building services) engineer. As mentioned in Section 1.3.2, the structural engineer's role is normally the design of the sub-structure and load-bearing elements. The M&E consultant will be responsible for specifying the required mechanical and electrical installations and the outputs and performance standards, while in more complex projects the detailed M&E design will be the responsibility of specialist M&E sub-contractors engaged directly by the main contractor.

A recent case has considered the extent to which a structural engineer was entitled to rely on advice obtained from a specialist subcontractor in producing a soil stabilization performance specification for the purposes of construction of a new supermarket. In the event, the floor slab suffered differential settlement as a consequence of the ground under the floor slab not being improved enough. The employer, the supermarket owner, contended that the engineers could not avoid liability by arguing that they were entitled to rely on the advice of the specialist subcontractors. However, the court held that it was not a question of whether the engineers had delegated their duty to the subcontractors but whether they had acted with reasonable skill and care in relying on advice from a specialist subcontractor instead of carrying out their own assessment. The court held that a construction professional could discharge its duty to take reasonable care by relying on the advice or design of a specialist provided that it acts reasonably in doing so. In the circumstances of the case the engineers had acted reasonably in doing so, and would not have been in breach of their duty of care even if even the advice of the specialist had been negligent (*Cooperative Group Ltd* v. *John Allen Associates Ltd* (2010)).

1.3.5 Specialist consultants

In large building projects, employers often employ specialist consultants to advise on specific areas, for example, planning consultant, environment and sustainability consultant, fire consultant, landscape architect, heating and ventilation consultant, lift consultant, interior and space consultants, transport consultant. In most cases a CDM co-ordinator will require to be appointed. See Section 1.7 in relation to a BIM Information Manager and Section 1.8 on GSL Lead or Champion in respect of 'Soft Landings'. The CIOB Complex Projects Contract 2013 referred to in Section 1.4.8 introduces a new role of 'Project Time Manager' to review progress, advise the employer's team on programming and work with the contractor to identify measures to recover delay. Additionally, the trend in recent years for building owners to demand that completed projects achieve a specified BREEAM rating has resulted in the emergence of the BREEAM consultant as a specialist. BREEAM (i.e.

the BRE Environmental Assessment Method) is a means of assessing the overall environmental performance of new and existing buildings. In order to achieve one of the ratings of Pass, Good, Very Good, and Excellent, a minimum number of points must be achieved.

1.3.6 Project manager

Depending upon the size of the building contract and the method of procurement, the employer may decide to engage a project manager. Although in the past a project manager in the construction context tended to be found only in a construction management contract (see Section 1.4.1), the role is nowadays found in all types of construction procurement and normally covers the management and coordination of the tender process and the works, including advising on procurement strategy and risks, coordinating the preparation of tender documents and assessment of tenders, producing contract documents, programming and monitoring of progress. Ultimately, however, the exact scope and extent of the duties will be determined by the terms of the appointment. The role of the Project Manager acting as contract administrator is recognized in NEC 3 and a project manager is often appointed by the employer to undertake the role of employer's agent in SBC/DB. The extent of the project manager's responsibilities to their client has been considered in a number of recent cases. In *Sweett (UK) Limited* v. *Michael Wight Homes Limited* (2012), the employer's agent's appointment imposed an express duty on it to 'prepare contract documentation and arrange for such documents to be executed'. The contractor was contractually obliged to provide a performance bond but went into liquidation after commencing work without having done so. The court rejected the employer's contention that the consultant was under an absolute obligation to arrange execution of the performance bond by the contractor and held that their duty was limited to using reasonable care to ensure it was provided. In the circumstances the consultant had fulfilled that duty. In contrast, in *The Trustees of Ampleforth Abbey Trust* v. *Turner & Townsend Project Management Limited* (2012), the project manager was held to have breached their duty to use reasonable care to ensure execution of the building contract by the contractor. The project manager had instead issued a series of letters of intent and had failed to warn the employer of the limited protection afforded by such letters as compared to a formal contract. As a consequence of the absence of a formal contract, the employer was unable to apply liquidated damages for delay against the contractor, and the court held the project manager liable to the employer for that loss.

1.3.7 Clerk of works

A clerk of works is generally employed on site by either the employer or the architect to act as construction inspector, oversee the execution of the works, and monitor compliance with the required contract standards. There is a professional body for clerks of works known as the Institute of Clerk of Works and Construction Inspectorate of Great Britain Inc (ICWCI). Clause 3.4 of SBC expressly permits the Employer to appoint a clerk of works whose duty is to act solely as inspector on behalf of the Employer under the directions of the Architect/Contract Administrator. The clerk of works cannot give any direction to the Contractor unless it is in regard to a matter in respect of which the Architect/Contract Administrator is expressly empowered by the Contract to issue instructions and the direction must be confirmed in writing by the Architect/Contract Administrator within two working days of being given. The corresponding role is carried out under NEC3 by the 'Supervisor' whose duties include the carrying out of tests and inspections, instructing the Contractor to search for defects, notifying the Contractor of defects, and issuing the defects certificate.

1.3.8 Contractor

Once an employer has decided upon the nature and extent of the work which they wish carried out (possibly with the assistance of the architect, quantity surveyor and/or project manager), they will usually invite one or more contractors to tender for the work. Where the employer is a contracting authority or a utility for the purposes of public procurement law (see Chapter 2), the mandatory rules in respect of competitive tendering will of course apply. The obligations of contractors under a building contract are considered in Chapter 5.

1.3.9 Sub-contractors

Often, and almost invariably in major projects, elements of the work are executed not by the main contractor itself but by sub-contractors or even sub-sub-contractors. These may include 'domestic sub-contractors' (chosen by the main contractor normally with the employer's consent); specialist contractors pre-named by the employer (which is not the same as nominated) or 'works contractors' (under a management contract). In the past these could also have included nominated sub-contractors.

The position of sub-contractors under a building contract is considered in Chapter 11.

In addition to the principal parties referred to above, other parties may have a role in a building project such as suppliers, insurers, funders, prospective tenants and purchasers of the building.

1.3.10 Experts

Some specialist consultants may be considered experts in their field. The use of expert witnesses in complex construction disputes is commonplace. An expert should act objectively and independently to avoid being viewed as no more than a hired gun. The expert witness or, to give him his proper title under Scots law, the 'skilled witness', is through practice or study, or both, specially qualified in a recognized branch of knowledge. A number of professional consultancy bodies, e.g. RIAS maintain a list of accredited expert witnesses.

The duties and responsibilities of experts in civil cases in England were set out in detail by Mr Justice Cresswell in *National Justice Compañía Naviera SA* v. *Prudential Assurance Co. Ltd ('The Ikarian Reefer') (No. 1)* (1993). These include the following:

- Expert evidence presented to the court should be, and should be seen to be, the independent product of the expert uninfluenced as to form or content by the exigencies of litigation, see *Whitehouse* v. *Jordan* (1981).
- An expert witness should provide independent assistance to the court by way of objective, unbiased opinion in relation to matters within his expertise, see *Polivitte Ltd* v. *Commercial Union Assurance Co. Plc* (1987) and *Re J (Child Abuse: Expert Guidance)* (1991). An expert witness should never assume the role of an advocate.
- An expert witness should state the facts or assumption upon which his opinion is based. He should not omit to consider material facts which could detract from his concluded opinion, see *Re J* (1991).
- An expert witness should make it clear when a particular question or issue falls outside his expertise.
- If an expert's opinion is not properly researched because he considers that insufficient data is available, this must be stated with an indication that the opinion is no more than a provisional one, see *Re J* (1991). In cases where an expert witness who has prepared a report could not assert that the report contained the truth, the whole truth and nothing but the truth without some qualification, that qualification should be stated in the report, see *Derby & Co. Ltd and Others* v. *Weldon and Others (No. 9)* (1990).
- If, after exchange of reports, an expert witness changes his view on a material matter having read the other side's expert's report or for any other reason, such change of view should be communicated (through legal representatives) to the other side without delay and when appropriate to the court.
- Where expert evidence refers to photographs, plans, calculations, analyses, measurements, survey reports or other similar documents, these must be provided to the opposite party at the same time as the exchange of reports.

These principles were subsequently approved by the Court of Appeal in *Stanton* v. *Callaghan* (2000).

In *Anglo Group plc* v. *Winther Browne & Co Ltd* (2000) it was said that the *Ikarian Reefer* analysis described above needed to be extended in accordance with the Woolf reforms of civil procedure in England. It set out the following analysis:

- An expert witness should at all stages in the procedure, on the basis of the evidence as he understands it, provide independent assistance to the court and the parties by way of objective unbiased opinion in relation to matters within his expertise. This applies as much to the initial meetings of experts as to evidence at trial. An expert witness should never assume the role of an advocate .
- The expert's evidence should normally be confined to technical matters on which the court will be assisted by receiving an explanation, or to evidence of common professional practice. The expert witness should not give evidence or opinions as

to what the expert himself would have done in similar circumstances or otherwise seek to usurp the role of the judge.

- He should co-operate with the expert of the other party or parties in attempting to narrow the technical issues in dispute at the earliest possible stage of the procedure and to eliminate or place in context any peripheral issues. He should co-operate with the other expert(s) in attending without prejudice meetings as necessary and in seeking to find areas of agreement and to define precisely areas of disagreement to be set out in the joint statement of experts ordered by the court.
- The expert evidence presented to the court should be, and be seen to be, the independent product of the expert uninfluenced as to form or content by the exigencies of the litigation.
- An expert witness should state the facts or assumptions upon which his opinion is based. He should not omit to consider material facts which could detract from his concluded opinion.
- An expert witness should make it clear when a particular question or issue falls outside his expertise.
- Where an expert is of the opinion that his conclusions are based on inadequate factual information, he should say so explicitly.
- An expert should be ready to reconsider his opinion, and if appropriate, to change his mind when he has received new information or has considered the opinion of the other expert. He should do so at the earliest opportunity.

In England, guidelines in respect of expert witnesses are now embodied in Part 35 of the Civil Procedure Rules 1998 and the relative Practice Direction. The 'Civil Justice Council Protocol for the Instruction of Experts to give Evidence in Civil Claims' contains guidance on best practice with regard to compliance with Part 35 of the Civil Procedure Rules and with the overriding objective that courts deal with cases justly.

In Scotland, there are presently no court rules in relation to the conduct of expert witnesses. The principles set out in the *National Justice Compañía Naviera SA* case have been referred to by the Scottish courts, with approval, in *Elf Caledonia Ltd* v. *London Bridge Engineering Ltd and Others* (1997) (one of the cases arising from the Piper Alpha disaster). Lord Caplan described the formulation of an expert's duties in *National Justice Compañía Naviera SA* as being helpful and correct. Certain observations, consistent with the principles set out in the *National Justice Compañía Naviera SA* case, were also made in *McTear* v. *Imperial Tobacco Ltd* (2005).

Certain professional bodies, for example, the RICS in Scotland, produce guidance for their members when acting as expert witnesses. The RICS Practice Statement 'Surveyors Acting as Expert Witnesses' applies to RICS members providing expert evidence to a tribunal. It follows similar principles to the court rules and procedures:

- The overriding duty is to the tribunal.
- That duty overrides any duty to the client.
- It involves setting out facts fully and giving truthful, impartial and independent opinions covering all relevant matters and whether or not they favour the client.

Up until recently it was thought that an expert was immune from any claim for negligence on the part of his client. However, this immunity (at least in certain cases) no longer applies following the decision of the Supreme Court in the English appeal of *Jones* v. *Kaney* (2011).

Some guidance as to the duty of an expert in a construction dispute litigation was recently given by the court in the English case of *Walter Lilly & Company Ltd* v. *Mackay and DMW Developments* (2012). The court in particular considered the correct approach to expert evidence on demonstrating delay and the meaning of practical completion. In addition, it was not for an expert to stray into legal matters, for instance, basing a report simply on the premise that the other side had not proved its case. See also *National Museums and Galleries on Merseyside (Trustees of)* v. *AEW Architects and Designers Limited* (2013), which reinforces the importance of an expert being truly independent.

1.4 Types of construction contract

Before issuing an invitation to tender or selecting a contractor, the employer will need to consider the appropriate type of construction contract and form of contract to be adopted. The contract documents are generally prepared on the advice of the project manager, architect or quantity surveyor, or in the case of complex procurement, by legal advisers. Where the building contract is being competitively tendered, the form of contract should be included in the invitation to tender issued to tenderers.

1.4.1 Procurement routes

There is a variety of different types of contractual arrangement which can be chosen. In smaller projects the contract may comprise simply a quotation by the contractor which is accepted, with or without qualification, by the employer. The quotation may have standard terms and conditions attached to it. The parties may negotiate over the incorporation of all, or part of, the standard terms and conditions into their contract. On the other hand, the perception of those within the construction industry is probably that the contracts regulating larger projects have become more and more complex over the years. The procurement route and type of contract chosen by the employer will to a large extent depend on the employer's key priorities, and the circumstances of the proposed project, such as speed, certainty of cost, and status of design.

Under traditional procurement, responsibility for the design process remains separate from the construction process. The contractor usually has no design responsibility and its obligations are limited to the execution of the works and the provision of materials to a design provided by the employer, unless it is contracted to design a specific part of the works, known in the JCT/SBC contracts as a 'Contractor's Designed Portion'. The contract is usually administered by the employer's professional team.

Nowadays, the most common type of procurement route in major projects is probably that of design and build, in which the contractor undertakes both the design and the construction of the works in return for a lump sum price. The contractor may appoint his own design team though often their appointments will initially have been made with the employer and then novated to the contractor.

A third common type of procurement route is that of 'management' which can be one of, or a combination of, management contracting, construction management and 'design and manage'. Under management contracting, the overall design of the works is the responsibility of the employer's design team. The employer appoints a management contractor who is responsible for managing the carrying out of the works by works contractors appointed by the management contractor under a number of works contracts for the various packages comprised in the works. The management contractor manages the overall process. Normally the management contracting route is used to allow a contract to be let where design is at an early stage so that design and construction can proceed in parallel along with the procurement of the works packages as and when appropriate in accordance with the programme. On the other hand, this usually means a loss of cost certainty for the employer, as even though the works packages will be procured competitively, the final prices will become known only after contract commencement. Under construction management, the employer appoints a design team and a construction manager. Unlike a management contractor, a construction manager does not appoint the contractors who actually carry out the works; instead those contractors (known as trade contractors) are appointed by the employer. The management of the construction process is performed by the construction manager on the employer's behalf. Under 'design and manage' procurement, the management contractor is not only responsible for managing the works packages but also for the design team.

1.4.2 Lump sum contracts

A lump sum contract is a contract in which the contractor agrees with the employer to carry out the building works for a pre-agreed price. The price is subject to adjustment only in certain limited circumstances specified in the contract, such as variations, employer default and other events (if any) which are expressly stated in the contract as being an employer risk, expenditure of provisional sums and (if applicable) inflation-related fluctuations in costs. The characteristics of a lump sum contract can apply to both design and build and traditional contracts. A contract using bills of quantities will be a lump sum contract if the bills are fully measured at the time the contract is entered into.

1.4.3 Measurement contracts

In this type of contract the sum payable by the employer to the contractor is determined by measuring the work done on completion of the project and by applying quantities to agreed rates or some other form of valuation. An example is a contract based on bills of approximate quantities, where the quantities cannot be accurately measured in advance of the contract being entered into.

1.4.4 Cost reimbursement and prime cost contracts

There are different types of such contracts such as cost contracts, cost plus contracts and prime cost contracts, but the common feature is that the sum which the employer pays the contractor is not a pre-agreed sum but a sum calculated by reference to the actual cost of the works carried out, generally with the addition of an amount to cover profit and a management fee. This may be a predetermined percentage of the costs, a predetermined fixed fee or a variable fee calculated according to a predetermined formula. An example of this type of contract is the NEC3 Option E (see Section 1.6). It is also common to combine the principles of a cost reimbursable contract with a 'target price' contract (e.g. the NEC3 Options C and D) under which costs will be reimbursable up to the level of a target figure If the final actual costs are less than the target, the contractor is entitled to a share of the savings calculated under a pre-agreed pain/gain share mechanism, whereas if the final actual costs exceed the target, that same pain/gain share mechanism will require the contractor to bear a share (which may be 100% if the contract so provides) of the excess.

1.4.5 Turnkey contracts

This term is sometimes used for certain types of design and build contracts, usually where certainty of final price and completion date are of particular importance, for example, where such certainty is demanded by private finance providers. Under this type of contract, the contractor undertakes a more onerous risk profile than under a standard design and build contract (though that does not mean that the contractor assumes all risk; risks such as force majeure will normally at least be shared). This type of contract is also commonly called an EPC contract (engineer, procure and construct). In the case of a process plant, power plant, or wind turbine project or the like, the contract will normally require the completion of commissioning and the achievement of performance tests to demonstrate compliance with the stipulated performance criteria in order to allow the operational phase to commence, backed up by performance liquidated damages in the event of failure. The European International Contractors (EIC) published the EIC Turnkey Contract in May 1994. FIDIC has also published turnkey contracts, beginning in 1995 with its 'Orange Book' entitled *Conditions of Contract for Design-Build and Turnkey*. This was replaced in 1999 by the *Conditions of Contract for Plant and Design-Build* (the 'Yellow Book') and the *Conditions of Contract for EPC/Turnkey Projects* (the 'Silver Book').

1.4.6 Two-stage tendering

This method of procurement is becoming more prevalent and its key characteristic is to involve the contractor at an early stage before completion of the design and before fully priced tenders have been obtained. The intention is to allow the contractor to collaborate with the employer and their consultants in the design and procurement process. This procurement method is normally only used with design and build contracts. The first stage tender is based on an outline design by the employer's professional team and the competitive element relates to the amount of preliminaries, the overhead and profit percentage and the pre-contract fee. Normally, the preferred bidder will enter into a pre-contract agreement to include the development of design and procurement of sub-contract packages. The SBCC has recently published two forms of pre-construction agreement. The Pre-Construction Services Agreement (General Contractor) for use in Scotland (PCSA/Scot 2013) is for a two-stage tendering process where the proposed main contract is SBC (Scottish Building Contract), SBC/DB, SBC Minor Works or JCT (Joint Contracts Tribunal) Major Projects. The Pre-Construction Services Agreement (Specialist) for use in Scotland (PCSA/SP/Scot 2013) is for a two-stage tendering process on a substantial or complex project where the party entering into the pre-construction agreement with the employer is a specialist who will in due course enter into a sub-contract with the main contractor, once appointed, and whose tender to the employer may subsequently be assigned to the main contractor.

Ideally the tenderers under a two-stage procedure should be committed to the amount of the tendered preliminaries, etc. and the work packages are then procured competitively and transparently so that there is no scope for negotiation. The risk for the employer is that the outstanding matters cannot be agreed during the second stage with the result that the process must be commenced from scratch. The two-stage tendering process may be combined with a form of guaranteed maximum price (GMP), where the contractor bids the GMP and takes the risk of the total amount of work packages subsequently exceeding that GMP. While two-stage tendering is sometimes used in public sector contracts, care needs to be taken to follow the EU public procurement rules, particularly if the restricted procedure applies, as the contracting authority cannot allow the second stage to develop into a negotiation. This would clearly infringe the requirements of the restricted procedure. Procurement is considered in Chapter 2.

It is possible that the growing popularity of two-stage tendering in time-critical major projects will increase in cases where BIM is used, in order to engage the contractor at an early stage in the development of the BIM model, see Section 1.7.

1.4.7 Joint ventures

This type of arrangement is now a very common method of procurement. A special purpose company or partnership is created by two or more parties (often a land owner and a developer), each contributing their respective assets, funds and/or skills with a view to procuring a construction project. The joint venture company/partnership will become the employer for the purposes of the building contract.

1.4.8 Partnering and alliancing

In the last twenty years or so there has been a perceived desire within the construction industry to move away from a confrontational and adversarial culture to a collaboration culture, with the objective of creating common goals between parties to the project and an understanding of each party's expectations and values. This is largely a product of two ground-breaking reports of the 1990s, *Constructing the Team* by Sir Michael Latham in 1993, and *Rethinking Construction* by a committee chaired by Sir John Egan in 1998. This has in turn led to a movement towards partnering contracts. Generally speaking, partnering aims to foster a sense of commitment to a project, to emphasize mutual goals and objectives, and to promote equity, trust, co-operation and fair dealing. The supposed consequential benefits of partnering include savings in time and cost, improved quality and fewer defects, and reduced risk of disputes. It is probably fair to say that partnering was initially regarded with some scepticism due to the arguably vague targets often included in partnering 'charters' and the somewhat aspirational wording used in these charters, which in most cases were non-binding. However, since the introduction in September 2000 of the ACA Standard Form of Contract for Project Partnering (PPC 2000), which was the first standard form of project partnering contract, there has been a growing tendency for partnering concepts to be incorporated as part of the contract itself rather than as a procedural overlay (as in the case of a partnering charter). A Scottish supplement has been published to accompany PPC 2000 and PPC 2000 was itself revised in 2003 and 2008, with a further amendment in 2011 to address the changes to the 1996 Act. The key features of PPC 2000 are as follows:

- PPC 2000 is a multi-party contract; not only do the client and the contractor enter into the contract, but also the client's representative and any consultants appointed by the client, and possibly certain specialists (who are sub-contractors appointed by the contractor).
- PPC 2000 contains various processes covering the period prior to construction on site, and assumes the selection of the contractor on the basis of quality rather than a lump sum price. Indeed, the parties are obliged to work together to arrive at an agreed maximum price, rather than a lump sum price being fixed at the very outset.
- PPC 2000 provides for the supply chain to be finalized, so far as possible, on an open book basis, encouraging partnering relationships with the specialist appointed by the contractor. Those specialists may themselves become full members of the partnering team, in which case they execute a joining agreement.
- The contractor is obliged to submit a business case to the client in respect of those parts of the work that it wishes to undertake directly by package or by the appointment of a specialist.
- PPC 2000 provides for a core group to be established, comprising key individuals representing partnering team members, who undertake regular previews of progress and performance and make decisions on certain matters.

Other standard form partnering contracts include:

- TPC 2005 (published by ACA and similar to PPC 2000, but for use with term contracts);
- Public Sector Partnering Contract (PSPC);
- NEC3 Secondary Option Clause X12: Partnering;
- JCT Constructing Excellence Contract 2011 and SBCC Constructing Excellence Contract 2006;

- CIOB Complex Projects Contract 2013. This new form published by the Chartered Institute of Building is intended for use in complex projects and imposes collaboration obligations on the parties in a more prescriptive manner than other forms.

In addition, the 2011 edition of the JCT and SBC contracts introduced a new optional supplementary condition as follows:

> The Parties shall work with each other and with other project team members in a co-operative and collaborative manner, in good faith and in a spirit of trust and respect. To that end, each shall support collaborative behaviour and address behaviour which is not collaborative.

It could of course be argued that the very fact that this is only an optional clause indicates that a collaborative culture in the construction industry has not yet been fully achieved, and indeed some might suggest that the period since 2008 has seen a slow-down in the trend towards partnering along with a desire by employers to exploit the increased competiveness in the market and the pressure on tender prices.

That said, a number of major employers have in recent years adopted a form of partnering contract, known as 'alliancing'. This form of procurement was originally adopted in the early 1990s in the UK oil and gas industry and was subsequently developed in major infrastructure projects in Australia, such as water projects, highways and dams, with the aim of avoiding the traditional client/contractor confrontational approach. The essence of an alliance contract is that the parties work together collaboratively in one integrated team and are bound by a risk/reward scheme which provides for a collective sharing of project risks and consequential savings or losses. This is in fact a step further than a partnering contract, since while the former focuses on cultural change in contractual relationships, the latter (at least in its true sense) aligns the parties' commercial and financial interests to the assumption of collective responsibility for the successful achievement of project goals.

A current UK example of alliancing is the Electricity Alliance operated by the National Grid, under which the gain share is calculated at the end of each year and divided among each regional partner based upon evidence of collaborative working, sharing best practice and driving efficiencies.

What is said to be the first UK rail sector example of a 'pure construction alliance' is the Stafford Area Improvement Programme on the West Coast Main Line, where a unified agreement was formed in early 2013 between Network Rail and the contractors and consultants, under which all parties share the benefits and risks under a 'one project, one organization' structure.

1.4.9 Framework contracts

Although framework contracts have been used for a number of years, their popularity, particularly in the public sector, has grown significantly since the late 2000s.

This is possibly linked to the removal of any lingering doubts as to the compliance of such contracts with public procurement rules by the express recognition of framework contracts in the EU Public Sector Procurement Directive (Directive 2004/18/EC, 31 March 2004) and in the Public Contracts (Scotland) Regulations 2006 (now consolidated under the Public Contracts (Scotland) Regulations 2012). Under a framework contract, the employer will procure either a single framework contractor or a panel of contractors (which, if subject to public procurement rules, must be a minimum of three). Individual contracts are then 'called off' under the terms governing the framework and if the framework consists of a panel of contractors, this is usually done by means of a 'mini-competition'. In cases where the employer is subject to the public procurement or utilities regime, the appointments to the framework are subject to the procurement regulations (and recent cases emphasize the need for care in compliance, see Section 2.16). However, the attraction to a public sector employer of a framework contract is that the mini-competitions are not subject to the full public procurement regime and procedures, such as publishing a notice in the *Official Journal of the European Union* (OJEU) (though certain criteria set out in the relevant regulations must still be met and the overriding principles of non-discrimination and equality of treatment must be followed). A further attraction is that the framework can be set up by a 'central purchasing body' for the benefit of a number of different employers or categories of employers described in the OJEU notice, who can then make individual call-offs under that framework.

Current examples of frameworks include the wide-ranging 'Buying Solutions' set up by the Government Procurement Service (which extends throughout the UK); ProCure21+ set up by the Department of Health for NHS projects in England and Wales and the corresponding NHS Frameworks Scotland set up by NHS Scotland in 2008 and re-tendered in 2013 with five 'principal supply chain partners' for construction-related services for both new-build and refurbishment. Construction frameworks have also been formed in Scotland and the rest of the UK by a number of local authorities and housing associations.

The publishers of the standard forms have recognized the growth in the use of frameworks and standard framework agreements are included in both the NEC3 and JCT/SBC suites.

1.4.10 Term contracts

As the name suggests, a term contract subsists for a specific period, and is usually for the provision of services during that period. However, these services can often be construction-related, most commonly for the carrying out of both planned and reactive repairs and maintenance, for example, on behalf of local authorities and housing associations. One of the most commonly used forms of term contract for this purpose is TPC 2005, published by ACA. Standard form term contracts are also comprised in the JCT/SBCC and the NEC3 suites of contract (the SBCC Measured Term Contract 2011 (and its JCT equivalent); the NEC3 Term Service Contract and the NEC3 Term Service Short Contract).

1.4.11 Other forms

This book focuses on the SBCC and the NEC3 forms of contract (see Sections 1.5 and 1.6). However, a number of other bodies within the construction and engineering industries produce their own standard forms of contracts and associated documentation. For example, the Association for Consultancy and Engineering (ACE) and the Civil Engineering Contractors Association (CECA) operate the suite of Infrastructure Conditions of Contract (ICC) for use in civil engineering-related contracts. This includes, among others, a Measurement version, a Design and Construct version, a Term version, a Target Cost version, and a Minor Works version (and it should be noted that these will require to be adapted for use in Scotland). These are based on, and effectively replace, the ICE Conditions of Contract, which were first published by the Institution of Civil Engineers in 1945. However the Institution withdrew its co-sponsorship of the ICE Conditions in 2011 as part of its decision to solely endorse the NEC3 suite of contracts. The 'GC/Works' family of contracts is now published by the Stationery Office and remains in gradually diminishing use in the public sector, including a recent major new prison project in Scotland. The longevity of these contracts is perhaps surprising given the lack of up-to-date revisions. GCWorks/1 comprises quantities and without quantities (1998) and two-stage design and build (1999) versions.

The Institution of Chemical Engineers (IChemE) publishes forms of contract primarily for use in the design and construction of process plants. Since such plants are usually performance-based, the passing of performance tests as a pre-condition of completion is a key element of these contracts. The suite of contracts comprises both UK and international versions and the most widely used form in the UK is the 'Red Book' for lump sum contracts, the fifth edition of which was published in 2013. Other main contract forms are the 'Green Book' for cost reimbursable contracts and the 'Burgundy Book' for target cost contracts. IChemE also publishes international versions of each of these contracts.

The Institute of Engineering and Technology (IET) and the Institution of Mechanical Engineers (IMechE) jointly publish a suite of contracts, known as Model Forms, in relation to electrical, electronic or mechanical plant. MF/1 Revision 5 (2010) is for use in home (i.e. the UK) or overseas contracts for the supply and erection of electrical, electronic or mechanical plant. MF/2 (1999) is for home or overseas contracts for the supply only of such plant; MF/3 (2001) is for home contracts for the supply only of mechanical and electrical goods where no initial design or subsequent installation or commissioning is required; and MF/4 is for home or overseas contracts for the engagement of a consulting engineer.

The FIDIC forms of contract (see Section 1.4.5) are also being used in major domestic projects, the most noteworthy being the adoption of the Silver Book as the basis for the fixed price lump sum design and build bespoke contract for the Forth Replacement Crossing (now known as the Queensferry Crossing) entered into in May 2011. In addition to the Yellow Book and Silver Book mentioned in Section 1.4.5, FIDIC also publishes the Red Book – the Construction Contract (Conditions of Contract for Building and Engineering Works, Designed by the Employer) published in 1999. This is essentially a re-measurement contract.

A more detailed examination of the forms of contract mentioned above is beyond the scope of this book but any of these forms may well be appropriate, depending upon the nature of the project in question. In any event, before using any standard form, users should ensure that it meets the employer's needs and is properly integrated with the other documents forming the contract.

1.4.12 PFI and PPP

Although the Private Finance Initiative (PFI) and Public Private Partnerships (PPP) (and the Scottish Government's variants, the Non-Profit Distributing model (NPD) and the 'hub' model) are beyond the scope of this book, it must be recognized that there has been a rapid growth in this method of procurement of public sector building and infrastructure works since the mid-1990s. The construction contracts forming part of the package of project documents are normally bespoke contracts, the form and terms of which are largely dictated by a pass-through of the obligations under the overarching project agreement. The funding and risk transfer features of such projects, coupled with the construction/operational interface, have led to the development of practices and principles peculiar to the construction contracts used in these projects. As a result, such contracts need to be regarded as a quite distinct category of construction contract.

1.5 The SBCC forms of building contract

In Scotland, many building contracts are entered into on the Scottish Building Contract Committee (SBCC) standard forms. The constituent bodies of the SBCC are currently the Association of Consultancy and Engineering, the Association of Scottish Chambers of Commerce, the Convention of Scottish Local Authorities, the National Specialist Contractors Council – Scottish Committee, the Royal Incorporation of Architects in Scotland, the Royal Institution of Chartered Surveyors in Scotland, Scottish Building, Scottish Casec, the Scottish Government – Building Division, and the Law Society of Scotland.

The SBCC has produced standard forms since 1964. Over the years, the number of standard forms has increased and numerous revisions and amendments have been issued. These can make it difficult to identify the precise terms upon which parties have contracted. Matters are often further complicated by the attempts of employers and contractors to modify the provisions of the standard form contracts.

The first forms of contract published by JCT were the 1963 Editions and a major revision to these (JCT 80), was carried out in 1980 (when a Design and Build form was also published for the first time). This was followed by numerous Amendments. The form was then reprinted as a 1998 edition (JCT 98), again followed by various Amendments. 2005 saw the launch of new editions of practically every JCT contract. This was the most comprehensive revision of the whole suite of JCT contracts for many years.

The overhaul of the JCT contracts in 2005 was in response to calls for change following market research. The JCT found that the industry expressed a preference for integrated documents for use rather than core documents with a series of supplements. There was uncertainty about which form to use and which supplement was appropriate to each form.

The JCT's aim was to present contracts in a user-friendly way. It achieves that by producing stand-alone contracts without the use of supplements. Each contract contains information about the circumstances in which it is suitable for use and many have their own Guides containing additional explanatory information.

The structure of the contracts has been substantially overhauled with clauses being grouped into sections such as 'Payment' and 'Control of the Works', being renumbered, and with the wording substantially revised in line with an aim to use plainer, non-legalistic language.

Many clauses have been shortened and simplified, sometimes by defining terms which tend to require long explanations (such as 'Interest' and 'Insolvency') and sometimes by incorporation by reference of statutory provisions (such as the CDM Regulations and VAT legislation) or procedural rules (such as the adjudication provisions in the Scheme for Construction Contracts and arbitration rules).

Articles and Contract Particulars are all located at the front of the contract so that all project-specific sections which require to be filled in are grouped together. A number of default provisions are contained so that if the particulars are not properly completed, this does not leave a gap but the default situation is automatically applied.

Certain provisions have been deleted – Nominated Sub-Contractors and Suppliers, Performance Specified Work, Contractors' Price Statement and Insurance for Employer's Loss of Liquidated Damages no longer appear in the standard form. There are optional provisions to be chosen as required, such as sectional completion and Contractor's Designed Portion. These are provided for within the wording of the standard form without the need for separate supplements to be read into the main contract form. The 2005 Edition of the various contracts and subsequent revisions have been consolidated into the 2011 Edition, which also takes into account changes required as a consequence of the amendments made to the 1996 Act by the Local Democracy, Economic Development and Construction Act 2009 with effect in England and Wales from 1 October 2011 and in Scotland from 1 November 2011.

The SBCC have produced, currently as 2011 Editions, Scottish versions of the majority of the JCT forms. The practice of publishing these as Scottish Supplements to be read into the JCT form ceased with the 2005 Edition when they were produced as stand-alone contract documents for the first time. This was a welcome user-friendly development and the differences between the JCT and SBCC versions are limited to those required to bring the JCT contracts in line with Scots law, terminology and procedure. All section and clause numbering is common between equivalent JCT and SBCC contract forms.

The principal areas of difference between JCT and SBCC are the third party rights and the arbitration provisions. Third party rights relate to the ability to confer benefits on a person who is not party to the contract, and would typically be used in a construction context where collateral warranties would otherwise be required. In England and Wales, this is governed by the Contracts (Rights of Third Parties) Act 1999. This

Act does not apply in Scotland. In Scotland, the equivalent is the *jus quaesitum tertio*, a common law right. The contract contains provisions to make an election as to whether warranties are to be provided or whether this will be dealt with by way of third party rights. For more detail on third party rights generally, see Chapter 13.

In relation to arbitration, the difference between JCT and SBCC relates to the procedural rules incorporated.

There is a wide variety of SBCC standard forms of contract available and the Standard Building Contract Guide for use in Scotland (November 2011) identifies the documents published by the SBCC.

New contracts have been added to the suite of standard forms for both the JCT and SBCC, including Minor Works with Contractor's Design 2013 and Framework Agreement. In addition, revised Sub-Contract Conditions (Sub/C/Scot 2011, Sub/D/C/Scot 2011 and DBSub/C/Scot 2011) have been published, respectively covering the situations where there is no design element in the sub-contract, where the main contractor is to design parts of the main contract works and the sub-contractor is to design all or part of the subcontract works, and for use with the design and build main contract, whether or not the sub-contract works include design by the sub-contractor.

The SBCC published in May 2012 a Named Specialist Update which contained optional provisions for incorporation into the SBC to enable the Employer, by means of an appropriate entry in the Contract Particulars to name individual specialists as domestic sub-contractors for identified parts of the Works.

In 2013, the SBCC published Project Bank Account Documentation for use in Scotland (PBA/Scot 2013) where the parties wish to adopt the use of a project bank account for payment purposes (see Section 8.7).

In the same year it published two versions of a Homeowner Contract – HOB/Scot 2013 for use by a homeowner/occupier who has not appointed a consultant to oversee the work, and HOC/Scot 2013 where a consultant has been appointed. See also Section 1.4.6 in relation to the new forms of Pre-Construction Services Agreement published by the SBCC.

In this book, references to clauses are (unless the text expressly specifies otherwise) to those in the Standard Building Contract With Quantities for use in Scotland, SBC/Q/Scot (2011 Edition). This contract is referred to in this book as 'the SBC'. In view of the predominant role of design and build as a method of procurement, this book also focuses on the relevant provisions of the Design and Build Contract for use in Scotland, DB/Scot (2011 edition), which is referred to for convenience as 'the SBC/DB'. In the sections which consider the relevant provisions of the SBC and the SBC/DB, we use the defined terms set out in those contracts.

1.6 The NEC3 forms of contract

The NEC forms of contract were first published in 1993 as the 'New Engineering Contract' followed by a second edition, the NEC2 in 1995, and by the NEC3 in 2005. Updated editions of the NEC3 suite, along with a new Professional Services Short Contract, were published in April 2013. The NEC family of documents has become increasingly popular and in particular is widely used by the public sector throughout the UK. The NEC3 suite of contracts now comprises the Engineering and Construction Contract; the Engineering and Construction Short Contract; the Engineering and Construction Subcontract; the Engineering and Construction Short Sub-Contract; the Professional Services Contract; the Professional Services Short Contract; the Term Service Contract; the Term Service Short Contract; the Supply Contract; the Supply Short Contract; the Framework Contract; and the Adjudicator's Contract. Of these, this book will focus only on the Engineering and Construction Contract. The term 'the NEC3' will be used in this book (except where otherwise stated) to refer to the NEC3 Engineering and Construction Contract, April 2013 edition.

NEC3 can be used for a wide range of different procurement routes and comprises nine Core clauses, with six Main Option clauses, two Dispute Resolution clauses and 18 Secondary Options clauses. The main options are as follows:

- Option A: Priced Contract with Activity Schedule
- Option B: Priced Contract with Bills of Quantities
- Option C: Target Contract with Activity Schedule
- Option D: Target Contract with Bills of Quantities
- Option E: Cost Reimbursable Contract
- Option F: Management Contract.

The Employer must choose one of these main options.

One of the Dispute Resolution Options W1 or W2 must be chosen, depending on whether or not the 1996 Act applies.

The Employer may then choose some or none of the Secondary Options numbered X1 to X20 (X8–X11 and X19 are not used).

These secondary Options range from those which are commonly used (e.g. X7 (Delay Damages); X13 (Performance Bond)) to the more esoteric (e.g. X3 (Multiple Currencies)), while others will have particular attraction to the Contractor (e.g. X15 (Limitation of the Contractor's liability for his design to reasonable skill and care); X18 (Limitation of liability)).

Option Y(UK)2 should be used where the 1996 Act applies since it incorporates Act-compliant payment provisions. The 2013 edition of the NEC3 reflects the changes to the 1996 Act.

Option Y(UK)3 should be chosen if the intention is to allow a third party to enforce a term of the contract under the Contracts (Rights of Third Parties) Act 2009 (though this would need express amendment for use in Scotland since the common law principle of *jus quaesitum tertio* applies in Scotland and not the 2009 Act).

The most significant changes to the NEC3 Engineering and Construction Contract introduced by the April 2013 edition were Project Bank Account provisions as new Option Y(UK)1 (see also Section 8.7) and also additional references to the CIC Building Information Modelling Protocol (see Section 1.7). There is also an expanded range of guidance booklets.

A striking feature of the NEC3 is its use of the present tense rather than the more familiar imperative mood. Clause 11 also makes a distinction between 'identified' and 'defined' terms. Terms which are identified in the Contract Data are used in italics throughout the contract and these are primarily contract-specific. Terms which are defined in clause 11.2 have initial capitals.

Another feature of the NEC3 is the provision for 'Z clauses' which allow for bespoke additional conditions of contract to be added. Since there is no equivalent NEC3 contract for use in Scotland, it is often necessary to use Z clauses to import changes necessary to reflect Scots law (e.g. third party rights, as mentioned above).

Another feature of the NEC3 is the incorporation into the contract of the Contract Data which sets out project-specific information, provided by the Employer in Part 1 and by the Contractor in Part 2.

A key document is the Works Information which is produced by the Employer and which needs to be drafted with care to ensure that all necessary information is included. For example, there is a single NEC3 form of Engineering and Construction Contract, with no separate design and build version. Clause 21.1 states that 'the Contractor designs the parts of the Works which the Works Information states he is to design'. Thus any design responsibilities to be imposed on the Contractor *must* be set out in the Works Information. This one example highlights the importance of ensuring the completeness of the Works Information.

The NEC has also published a comprehensive set of guidance notes and flow charts for the various forms of contract within the NEC3 suite.

1.7 Building Information Modelling (BIM)

At the time of writing, the potential impact on contractual relationships of Building Information Modelling (BIM) is still not fully known, but there is no doubt that the construction industry is moving towards support for the use of BIM. The UK Government Construction Strategy, published in May 2011, confirmed the commitment to using BIM, with the intention that by 2016 central government projects above a certain level will be executed using BIM Level 2, with the ultimate goal of moving to BIM Level 3. A similar approach was taken in the Scottish Government's 'Review of Scottish Public Sector Procurement in Construction', published in October 2013, which recommended the use of BIM for central Scottish Government projects with the objective that construction projects across the public sector adopt a BIM Level 2 approach by April 2017.

The annual BIM construction industry-wide survey carried out by the NBS in the three months to February 2013 found that, of 1350 professionals participating, 39% were using BIM, compared with a corresponding figure in 2010 of 13%, and that 71% agreed that BIM represented the 'future of project information' (though the fact that 74% agreed that 'the industry is not yet clear enough on what BIM is yet' might suggest that much educational work still needs to be done).

The most commonly used definition of BIM is that given by the Construction Project Information Committee (CPIC): 'a digital representation of physical and functional characteristics of a facility creating a shared knowledge resource for information about it forming a reliable basis for decisions during its life cycle, from earliest conception to demolition'. In broad terms, BIM Level 2 is a process involving a managed 3D environment achieving integration by proprietary interfaces and 4D program data, while BIM Level 3 involves full integration of data managed by a single collaborative model server. Each member of the design team and also design sub-contractors contribute to the project model.

It is envisaged that a BIM Manager (also referred to in some contracts as the BIM Information Manager) will be appointed with responsibility for management of the model. They may be the lead designer or this may evolve as a separate discipline. Their role will include responsibility for user access to the model and coordinating and integrating the individual designs into the model.

This leads to a number of potential legal issues, such as intellectual property rights in and ownership of the integrated model, insurance, and also the responsibilities and liabilities of each designer who contributes to the model, as well as that of the BIM Manager for such contributions (for example, for not identifying clashes), particularly bearing in mind that there is likely to be a need to use the model for many years after construction completion.

A key document for any project using BIM is the BIM Protocol, which in most cases will be a contract document, for both professional appointments and the building contract, and will set out the roles and responsibilities of the design team in relation to the creation and updating of the model; access rights to the model; and the role of the BIM Manager.

It may well be that concerns as to the uncertainty of legal responsibilities for the model are exaggerated and that it is no more complicated than the designer being responsible for his own discipline-specific contribution in accordance with the usual principles of traditional design. Even so, it is possible that the growth in the use of BIM will see a corresponding increased use in major projects of integrated design teams led by a lead designer with a number of sub-consultants, so as to avoid the employer being exposed to any gaps in responsibility.

The industry standard form contracts have addressed BIM in different ways. JCT published a Public Sector Supplement in September 2011 which they then updated in December 2011. In Scotland, the SBCC published a Public Sector Supplement for use in Scotland, in November 2011. In relation to BIM, the Supplements include amendments to the main JCT and SBC and SBC/DB forms which provide for the inclusion of 'any agreed Building Information Modelling protocol' as a contract document, thus imposing a duty on the Contractor to comply with the protocol. The April 2013 edition of the NEC3 includes references to the CIC Building Information Protocol. The NEC has also published a new Guide: *How to Use BIM with NEC3 Contracts*, which sets out some practical steps on using BIM, dealing with the contractual and technical matters that arise.

These recent amendments made to the JCT/SBCC and the NEC3 forms of contract to address BIM are of a fairly 'high level' nature, marked by the absence of substantial amendments to the contract clauses. The NEC3 Guide on BIM referred to above includes guidance on using the BIM Protocol and provides suggested additional clauses to use with the NEC3, the intention being that these be framed as additional clauses under Option Z, rather than as part of a new stand-alone Secondary Option. These clauses would include additional compensation events, i.e. where a party is unable to provide its contribution to the model due to events outside its control, and where the Employer is obliged to revoke any sub-licence that may have been provided to use information provided by others. The Guide also emphasizes the importance of including the BIM Protocol in the Works Information.

The BIM Protocol intended for use with the NEC3 is the standard form protocol published in March 2013 by the Construction Industry Council (CIC), which is for UK contracts using Level 2 BIM. At the same time the CIC also published two other BIM-related documents, namely, Best Practice Guide for Professional Indemnity Insurance when using BIM; and Outline Scope of Services for the Role of Information Management. The CIC BIM Protocol sets out the contractual basis for the use of BIM and the relevant obligations of the project team members, including the production of models, and the employer's obligation to appoint a BIM Manager (termed the 'BIM Information Manager'). It is intended that the BIM Protocol be part of each of the relevant contract documents, i.e. the building contract and the professional appointments.

Although, unlike the NEC3, the JCT and SBCC BIM-related amendments contained in the Public Sector Supplement do not provide for a specific form of BIM Protocol, there seems no reason why the CIC BIM Protocol could not be used with that Supplement, and indeed it does appear to be accepted that incorporation into the building contract of a standard form BIM Protocol can avoid the need for substantial amendments to the contract, at least where Level 2 is used, other than to dovetail with the Protocol requirements.

Other recent developments signifying the industry move towards BIM are the publication of the PAS (Publicly Available Standard) 1192-2:2013, which is a specification for information management for the capital/delivery phase of construction projects using BIM (with full upgrade to a British Standard expected before 2015), the imminent publication by BIS of BS 7000-4 (Design Management Systems), a guide to managing design in construction which will align with PAS 1192-2:2013 and will take into account the development of BIM and the release by BSI of PAS 1192-3 'Specification for Information Management for the Operational Phase of Construction Projects Using Building Information Modelling' for public consultation which closed early in December 2013. This is a partner document to PAS 1192-2 focussing on the operational phase of assets. See also Section 1.3.2 in relation to the new RIBA Plan of Work 2013, which specifically takes BIM into account.

1.8 *Soft Landings*

Soft Landings is a concept that seeks to address issues arising subsequent to the handover of a building to its users, and can be seen as part of the move towards more sustainable buildings. In June 2009, an industry task group convened by the Building Services Research and Information Association (BSRIA) produced the Soft Landings Framework for better briefing, design, handover and building performance in-use (BSRIA BG 4/2009).

BSRIA's report, *Introducing Soft Landings* describes Soft Landings as:

> Soft Landings means designers and constructors staying involved with buildings beyond practical completion to assist the client during the first month of operation and beyond to help fine-tune and de-bug the systems, and ensure the occupiers understand how to control and best use buildings.

The purpose of Soft Landings is described in the Framework as being 'to smooth the transition into use and to address problems that post-occupancy evaluations show to be widespread'.

Soft Landings subsists throughout the whole life of a construction project from design brief and feasibility through construction and commissioning into the period immediately after handover and finally into the early years of operation of the building. The Framework splits these into stages as follows:

Stage 1 Inception and briefing
Stage 2 Design development
Stage 3 Pre handover
Stage 4 Aftercare
Stage 5 Years 1 – 3 extended aftercare.

For each of these stages, the Framework sets out a Checklist and Supporting Notes with details of key matters to be considered.

At the same time, the UK Government identified the need to align design and construction with operational asset management, and for that purpose Government Soft Landings (GSL) was developed. The Government Construction Strategy of May 2011 identified that integration of the design and construction of an asset with the operation phase should lead to improved asset performance, while the Government Soft Landings Policy of September 2012 recommended that the policy should apply to all new central government projects and major refurbishments and should be implemented by central government departments during 2013, working towards a mandate in alignment with BIM in 2016.

The Government Construction Strategy One Year On Report and Action Plan Update of July 2012 reports that ten trial projects have been introduced to test GSL. A draft policy document on GSL to reflect findings from the trial projects will be produced and proposals will then be developed for roll-out of the strategy.

The responsibility for GSL from October 2012 onwards moved to the BIM Task Group in order to ensure that BIM and GSL are in alignment and to allow work towards the combined mandate in 2016.

It is clear that, for Soft Landings to operate as intended, the end users, operators, contractors and designers must all be involved from a very early stage in the project. The BIM Task Group suggests that parties involved engage a GSL Lead or Champion.

Contractual documentation to take into account the needs of Soft Landings is being prepared by the BIM Task Group, recognizing that the post-handover requirements of GSL require an involvement beyond that conventionally provided for within standard construction contracts and professional appointments and that a contractual arrangement to engage these parties for the post-completion period will be necessary.

The BIM Overlay to the RIBA Outline Plan of Work (see Section 1.3.2) contains various references to Soft Landings-related activities, which indicates some movement already towards inclusion of this within architects' work scopes.

The BSRIA Core Principles suggest a light touch approach to the contractual documentation in respect of Soft Landings with a commitment to use the Soft Landings process as an overall aim. The philosophy is that Soft Landings will work best where parties work together in a collaborative manner, sharing risks and rewards; and there is a concern that if contractual obligations and responsibilities are too tightly specified, gaps can occur.

There is clearly also a close interaction between Soft Landings and BIM, as illustrated by the integration of GSL within the BIM Task Group's remit. Therefore, contractual arrangements for Soft Landings need to be considered along with those required for BIM and a common and integrated approach taken to achieve the aims.

Chapter 2
Procurement

2.1 Introduction

This chapter describes an important element of law relating to construction contracting, namely the requirements in relation to contracts which require to be awarded, in accordance with the Public Contracts (Scotland) Regulations 2012 as amended (in this chapter, referred to as 'the Regulations'), by public bodies which are subject to those Regulations. The Public Contracts (Scotland) Regulations 2012 came into force on the 1st of May 2012 and update Regulation 23, taking into account offences under the Criminal Justice and Licensing (Scotland) Act 2010, the Bribery Act 2010, the Proceeds of Crime Act 2002 and the Drug Trafficking Act 1994; clarify that the award criteria under Regulation 30(2) are indicative; and reduce the time limit in Regulation 47 for the bringing of Court proceedings from 3 months to 30 days.

This chapter also briefly introduces the European Union's public procurement regime, and explains some features which are significant in relation to construction contracting.

The Regulations implement the public contracts Directive 2004/18/EC in Scotland ('the Directive') which is known as the Classic Directive. The Directive modernized and updated the procedures in relation to the awarding of contracts in the public sector originally established by Directives 92/50/EEC, 93/36/EEC and 93/37/EEC. There are corresponding Regulations for England, Wales and Northern Ireland. This chapter refers to the provisions of the Regulations, unless otherwise indicated, and it should be noted that there are a number of differences between the Regulations, and The Public Contracts Regulations 2006 (as amended), notably in relation to remedies.

Procurement by entities operating in the water, energy, transport and postal services sectors has, from 31 January 2006, been subject to Directive 2004/17/EC, which is now implemented into Scots law by the Utilities Contracts (Scotland) Regulations 2012. These Regulations apply to the award of relevant contracts entered into by utilities in relation to the activity for which they are a specified utility.

The public procurement legislation referred to above essentially comprises procedural rules founded with economic purpose, to ensure that a level playing field across the EU is put in place for those competing for public sector contracts.

MacRoberts on Scottish Construction Contracts, Third Edition. MacRoberts.

2.2 *The EU public procurement regime and EU economic and legal principles*

The founding economic objectives of the European Union include providing for the free movement of labour, capital, goods and services throughout EU member states in a free internal market. Accordingly, barriers to trade such as restrictions on the use of foreign products, quota systems and subsidies to domestic industry, are contrary to EU (and national) law. Various Articles of the Treaty on the Functioning of the European Union (TFEU) are relevant to public procurement law, but (i) Article 34 TFEU prohibits quantitative restrictions on imports and all measures having equivalent effect between member states (the free movement of goods); (ii) Article 49 TFEU prohibits restrictions on freedom to provide services within the Community in respect of nationals of member states (the right of establishment); and (iii) Article 18 TFEU prohibits discrimination on grounds of nationality. In addition, and depending on how it is conducted, the award by government bodies of contracts to third parties (public procurement) may also act as a barrier to trade by hindering equal market access and fair competition between all EU undertakings to such contracting opportunities.

In response, EU legislation has been introduced to co-ordinate and converge the public procurement procedures in EU member states. The primary objectives of these rules are economic, to create circumstances where economic operators may compete for public contracts on a level playing field in economic terms. Corresponding economic principles predict the benefits of open competition; they dictate that the number of market participants competing for such contracts should thereby increase, so creating competitive tension during public procurement competitions and causing bidders to reduce prices and increase the quality of their proposals.

The framework of legislation governing public procurement which has been enacted to achieve these economic aims is quite complex. In addition, the case law of the European Court of Justice has drawn on the fundamental rules of the EC Treaty in establishing legally binding principles relevant to public procurement, the most important of which are prescribed by regulation 4 of the Regulations, which requires (when those Regulations apply) that a contracting authority (i) must not treat a person who is not a national of a relevant State and established in a relevant State more favourably than one who is; (ii) must treat economic operators equally and without discrimination; and (iii) must act in a transparent and proportionate manner. Each of these principles merits some further comment, as follows.

2.2.1 Transparency

Contracting authorities must ensure that information on procurement opportunities and on relevant rules, policy and practice is made available to all interested parties, notably potential works contractors, suppliers and service providers (known as 'economic operators'). These interested parties have extensive rights of access to such information and, according to the European Court of Justice, a company which is closely involved in the tendering procedure (including the successful tenderer) must receive, without delay, precise information concerning the conduct of the entire procedure, see *Embassy Limousines & Services* v. *European Parliament* (1998) and also *Aquatron Marine, t/a Aquatron Breathing Air Systems v Strathclyde Fire Board* (2008) where the authority was not qualified to evaluate the bids so had excluded the bid erroneously. In the *Serrantoni* case (2008), the Court found that it is not lawful to exclude bidders automatically.

2.2.2 Non-discrimination

Contracting authorities must not discriminate against contractors or providers from other EU countries compared to domestic undertakings, and must not discriminate between domestic and imported products or services. Contracting authorities should not impose conditions on non-domestic bidders which are different to or more demanding than conditions imposed on domestic bidders. In relation to conditions of tendering, the European Court of Justice has declared that observance of the principle of equal treatment of tenderers requires that all the tenders comply with the tender conditions so as to ensure an objective comparison of the tenders submitted by the various tenderers, see *European Commission* v. *Kingdom of Denmark* (1993).

2.2.3 Equality

Contracting authorities may not impose conditions on some bidders and not on others unless (exceptionally) there is reasonable justification for such treatment. All enquiries and requests for information or other assistance must be treated fairly and equally.

2.2.4 Mutual recognition

Contracting authorities must accept technical specifications, diplomas and qualifications if supplied by undertakings from other EU countries when they are generally recognized as being equivalent to those required or recognized in the UK, see regulation 9(15) of the Regulations.

2.2.5 Proportionality

This principle requires that a contracting authority's definition of performance and technical specifications is necessary and appropriate in relation to the objectives to be reached by the awarding body, i.e. that contracting authorities do not apply excessive and disproportionate technical, professional or financial conditions when selecting candidates for a procurement. See *Azam & Co Solicitors v Legal Services Commission* (2010) in relation to the proportionality of a decision not to accept a late bid.

It is important to note that the Regulations implement the Directive into Scots law and must be interpreted purposively in accordance with that Directive and other applicable EU legislation and case law, notably the EU legal principles referred to above, see *Von Colson* v. *Land Nordrhein-Westfalen* (1984) and *Marleasing SA* v. *La Comercial Internacional de Alimentación SA* (1990). This, in turn, means that even if the terms of the Regulations are clear, they may need to be interpreted in a manner not strictly in accordance with that clear meaning. It is possible that a court applying a purposive interpretation of the Regulations may re-write all or part of them, changing their meaning and the implications, to an extent consistent with the purposes of the relevant EU rules.

The legal principles outlined above are also important from the contracting authority's perspective, because complying with them should enable it to create strong competitive conditions.

2.3 *Beyond the EU*

The EU public procurement regime offers protection not just to nationals (both legal and natural persons) of EU member states, in respect of contracts being awarded by EU public bodies. By virtue of certain international agreements, there are also some protections for EU nationals in tendering for contracts outside the EU, and for non-EU nationals tendering within the EU.

First, the European Economic Area Agreement ('the EEA Agreement') which established a single market in public procurement and is worth over € 2,150 billion per year, is intended to promote trade between the European Free Trade Association (EFTA) and the EU, and application of the EU public procurement procedures is extended to four EFTA member states (Iceland, Liechtenstein, Norway and Switzerland). The EEA Agreement establishes public procurement principles and procedures in these EFTA States which are similar to the EU public procurement position. Application of the EEA Agreement by these EFTA States is monitored by the EFTA Surveillance Authority and the EFTA Court. The European Commission and the EU General Court, formerly the European Court of Justice (ECJ), monitor application within the EU.

Second, the EU and EU member states entered into a number of 'Europe Agreements' with certain States of Central and Eastern Europe, many of who have since become EU member states. These agreements provide for access by undertakings of the relevant State to EU contracts on terms no less favourable than those applied to EU nationals.

Third, the Government Procurement Agreement (GPA) has coverage in parallel with the EU procurement provisions. As well as giving rights to EU-based tenderers, it applies to contracting authorities in GPA States outside the EU, thereby affording EU and non-EU tenderers protections outside the EU as well as within. As part of wider international trade negotiations through the World Trade Organisation on 30 March 2012, GPA States adopted a revised GPA, which extends coverage to a number of additional government entities. The revision is designed to clarify and modernize the GPA, increasing transparency and making it easier for developing countries to access such contract opportunities.

The GPA largely conforms to the same principles as the EU procurement regime, and signatories undertake to treat each other on the basis of mutual reciprocity and provide guaranteed market access to specified listed areas. Similarly to the EU procurement regime, the GPA is designed to make laws, regulations, procedures and practices relating to government procurement more transparent and to guard against discrimination against foreign products or suppliers. As with the EU public procurement regime, the GPA imposes deadlines, prohibits the splitting of contracts and establishes detailed rules on the content of tender documentation and the contract award process. The revised GPA now allows for procurement notices to be published electronically, provided that they are readily accessible to potential bidders. The post-award information and publication requirements in the GPA require parties to publish award notices and for suppliers from GPA States to receive prompt information on contracting authorities' procurement practices, an explanation of why the supplier's application to qualify was rejected, why its existing qualification to tender was brought to an end, and information on the characteristics and relevant advantages of the tender selected, see Article XVII of the GPA. The GPA requires States bound by the GPA to provide a timely, effective, transparent and non-discriminatory administrative or judicial review procedure to challenge alleged breaches of the GPA, see Article XIII of the GPA. Disputes between GPA Parties are subject to the procedures of the WTO Understanding on Rules and Procedures Governing the Settlement of Disputes. They constitute separate requirements, but the similarities between the EU and GPA procedures are such that by complying with the Directive and the Regulations, a contracting authority should also be in compliance with the GPA.

2.4 *Conditions for application of the procurement rules*

The Regulations only apply to the procurement and award of contracts if certain pre-conditions are met. In summary, these pre-conditions are as follows:

- that the body awarding the contract is a contracting authority;
- that the object of the contract falls within the scope of 'works', 'services' or 'supplies' as defined in the Regulations;
- that the value of the works, services or supplies under the proposed contract is in excess of the relevant financial threshold set out in the Regulations (though regard should be had to Section 2.18 on the obligations relating to lower-value contracts).

2.5 *Who must comply with the Regulations?*

A body is required to follow procedures under the Regulations for the award of relevant contracts if it falls within the definition of 'contracting authority' in regulation 3 of the Regulations. Regulation 3 lists a number of central and local government bodies, and includes within that definition a wider category of what are known as bodies governed by public law, being a corporation established, or a group of individuals appointed to act together, for the specific purposes of meeting needs in the general interest, not having an industrial or commercial character, and

- financed wholly or mainly by another contracting authority, or
- subject to management supervision by another contracting authority, or
- more than half of the board of directors or members of which, or, in the case of a group of individuals, more than half of those individuals, are appointed by another contracting authority.

There is considerable case law in relation to the various elements of this definition. In *Mannesmann Anlagenbau Austria AG* v. *Strohal Rotationsdruck GmbH* (1998), the ECJ held that the body in question had to have been established to meet needs in the general interest not having an industrial or commercial character and the fact that it also carried out other, commercial, activities was irrelevant, but see also *Universale-Bau AG* v. *Entsorgungsbetriebe Simmering GesmbH* (2002), *Adolf Truly GmbH* v. *Bestattung Wien GmbH* (2003) and *Arkkitehtuuritoimisto Riitta Korhonen Oy* v. *Varkauden Taitotalo Oy* (2003).

The requirement of being financed wholly or mainly by another contracting authority was clarified in *R* v. *HM Treasury, ex parte University of Cambridge* (2000). The expression 'for the most part' has its ordinary meaning of 'more than half', and that the decision as to whether a university is a contracting authority should be made annually, with the budgetary year during which the procurement procedure was begun being the most appropriate period for calculating how it was financed.

It may be clear that certain purchasing bodies constitute contracting authorities for the purposes of the Regulations, but for other bodies the procurement position may need to be considered more closely. In a construction and major projects context, care may need to be taken, for example, where a joint venture includes a public sector party, to ensure that where the joint venture body does fall within regulation 3(1)(bb) of the Regulations, it complies with the Regulations in awarding relevant contracts.

2.6 Treatment of a proposed contract as a works contract, supply contract or a services contract

The original EU public procurement legislation, which was consolidated into the Directive, comprised separate Directives in respect of works contracts, supply contracts and services contracts. Accordingly, central to the scope of application of each of these Directives, and also the UK implementing Regulations, were the definitions of a public works contract, a public supply contract and a public services contract. In addition, there are specific rules in relation to the treatment of contracts for a combination of works, supplies and/or services (mixed contracts).

2.6.1 What is a works contract?

A contract will be a works contract to be awarded under the Regulations if it is a 'public works contract', defined by regulation 2(1) of the Regulations as a contract, in writing, for consideration (whatever the nature of the consideration):

- for the carrying out of a work or works for a contracting authority, or
- under which a contracting authority engages a person to procure by any means the carrying out for the contracting authority of a work corresponding to specified requirements.

A subsidised public works contract is one for which a contracting authority undertakes to contribute more than half of the consideration to be, or expected to be, paid under a contract, but which has been or is to be entered into by another person (the subsidised body), see regulation 34(1) of the Regulations. Depending on the subject matter, there is a requirement on the contracting authority to impose a condition of making such a contribution that the subsidised body complies with the Regulations in relation to that contract as if it were a contracting authority itself, and either to ensure that the subsidised body does so comply or to recover the contribution, see regulation 34(2) of the Regulations. This requirement applies to public works contracts for any of the civil engineering activities specified in Schedule 2 to the Regulations, and to building work for hospitals, facilities intended for sports, recreation and leisure, school and university buildings or buildings for administrative purposes, see regulation 34(2)(a) of the Regulations. It also applies to public services contracts covered by the Regulations for providing services in connection with such subsidised public works contracts, see regulation 34(2)(b) of the Regulations.

2.6.2 What are works?

Works are defined under the Regulations as any activities specified in Schedule 2 to the Regulations. This long list specifies works on a broad basis and includes:

- construction of new buildings and works and restoring and common repairs;
- site preparation (demolition and wrecking of buildings, earth moving and site clearing, building site drainage and drainage of agricultural or forestry land);
- test drilling, boring and core sampling for construction, geophysical, geological or similar purposes;
- general construction of buildings and civil engineering works (including bridges, pipelines, power lines and assembly and erection of prefabricated constructions on site);
- construction of highways, roads, airfields, sports facilities and water projects;
- other construction work involving special trades (pile driving, water well drilling and construction, shaft sinking, steel bending, bricklaying and stone setting, scaffold erecting and dismantling, including renting of scaffolds);

- installation of electrical wiring and fittings (including telecommunications and electrical heating systems, residential antennas and aerials, fire alarms, burglar alarm systems, lifts and escalators);
- insulation work activities (thermal, sound or vibration);
- plumbing (installation of plumbing and sanitary equipment, gas fittings, heating, ventilation, refrigeration or air-conditioning equipment and sprinkler systems);
- other building installation (illumination and signalling systems for roads, railways, airports and harbours);
- building completion (plastering, joinery installation, floor and wall covering, painting and glazing); and
- renting of construction or demolition equipment with operator.

2.6.3 What is a supply contract?

A contract will be a supply contract to be awarded under the Regulations if it is a 'public supply contract', defined by regulation 2(1) of the Regulations, as a contract, in writing, for consideration (whatever the nature of the consideration):

- for the purchase of goods (whether or not the consideration is given in instalments and whether or not the purchase is conditional upon the occurrence of a particular event); or
- for the hire of goods by a contracting authority (both where the contracting authority becomes the owner of the goods after the end of the period of hire and where it does not); and for any siting or installation of those goods, but where under such a contract services are also to be provided, the contract is only a public supply contract where the value of the consideration attributable to the goods and any siting or installation of the goods is equal to or greater than the value attributable to the services.

2.6.4 What is a services contract?

A contract will be a services contract to be awarded under the Regulations if it is a 'public services contract', defined in regulation 2(1) of the Regulations as a contract, in writing, for consideration (whatever the nature of the consideration) under which a contracting authority engages a person to provide services but does not include a public works contract or a public supply contract.

2.6.5 Mixed contracts

The Regulations address the possibility of contracts being for mixed requirements. Thus a contract both for supply of goods and of services shall be considered to be a public services contract if the value attributable to those services exceeds that of the goods covered by the contract.

A contract for services which includes works elements (i.e. activities specified in Schedule 2 to the Regulations) that are only incidental to the principal object of the contract shall be considered to be a public services contract. This is similar to the test developed by the ECJ in its judgments on certain public procurement cases under the preceding Directives, see *Telaustria Verlags GmbH* v. *Telekom Austria AG* [2000], and *Gestión Hotelera Internacional SA* v. *Comunidad Autonama de Canarias* (1994).

2.6.6 Part A and Part B services

The procedures which must be followed under the Regulations differ according to whether the contract to be awarded is a contract for Part A services (commonly referred to as 'priority' services) or for Part B services (commonly referred to as 'residual' services). In terms of the Regulations, a Part A services contract is a contract under which services under Part A of Schedule 3 are to be provided, such as financial services, computer and related services and architectural services. Similarly, a Part B services contract is a contract under which services specified in Part B of Schedule 3 are to be provided, such as health and social services and legal services. The full procedures apply in respect of Part A services, but not in respect of Part B services (though this distinction is likely to be abolished in the future, see Section 2.19). The general principles of transparency, equal treatment and non discrimination, and the regulations in relation to the enforcement of obligations do apply in relation to Part B services.

A single contract for services specified in both Parts A and B of Schedule 3 to the Regulations is required to be treated as:

- a Part A services contract if the value of the consideration attributable to the services specified in Part A is greater than that attributable to those specified in Part B; and
- a Part B services contract if the value of the consideration attributable to the services specified in Part B is equal to or greater than that attributable to those specified in Part A.

2.7 *What are the relevant financial thresholds?*

The Regulations only require to be followed by contracting authorities for certain contracts whose estimated value (net of VAT) exceeds a particular threshold amount, see regulation 8(1) of the Regulations.

In respect of the award of public works contracts and subsidised public works contracts, the Regulations apply to contracts to be awarded by a contracting authority which have an estimated value which is expected to exceed a threshold currently of £4,348,350 (€5,000,000), see regulation 8(2) of the Regulations.

For public supply contracts or public services contracts a distinction must be made between two categories of public sector bodies, see regulation 8(3) and regulation 8(4) of the Regulations. Schedule 1 to the Regulations lists central government bodies which are subject to the World Trade Organisation Government Procurement Agreement. The Regulations apply to contracts to be awarded by these bodies with an estimated value which is expected to exceed a threshold currently of £113,057 (€130,000) in relation to public supply contracts or public services contracts, with the exception of Part B (residual) services, Research & Development Services, certain Telecommunications services in Category 5 and subsidised services contracts. These have a threshold currently of £173,934 (€200,000). A higher threshold applies in respect of other public sector contracting authorities; the Regulations apply where public supply contracts and public services contracts are to be awarded by a contracting authority which is not listed in Schedule 1 to those Regulations, if the estimated value is expected to exceed currently £173,934 (€200,000).

The estimated value of a public contract shall be the value of the total consideration payable, net of value added tax, which the contracting authority expects to be payable under the contract, and any form of option, renewal of the contract, fees or commissions which are to be included in the calculation, see regulation 8(7) and regulation 8(8) of the Regulations.

A specific aggregation rule requires a contracting authority which has a single requirement for goods, services or works and enters (or proposes to enter) into a number of contracts, to aggregate for the purposes of regulation 8(1) the consideration expected to be payable under each of those contracts, see regulation 8(11) of the Regulations.

Regulation 8 of the Regulations contains a number of other relevant considerations and requirements in relation to contract value thresholds, but importantly provides that a contracting authority shall not enter into separate contracts nor exercise a choice under a valuation method with the intention of avoiding the application of the Regulations, see regulation 8(19) of the Regulations.

2.8 *Is there an applicable exclusion?*

The Regulations do not apply to the seeking of offers in relation to a proposed public contract, framework agreement or dynamic purchasing system where the contracting authority is a utility within the meaning of regulation 3 of the Utilities Contracts (Scotland) Regulations 2012, nor to other contracts which meet certain other conditions, see regulation 6(1) of the Regulations. Regulation 6(2) contains certain other exemptions, for contracts:

- in relation to telecommunications;
- which are secret or require special security measures;
- for the acquisition of land, including existing buildings, land covered with water and any estate, interest, easement, servitude or right in or over land;
- in relation to broadcasting;
- for arbitration or conciliation services;
- for certain financial services;
- for central banking services;

- for research and development services (unless certain conditions are fulfilled);
- under which services are to be provided by a contracting authority because that contracting authority or person has an exclusive right to provide the services; or which is necessary for the provision of the services; and
- for a services concession, subject to regulation 46 which provides a duty of non-discrimination in certain circumstances.

2.9 *Types of procurement procedure*

The Regulations provide for four main types of competitive procedure which a contracting authority may follow. Three of these, the open procedure, the restricted procedure, and the competitive negotiated procedure, were established in Directives 92/50/EEC, 93/36/EEC and 93/37/EEC. The fourth procedure, the competitive dialogue procedure, was introduced by the Directive.

The detailed rules applying to each of these procedures differ. The competitive dialogue and competitive negotiated procedures may only be used in specified limited circumstances. A contracting authority should carefully consider which procedure is appropriate and document its reasons for the decision it takes in this regard.

2.10 *The open procedure*

The stages of the open procedure (in which all interested parties may submit proposals), are set out in regulation 15 of the Regulations and are summarized below.

2.10.1 Step 1: Advertising

The contracting authority must send to the Official Journal a contract notice submission in prescribed form under regulation 15(2). The contracting authority must send this as soon as possible after forming the intention to seek offers, and must use the form of the contract notice in Annex II to Commission Regulation (EC) No. 1564/2005, colloquially known as an OJEU Contract Notice (OJEU being *The Official Journal of the European Union* (henceforth 'the Official Journal')). This advertises the contracting authority's requirement and gives interested potential bidders details of how to obtain further information. In completing a contract notice submission, the contracting authority is required to provide a range of information about its contract requirements and about the procedure being conducted. Although there is no pre-qualification stage under the open procedure, the contracting authority may require an economic operator to satisfy minimum levels of economic and financial standing and/or technical or professional ability if they are specified in the contract notice and are related and proportionate to the subject matter of the contract, see regulation 15(12).

2.10.2 Step 2: The tender period

A period of not less than 52 days (the tender period) must be allowed to enable interested parties to prepare and submit tenders, though that time limit can be reduced further if a Prior Information Notice has been issued, if the award notice has been sent electronically and if the contracting authority has given unrestricted and full direct access by electronic means to the contract documents.

Periods for taking action under the Regulations run from the day after the day on which the action is taken, and shall be extended where necessary to include two working days or to end on a working day, see regulation 2(6).

In the open procedure, except for minor clarifications, the contracting authority must choose between tenders as they are bid, though in limited circumstances bidders may be asked to re-tender.

2.10.3 Step 3: Evaluation

The contracting authority must evaluate all bids either on the basis of lowest cost or most economically advantageous tender, see regulation 30(1).

2.10.4 Step 4: Contract award

Following evaluation, a contracting authority may decide to award a contract (as described in more detail below in relation to the restricted procedure) and must publish a contract award notice in the Official Journal within 48 days, see regulation 31(1).

2.11 *The restricted procedure*

The stages of a restricted procedure are set out in regulation 16 of the Regulations and include an additional stage, namely the contracting authority making a preliminary assessment of those who express an interest in the procurement, as to whether they should pre-qualify for the tender competition (pre-qualification). In summary, the stages in a restricted procedure are as follows.

2.11.1 Step 1: Advertising

The contracting authority must send to the Official Journal a contract notice application in prescribed form, see regulation 16(2). The contracting authority must send this as soon as possible after forming the intention to seek offers. As in the open procedure, this advertises the contracting authority's requirement and gives details of how interested parties may obtain further information.

2.11.2 Step 2: Pre-qualification stage

Interested parties must have (at least) 37 days from the date of dispatch of the contract notice, within which to notify the contracting authority that they wish to be invited to tender, see regulation 16(3) and regulation 16(5). However, that limit can be reduced further if the award notice has been sent electronically or where the minimum time limit is rendered impractical by reason of urgency. See as follows:

- regulation 16(5): where the contract notice is submitted by electronic means in accordance with Annex VIII of the Directive, the time limit may be reduced by seven days;
- regulation 16(6)(a): a time limit of not less than 15 days from dispatch of the OJEU Contract Notice for reasons of urgency;
- regulation 16(6)(b): a time limit of not less than 10 days where the contract notice has been submitted by electronic means in accordance with regulation 16(5) and compliance with the minimum time limit of 37 days is rendered impractical for reasons of urgency.

The contracting authority must select tenderers in accordance with regulations 23–26 of the Regulations. An economic operator can only be excluded from the group of economic operators from which a contracting authority is to select those to be invited to tender, on the grounds for exclusion set out in regulation 23 (such as insolvency or conviction of a criminal offence), or if the economic operator fails to satisfy minimum standards of economic and financial standing or technical or professional ability, see regulation 16(7).

The number of persons which the contracting authority can invite to tender must be sufficient to ensure genuine competition and must at least be equal to any minimum number which may have been specified in the contract notice, see regulation 16(10). Where there is a sufficient number of economic operators suitable to be invited to tender, the contracting authority may limit the number which it intends to invite, but the contract notice must have specified the objective and non-discriminatory criteria which would be applied in so doing, and must also have specified the minimum number (which shall be not less than five) and (where appropriate) the maximum number of economic operators which the contracting authority intends to invite to tender.

No price or other bidding indications may be asked for at this stage, nor considered if voluntarily provided by an economic operator. The criteria used for the selection of tenderers do not apply to the award of a public contract. In pre-qualification, the contracting authority is selecting tenderers whereas after pre-qualification and during the tender stage it is evaluating tenders. This is an important distinction to maintain during public procurements.

2.11.3 Step 3: The tender period

Those selected to tender by the contracting authority will usually be sent a formal invitation to tender (or 'ITT') by the contracting authority which must be accompanied by the contract documents, see regulation 16(13). A period of not less than 40 days (the tender period) must be allowed to enable interested parties to prepare and submit tenders, see regulation 16(16). However, that time limit can be reduced further for reasons of urgency, if a Prior Information Notice has been issued and if the contracting authority has given unrestricted and full direct access by electronic means to the contract documents. See as follows:

- regulation 16(17): a time limit of not less than ten days from the date of dispatch of the invitation where the minimum time limit of 40 days is rendered impractical for reasons of urgency;
- regulation 16(18): generally to be not less than 36 days but in any event not less than 22 days, if a Prior Information Notice was submitted to the Official Journal at least 52 days and no more than 12 months prior to the dispatch of the OJEU Contract Notice;
- regulation 16(19): the time limits may be reduced by five days, provided that the authority offers unrestricted and full direct access to the contract documents by electronic means and the contract notice specifies the internet address at which the contract documents are available.

In the restricted procedure, and in common with the open procedure, the contracting authority must choose between tenders as they are bid, though it is possible to seek clarification of the terms of bidders' proposals, and in certain circumstances bidders may be asked to submit further tenders.

2.11.4 Step 4: Evaluation

The contracting authority must evaluate all bids either on the basis of lowest cost or most economically advantageous tender, see regulation 30(1). When using the latter basis of evaluation, the contracting authority must state the weighting which it gives to each of the criteria chosen in the contract notice or in the contract documents, which is one of the requirements of transparency and helps bidders to understand how they might make their proposals as attractive as possible, see regulation 30(3).

In *Lianakis and Others v Dimos Alexandroupolis and Others* (2008), it was held that an authority cannot apply weightings and sub-criteria to award criteria set out in tender documents unless those weightings or sub-criteria have been previously brought to the bidder's attention. Furthermore, an authority cannot take account of a bidder's experience, manpower, equipment or ability to perform the contract by an anticipated deadline as part of the award criteria. Such criteria can only be taken into account at the pre-qualification stage.

2.11.5 Step 5: Contract award

Following evaluation, a contracting authority may make a contract award decision, and on doing so must provide bidders and those who applied to be selected to tender, as soon as possible after the contract award decision is made, with relevant information. Section 2.15 describes these requirements in more detail. Subject to those requirements, a contracting authority may proceed to award a contract after evaluating bids, and must publish a contract award notice in the Official Journal within 48 days, containing specified details about the contract awarded, the successful contractor and also about the procurement competition, see regulation 31(1).

2.12 The negotiated procedure with advertisement

2.12.1 Significance of the negotiated procedure with advertisement

A significant feature of the negotiated procedure with advertisement is that it allows a contracting authority to negotiate commercial and pricing proposals and contract terms with bidders, which is not possible under either the open procedure or the restricted procedure.

The negotiated procedure with advertisement is set out in regulation 17 of the Regulations. In summary, the stages in a negotiated procedure with advertisement are as follows:

Step 1: Advertising

The contracting authority must send to the Official Journal a contract notice submission in prescribed form (see under regulation 17(3) of the Regulations), and the contracting authority must send this as soon as possible after forming the intention to seek offers, and must use the form of the contract notice in Annex II to Commission Regulation (EC) No. 1564/2005. As in the open and restricted procedures, this advertises the contracting authority's requirement and gives details of how interested parties may obtain further information.

Step 2: Pre-qualification stage

Interested parties must have (at least) 37 days from the date of dispatch of the contract notice within which to notify the contracting authority that they wish to be selected to negotiate, see regulation 17(5). That limit can be reduced further if the award notice had been sent electronically or where the minimum time limit is rendered impractical for reasons of urgency. See as follows:

- regulation 17(5): where the contract notice is submitted by electronic means in accordance with Annex VIII of the Directive, the time limit may be reduced by seven days;
- regulation 17(8)(a): a time limit of not less than 15 days from dispatch of the OJEU Contract Notice for reasons of urgency; and

- regulation 17(8)(b): a time limit of not less than ten days where the contract notice has been submitted by electronic means in accordance with regulation 17(5) and compliance with the minimum time limit of 37 days is rendered impractical for reasons of urgency.

The contracting authority must select tenderers in accordance with regulations 23–26 of the Regulations. An economic operator can only be excluded from the group of economic operators from which a contracting authority selects those to be invited to tender on the grounds for exclusion set out in regulation 23 (such as insolvency or conviction of a criminal offence), or if the economic operator fails to satisfy minimum standards of economic and financial standing or technical or professional ability, see regulation 17(7).

The number of persons selected to negotiate must be sufficient to ensure genuine competition and must at least be equal to any minimum number which may have been specified in the contract notice, see regulation 17(12). Where there is a sufficient number of economic operators suitable to be selected to negotiate (see regulation 17(11)), the contracting authority may limit the number which it intends to select to negotiate, but the contract notice must have specified: (i) the objective and non-discriminatory criteria which would be applied in so doing; (ii) the minimum number, to be not less than three; and (iii) where appropriate, the maximum number of economic operators which the contracting authority intends to invite to negotiate.

In pre-qualification, the contracting authority is selecting parties to negotiate, whereas after pre-qualification and during the negotiation tender stage, it is evaluating proposals. No price or other bidding indications may be asked for or taken into account at the pre-qualification stage, nor may they be considered if voluntarily provided to the contracting authority, see regulation 17(9). The criteria used for selection of tenderers do not apply to the award of a public contract, see regulation 30(1).

Step 3: Tender and negotiation period, leading to evaluation and contract award

Those selected to negotiate by the contracting authority will usually be sent a formal invitation to negotiate (or 'ITN') by the contracting authority which must be accompanied by the contract documents, see regulation 17(15).

Conduct of this phase has a very significant bearing on the outcome of the procurement and, as a minimum, the following key points should be borne in mind in relation to bid preparation and communication during the tender period.

For the contracting authority, the priority should be to ensure that all tenderers clearly understand in detail the requirements which it wishes to be met through the contract it proposes to award. Public sector bodies may have multiple policy objectives. For example, a national health service contracting authority may have clinical, financial and health and safety objectives for the same procurement. These may compete in priority terms, or even conflict. In such circumstances, it may not be possible for tenderers fully to understand the contracting authority's requirements unless they are told the relative importance of such competing or conflicting requirements. Economic operators preparing proposals will be particularly concerned to understand how these proposals will be evaluated, hence the important requirements for the contracting authority to provide evaluation criteria and weighting information. It will also be important to the contracting authority to ensure proposals received meet its expectations and requirements. Providing tenderers with clear details in relation to evaluation should assist in this.

The contracting authority must evaluate all bids either on the basis of lowest cost or most economically advantageous tender, see regulation 30(1). In the negotiated procedure, as well as seeking clarification of proposals, the contracting authority is permitted also to negotiate with tenderers. The Regulations provide that, where it needs to identify the best tender in order to award the public contract, the contracting authority is obliged to negotiate with economic operators which have submitted tenders with the aim of adapting the tenders to the requirements specified in the contract documents, see regulation 17(21).

Step 4: Contract award

Following evaluation, a contracting authority may decide to award a contract (as described in more detail above in relation to the restricted procedure), and must publish a contract award notice in the Official Journal within 48 days, see regulation 31(1).

2.12.2 Restrictions on using the negotiated procedure

The negotiated procedure with advertisement can only be used in specified circumstances, namely: (i) because of irregular or unacceptable tenders pursuant to an open procedure or restricted procedure, but only if the terms of contract used for that earlier procurement are not substantially altered in the negotiated procedure; (ii) exceptionally, when the nature of the work or works to be carried out, the goods to be purchased or the services to be provided or the risks attaching to them, are such as not to permit overall pricing; (iii) for a public services contract, when the nature of services to be provided (in particular intellectual services) is such that specifications cannot be drawn up with sufficient precision to permit award of the contract using the open or restricted procedure; and (iv) for a public works contract, when the work or works are to be carried out under the contract solely for the purpose of research, testing or development, but not with the aim of ensuring profitability or to recover research and development costs.

The precise conditions in which a contracting authority should regard itself as being permitted to use the negotiated procedure have been under close scrutiny. The European Commission developed the competitive dialogue procedure described below in preference to the negotiated procedure and contracting authorities should only use the negotiated procedure where there is clear justification for doing so. Choice of procurement procedure is discussed further in Sections 2.13.1 and 2.13.2.

2.13 *The competitive dialogue procedure*

2.13.1 Nature of competitive dialogue procedure

The competitive dialogue procedure is intended for use in the award of 'particularly complex' contracts, where there is a need for contracting authorities to discuss their requirements with shortlisted candidates before final written tenders are received. The Recitals to the Directive describe the purpose as providing 'a flexible approach', preserving competition between operators and permitting discussion of 'all aspects' of the contract with each candidate.

The stages of the competitive dialogue procedure are set out in regulation 18 of the Regulations. In summary, the stages are as follows:

Step 1: Advertising

The contracting authority must send to the Official Journal a contract notice submission in the prescribed form, see regulation 18(4) of the Regulations. The contracting authority must submit this as soon as possible after forming the intention to seek offers, and must use the form of the contract notice in Annex II to Commission Regulation (EC) No. 1564/2005 to do so. As in the open, restricted and negotiated procedures, this advertises the contracting authority's requirement and gives details of how interested parties may obtain further information.

Step 2: Pre-qualification

Interested parties must have (at least) 37 days from the date of dispatch of the contract notice, within which to notify the contracting authority that they wish to be selected to participate, see regulation 18(7). However, that limit can be reduced further if the award notice had been sent electronically, see regulation 18(9).

The contracting authority must select participants for the competitive dialogue in accordance with regulations 23 to 26 of the Regulations. An economic operator can only be excluded from the group of economic operators from which a contracting authority selects those to be invited to tender on the grounds for exclusion set out in regulation 23 (such as insolvency or conviction of a criminal offence), or if the economic operator fails to satisfy minimum standards of economic and financial standing or technical or professional ability, see regulation 18(10) and regulation 18(11).

The number of persons selected to participate in the dialogue must be sufficient to ensure genuine competition and must at least be equal to any minimum number specified in the contract notice, see regulation 18(13). Where there is a sufficient number of economic operators suitable to be selected to participate in the dialogue (see regulation 18(12)), the contracting authority may limit the number which it intends to invite to participate in the dialogue, but the contract notice must have specified: (i) the objective and non-discriminatory criteria to be applied to limit that number; (ii) the minimum number, to be not less than three; and (iii) where appropriate, the maximum number that the contracting authority intends to invite to participate in the dialogue.

No price or other bidding indications may be asked for or taken into account at this stage, or considered if voluntarily provided, see regulation 18(11). The criteria used for selection of tenderers do not apply to the award of a public services contract, see regulation 30(1).

Step 3: Dialogue phase

Tenderers participate in competitive dialogue with the contracting authority in response to an invitation to participate. The contracting authority's required aims during the dialogue are to identify and define how its needs can best be satisfied, in consultation with the participants, see regulation 18(21).

The contracting authority may discuss 'all aspects of the contract', but shall ensure equality of treatment among all participants, see regulation 18(22)(a) and (b). A concern for participants is that others could acquire and exploit their proprietary ideas, and so proposed solutions or confidential information are not to be divulged to other candidates without consent, see regulation 18(22)(c). The competitive dialogue procedure is to continue until one or more comparable solutions can be identified which are capable of meeting the contracting authority's needs, see regulation 18(25).

The procedure may be conducted in successive stages, permitting stage by stage the reduction of solutions, but a sufficient number of bidders must remain to ensure genuine competition at tender stage, see regulation 18(24).

Step 4: Post-dialogue tender stage

The contracting authority may continue the dialogue until it can identify one or more solutions capable of meeting its needs, if necessary after comparing them, and should formally declare the dialogue concluded, see regulation 18(25) and regulation 18(26). Contracting authorities should consider carefully when this declaration should be made as it marks an important transition during the procedure.

Once the dialogue is concluded, the contracting authority invites each participant to submit a final tender on the basis of any solution or solutions presented and specified (not necessarily by that tenderer) during the dialogue, see regulation 18(26). The contracting authority cannot invite fewer than three tenderers to do so, provided a sufficient number of candidates satisfy the qualitative selection criteria.

Step 5: Evaluation and fine tuning to award

Tenders are then evaluated on the basis of the award criteria and the contracting authority shall award the contract to the participant which submits the most economically advantageous tender, see regulation 18(28). Importantly, at this stage there is no further scope for dialogue or negotiation and tenderers can only be asked to clarify, specify or fine-tune their proposals, see regulation 18(27).

2.13.2 How complex is particularly complex?

The competitive dialogue procedure is confined to 'particularly complex' contracts, defined as contracts where the contracting authority is not objectively able:

- to define the technical means capable of satisfying its needs or objectives; or
- to specify either or both of the legal and financial make-up of a project, see regulation 18(1).

This definition lacks clarity. The first limb refers to an inability to define the 'technical means' for meeting the contracting authority's needs. Although not clearly defined in the Regulations, this may be interpreted as referring to the skills, knowledge, technology or methods capable of realizing the contracting authority's overall objectives.

The second limb of the definition of 'particularly complex' is equally unclear. The term 'legal and/or financial make-up' is not defined further, though it may refer to difficulties in predetermining the contractual structure and terms (including funding arrangements). In the initial stages of major projects there is likely to be some uncertainty in this regard, but it is not clear to what degree the contracting authority must be unable to specify either the legal or financial make-up of a project in order to justify using the competitive dialogue.

A competitive dialogue is permitted where contracting authorities consider that the use of the open or restricted procedure will not allow award of the contract. This suggests a degree of discretion but a contracting authority would need to establish that it is not 'objectively able' to define or specify the required information. Recital 31 to the Directive describes the circumstances for which competitive dialogue is intended as being when it is objectively impossible to define the means of satisfying the contracting authority's needs, or of assessing what the market can offer in the way of technical solutions and/or financial/legal solutions. Those recitals do state that projects for integrated transport infrastructure or for large computer networks may be regarded as 'particularly complex' and do refer to projects involving 'structured financing', which may include most PPP/PFI and NPD projects.

2.13.3 Interaction between the competitive dialogue and the open, restricted and negotiated procedures

Dialogue between the contracting authority and bidders to identify and define how the authority's needs can best be satisfied is not part of the open procedure, nor is it part of the restricted procedure. It is presupposed, for both of these procedures, that the contracting authority's requirement has been accurately described in advance. In contrast, two existing grounds for using the negotiated procedure assume the contracting authority's requirement is not entirely clear, namely, when:

- exceptionally, the nature of the works or services or the risks attaching to them are such as not to permit overall pricing, see regulation 13(b); or

- for public services contracts, the nature of the services is such that the specification cannot be established with sufficient precision to permit award by open or restricted procedures, see regulation 13(c).

Accordingly there is some overlap between when the competitive dialogue procedure and the competitive negotiated procedure can be used. An inability to predetermine technical means (competitive dialogue) is similar to an inability to establish contract specifications with sufficient precision (negotiated procedure), and difficulty in specifying the 'financial make-up' of a project (competitive dialogue) may be a reason why 'prior overall pricing' is not possible (negotiated procedure).

2.13.4 Practical issues in conducting a competitive dialogue

Strategies for dialogue

Contracting authorities must not reveal one participant's proposed solution to another participant without consent, see regulation 18(22)(c). That right to confidentiality sits uneasily with the intention through competitive dialogue to encourage innovation towards best solutions. Participants may take different approaches to the dialogue, including on how early and in how much detail to share their proposals with the contracting authority. As a result, the contracting authority may receive rudimentary proposals with high potential as well as better developed proposals with less potential; if so, it may be difficult for a contracting authority to compare them fairly and objectively.

No 'cherry-picking'

A participant may fear other participants exploiting or 'cherry-picking' its ideas, and may object to the contracting authority disclosing to other participants a solution it proposes. The concern may be particularly great in relation to a solution which includes proprietary technology or intellectual property. The perceived risk of participants having such concerns may lead a contracting authority to require each bidder to submit a tender based on its own proposal for a solution, see regulation 18(26). However, that may prevent or restrict participants from competing on the basis of the contracting authority's favoured solution, in turn potentially impairing the contracting authority's ability to obtain best value for money from the procedure.

Clarification and fine-tuning

Final tenders can be clarified and fine-tuned but this shall 'not involve changes to the basic features of the tender … when those variations are likely to distort competition or have a discriminatory effect', see regulation 18(27).

Those familiar with PPP/PFI/NPD and major construction projects will appreciate how hard it would be in practice to avoid negotiation completely in the lead-up to conclusion of contract or financial close. But according to the Office of Government Commerce (OGC) Guidance, negotiation can be undertaken between the contracting authority and the bidders when working on final tenders at preferred bidder stage (just as under the negotiated procedure), so long as (as detailed above) the amendments do not distort competition or have a discriminatory effect. The OGC Guidance contends that this type of fine-tuning should not be regarded as distorting competition, on the basis that this period of negotiation would need to be done with whichever bidder is appointed as preferred bidder.

The Scottish Procurement Directorate has issued a number of Notes in relation to public procurement. Although not legally binding, this and other central government guidance are relevant to public procurement, and compliance may be a precondition for contracting authorities to receive funding and approval for contract award.

In reality, it may be hard for one participant closely to scrutinize others' dialogue with the contracting authority on specifications or terms. Rights of confidentiality may make it difficult for participants to know precisely what was discussed with the successful tenderer, both before and after tenders were invited.

2.14 Awarding the contract

It is essential that a contracting authority develops a robust evaluation model including evaluation criteria and evaluation weightings to enable it to assess bids, just as with any public procurement procedure.

2.14.1 Criteria for the award of the contract

The contracting authority shall award a public contract on the basis of the offer which:

- is the most economically advantageous to the contracting authority, or
- offers the lowest price.

This is intended to ensure that contracting authorities take decisions to award contracts on the basis of objective (commercial) criteria.

Using the lowest price criterion may be unduly restrictive for a procurement if proposals received offer economic advantages for the contracting authority other than in terms of the price payable. In choosing to make its evaluation assessment based on what is most economically advantageous, the contracting authority can take into consideration other factors as well as price, e.g. programme to completion or delivery, quality, environmental characteristics, design and aesthetic characteristics, functional features and technical assistance.

2.14.2 The most economically advantageous tender

The Regulations provide a non-exhaustive list of the type of factors that may be taken into account in assessing what is most economically advantageous. The Regulations provide that the criteria must be 'linked to the subject matter of the contract', see regulation 30(2).

An authority must disclose its chosen evaluation criteria for assessing what is most economically advantageous in the OJEU Contract Notice or contract documents, or in the case of the competitive dialogue procedure, in the descriptive document, see regulation 30(3). Further, the contracting authority should state the weighting which it gives to each of the chosen criteria. If the contracting authority does not believe that it is possible to allocate weightings, it must indicate the criteria in descending order of importance, see regulation 30(5).

If a contracting authority fails to state the relevant award criteria or fails to state them clearly in the OJEU Contract Notice (or in the contract documents), it will be required to award the contract on the basis of the lowest price, see *R* v. *Portsmouth City Council, ex parte Coles Colwick Builders Ltd and George Austin Ltd* (1997).

2.14.3 Abnormally low bids

Contracting authorities are under a duty to investigate when they receive an abnormally low bid, see regulation 30(6), *Amey LG Limited v The Scottish Ministers* (2012) and *Morrison Facilities Services Ltd v Norwich City Council* (2010).

2.15 *The* Alcatel *mandatory standstill period*

Under the EC Directives providing for procurement remedies ('the Remedies Directives'), EU Directives 89/665 and 92/13, once a public contract had been awarded to a successful tenderer the only statutory procurement remedy available to unsuccessful tenderers was to seek damages from the contracting authority for any alleged breach. Prior to contract award, interim remedies other than damages were (and still are) available, as described in Section 2.17. The ECJ decision in *Alcatel Austria* v. *Bundesministerium für Wissenschaft und Verkehr* (1999) found that the Remedies Directives are intended to protect tenderers against arbitrary decisions by contracting authorities, in particular at the stage where infringements can still be rectified, but such protection cannot be effective if the tenderer is not able to rely on these rules against the contracting authority. Instead effective legal protection presupposes, first, an obligation to inform all tenderers of the award decision, so that each has a genuine possibility of raising proceedings and exercising their remedies. In addition, it must be possible for the unsuccessful tenderer to examine in sufficient time the validity of the award decision; a reasonable period must therefore elapse between the time when the award decision is communicated to unsuccessful tenderers and conclusion of the contract.

2.15.1 The ten-day standstill period

In order to give effect to the terms of the ECJ judgment in *Alcatel*, the Regulations provide for a 'standstill' period between when an award decision for a contract awarded under those Regulations is notified to bidders, and the date on which that contract is to be entered into, see regulation 32(3).

Contracting authorities are required to issue a notice of their award decision to unsuccessful tenderers and candidates in writing at least 10 days prior to their entry into the contract (or such longer or extended period as described below). The standstill period is designed to enable an aggrieved bidder to review the procurement and, if necessary, to raise legal proceedings against the contracting authority before the contract is entered into. The standstill notice provides an explanation of the reasons for the contract award at the start of the standstill, by containing the following detailed information:

- the contract award criteria;
- the name of the winning tenderer;
- the score obtained by the unsuccessful tenderer receiving the notice as well as that of the tenderer awarded the contract;
- the relative characteristics and advantages of the winning bid compared to that of the unsuccessful tenderer (or a summary of why a candidate was not successful);
- a precise statement of the effect of the standstill arrangement on that tenderer or candidate; and
- how long the standstill period will be.

The standstill period commences on the date the last notice is sent to the relevant tenderers and candidates. If the standstill period is to end on a non-working day, it must be extended to the next working day and according to the general rules on notice periods in regulation 2(6), the date of sending the notice is not counted in the standstill period. The standstill period does not apply to below threshold procurements, Part B services or where there was only one tenderer. If notification is in writing and not sent by fax or email, the period is extended to 15 days. See the definition of 'relevant standstill period' in regulation 2(1).

If an unsuccessful economic operator makes a written request, the contracting authority must within 15 days of such request inform that economic operator of the reasons why it was unsuccessful, and of the characteristics and relative advantages of the winning tenderer and their name, if it has not already provided that information, see regulation 32(6).

Requests for additional de-briefing within the mandatory standstill period may lead the contracting authority to alter the duration of the standstill period. To reduce uncertainty as to the contract commencement date, when calculating the standstill period, authorities may wish to make allowance for time needed to conduct additional de-briefing(s), or to provide the additional de-brief information to all unsuccessful tenderers at the time of contract award decision notification.

2.16 The use of framework agreements in public procurement

The Directive, for the first time, introduced provisions covering the use of framework agreements for public contracts, which it defines as being:

> an agreement between one or more contracting authorities and one or more economic operators, the purpose of which is to establish the terms governing contracts to be awarded during a given period, in particular with regards to price and, where appropriate, the quantity envisaged.

This followed a series of consultations and negotiations on the subject going back to 1996 and a period during which frameworks were permitted for utilities procurement.

Framework agreements establish the terms which will apply if the parties subsequently conclude a contract for goods, services or works. It is important to note that a framework agreement itself does not constitute a public contract and as such the framework agreement does not create an obligation on either party to perform (i.e. the parties to the framework agreement are not obliged to enter into any contract for the provision of goods, works or services). Instead the framework agreement is facilitative and is used to set out in advance the terms and conditions that would apply should the parties wish to 'call off' during the period of the framework contracts of a type to which the framework relates.

Since a framework agreement is defined in the Directive as being 'between one or more contracting authorities and one or more economic operators', it is clear that there can be more than one contracting authority and/or more than one economic operator involved in any particular framework agreement. This means that a contracting authority can enter into a framework agreement for works, goods or services intended for other contracting authorities by acting as a 'central purchasing body' (see the definition in regulation 2(1)), and a contracting authority can enter into a framework agreement with more than one economic operator. The latter is known as a multi-supplier framework and must consist of at least three suppliers. Single-supplier frameworks are, as the name suggests, framework agreements entered into by one or more contracting authorities with only one economic operator.

Framework agreements are covered by regulation 19 of the Regulations. This provides that 'the contracting authority must not conclude a framework agreement for a period which exceeds 4 years except in exceptional circumstances'. These exceptional circumstances might include, for example, a situation where the economic operator is required to make a substantial up-front investment which may take some time to yield a return.

Regulation 19 also deals with the process for entering into a framework agreement. This process, similar to that involved in entering into a public contract, begins with the publication of a contract notice in the Official Journal. Thereafter the contracting authority must follow one of the permitted award procedures outlined above to be used by contracting authorities in the procurement of a public contract.

Framework agreements can, if used in appropriate circumstances, have considerable advantages. If a contracting authority is likely to have a number of separate but similar contract requirements over a period, a framework agreement avoids the need for repeated procurement, or, in the case of multi-supplier frameworks, will limit procurements to 'mini-competitions' based on terms and conditions already set up under the framework agreement, with resultant savings in cost and time.

On the other hand, there are also a number of potential disadvantages to using framework agreements; they can complicate the award of 'straightforward' contracts and can, if the legal procurement requirements are not met, result in lengthy and costly litigation by parties concerned over 'foreclosure' of the market. Two Northern Ireland cases highlighted the potential pitfalls in procuring framework agreements. In *McLaughlin & Harvey Ltd v Department of Finance & Personnel* (2008), the contracting authority was held to be in breach of the Regulations for failing to disclose 39 sub-criteria used in assessing bids for its framework and the weighting to be given to each of the criteria in the assessment. In *Henry Brothers (Magherafelt) Ltd & Ors v Department of Education for Northern Ireland* (2008), the contracting authority was held to have erred in the pricing mechanism used to assess the bids. In both these cases the court set aside the framework agreements entered into.

It also needs to be borne in mind that, though mini-competitions for contracts called-off under the framework are not subject to the full public procurement regime under the Regulations, in particular, the publication of an OJEU notice, certain criteria set out in regulation 19 must still be met and the overriding obligations set out in regulation 4(3) of treating economic operators equally and without discrimination and acting in a transparent and proportionate manner must be followed. In addition, the award criteria for each mini-competition must follow those set out in the framework agreement (which in turn means that these need to form part of the contract documents in the original tender process for the framework), see regulation 19(9)(d). The terms and conditions for the call-off contracts also need to be included as part of the original framework agreement and there are limitations to the extent to which these can be modified in the subsequent mini-competitions, since regulation 19(4) provides that the terms of the call-off contract awarded pursuant to a framework must not be 'substantially amended' from those laid down in the framework agreement. See also regulation 19(8), which states that any mini-competition must be 'on the basis of the same or, if necessary, more precisely formulated terms, and where appropriate other terms referred to in the contract documents based on the framework agreement'.

2.17 Remedies against contracting authorities (bidder grievances and complaints)

The number of formal complaints and successful court actions concerning claimed breaches of EU public procurement law has increased since reforms of the remedies available to aggrieved economic operators took effect. This section explains the legal position in some detail. Increasingly contractors are taking advantage of the award decision information that must be provided at the start of the standstill period to seek further information about particular aspects of evaluation that are unclear or of concern, without taking the expensive and difficult step of raising legal proceedings. Equally, the incidence is increasing of contracting authorities electing to re-tender a procurement which is subject to substantive or serious bidder complaints, thereby reducing the risk of legal proceedings being raised.

2.17.1 The Remedies Directives

The current EU Directive (Directive 2007/66/EC) was adopted by the Council in December 2007. The new Directive implemented two key changes: introduction of an ineffectiveness remedy and new standstill obligations as explained below.

Each EU member state is required to ensure that effective remedies and means of enforcement are available to suppliers, contractors and service providers who believe that they have been harmed as a consequence of a breach of their respective public procurement rules.

The rights of action laid down in the Regulations are available to any person who sought to tender for a relevant contract and potentially are available to any economic operator who had an interest in being engaged to perform the contract in question. The complainant must be an economic operator from an EU country or from a country which is a signatory to the GPA. See *Federal Security Services Ltd v Chief Constable for the Police Service of Northern Ireland* (2009) for the position with regard to Part B contracts.

2.17.2 Contracting authority duty to comply with the Regulations

The statutory rights and remedies for economic operators are founded upon regulation 47(1) of the Regulations which provides that a contracting authority owes a duty to 'economic operators' (including bidders, would-be bidders and interested parties), to comply with the provisions of those Regulations and with any enforceable EU obligation in respect of a public contract. For breach of that duty, economic operators suffering loss may pursue statutory remedies of: (i) interim suspension of the procurement procedure or decisions under it; (ii) setting aside of the procurement procedure or decisions under it; and/or (iii) damages. See regulation 48(1), as well as an order of contract ineffectiveness under regulation 49, as explained below.

2.17.3 What statutory remedies are provided?

Interim measures

The complainant may ask the court to issue an interim order, which suspends the allegedly defective award or suspends the implementation of any decision or action taken by the awarding authority in the course of such a procedure, such as the inclusion of a discriminatory or unfair contract term.

Set-aside and amendment orders

The court may also set aside any decision or act taken unlawfully in a procurement procedure and may order the awarding authority to amend any documents. Set-aside and amendment orders, like interim measures, may only be granted if the contract in question has not been entered into. See *Exel Europe Ltd v University Hospitals Coventry and Warwickshire NHS Trust* (2010) where the court applied the American Cyanamid Test and considered the balance of convenience, taking into consideration the public interest in the efficient and economic running of the NHS.

Damages

Under regulation 48(1) of the Regulations, a remedy in damages is also available to a complainant, regardless of whether or not the contract in question has been entered into.

Damages are available to an economic operator who has suffered loss or damage as a consequence of a breach of the Regulations. Although regulation 48(1) does not expand upon the principles governing the availability and amount of damages in proceedings under Scots law, successful proceedings may potentially lead to significant financial liabilities for the contracting authority.

Ineffectiveness

A new remedy of ineffectiveness was introduced by Directive 2007/66/EC, which allows the courts to set aside contracts which have been entered into by making a declaration of ineffectiveness. The relevant provisions in respect of this new remedy are contained in regulation 49 of the Regulations. There are three grounds on which a declaration can be awarded:

1. If a contract notice was not sent to the Official Journal, i.e. there has been an illegal direct contract award, see regulation 49(5).

2. A contract has been awarded in breach of the standstill rules or proceedings have been served on the contracting authority and not disposed of by the court, coupled with another breach of the Regulations which has had an effect on the bidder's chance of winning the contract, see regulation 49(7); or

3. There has been a breach of the procedural rules for the award of a contract under a framework or dynamic purchasing system, being an above threshold contract, see regulation 49(8).

The court has discretion not to grant a remedy of ineffectiveness even if ineffectiveness has been established. Regulation 48(2) provides that in any interim proceedings under regulations 47–50 the court may decide not to grant an interim order when the negative consequences of such an order are likely to outweigh the benefits, having regard to the following considerations:

(a) that decisions taken by a contracting authority must be reviewed effectively and, in particular, as rapidly as possible;
(b) the probable consequences of an interim order for all interests likely to be harmed; and
(c) the public interest.

If a court decides to grant a declaration of ineffectiveness, it may shorten the duration of the awarded contract and/or award a civil financial penalty against the contracting authority. Regulation 48(3) states that where the court is satisfied that regulation 49(7)(a) applies but the second ground for ineffectiveness is not otherwise met (i.e. the other pre-conditions for ineffectiveness under paragraphs (b), (c) and (d) of regulation 49(7)), it must order:

- the contracting authority to pay a financial penalty to the Scottish Ministers; or
- where the contract in relation to which the breach occurred has been entered into, or the framework agreement in relation to which the breach occurred has been concluded, the shortening of the duration of the contract or framework agreement, and ensuring that the terms of the order are effective, proportionate and dissuasive.

European remedies: corrective procedure

As well as (or instead of) bringing an action before a national court, aggrieved parties may also lodge complaints with the European Commission. Once such a complaint is lodged, the European Commission may initiate what is known as a 'corrective' procedure if it is satisfied that a clear and manifest breach of public procurement rules has been committed during a contract award procedure, see EC Directive 92/13, Chapter III.

2.17.4 Bringing proceedings under regulation 47

The Regulations provide that proceedings may not be brought against a contracting authority unless that contracting authority is informed of the breach or apprehended breach and of the complainant's intention to bring proceedings, see regulation 47(6) and *Amaryllis Ltd v HM Treasury (sued as OGC buying solutions)* (2009). If seeking an ineffectiveness order, such proceedings must be brought within thirty days beginning with the date when the economic operator first knew or ought to have known that grounds for starting the proceedings had arisen, see regulation 47(7)(b).

The Regulations were amended in this respect on account of case law which indicated that the long-established principle that public procurement proceedings must be commenced 'promptly' is contrary to European Community law, see *Uniplex (UK) Ltd (Uniplex) v NHS Business Services Authority (NHS)* (2010). To ensure there is effective review, the ECJ said limitation periods for actions under the Regulations may not start to run until such time as the claimant knew or ought to have known of the alleged breach of procurement law. The judgment indicated that a concerned bidder can only come to an informed view as to whether there has been an infringement after it has been informed of the reasons for its elimination, and referring in this regard to economic operators' need for legal certainty. National limitation periods must not make it uncertain, impossible or excessively difficult to exercise any rights stemming from EC law, such as a right to effective review.

2.18 Awarding low-value contracts fairly

The Regulations and the preceding legislation set out the procedures to be followed in the award of public contracts valued above a certain threshold, but for some time it has been unclear as to what obligations apply to contracting authorities regarding the award of certain public contracts to which those Regulations do not apply. On 24 July 2006, the European Commission issued an interpretative communication on the Community law applicable to contract awards not (or not fully) subject to the provisions of the Public Procurement Directives ('the Communication'). This seeks to clarify the rules which apply to such public contracts, e.g. contracts valued at below the threshold for application of the Regulations and Part B contracts. The terms of the Communication may assist in interpreting one novel requirement in the Regulations which was not contained in Directives 92/50/EEC, 93/36/EEC and 93/37/EEC, as follows. A contracting authority proposing to award a public contract with an estimated value below the relevant threshold, or a proposed public contract which is otherwise exempt from the requirement to be advertised must, if required by its general EU obligations, for the benefit of any potential economic operator, ensure a degree of advertising sufficient to enable open competition and meet the requirements of the principles of equal treatment, non-discrimination and transparency, see regulation 8(21).

The Communication does not contain new legislative rules but provides Commission guidance on the application of the minimum standards of equal treatment, non-discrimination and transparency derived from the EC Treaty in the award of below-threshold public contracts. The requirement for transparency consists in ensuring, for the benefit of any potential tenderer, a degree of advertising sufficient to enable the services market to be opened up to competition and the impartiality of the procedures to be reviewed.

Importantly, the contracting authority must first decide if the public contract in question would be of interest to economic operators located in other EU member states. The Communication proposes that in making such a decision a contracting authority must be guided by an assessment of the relevance of that contract to the internal market on the basis of its subject matter, value and customary practices in the relevant sector.

Should the contracting authority decide that a public contract might be of interest to economic operators located in other EU member states, the Communication provides guidance under three distinct heads – advertising, contract award and review procedures.

In relation to advertising, an undertaking located in an EU member state must have access to appropriate information regarding a public contract before it is awarded, to allow it to be in a position to express its interest in that contract. In order to satisfy this requirement the Commission is of the view that a contracting authority must publish a sufficiently accessible advertisement prior to the award of the contract, and lists optional means of publication. The Communication states that the greater the interest of the contract to potential bidders from other EU member states, the wider the coverage should be. The advertisement should provide as much information as an economic operator from another EU member state will reasonably need to make a decision on whether to express interest in the procurement procedure.

It can be seen from *Sidey v Clackmannanshire Council* (2011) that rights and obligations of contracting authorities and bidders in a below threshold process fall into three tiers:

- Above threshold: the Regulations/Directives apply to procurement processes, remedies are available in terms of the Regulations/Directives and obtainable by way of sheriff court, commercial or ordinary action;
- Below threshold with cross-border interest: general principles of European law and public law apply to procedures followed and decisions taken but statutory remedies are not available in terms of the Regulations/Directives. In principle, contractors may bring Judicial Review proceedings against a contracting authority;
- Below threshold with no cross-border interest: European legal rules are not engaged under the Directive/Regulations but public law would apply in relation to contracting authorities and, in principle, contractors may bring Judicial Review proceedings against a contracting authority in respect of its acts and decisions it has taken.

2.19 Forthcoming changes in the Procurement law landscape

The law as described in this chapter is intended to be correct at the time of writing. However, the law in this area is currently undergoing significant transformation. An overhaul of EU public procurement rules has been progressing for some time and new Public Procurement Directives were adopted by the European Parliament on 15 January 2014 published in the Official Journal of the EU on 28 March 2014 and come into force on 17 April 2014. Implementation is require by member states within two years. These replace the existing Public Sector Directive (2004/18/EC), the new Public Sector Directive (2014/24/EU), the Utilities Directive (2004/17/EC) and introduce new rules on the award of concession contracts with the introduction of the Concessions Directive (2014/23/EU). The provisions of the new Directives include:

- self-certification of compliance with prequalification requirements;

- the introduction of a procedure known as Innovation Partnerships to allow contracting authorities to procure currently unavailable innovative solutions;
- greater scope to exclude potential applicants, including for poor performance under previous contracts;
- the abolition of the distinction between Part A and Part B services and a new 'light touch' regime;
- the right on the part of utilities to use competitive dialogue;
- the introduction of measures to encourage the division of larger contracts into lots;
- the introduction of further rules governing permissible changes to existing contracts.

There are no changes to the existing remedies regime (see Section 2.17).

The Scottish Parliament is also proceeding with changes in the area of procurement law and long-awaited Procurement Reform (Scotland) Act 2014 has now received Royal Assent and commenced on 17 June 2014. The A includes provisions obliging or requiring contracting authorities:

- to prepare and publish procurement strategies and annual procurement reports (for contracting authorities with a procurement annual spend over £5 million);
- to consider including community benefit requirements (for major procurement contracts valued above £4,000,000); and
- to take into account guidance to be issued by the Scottish Ministers, when preparing pre-qualification questionnaires.

The Act also enables the Scottish Ministers to make regulations as to how Scottish public bodies assess bidder suitability to tender for public contracts, in the light of concerns over blacklisting and related practices.

Chapter 3
Entering into a Construction Contract

3.1 *Introduction*

Many people may not appreciate the frequency with which they enter into a contract while going about their everyday business. For example, purchasing a train ticket constitutes the formation of a contract between the railway company and the passenger. Few people are aware that this is a formal legal arrangement which imposes rights and obligations on both the passenger and the railway company. In reality, contractual relationships of one nature or another hold the very fabric of the commercial world together, including the construction industry. Without the certainty that a contract provides, the resulting chaos would inevitably render the conduct of business, in any meaningful sense of the term, impossible.

3.2 *Essentials of written and oral contracts*

3.2.1 Agreement

A contract is essentially an agreement, expressed either in writing or verbally, between a number of parties (not necessarily restricted to two) regarding the same subject matter. The law relating to the formation of contracts is of general application, notwithstanding the diversity of subject matter which may constitute the agreement between the parties. In this regard there is little distinction between, for example, a contract for the sale of goods and a building contract. The essentials of formation for both are identical.

A contract is formed when the parties to it reach agreement as to the essential elements of the transaction. There must be what is termed *consensus in idem*, the literal meaning of which is 'agreement in the same thing'. There is no need for consensus between the parties in relation to every detail of the transaction – the test is an objective one. It was held in the case of *Muirhead & Turnbull* v. *Dickson* (1905) that: 'Commercial contracts cannot be arranged by what people think in their inmost minds. Commercial contracts are made according to what people say.' This does not mean that the courts will always disregard the presumed intention of the parties at the time the contract was entered into. In *Bank of Scotland* v. *Dunedin Property Investment*

MacRoberts on Scottish Construction Contracts, Third Edition. MacRoberts.

Co. Ltd (1998), the court held that in order to interpret a contract, the court was entitled to have regard to discussions between the parties in order to establish the parties' knowledge of the circumstances with reference to which particular terms were used in a contract. Only if both parties had something in their contemplation at the time the terms of the contract were agreed would the court have regard to such matters (but see Section 3.7 for more detail on contract interpretation).

There are certain exceptions to the general principle that a contract can only be construed by reference to what it actually says. Implied terms are considered in Section 3.4. There have also been statutory inroads. Sections 8 and 9 of the Law Reform (Miscellaneous Provisions) (Scotland) Act 1985 provide that the court may rectify a document which fails to express accurately the common intention of the parties at the time the agreement was made. It is important to note that for an application for rectification to be successful, an underlying agreement and common intention must be demonstrated. In other words, *consensus in idem* must be established. Further, in terms of section 1 of the Contracts (Scotland) Act 1997, contrary evidence may be led to show that there are additional terms to an apparently complete document (save where the document expressly states that it comprises the whole terms of the contract).

Even where the parties to a contract do not think that they have reached agreement, the court may consider that they have. In *Uniroyal Ltd* v. *Miller & Co. Ltd* (1985), it was held that, in establishing whether parties had entered into a contract, the fundamental principle to be applied is to consider whether or not, and when, there has been *consensus in idem* between parties. In that case there was no agreement regarding the price of goods to be supplied under the contract and it was held that such lack of consensus was fatal. The court held that there was no contract between the parties as, in this particular instance, price was a fundamental and essential part of the contract.

Performance of a purported contract in the mere belief by the parties to it that it was binding and where none of the essentials had been agreed will generally not be sufficient to enable the courts to conclude that there is *consensus in idem*. In *Mathieson Gee (Ayrshire) Ltd* v. *Quigley* (1952), it was held that it is not enough for the parties to agree that there was a concluded contract if there was otherwise lack of agreement on essentials.

Even where there appears, on the face of it, to be no agreement there may still be a concluded contract as a result of the actions of the parties. In *Roofcare Ltd* v. *Gillies* (1984), the pursuers submitted a tender to carry out repairs to the roof of the defender's property. Their offer was made subject to the condition that the quotation was 'subject to the undernoted terms and conditions and no alterations, exclusions, additions, or qualifications to the quotation and specification will be made unless confirmed in writing by Roofcare'. The defender accepted the quotation, confirming that the pursuers should proceed with the repair work to the roof 'making same wind and watertight'. The pursuers did not reply to this qualification. The defender then allowed the work to proceed and the pursuers sued the defender for payment. The defender contended that there was no *consensus in idem* due to the wind and watertight qualification not being accepted by the pursuers, and so there was no contract. The Sheriff

Principal held that there had indeed been consensus. The pursuers had presented an offer that was the only basis upon which they would carry out the contract unless otherwise agreed in writing. The defender, who knew of that condition, added his qualification knowing that if it was to be accepted the pursuers would do so in writing. The pursuers did not accept the condition in writing. Despite this, the defender allowed the work to proceed. By his actions the defender was held to have accepted that the wind and watertight qualification did not apply and that there was a contract between him and the pursuers.

3.2.2 Offer and acceptance

Offers should be contrasted with 'invitations to treat', where a party demonstrates by words or by conduct a willingness to negotiate a contract. In a construction context, where tenders are invited, these constitute invitations to treat. The tender that is submitted by the contractor in response constitutes an offer which is available for acceptance by the employer.

An offer must be communicated to the party to whom it is made, see *Thomson* v. *James* (1855). It is thought that where an offer is communicated by a third party (unless an authorized agent of the offerer, for example, a solicitor), it cannot be accepted. Only where the offer is communicated by the party making it can it be accepted.

A simple, unconditional offer may be revoked at any time before acceptance, see *Thomson*. The revocation must be communicated to the recipient of the offer before it has any effect. Thus, offerers may change their mind at any time prior to acceptance. On the other hand, if an offer is stated to be irrevocable for a certain period, it cannot be withdrawn during that time. However, the period during which a firm offer is to be kept open cannot be vague or it is unenforceable, see *Flaws* v. *The International Oil Pollution Compensation Fund* (2002).

Where a time limit for acceptance is specified within the offer and no acceptance is received within that time, the offer will fall unless the offerer extends the time limit for acceptance. In *Thomson* it was held that an offer, pure and unconditional, puts it in the power of the party to whom it was addressed to accept the offer, until by the lapse of reasonable time he has lost the right. What constitutes a reasonable time will depend on the facts and circumstances of each case.

A simple offer made without limit of time may lapse where there is a material change of circumstances after the offer has been made. In *McRae* v. *Edinburgh Street Tramways Co.* (1885), it was held by Lord President Inglis that the change of circumstances must render the offer 'unsuitable and absurd' before it will lapse.

Where an offer is made and has not lapsed due to any of the above factors the contract will be concluded, provided there is agreement between the parties, when the offer is accepted. Acceptance can be express or it can be implied from the actions of the recipient of the offer. Again, it is essential that the acceptance is communicated to the offerer.

There is a general rule in Scots law that silence by the recipient of the offer does not imply acceptance of the offer, subject to two exceptions:

- unilateral or 'if' contracts where uncommunicated acts of the party accepting the offer may be sufficient to conclude the contract (see *Carlill* v. *Carbolic Smokeball Company* (1893)); and
- the postal acceptance rule (see Section 3.2.3).

If an offeror stipulates that silence will amount to acceptance, that offeror will be personally barred from later arguing against the offeree that no contract exists as a result of the absence of a communicated acceptance.

In *Wylie & Lochhead* v. *McElroy & Sons* (1873), it was held that the contention by the pursuers that the offerees' silence inferred acceptance was a most unreasonable one. Actions on the part of the offeree may be sufficient to infer that they have accepted the offer. In *Gordon Adams & Partners* v. *Jessop* (1987), the defender instructed the pursuers to place his property on a list of properties for sale. The pursuers, after inspection of the premises, wrote to the defender stating that the property was placed with them on a 'sole agency' basis. The defender's solicitor wrote to the pursuers stating that while the pursuers were instructed to place the property on the list, they were not appointed as sole agents. The pursuers responded that they would not accept property unless it was on a sole agency basis. The defender did not respond to that but allowed the pursuers to continue to place the property on their list. It was held that a contract existed between the parties. The defender, in the full knowledge that the pursuers were insisting that they were sole agents, allowed the pursuers to place the property on their list. In the light of the defender's actions the pursuers' belief that there was a contract between the parties was a reasonable one, induced by the defender's behaviour.

Acceptance can be verbal, written or implied from the conduct of the parties. Above all, the acceptance must meet the offer. An acceptance which does not accept all of the parts to the offer or which tries to incorporate conditions or qualifications into the offer is not an acceptance at all but a counter-offer, see *Wolf & Wolf* v. *Forfar Potato Co.* (1984). In general, the effect of a counter-offer is to refuse the original offer, which will then fall and can no longer be accepted. Where a counter-offer is accepted unconditionally by the original offerer, then the contract will be concluded.

Acceptance of an offer must be communicated to the offerer before the contract is concluded. There are exceptions to this general rule, for example, where the contract is concluded as a result of the actions of the parties and where acceptance is made by post. The offer may stipulate the method of acceptance, for example, by post, email, fax or telephone. Where the method of acceptance is stipulated, communication of the acceptance must be made by that method or it will be invalid. Where no method of acceptance is stipulated, the acceptance is valid provided it is made in a competent manner. Particular rules apply regarding postal communications, see Section 3.2.3.

It is trite law, but worthy of reinforcing, that the sending of contract documents by an employer to a contractor and subsequent signature by the contractor does not amount to an offer by the employer that has then been accepted by the contractor – the contract signed by the contractor constitutes the offer which to be binding must then be accepted by the employer (*Liberty Mercian Ltd* v. *Cuddy Civil Engineering Limited and Cuddy Demolition and Dismantling Limited* (2013)).

As regards the use of emails, case law has confirmed that where the terms of a contract have been negotiated by email, and agreement has been reached, this could create a binding contract, in writing and signed, even if the actual physical document was not prepared and/or signed, see *Golden Ocean Group Ltd* v. *Salgaocar Mining Industries PVT Ltd and another* (2011). If there is a reply accepting the terms, then the contract can be formed. This is especially so if the reply is not in a fresh email, but is part of a chain of emails, see *Nicholas Prestige Homes* v. *Neal* (2010). Of course, in terms of a variation to a contract, any variation must follow the form (if any) stipulated by the original contract. In the absence of such a stipulation (and subject to any statutory requirements such as those in the Requirements of Writing (Scotland) Act 1995), there is no set form for a variation, i.e. it need not match the form of the original contract.

The Scottish Government is, at the time of writing, considering legislation which would allow the English concept of execution of contracts by counterparts to be valid for Scots law contracts as well as conclusion of contracts by email, (the Legal Writings (Counterparts and Delivery) (Scotland) Bill).

3.2.3 The postal acceptance rule

An offer, withdrawal or rejection of an offer is only valid when it has been received by the other party. Conversely, a contract is formed when an unqualified acceptance is posted, and not when it is received, even if a specified time limit for acceptance has passed by the time the acceptance is received (but not prior to the acceptance being posted), see *Jacobsen Sons & Co.* v. *E Underwood & Son Ltd* (1894).

The case of *Thomson* v. *James* (1855) sets out the principles of the postal acceptance rule. In that case the offer was posted to the offeree. The offeree posted his acceptance and, on the same day, the offerer posted a letter withdrawing the offer. Both letters arrived at their respective destinations on the same day. The question for the court was which letter took effect first – was there a concluded contract or did the letter withdrawing the offer take effect before the letter accepting the offer? Obviously, if the retraction was effective first, then the offer no longer existed and could not be accepted.

It was held by the court that the acceptance was effective and that, therefore, there was a concluded bargain between the parties which could not be affected by the letter of revocation. The rationale was that an acceptance is effective when physically posted whereas a letter revoking an offer is not effective until it actually becomes known to the offeree.

Where the offer specifies a time limit within which it must be accepted, acceptance will be effective provided the acceptance is posted within the time limit. It is of no consequence to establishing whether there is a concluded contract if the acceptance is not actually received until a few days after the time limit expires, provided it is posted before the time limit expires.

The courts may, in very exceptional circumstances, depart from a strict application of the postal acceptance rule if to apply it would lead to an absurd result, see, for example, *Burnley* v. *Alford* (1919).

Although telex may be largely obsolete now, the case law is relevant when considering more modern forms of communication. Where acceptance is made by telex transmission, it has been held in the English case of *Brinkibon Ltd* v. *Stahag Stahl* (1983) that the postal rule does not apply – telex is a method of instantaneous communication and is therefore treated in the same way as an oral communication. A telex acceptance was held to be effective when printed out at the offerer's end.

Facsimile transmissions are accepted as being binding, provided that the type of contract is not one which is required to be in writing by the Requirements of Writing (Scotland) Act 1995 (for example, missives in relation to the sale of land), see *EAE (RT) Ltd* v. *EAT Property Ltd* (1994) and *Park, Petitioners (No. 2)* (2009). It would always be prudent to follow up faxed documents with hard copies. The Court's concern in *Park* was that transmission by fax did not put the relevant document beyond the control of the party sending the fax, and there was no indication that the document was being held to the other party's order.

The position regarding emails is less certain. The courts in England have said that the postal rule is not applicable to email as it is instantaneous, see *Thomas and Gander* v. *BPE Solicitors (a firm)* (2010). That said, email communications are not necessarily instantaneous and may take some time to reach their recipient. This might lead to a conclusion that the general postal acceptance rule should apply. The 'control' argument also favours this conclusion; once the email has been sent, the sender has no control over ensuring it reaches its recipient.

There is, however, the counter-argument that, generally speaking, emails do reach their recipients quickly and it is often possible to track receipt of emails whether electronically or by telephoning the recipient to ensure successful receipt. The origin of the postal rule was to create certainty at a time when post was the only form of communication and was generally slow. It may be prudent to set out in an order when acceptance will be deemed to have taken place, in particular where email is used.

There is also uncertainty over how the courts would treat the recipient of a fax or email who does not activate the means of accessing the communication within a reasonable time. It may be that the courts will consider this to be similar to the circumstances in *Burnley* v. *Alford* (1919), holding that the communication becomes effective when it ought, in the ordinary course of business, to have been read. It is immaterial if one party did not read the communication either fully or at all, see *Thomas and Gander* (2010) in which it was held (reiterating *Brinkibon*) that the question should be 'resolved by reference to the intentions of the parties, by sound business practice and in some cases by a judgment made where the risks should lie'. See also *Bernuth Lines Limited* v. *High Seas Shipping Limited* (2005), where notice of arbitration was held to have been validly served notwithstanding it might not have reached the relevant staff of the recipient company. However, this is open to contrast with the Scottish Law Commission's view that 'in a context where the concept of "business hours" is relevant, a communication that reaches the addressee's system outside those hours will become accessible … when the next period of business hours opens' (Scottish Law Commission Discussion Paper on Formation of Contract (Disc Paper No. 154, 2012, paragraph 2.17)).

Regulation 11 of the Electronic Commerce (EC Directive) Regulations 2002, while not resolving the point completely, states that orders (including contractual offers) and acknowledgements of receipt 'will be deemed to be received when the parties to whom they are addressed are able to access them'. Regulation 11 applies to parties who are not consumers.

Beware also relying on the use of email disclaimers as they will not always be sufficient; ideally they would refer to both the email and any attachments. It may be advisable to spell out the position in the communication itself (whilst there is little case law on this point, it was briefly considered in *Baillie Estates Ltd* v. *Du Pont (UK) Ltd* (2009), on which see Section 3.2.4).

3.2.4 Battle of the forms

It is common within the construction industry for offers to be made subject to the offerer's standard conditions of contract (frequently printed on the reverse side of the offer or appended to it). Difficulties arise where the offeree accepts the offer subject to the qualification that the offeree's standard conditions will apply. In the ordinary course of events this would undoubtedly constitute a counter-offer requiring the offerer's acceptance. Where, however, work is commenced prior to the counter-offer being accepted, a question arises as to whether there was, in fact, a contract and, if so, on whose terms. While it has often been said of this scenario that the person firing the last shot will be successful, it has also been commented that it may be more helpful to look at the documents as a whole to determine whether the parties have reached agreement on essential points, notwithstanding differences between the forms, see *Butler Machine Tool Co. Ltd* v. *Ex-Cell-O Corporation* (1979).

The Scottish case of *Baillie Estates Ltd* v. *Du Pont (UK) Ltd* (2009) is interesting regarding this point. Du Pont sent an email with its commercial proposal on pricing and delivery, to which Baillie replied 'go ahead'. It was at this point that the contract was formed, and not a few days later when Du Pont sent its standard terms and conditions to Baillie. As these were sent subsequent to the contract being formed, they were not applicable.

It is of course possible that neither set of conditions will apply. The High Court in England so held in *GHSP Inc* v. *AB Electronic Ltd* (2010), as it was clear that neither party would accept the other's standard terms. Notwithstanding the 'last shot' doctrine, no formal contractual terms were ever concluded and, in that case, the implied terms of the Sale of Goods Act 1979 applied. This does not of course mean that this will always be the outcome of such a battle, but it does highlight a need for greater certainty over contractual terms prior to commencing works. Another example of this is *AE Yates Trenchless Solutions Limited* v. *Black and Veatch Limited* (2008). In this case an invitation to tender specifying a particular standard form (the IChemE Form of Contract: Subcontract for Civil Engineering Works (the Brown Book)) had been issued by a contractor, with the sub-contractor issuing a tender in response specifying a different standard form. The court found that the contractor had, at a pre-contract meeting, rejected the sub-contractor's offer (and its specified terms) by stating that the Brown Book would apply. It appeared from the evidence that the sub-contractor had not taken issue with this, and in particular had not insisted at this stage, or at any time thereafter, that its required ground conditions clause must be incorporated into the sub-contract. Despite the sub-contract prepared by the contractor not being signed, the parties proceeded on the basis of the Brown Book and the court therefore held (following long-established case law starting with *Brogden* v. *Metropolitan Railway* (1877) whereby an objectively construed course of conduct and dealing can amount to acceptance of an offer) that it was this form which applied. Although the sub-contractor's tender had been incorporated into the sub-contract, it fell lower down the order of precedence than the standard form, which therefore took priority.

Ultimately, the issue will be decided on the basis of an objective assessment of what the parties agreed, looking at the evidence in the particular circumstances of the case. Clearly in a world of faster electronic communication, and in particular the prevalence of email, parties will need to exercise care over when contracts are formed and their terms.

3.3 *Capacity to contract*

Special rules apply to the capacity of certain categories of persons to enter into contracts. The main categories are as follows.

3.3.1 Young persons

The Age of Legal Capacity (Scotland) Act 1991 makes a distinction between two groups of young people, namely, those under the age of 16 and those aged between 16 and 18. With limited exceptions, a person under the age of 16 has no legal capacity to enter into any transaction (s.1(1)(a)).

A person over the age of 16 has legal capacity to enter into contracts. A person who enters into a contract between the ages of 16 and 18 can, in certain circumstances, apply to the court to set aside the contract if it is shown to be prejudicial and provided that such an application is made before the person concerned attains the age of 21 (s.3).

3.3.2 Insanity

It is a general principle of Scots law that an insane person has no power to contract and any contracts which such a person purports to enter into are void, see, for example, *Gall* v. *Bird* (1855). In addition, such contracts are generally void even although the other party may not have known that he was dealing with a person of unsound mind at the time that the contract was entered into, see *John Loudon & Co* v. *Elder's Curator Bonis* (1923). However, continuing contracts into which a party has entered while they were sane are not necessarily rendered void by that party's subsequent insanity, see *Howie* v. *CGU Insurance plc* (2005).

3.3.3 Aliens

It is the position in Scots law that, conforming to the Rome Convention, a contract made during a period of residence in Scotland cannot be set aside on the ground that one of the parties was an alien who lacked contractual capacity under his or her own legal system unless it is proved that the other party knew of the incapacity or was negligently unaware of it. This principle is incorporated into the Scottish legal system by s.2 of the Contracts (Applicable Law) Act 1990.

3.3.4 Corporate bodies

A corporate body is a distinct legal entity which is entirely separate from the members of the corporation. A corporate body can enter into contracts and can sue and be sued. Corporate bodies will contract through their agents. The agent must have express or ostensible authority to bind the corporation to the contract he purports to make. Directors of companies have ostensible power and authority to bind the company in transactions.

A corporate body created by statute, or exercising statutory powers, cannot enter into any contract or dispose of its funds in any way which is not sanctioned by the statute or reasonably incidental to the powers conferred. To do so would be *ultra vires*, i.e. beyond its powers. Where a party is dealing with a company incorporated under the Companies Acts, the position regarding *ultra vires* has been simplified so that both new and existing companies have one main constitutional document, that being articles of association. A deeming provision will transfer an existing company's memorandum of association into its articles (see s.28 Companies Act 2006).

In terms of s.31 Companies Act 2006, unless a company's articles specifically restrict the objects of the company, its objects are unrestricted. Where a company deals with a person in good faith, the power of the directors to bind the company, or to give permission to others to do so, is currently deemed to be free of any limitation under the company's memorandum and articles of association, see s.40 Companies Act 2006. A person is not to be regarded as acting in bad faith by reason only of their knowing that an act is beyond the powers of directors under the memorandum and articles of association of the company. In addition, a person is presumed to have acted in good faith unless the contrary is proved, see s.40(2) Companies Act 2006. Further, a party to a transaction with a company is not bound to enquire as to whether the transaction is permitted by the company's memorandum or as to any limitation on the powers of the board of directors to bind the company or to sanction others to do so, see s.40(2) Companies Act 2006.

3.3.5 Limited liability partnerships

Limited liability partnerships (LLPs) were created by the Limited Liability Partnership Act 2000 and are designed to offer the structural flexibility and tax status of partnerships combined with limited liability for its members. An LLP is a body corporate, and is a separate legal body from its members. This allows the LLP to enter into contracts and hold property. Unless provided under statute, usual partnership law shall not apply to an LLP, as LLPs are more akin to companies than partnerships. Unlike a partnership, if a member leaves the LLP, the LLP continues to exist.

Where a party contracts with the LLP rather than an individual member, any claim for breach of contract lies against the LLP, and only a claim in delict may be brought against the individual member. The members will no longer be liable jointly (and in Scotland severally also) for the debts and obligations of the LLP.

Members of the LLP may represent and act on behalf of the LLP in all its business. However, the LLP will not be bound by the actions of a member where that member has no authority to act for the LLP and the person dealing with the member is aware of this or does not know or believe that the member was in fact a member of the LLP. Where a person has ceased to be a member of the LLP, he will still be regarded as a member of the LLP in a question with a person dealing with that former member unless (a) the person dealing has had notice that the former member has ceased to be a member of the LLP or (b) notice that the former member has ceased to be a member of the LLP has been delivered to the Registrar of Companies.

3.4 *Implied terms*

Implied terms are those which may be implied into a contract to reflect the presumed though unexpressed intention of the parties or which may be implied by statute or other rule of law irrespective of the intention of the parties, see *William Morton & Co* v. *Muir Brothers & Co* (1907). They may not have been expressly agreed by the parties, but apply anyway.

3.4.1 Factual implication

A number of tests have been devised by the courts to determine whether the implication of terms may be permissible. It goes without saying that factual implication is heavily dependent upon the facts and circumstances of the case in question.

In *The Moorcock* (1889), Lord Justice Bowen formulated a test for implying terms, which has become almost universally known as the 'business efficacy' test. A comprehensive re-statement of this can be found in the analysis of Lord Simon of Glaisdale in *BP Refinery (Westernport) Pty Limited* v. *Shire of Hastings* (1977):

> 'for a term to be implied, the following conditions (which may overlap) must be satisfied: (1) it must be reasonable and equitable; (2) it must be necessary to give business efficacy to the contract, so that no term will be implied if the contract is effective without it; (3) it must be so obvious that 'it goes without saying'; (4) it must be capable of clear expression; (5) it must not contradict any express term of the contract.'

Another approach adopted by the courts is to apply what has become known as the 'officious bystander' test. The test derives its name from the judgment in the English

Court of Appeal case of *Southern Foundries (1926) Ltd* v. *Shirlaw* (1939). The test is essentially whether an 'officious bystander', a fictitious person who was privy to the discussions of the contracting parties, upon proposing the inclusion of a term would be 'testily suppressed with a common "Oh of course" by the parties'. Put more simply, is the term proposed by the 'officious bystander' so obvious that its intended inclusion goes without saying? Would a reasonable person, given the background knowledge available to the parties at the time, have understood the parties' intentions to be to include such a term?

In either instance it is clear that the factual implication of a term must be reasonable in the circumstances, see *Morton*. However, not all terms that are reasonable may be capable of being implied. The courts will not imply a term merely because to do so would be fair. The terms of a contract can be harsh and unfair, but terms will be implied only if they are necessary to make the contract work, see *Mediterranean Salvage and Towage Ltd* v. *Seamar Trading and Commerce Inc* (2009), in which the Court of Appeal noted that it is not sufficient that a clause is merely reasonable; and also *Leander Construction Ltd* v. *Mulalley and Company Ltd* (2011) where the contract operated 'perfectly satisfactorily' without the implied term argued for. The court must consider whether the term in question would 'spell out in express words what the instrument, read against the relevant background, would reasonably be understood to mean' (Lord Hoffman, in *Attorney General of Belize* v. *Belize Telecom Limited* (2009)). The Privy Council in *Belize* went further and confirmed that the conditions set out in *BP Refinery (Westernport) Pty Ltd* v. *Shire of Hastings* (1977) noted above were a 'collection' of different ways of saying the same thing, i.e. that a proposed implied term must spell out what the contract actually means. However, although the *Belize* case has been followed and is regarded as the leading case on the issue of implied terms, the courts have stressed that the business efficacy test should still be applied (see *Mediterranean Salvage*).

In relation to the implication of a contract following expiry of letters of intent, the test for implication is necessity (see *The Trustees of Ampleforth Abbey Trust* v. *Turner & Townsend Project Management Limited* (2012) citing Tomlinson LJ in *JD Cleverly Limited and Cwmbran Motors Limited* v. *Family Finance Limited* (2010)).

3.4.2 Course of dealings and custom and usage

Another route to inclusion of an implied term is where the parties have consistently used certain terms in previous dealings. A court may imply a term if it can be shown that it was the reasonable expectation of the parties that a particular term would apply, notwithstanding it is not expressly included, and on the basis that there is no express term to the contrary in the relevant contract. Of course, there would need to be consistent, regular trading over a period of time. This may be a difficult argument to win; a limited course of dealing may not be sufficient for a court to decide that an impartial observer would have concluded that the parties intended particular terms to apply, see *Capes (Hatherden) Ltd* v. *Western Arable Services Ltd* (2009).

Furthermore, a term may also be implied on the basis of custom and usage, particularly in a district or trade or other context, see *Morton*. Implication under this head still requires evidence of necessity and to some extent overlaps with the officious bystander test, albeit it in a restricted sense. Usage must be notorious, certain, reasonable, and not contrary to law, see *Yates* v. *Pym* (1816). Further, it must be more than a mere trade practice (*Cunliffe-Owen* v. *Teather and Greenwood* (1967)), and it must not be contrary to the express terms of the contract.

3.4.3 Rule of law

Implication arising by operation of a rule of law manifests itself in a number of ways. The implied duties that are incumbent upon a seller as to quality have long been elevated to statutory form and indeed still are, see the Supply of Goods and Services Act 1982. However, there are many instances where the present legislation applicable to sale is inapplicable. In such instances the courts have been willing to imply almost identical duties on the seller as those imposed under statute, simply because they are legal incidents to contracts of sale. However, in the absence of a precedent, implication on this basis will seldom arise, see *Scottish Power plc* v. *Kvaerner Construction (Regions) Ltd* (1999) where the court found that any implication of terms should be exceptional. There are also of course a number of terms which are implied by law into construction contracts by virtue of the Housing Grants, Construction and Regeneration Act 1996 (as amended) regarding payment terms (though not actually a right to payment, see *Mowlem plc* v. *Phi Group Limited* (2004)) and adjudication.

3.4.4 Implied terms in the construction industry

The following part of this chapter is restricted to a consideration of the implication of terms in contracts within the construction industry. A more detailed appraisal of implied terms lies outwith the scope of this book.

It would appear that the presence of an alternative remedy under a contract (albeit less attractive than the one sought under the implied term) will generally preclude the implication of an implied term. Thus, where a contract provided for works to be carried out in phases and only one phase provided for an extension of time in the event of delay, the House of Lords refused to imply an extension of time clause into another phase where delay could have been dealt with under alternative provisions of that contract, see *Trollope & Colls Ltd* v. *North West Metropolitan Regional Hospital Board* (1973). In *F Brown plc* v. *Tarmac Construction (Contracts) Ltd* (2000), the court refused to imply a term where the claim rested on express provisions which sat alongside the proposed (broader) implied term. It should be noted that a 'more tightly and precisely formulated implied term' might have been justified, if pleaded. A more recent example is *Leander Construction Ltd* v. *Mulalley and Company Ltd* (2011) (on which more later) where the employer sought to withhold monies otherwise due to the contractor on the basis that the contractor had breached an implied term to proceed regularly and diligently with the works. The employer was ultimately unable to rely on this as the court refused to imply the term, notwithstanding there was an express right of termination for failure to so proceed. Indeed, an express right to terminate pointed away, rather than to, an implied term, because the parties had clearly already considered the consequences of a failure to proceed regularly and diligently. Generally, priority was given to the principle that 'provided that the main contractual obligation was an obligation to complete by a certain date, it was unnecessary and unhelpful to impose other interim progress obligations on the Contractor'.

The most common examples in construction contracts of implied obligations relate to the standards to be achieved by a contractor (see Section 5.3).

Over the years, the courts have become increasingly willing to imply terms which, in their most general form, have tended to require the employer and his agents (e.g. the architect or engineer) to fulfil their obligations timeously to allow a contractor to progress their works. While each case will necessarily turn on its own facts, the following terms have been implied in main contracts by the courts:

- that the employer and its agent (in this case the engineer) were obliged to provide the contractor with all necessary details and instructions in sufficient time to enable the contractor to execute and complete the works in 'an economic and expeditious manner and/or in sufficient time to prevent the claimants being delayed in such execution and completion', see *Neodox Ltd* v. *The Mayor, Aldermen and Burgesses of the Borough of Swinton and Pendlebury BC* (1958). Similar duties to the foregoing would also appear to be incumbent upon a contractor when supplying necessary information and not hindering completion by a sub-contractor, see *J & J Fee Ltd* v. *The Express Lift Co. Ltd* (1993). Where the contractor has prepared a contract programme such that the programmed completion date is earlier than the date for completion stated in the contract, the contractor is entitled to complete the works in accordance with the programme. However, the employer is only obliged to act within timescales that allow the contractor to complete the works in accordance with the contractual completion date and not to the contractor's accelerated programme, and the court will not imply a term to the contrary, see *Glenlion Construction Ltd* v. *The Guinness Trust* (1987).
- not to hinder or prevent the contractors from carrying out their obligations in accordance with their contract or from executing the works in a regular and orderly manner, and to give possession of the site within a reasonable time, see *London Borough of Merton* v. *Stanley Hugh Leach Ltd* (1985).
- the employer must take all reasonable steps to enable the contractor to discharge its obligations and to execute the works in an orderly and regular manner, including things which the architect is obliged to do to facilitate this, see *Mackay* v. *Dick and Stevenson* (1881) and *Lubenham Fidelities and Investments Co. Ltd* v. *South Pembrokeshire DC and Another* (1986).
- an employer, if he becomes aware that a member of his professional team is not performing correctly, has a duty to advise that team member accordingly, akin to the duty discussed by the Court of Appeal in *Panamena Europea Navigación (Compañía Limitada)* v. *Frederick Leyland & Co Ltd* (1947) and also discussed (but held not to be implied in the circumstances of that case) in *Lubenham Fidelities Ltd.*

In *Neodox* the court considered 'reasonable time' in the context of the provision of information. It said:

> 'What is reasonable time does not depend solely upon the convenience and financial interest of the claimant. No doubt it is to their interest to have every detail cut and dried on the day the contract is signed, but the contract does not contemplate that. It contemplates further details and instructions being provided, and the engineer is to have a time to provide them which is reasonable having regard to the point of view of him and his staff and the point of view of … [the employer], as well as the point of view of the contractors.'

The courts have been unwilling to imply additional terms as to the timing of a contractor's performance prior to the specified completion date, despite numerous claims suggesting there should be an implied obligation to proceed regularly and diligently with the works. The decision in *Greater London Council* v. *The Cleveland Bridge and Engineering Company Limited and Anor* (1986) suggested that a contractor was 'free to plan and perform the work as he pleases, provided always that he finishes it by the time fixed in the contract' (and of course meets any specified key dates/milestones). This proposition has generally been upheld in recent cases, and was considered in detail in *Leander* in which the court considered that it would be 'unnecessary and unhelpful' to impose interim progress obligations. Of interest in this case are the court's justifications for rejecting the implied term: (a) a right of termination for failure to proceed regularly and diligently could exist independently of an obligation to so proceed; (b) the inclusion of the termination right suggests that such an obligation was considered but rejected; and (c) though the contract included a completion date, if the implied term were allowed, further terms would need to be implied to provide a contractual mechanism to operate the implied term. See also the discussion of this topic in Sections 5.5.1 and 6.4.1.

Where the parties have clearly contemplated a risk, legal implication will not be sufficient to imply a term unless it satisfies the additional test of necessity, see *Martin Grant & Co. Ltd* v. *Sir Lindsay Parkinson & Co. Ltd* (1984). In addition, implication will generally be precluded where the term seeks to impose liability on a party for matters over which they have no control, see *Ductform Ventilation (Fife) Ltd* v. *Andrews-Weatherfoil Ltd* (1995). Finally, the courts have been unwilling to imply a term where the implied terms sought were at variance with the express provisions of the contract, see *Scottish Power plc* v. *Kvaerner Construction (Regions) Ltd* (1998).

Regard should also be had to any entire agreement or exclusion clauses which may preclude the inclusion of implied terms, see *Axa Sun Life Services plc* v. *Campbell Martin Ltd* (2011). Whether a term is implied will, of course, depend on the precise words of the relevant entire agreement or exclusion clauses. However, the *Axa* case did suggest that terms implied for business efficacy were 'intrinsic' provisions of the contracts and were not therefore excluded by the entire agreement clause.

3.5 Letters of intent

The Technology and Construction Court (TCC) in England has commented (in *The Trustees of Ampleforth Abbey Trust* v. *Turner & Townsend Project Management Limited* (2012)) that 'it is extremely rare for construction projects of any significance to be completed under letters of intent' and that 'completion of such a project under letters of intent is a mark of something having gone wrong'. Nevertheless, though best avoided, commercial necessity may demand that certain works or services are carried out before the parties are in a position to enter into the formal building contract. It is possible for design, supply and even construction to be started and sometimes completed on the basis of a letter of intent. The term 'letter of intent' covers a wide variety of pre-formal contract arrangements, both formal and informal, between parties. Indeed, it is probably a misnomer for the type of document that generally goes under that description in construction projects. In certain cases, such letters may be intended to fall short of establishing a legal relationship and merely to provide comfort to the recipient. However, they are normally intended to create a binding relationship, albeit for a restricted purpose and a limited period.

The expression of a future intention to contract under a letter of intent is capable of being construed as a legally enforceable promise under Scots law. Whether a letter of intent will fall short of establishing a legal relationship or constituting an enforceable promise will depend largely on the form of wording used in each case, a view supported by the court in *ERDC Group Ltd* v. *Brunel University* (2006). The phrase 'letter of intent' was said not to be a term of art, but one whose meaning and effect depend on the circumstances of each case.

A number of relatively recent cases have reinforced the proposition that valid and enforceable contracts can be created by letters of intent, see *Hackwood Ltd* v. *Areen Design Services Limited* (2005), and *Robertson Group (Construction) Ltd* v. *Amey-Miller (Edinburgh) Joint Venture and Others* (2005). Although in the latter case the letter of intent was agreed as being a 'stop-gap' arrangement in the context of on-going negotiations towards a formal contract, once the contractor had commenced work on the basis of the letter, it was contractually obliged to proceed with the works.

In many instances, where work has been carried out by one party pursuant to a letter of intent, the analysis of whether a contract has been formed is somewhat irrelevant; if there is no contract, the party who has tendered performance will have a claim based on *quantum meruit*, i.e. reasonable payment for work done. In *ERDC Group Ltd*, work proceeded on the basis of a letter of intent, with the scope of the works and the financial authority being increased by four further letters of intent. The authority in the final letter of intent expired on 1 September 2002, but the contractor continued working until the works were completed in November 2002. A formal contract was never signed, because the contractor argued that the scope of the works had changed significantly from that anticipated at the outset, and that it was therefore entitled to be paid on a *quantum meruit* basis. The court found that until 1 September 2002, there was a valid contract and payment was to be made (and had been made) in accordance with that contract. For work done after 1 September 2002, the contractor was to be paid on a *quantum meruit* basis but on the basis of the tender rates and prices as there was no reason why this original basis of payment should be different (the contractor had argued that it should be paid on a cost plus basis). The court noted the decision in *Sanjay Lachhani* v. *Destination Canada (UK) Ltd* (1997) that a contractor should not be better off as a result of the failure to conclude a contract than they would have been if their offer had been accepted.

However, where a party who has received performance seeks damages for breach of contract, the issue is likely to be highly important. Where a contract has been formed, it may be difficult to determine what form the contract takes. Unsurprisingly, this will depend on the relevant facts and circumstances. See, for instance, *RTS Flexible Systems Limited* v. *Molkerei Alois Müller Gmbh & Co KG* (2010), where three courts came to three separate conclusions as to what, if any, contract had been formed.

Where a letter of intent anticipates that, for example, a standard form of contract will be entered into by the parties at some future date and no such contract is subsequently entered into, the party who has received performance will be deprived of the protection which the terms of the standard form might otherwise have provided. The TCC has recently held that liquidated damages provisions in a standard form contract were not incorporated where the works were carried out and completed under letters of intent which referred to that standard form: '[t]he fact that the period mentioned in the final letter of intent had expired does not make it necessary to imply a full contract' (*The Trustees of Ampleforth Abbey Trust* v. *Turner & Townsend Project Management Limited* (2012)). Although in that case the liquidated damages provisions were not enforceable, the employer was awarded damages against the project managers based on how it would have benefitted from an executed building contract, taking into account the risk that the contractor would not have signed the contract. A slightly older example is the case of *Wescol Structures Ltd* v. *Miller Construction Ltd* (1998), where negotiations between the sub-contractor and both the main contractor and the employer's representatives proceeded on the basis of letters of intent. The sub-contractor insisted in its replies to the letter of intent that the standard DOM/2 form of sub-contract would apply, whereas the employer's representatives, who wrote the letters of intent, stated that the sub-contract would be 'back to back with the main contract' but failed to detail any specific terms. The case, which related to payment terms, was decided on the basis of the standard form even though the standard form was never entered into, partly because the employer had never challenged the sub-contractor's assumption.

English law is more developed in relation to letters of intent. The leading English authority on this point is *British Steel Corporation* v. *Cleveland Bridge & Engineering Co.* (1984). Here, where a party commenced work on the basis of the words 'pending the preparation and issuing to you of the official form of sub-contract' contained in a letter of intent, it was held that it was 'very difficult to see how [the plaintiff], by starting work, bound themselves to any contractual performance'. Among other things, neither the price, the delivery dates, nor the applicable terms of contract had been agreed. The *RTS Flexible Systems* case came to a similar conclusion that essential agreement had been reached, despite there remaining notes as to other clauses the parties wished and not all schedules having been agreed. None of the remaining items was considered essential and requiring agreement before the contract could be signed. It is worth noting that the omission of agreement as to price need not be fatal, see *Amec Capital Projects Ltd* v. *Whitefriars City Estate Ltd* (2003) and *Hackwood Ltd.* The use of the word 'pending' was indicative of a state of preparation only.

In trying to establish whether a contract does in fact exist, the courts look at correspondence between the parties and the conduct of the parties, see the Supreme Court's comments in *RTS Flexible Systems*. The conduct of the parties could even lead to a conclusion that an agreed prerequisite to the contract being formed had been waived (such as a clause in the letter of intent requiring the parties to sign the contract as in *RTS Flexible Systems*). This was effectively confirmed in *Ampleforth* where the TCC said that even

> '[the] fact that the parties have been dealing on a 'subject to contract basis' … does not of itself exclude the possibility that the time will come when the necessary implication of their conduct is that they have waived the requirement of a formal written contract.'

In *Hackwood Ltd* the letter of intent stated that the Joint Contracts Tribunal's (JCT) conditions 'will be' the basis of the contract, and the TCC held that such a reference was sufficient to incorporate the standard form save to the extent that such terms were inconsistent with the terms of the letter of intent, notwithstanding that the appendix and other project-specific data were not yet agreed. If relevant information was missing (for example, the amount of liquidated damages), then that mechanism only would fall away.

Notwithstanding the analogy with promise, a letter of intent may also be construed as an offer capable of being accepted, depending on its terms, see *Uniroyal Ltd* v. *Miller & Co. Ltd* (1985) and *Mowlem plc (t/a Mowlem Marine)* v. *Stena Line Ports Ltd* (2004). In determining whether an offer has been made, the terms of the letter itself are crucial.

3.6 Incorporation of terms by reference to another document

Generally, reference to a particular form of contract will be sufficient to incorporate its terms. Reference can be oral, but it is preferable that it is in writing (though not conclusive, see *Sidney Kaye, Eric Firmin & Partners (a firm)* v. *Bronesky* (1973)). An agreement that a contractual relationship will be governed by reference to a particular form of contract will be sufficient to incorporate those terms into that contract, subject to the conditions referred to being readily identifiable or at least identifiable with reference to common industry knowledge, see *Modern Building Wales Ltd* v. *Limmer & Trinidad Co. Ltd* (1975). Thus, reference to a sub-contractor's order being 'in accordance with the appropriate form for nominated sub-contractors RIBA 1965 edition' was sufficient, after evidence had been led to show that, while a contract formally called 'RIBA 1965 edition' did not exist, the term was commonly used in the building trade to refer to the 'green form'. See also *Aqua Design and Play International Limited* v. *Kier Regional Ltd* (2002), as to whether an amended or unamended standard form had been incorporated.

The courts will consider if the words used on the face of a document are reasonably capable of being understood as intended to incorporate the particular terms, see *Rooney and another* v. *CSE Bournemouth Ltd* (2010). Adequate notice of terms must be given, and the courts will consider if the parties intended that particular terms should be incorporated. See *Cubitt Building and Interiors Ltd* v. *Richardson Roofing (Industrial) Ltd* (2008), where terms which were said to be attached to the order were not in fact so attached, and it was held that this meant the parties did not intend for them to be incorporated. Of particular note are cases where documents have been sent by fax but terms and conditions printed on the back of those documents have not also been faxed. In these circumstances, the courts are unlikely to conclude that these have been sufficiently incorporated, see *J Murphy & Sons Ltd* v. *Johnston Precast Ltd* (2008).

The TCC case of *Allen Fabrications Limited* v. *ASD Ltd* (2012) is interesting for its summary of incorporation of standard terms and conditions. The court set out the two ways this may happen: either they may be on or referred to in a document that is provided to the other party prior to or at the time the contract is entered into, or they may be in or referred to in a post-contractual document (e.g. on an invoice) where there is a prior course of dealing between the parties using those documents, so that it can be inferred that the parties intended to contract on those terms.

The foregoing scenario envisages, however, that the terms of any conditions referred to will be suitable in the circumstances, for example, that a sub-contract relationship will be governed by known sub-contract terms. What is more problematic and, indeed, a relatively common occurrence in the construction industry, is where party A attempts to impose the terms and conditions to which it is subject, for example, under a main contract, into a sub-contract which they have entered into with party B. This is commonly referred to as a 'back to back' arrangement.

This issue was considered by the Outer House of the Court of Session in *Parklea Ltd* v. *W & J R Watson Ltd* (1988). Here a sub-contract purported to incorporate the main contract conditions into the sub-contract that also contained other express terms. A dispute arose as to whether the arbitration clause in the main contract was applicable to the sub-contract. A number of principles emerge from this case that are of guidance in assessing whether such terms are capable of incorporation:

- The starting point must be to consider whether the parties have incorporated the whole of the main contract conditions; it is irrelevant that some (and not others) of the conditions would have fitted very neatly into the sub-contract conditions.
- Do the words incorporating the sub-contract conditions make clear that they are applicable to the exclusion of all other provisions? It was held that a reference to the main contract conditions 'solely regulating' the relationship between the parties was not indicative of an exclusion of all other conditions; the subsequent reference to the applicability of the main contract provisions being excluded where they conflicted with other express terms of the sub-contract mitigated against such a construction.
- Where the purportedly incorporated terms conflict or duplicate other express terms of the sub-contract or duplicate the terms of the main contract, this will militate against the conclusion that the main contract terms will exclusively regulate the parties' contractual relationship.

Parklea Ltd follows a line of authority whereby the Scottish courts have been reluctant to apply the terms of an arbitration clause in similar circumstances. See also the later case of *Babcock Rosyth Defence Ltd v. Grootcon (UK) Ltd* (1998) discussed in Section 11.4. It is interesting to note that in *Parklea Ltd* it was a matter of agreement between the parties that only wholesale incorporation of the terms of the main contract would be sufficient to incorporate the arbitration clause. It would appear that the English courts might be prepared to adopt a broader approach. They have held that where the main contractor's terms are not inconsistent with the sub-contractual relationship, they could be incorporated. See *Brightside Kilpatrick Engineering Services* v. *Mitchell Construction* (1973) Ltd (1975), and, with particular regard to arbitration clauses, also *Giffen (Electrical Contractors) Ltd* v. *Drake & Skull Engineering Ltd* (1993) and *Roche Products Ltd and Another* v. *Freeman Process Systems Ltd and Another* (1996). However, where only 'general words of incorporation' are used, these will only be sufficient to incorporate terms of a main contract that are germane to the sub-contract and not those that are merely ancillary (see *Siboto K/S* v. *BP France SA* (2003)), but contrast this English case with the Scottish case of *Cameron (Scotland) Ltd* v. *Melville Dundas Ltd* (2001), where general incorporation wording was unsuccessful though that did not mean that specific elements of the main contract could not be relevant, in particular clauses regarding the quality of the work and materials, and performance by the sub-contractor not giving rise to conflict under the main contract.

More in keeping with the approach of the Scottish courts is the Canadian case of *Smith and Montgomery* v. *Johnson Brothers & Co. Ltd* (1954) where it was held that where a contract made reference to the terms of another contract and expressly incorporated a number of those terms, then only those expressly included would form part of that contract. Here the incorporation of an express term of a main contract, which made reference to the payment provisions in respect of nominated sub-contractors, and which was expressly incorporated into the sub-contract, was held to be valid. The English courts have also addressed the issue as to whether the words 'shall be deemed to have notice of all the provisions of the main contract' are sufficient to incorporate those terms and have answered in the negative, see *The Jardine Engineering Corporation* v. *The Shimizu Corporation* (1992).

Incorporation of terms may be relevant when considering letters of intent as these often refer to the terms of a building contract which has not yet been finally agreed. This can be a subject of disagreement if the final contract is not executed, as often one party argues that it cannot have been the intention of the parties that the terms of a contract that was still being negotiated should govern the relationship between the parties. In *Hackwood Ltd*, the TCC found that the object of the letter of intent was to establish an interim contract that would govern the relationship between the parties until the final contract was agreed, on terms that both parties appreciated could govern the whole of the project. In particular, the use of the future tense ('the basis of the contract will be ... ') did not indicate that the referenced terms should only apply to the final agreed contract and not the interim contract. In this case, the standard form of contract was incorporated into the interim contract, save for those terms that were inconsistent with the terms of the letter of intent.

The English courts have taken a fairly strict approach in relation to the incorporation of arbitration and dispute resolution clauses, requiring express reference to such clauses in the incorporation clause. This is because agreements regarding dispute resolution are regarded as personal to the parties, and collateral to the main obligations in the contract, see *Yorkshire Water Services Ltd* v. *Taylor Woodrow Northern Ltd* (2002), following the reasoning in *Aughton Ltd* v. *M F Kent Services Ltd* (1992). Jurisdiction clauses are usually treated in a similar manner.

3.7 Contract interpretation

The contract that is entered into records the bargain made by the parties, whether or not that is different from what was previously stipulated. When considering the express terms of a contract and trying to ascertain the parties' intentions, the courts have set out some broad principles of interpretation. The starting point is the parties' 'language interpreted in accordance with conventional usage' (*Bank of Credit and Commercial International SA (in compulsory liquidation)* v. *Ali and others* (2001)), from the standpoint of a reasonable businessperson.

The leading statement of the law in this area is now to be found in the recent decision of the Supreme Court in *Rainy Sky SA and others* v. *Kookmin Bank* (2011). In that case, Lord Clarke said that 'the ultimate aim of interpreting a provision in a contract, especially a commercial contract is to determine what the parties meant by the language used, which involves ascertaining what a reasonable person would have understood the parties to have meant'.

The House of Lords in *Investors Compensation Scheme Ltd* v. *West Bromwich Building Society* (1998) gave a summary of the principles of interpretation that should be adopted. In that case, Lord Hoffman stated that the process of interpretation, where there is a clear mistake in the language of the contract, is to decide 'what a reasonable person would have understood the parties to have meant by using the language which they did'. A reasonable person is one who has all the background knowledge which would reasonably have been available to the parties in the situation in which they were at the time of the contract (*Rainy Sky*). The court in *Chartbrook Limited* v. *Persimmon Homes Limited and others* (2009) stated that a court should only depart from these principles if it were confident that either the rule was impeding the proper development of the law or it was contrary to public policy.

The court will try to achieve an interpretation consistent with business common sense, considering the commercial purpose of the contract and the contract as a whole (and not just the particular clause in question). If there is no sensible commercial justification, this is unlikely to be the preferred interpretation.

Generally, if the language of the contract leads to a conclusion that one or other interpretation is correct, then the court must give effect to that interpretation. Where there is more than one possible interpretation of a clause, the court will normally adopt the one most consistent with business common sense (*Rainy Sky*), though there is no requirement for a court to 'attribute to the parties an intention which they plainly could not have had' (*Investors Compensation Scheme*).

In considering the terms of a contract, the court should take into account:

- the actual wording of the contract, as well as the wider (relevant) background knowledge and context to which the parties would have had access at the time of agreeing the contract;
- the general position in the market (provided there is evidence in support) (*Thomas Crema* v. *Cenkos Securities plc* (2010)).

This does not, however, extend to allowing evidence of what was said or done during pre-contractual negotiations as an aid to construction or for the purpose of drawing inferences about what the contract meant (though it may be useful background as to what the parties meant or for establishing what facts were known to the parties at the time).

The outcome of this exercise by the court may mean that the parties are held to terms which, had a full investigation been undertaken, a reasonable person would not have concluded was intended. Further, the mere fact that a contract may appear to be unduly favourable to one party is not a sufficient reason to decide that wording has a different meaning; clearly, a contract is not always a fair balance between the parties.

3.8 Signing a building contract

3.8.1 General

The requirements of Scots law in relation to the signing of documents are set out in the Requirements of Writing (Scotland) Act 1995 ('the 1995 Act').

Writing is not required for the constitution of a contract except where the contract relates to the creation, transfer, variation or extinction of an interest in land (s.1). Although writing is not required for other forms of contract, the parties may execute their contract in such a way as to render the contract self-proving.

The 1995 Act distinguishes between a document which has been validly signed and a document which has self-proving status. A validly signed document is one which has been subscribed by the grantor (s.2). Here extrinsic evidence is necessary to confirm the validity of the signatures. However, if the requirements of the 1995 Act regarding witnessing (which will be discussed in more detail later) have been followed, then the signatures of the parties will be afforded self-proving status. In effect, this means by virtue of the means of execution the signatures of the parties are presumed valid and need not be proved (s.3).

Where a contract has schedules annexed to it, the schedules will be incorporated into the contract if they are referred to in the body of the contract and it is identified on the face of the schedules that they are the schedules referred to in the contract. If this is done, there is no need for the schedules to be signed (s.8(1)). It is only where a contract relates to land and any attached schedule describes or shows all or any part of that land, that the schedule in question requires to be signed (s.8(2)); and depending on the nature of such schedule, it may require to be signed either on each page or on the last page (s.8(2)(c)).

The precise requirements of subscription vary depending upon the designation of those signing the contract.

3.8.2 Individuals

If a party to a building contract is contracting as an individual (which includes a sole trader), then that person must subscribe the contract and have their signature witnessed by one witness. If this is done, subscription by that person will be self-proving (s.3(1)).

3.8.3 Partnerships

In the absence of specific internal signing requirements, a contract will be validly executed on behalf of a partnership if it is signed by one partner or another person with authority (Schedule 2, paragraph 2(1)(3)). The signatory can either sign his own name or the name of the firm (Schedule 2, paragraph 3). The law regarding the power of a partner to bind a firm is set out in the Partnership Act 1890. The signature of the partner, or the authorized person, must be witnessed by one witness in order to make their subscription self-proving.

3.8.4 Companies

A company will validly execute a contract if it is signed by one director, the company secretary or by a person authority to sign the contract on the company's behalf (Schedule 2, paragraph 3). Again, for the subscription to be self-proving, a single witness must witness it. The contract will also be self-proving if it is signed by two directors, or a director and the company secretary or by two persons with authority (Schedule 2, paragraph 2(5)). In these circumstances there is no need for the signatures to be witnessed.

3.8.5 Limited liability partnerships

Signature of a document by a limited liability partnership (LLP) will be self-proving if it has been signed by a member of the LLP in front of a witness, or by two members of the LLP (Schedule 2, paragraph 3A(5)), though the document will be validly executed on behalf of the LLP if it is signed by a member of the LLP (Schedule 2, paragraph 3A(1)).

3.8.6 Local authorities

A contract will be validly executed by a local authority if it is signed by the proper officer, usually the chief executive, see Schedule 2, paras 4(1) and (3). A person purporting to sign as the proper officer is presumed to be the proper officer, see Schedule 2, paragraph 4(2). For the subscription to be self-proving, the contract must be subscribed by the proper officer on the local authority's behalf and either (a) the signature is witnessed by one witness or (b) the contract is sealed with the local authority's seal, see Schedule 2, paragraph 4(5).

3.8.7 Witnesses

The 1995 Act reduced the requisite number of witnesses from two to one. If more than one signatory is signing at the same time, one independent person can competently witness all signatures.

Witnesses must be independent with no direct interest in the contract. In addition, witnesses must be over the age of 16; be of sound mind; be able to write; and not be blind.

The witness must see the signatory sign the contract or, alternatively, the signatory can sign the contract outwith the presence of the witness and thereafter show their signature on the contract to the witness and acknowledge to the witness that the signature is in fact his. The witness must know the signatory but all that is required in that regard is a reliable introduction prior to signing or acknowledging, see *Brock* v. *Brock* (1908).

It is the practice for witnesses to sign opposite the signatory's signature and customary, though not strictly necessary, for witnesses to write the word 'witness' after their signature.

3.8.8 Electronic signature

The Electronic Communications Act 2000 means that electronic signatures and certificates supporting them are admissible as evidence 'in relation to any question as to the authenticity of the communication or data or as to the integrity of the communication or data' (s.7(1)). The question of whether electronic form is permissible in specific areas will be addressed on a case-by-case basis by statutory instrument; however, it is not yet possible to dispose of heritable property electronically (but see our comments at Section 3.2.2).

Chapter 4
Employers' Obligations

4.1 Introduction

A construction contract will usually set out, in express terms, the obligations owed by the employer to the contractor. Where these are not set out, certain terms will be implied due to the nature of the contract. Implied terms have been considered in Section 3.4.

The employer's obligations, be they express or implied, broadly fall into two main categories. First, an obligation of co-operation or the requirement to do certain things to put the contractor in a position of being able to carry out their own obligations under the contract. Second, an obligation to make payment for the work carried out by the contractor.

Although these will be referred to as the employer's obligations, the employer commonly employs others to perform certain of these functions on their behalf, for example, the architect/contract administrator under the SBC, the employer's agent under the SBC/DB, and the project manager and the supervisor under the NEC3. In other engineering contracts (e.g. the MF/1 form published by the Institution of Engineering and Technology) that role will usually be carried out by the person designated as 'engineer'. It follows that any reference to the employer's obligations will include obligations to be performed by other parties on the employer's behalf. Breach by these parties will lead to the employer in turn being in breach of its obligations to the contractor, see *Neodox Ltd* v. *Swinton and Pendlebury BC* (1958).

We will consider first the duties to do certain things necessary to enable the contractor to carry out his works. In *Mackay* v. *Dick and Stevenson* (1881), Lord Blackburn stated that it was:

> a general rule that where in a written contract it appears that both parties have agreed that something shall be done, which cannot effectually be done unless both concur in doing it, the construction of the contract is that each agrees to do all that is necessary to be done on his part for the carrying out of that thing, though there may be no express words to that effect. What is the part of each must depend on circumstances.

While this might be described as the positive duty flowing from the employer's obligation to do everything necessary to enable the contractor to carry out their works,

MacRoberts on Scottish Construction Contracts, Third Edition. MacRoberts.

another term which has been frequently implied is the obligation not to do anything which will hinder the contractor from carrying out their obligations under the contract or from executing work in a regular and diligent manner, see *London Borough of Merton* v. *Stanley Hugh Leach Ltd* (1985). In other words, the employer cannot do anything to prevent the contractor from performing their obligations under the contract.

Examples of the general obligation to do all that is necessary to enable the contractor to carry out their works are the obligation to give the contractor possession of the site; the obligation to administer the site; and the obligation to issue instructions and to provide information.

4.2 Possession of the site

4.2.1 General

In certain contracts it is an express term that the employer will give possession of the site, or the relevant part of it, to the contractor to enable him to carry out the works, see, for example, clause 2.4 of the SBC, clause 2.3 of the SBC/DB and clause 33.1 of the NEC3. Where this is not expressly stated, it will be implied as, in the majority of contracts, a contractor cannot carry out their works unless they actually have possession of the site, see *R* v. *Walter Cabott Construction Ltd* (1975). In referring to the requirement for the employer to give the contractor possession of the site, this is not possession in its legal sense but more a right of entry or control falling short of literal possession. Indeed, clause 33.1 of the NEC3 refers to 'access to and use of' the site rather than possession.

The questions which then arise are: when is the employer required to give the contractor possession?; what is the nature and extent of the possession which the employer requires to give the contractor (including consideration of whether or not the giving of possession implies that the contractor will have uninterrupted access to and possession of the site)?; and, finally, what is the duration of this obligation?

4.2.2 Time of possession

As stated above, the contract will often expressly state when the employer is required to give the contractor possession. Under the SBC and the SBC/DB this is the Date of Possession stipulated in the Contract Particulars, while under the NEC3 this is the 'access date' stipulated in the contract data or the accepted programme. Where the contract does not expressly provide for such a date, it will be implied that possession must be given within a reasonable time to enable the contractor to complete the works by any required completion date, see *T & R Duncanson* v. *The Scottish County Investment Co. Ltd* (1915). Under the SBC and the SBC/DB the Employer may defer the giving of possession for a period not exceeding six weeks from the Date of Possession, or such lesser period as is stated in the Contract Particulars, see clause 2.5 of the SBC and clause 2.4 in the SBC/DB. Any such deferment is a Relevant Event giving rise to a claim for an extension of time under clause 2.29.3 of the SBC and 2.26.3 of the SBC/DB. Clause 4.23 of the SBC and 4.20 of the SBC/DB provide for deferment of possession as a ground for claiming loss and/or expense.

Under clause 60.1(3) of the NEC3 a failure by the Employer to allow access to the site by the access date is a compensation event which may give rise to a claim by the Contractor for both an extension of time and costs.

Failure by or on behalf of the employer to grant an extension of time where they have delayed giving the contractor possession may result in the employer being unable to apply liquidated damages where the contract is prevented from achieving the completion date, see *Wells* v. *Army & Navy Co-operative Society Ltd* (1902).

4.2.3 Nature and extent of possession to be given

The contractor is normally entitled to possession of the whole site, subject to any express provision otherwise, such as in the case of sectional completion. That would appear to be what is meant by clause 2.4 of the SBC and clause 2.3 of the SBC/DB which refer to the Contractor being given possession of the 'site' on the specified date of possession. 'Site' is not a word that is defined by clause 1.1. In any case, it will be implied that the employer must make available the entire area that is necessary to enable the contractor to carry out their contract works. Clause 33.1 of the NEC3 is probably closer than the SBC to what would otherwise be the implied position in that the access and use which are to be given are to 'each part of the Site ... which is necessary for the work'. In some cases the necessary area has been held to extend beyond the actual area which will be occupied by the completed structure into other areas, for example, to provide working space and to enable the contractor to work efficiently and in accordance with generally accepted construction practices, see *R* v. *Walter Cabott Construction Ltd* (1975).

It is not an implied term, however, that the employer must provide work to the contractor in such a way as to enable them to carry out the work on an economic basis, see *Martin Grant & Co. Ltd* v. *Sir Lindsay Parkinson & Co. Ltd* (1984). A similar consideration arose in *Scottish Power plc* v. *Kvaerner Construction (Regions) Ltd* (1998). In that case the contract stipulated a period of 24 weeks for the work. There was no guarantee of continuous working. The Lord Ordinary held that the employer had power to interrupt the continuity of the period of 24 weeks. The reasoning behind that decision turns very much upon the provisions of the particular contract in question. We would suggest that the decision, which is to the effect that where there is a specified contract period and no guarantee of continuous working the employer has the right to interrupt, is not one which should be followed as a general principle. If this were an absolute right, an absurd situation could arise with the employer being entitled to commence then stop the works at will.

The employer does not have any implied right to come on to site after possession has been given to the contractor. If they wish to retain the right to do so, the contract should expressly provide for this. An example of this is to be found in clause 2.6 of the SBC and 2.5 of the SBC/DB which, with the consent in writing of the Contractor (not to be unreasonably withheld and subject to insurers' confirmation that insurance will not be prejudiced), allows the Employer to use or occupy the site or the Works or part of them prior to completion, whether for storage or otherwise. Such consent, where given, appears to provide the Employer with the equivalent of a licence from the Contractor to use or occupy part of the site to the extent necessary for the particular purpose falling within the scope of the clause, see *Impresa Castelli SpA* v. *Cola Holdings Ltd* (2002).

Further examples can be found in clauses 3.1 and 3.4 of the SBC, which provide for the presence on site of the Architect and clerk of works and in clause 3.1 of the SBC/DB in relation to the Employer's Agent. Likewise, clause 27.2 of the NEC3 requires the Contractor to provide access to the Project Manager and the Supervisor.

Regardless of such express provisions, it is submitted that the contractor's right of possession should be subject to the implied qualification that the employer, or those employed by him, should have a right of reasonable access for the purposes of inspection, supervision and administration of the contract. Where nothing is said about possession, the contractor must be allowed use and possession of the site as required for the purposes of carrying out their works, see *Ductform Ventilation (Fife) Ltd* v. *Andrews Weatherfoil Ltd* (1995).

The NEC3 provides separate definitions for Working Areas and the Site. The Working Areas include the Site, and the areas which are both necessary for providing the works and are used only for work in the contract (unless that is changed in accordance with the contract). An example of where this might be useful is a city centre development, where the provision of the works may require the use of land outwith the Site itself. The NEC3 provides that materials may be brought to the Working Areas provided that any tests or inspection have been passed (clause 41.1), and makes provision for transfer of title on delivery (clause 70.2). The NEC3 clause 15.1 allows the Contractor to submit a proposal to the Project Manager for adding an area to the working areas stated in the Contract Data, and a reason for refusing such a proposal is that the proposed area is either not necessary for providing the works or used for work not in the contract.

After starting on site, the contractor may be denied undisturbed occupation for a variety of reasons, only some of which may be the responsibility of the employer. Clause 2.7 of the SBC and 2.6 of the SBC/DB provide that where another contractor or supplier (defined as Employer's Persons) is employed by the Employer to carry out works outwith the Contractor's scope of works, the Contractor is under an obligation to permit the execution of such other work where the contract provides sufficient information to allow the Contractor to do so and still carry out the works in accordance with the contract. Where such information has not been made available to the Contractor, the Employer's right to have access to, and to instruct other contractors to execute works on the site, is subject to the consent of the Contractor, not to be unreasonably withheld or delayed.

In the case of the NEC3, any right on the part of other contractors engaged by the employer (i.e. 'Others') to share the Contractor's working areas should be specified in the Works Information, see clause 25.1. Any departure from this may be a compensation event under clause 60.1(5).

Unless the contract provides otherwise, an employer will not be liable where there is unauthorized occupation by a third party such as picketers, unless they have induced or condoned the obstruction, see *London Borough of Merton* v. *Stanley Hugh Leach Ltd* (1985). There is no implied warranty by the employer that access for the contractor will not be prevented by a third party, such as a picketer, see *LRE Engineering Services Ltd* v. *Otto Simon Carves Ltd* (1981). Notwithstanding the right of possession of the site, the contract may impose restrictions on the means of access to the site. Clause 2.13.1 of the SBC provides that the bills of quantities (unless otherwise specifically stated in the bills in relation to any specified item) are to have been prepared in accordance with the standard method of measurement (SMM 7), which in turn requires that any conditions relating to access should be stated in the bills of quantities. Where the Employer fails to provide access to the site as provided for in the contract, this may be an act of 'impediment, prevention or default', giving rise to a claim for extension of time under clause 2.29.7 of the SBC and 2.26.6 of the SBC/DB and for loss and expense under clause 4.24.5 of the SBC and 4.21.5 of the SBC/DB.

Under the NEC3, failure on the part of the Employer to allow access in accordance with the contract would be a compensation event under clause 60.1(2).

It should be noted, however, that clauses 3.10 and 5.1.2.1 of the SBC and clauses 3.5 and 5.1.2.1 of the SBC/DB allow the Employer to impose obligations or restrictions in regard to access to the site, subject to the Contractor's right of reasonable objection. Such imposition will be treated as a Variation.

Under the NEC3 this could be dealt with as a change to the Works Information.

4.2.4 Duration of the obligation to give the contractor possession

The obligation to give possession of the site will normally subsist until completion of the works. This is subject to any provision for sectional completion and handover to the employer, for example, as provided for by the Sixth Recital of the SBC Articles of Agreement and the Fifth Recital of the SBC/DB Articles of Agreement. Under the NEC3, sectional completion requires the use of secondary Option X5.

Where there is no express provision for the contractor to give up possession of the site, they will be entitled to possession for so long as is necessary to allow them to perform their obligations under the contract, see *Castle Douglas and Dumfries Railway Company* v. *Lee, Son and Freeman* (1859).

Under clause 2.33 of the SBC and 2.30 of the SBC/DB, the Employer may, with the consent of the Contractor (which consent should not be unreasonably withheld), take possession of any part of the Works prior to practical or sectional completion (if applicable). The taking of partial possession will have important consequences for that part of the Works in respect of the practical completion date, the Rectification Period, insurance and liquidated damages, see clauses 2.34–2.37 of the SBC and 2.31–2.34 of the SBC/DB.

In relation to early possession under the NEC3, clause 35.2 provides that the Employer may use any part of the works before completion has been certified, and if he does so, he takes over such part when he begins to use, except if the use is for a reason stated in the Works Information or to suit the Contractor's method of working. This 'deeming' of taking over is important as early taking over will result in the Contractor ceasing to have liability for delay damages applicable to the relevant part of the works (see clause X7.3) and the Employer assuming the risk of loss or damage to that part (see clause 80.1).

An intervening event may occur which allows the employer to take back possession of the site prior to practical completion, where, for example, they are entitled to terminate the contractor's employment under the contract.

The Employer's grounds of termination are to be found in clauses 8.4, 8.5 and 8.6 of the SBC and also of the SBC/DB. In those circumstances, the Employer may take possession of the site under clause 8.7.2.1 and the Contractor may be obliged to remove or have removed from the site any temporary buildings, plant, tools, equipment, goods and materials.

Under the NEC3, similar provisions are contained in clause 92.2.

Termination generally is considered in Section 9.4.

4.3 Administration

4.3.1 General

The employer is under an obligation to administer the site in such a way as to ensure that the contractor can meet their obligations under the contract. This section deals with the obligation incumbent on the Employer under the SBC, the SBC/DB and the NEC3 to appoint professional consultants to act on its behalf in administering the contract, together with the obligation not to interfere with the certifying process where a certifier, such as an architect, has been appointed and the obligation incumbent upon an employer to use his best endeavours to ensure that the architect carries out his required functions, where it appears that he may be failing to do so.

4.3.2 Appointment of an architect and other professionals

The SBC requires the appointment of an Architect (or Contract Administrator, as the case may be) together with a Quantity Surveyor if appropriate, see articles 3 and 4 of the Articles of Agreement. If the Architect is not also the CDM Co-ordinator for the purposes of the CDM Regulations, then that person is specified in article 5. The Employer also has the option of appointing an employer's representative and/or clerk of works, see clauses 3.3 and 3.4. Under article 3 of the Articles of Agreement of the SBC/DB, the Employer will specify the Employer's Agent. The CDM Co-ordinator is specified in article 5.

Under the NEC3 the Employer should specify the details of the Project Manager and of the Supervisor in part 1 of the Contract Data.

Under the SBC the Architect acts in all respects as the agent of the Employer. Normally, the architect will have been appointed prior to the contractor tendering for the contract. Failure to appoint an architect where the employer is contractually obliged to do so is a breach of contract by the employer, see *London Borough of Merton* v. *Stanley Hugh Leach Ltd* (1985). Where the contract calls for the appointment of an architect, it may well be that this is a condition precedent to the contractor's obligation to perform the work. Contractors may, however, be personally barred from insisting on the appointment of the architect if, for example, they commence work and the contract proceeds without the appointment of an architect.

If for any reason the architect, contract administrator or quantity surveyor becomes unable to act, the employer has a duty to appoint a replacement, see clause 3.5 of the SBC. They should do so within a reasonable time and their refusal to do so may amount to a repudiation of the contract entitling the contractor to rescind. Repudiation and rescission are considered in Chapter 9.

Unless there is an express term to the contrary, the employer cannot appoint itself to perform the architect's, or any other professional's, certification or decision-making functions part way through a building contract, see *Scheldebouw BV* v. *St James Homes (Grosvenor Dock) Ltd* (2006). The position is different under a design and build contract. There is no reason in principle why the employer cannot undertake such functions, and indeed the wording of the SBC/DB suggests that it is the Employer who has the primary obligation to undertake such roles, with the Employer's Agent acting in a delegated capacity.

The NEC3 provides for specific roles to be undertaken by both the Project Manager and the Supervisor. The Project Manager's role includes assessing amounts due for payment, certifying completion and determining the cost and time consequences of compensation events. The Supervisor's role includes supervising tests and inspections, notifying the Contractor of defects and issuing the defects certificate. The Project Manager's focus is therefore the management of the contract, and the Supervisor deals with inspections and quality. However, in practice, the two roles are often carried out by the same organization. Clause 10.1 provides that the Employer, the Contractor, the Project Manager and the Supervisor shall act as stated in the contract and in a spirit of mutual trust and co-operation. This suggests that the Employer under an NEC3 contract is under an obligation to complete the appropriate entries for such appointments in the Contract Data in order to make the contract operable. However, there is nothing in the NEC3 which prevents the Employer from naming himself as project manager or supervisor, or both, and there seems to be no reason in principle why the Employer should not do so.

While in most contracts the identity of the architect will be expressly stated, it would be wise for the contract to be worded to refer to the appointment of the individual architect 'or such other person as may be nominated by the employer'. This is to avoid a situation where there may be confusion surrounding the existence of an obligation to appoint a successor should the appointment fail for any reason. In *Croudace Ltd* v. *London Borough of Lambeth* (1986), it was held that there had been a breach of contract on the part of the council where the architect employed by them on a contract, and who had been dealing with the contractor's claim for loss and expense, retired and the council delayed in appointing a successor. In that case, the architect named in the contract had a responsibility to ascertain the contractor's claims. There would appear to be an obligation on the employer to ensure that the successor to the original architect is reasonably competent to perform the job, see *London Borough of Merton* v. *Stanley Hugh Leach* (1985).

It should be noted that clause 3.5.1 of the SBC allows the Contractor, except where the Employer is a local authority and the nominated replacement is one of its officials, to object to the Employer's nominated replacement as Architect, contract administrator or Quantity Surveyor, and any dispute is referable to the dispute resolution procedure under the contract. There are no equivalent restrictions to the Employer's replacement of the Employer's Agent under the SBC/DB (see article 3).

The only condition of the replacement of the Project Manager or the Supervisor under the NEC3 is that the Employer notifies the Contractor first (see clause 14.4). Clause 3.5.2 of the SBC provides that no such replacement who is appointed shall be entitled to disregard or overrule any certificate, opinion, decision, approval or instruction by any predecessor unless that predecessor would have had power under the contract to do so.

4.3.3 Nomination and naming of sub-contractors and specialists

Sub-contractors and suppliers are considered in Chapter 11. It used to be the case that an employer could decide to nominate a sub-contractor or supplier where, for example, they wish to ensure the quality of certain work that is to be performed, or the quality of certain materials that are to be supplied, or to avoid the price constraints which the contractor may be under. In these circumstances the sub-contractor/supplier was termed a nominated sub-contractor/supplier. Unlike its predecessors, the SBC does not contain any provisions allowing for nominated sub-contractors. For a summary of the differences between nominated and domestic sub-contractors under earlier editions of the SBC and the SBC/DB, see Section 11.4.

Clause 3.8 of the SBC gives the option to the Employer of listing no less than three persons in the Contract Bills to provide certain work measured or described there, but the Contractor ultimately has the final say as to which of those persons on the list carries out that work, and clause 3.8.4 makes it clear that that person remains a domestic sub-contractor as opposed to a nominated sub-contractor. The SBC/DB at paragraph 2 of Schedule Part 2, Part 1 provides an option for the contract to state that work is to be executed by a named sub-contractor. See Section 11.5 for details of these provisions. The Named Specialist Update 2012 to the SBC allows the Employer to name specialists as domestic sub-contractors (see Section 11.5).

4.3.4 Obligation of non-interference

The architect's role is *quasi* arbitral in nature. This is considered in Chapter 7. While he is a professional person, he is not independent. He is an agent of the employer, see *Beaufort Developments (NI) Ltd* v. *Gilbert-Ash NI Ltd and Another* (1998). The employer is nevertheless under an implied obligation not to interfere with the operation of the certification process by the architect. Employers may be open to a claim for damages should they attempt to do so. Employers owe a duty to ensure that the architect discharges his obligations properly, see *London Borough of Merton* v. *Stanley Hugh Leach Ltd* (1985). They may also owe a duty to the contractor to replace an incompetent architect where they become aware that the architect is failing to perform his functions under the contract, or is taking into account things he ought not to, having regard to the contract, see *Panamena Europea Navigación Compañía Limitada* v. *Frederick Leyland & Co. Ltd* (1947).

4.4 *Information and instructions*

The SBC provides that where not included in the Information Release Schedule, the Architect shall from time to time provide the Contractor with such further drawings or details as are reasonably necessary to explain and amplify the Contract Drawings and shall issue such instructions as are necessary to enable the Contractor to carry out and complete the Works in accordance with the Contract, see clause 2.12.1. Should the Contractor not receive information and/or instructions within the necessary time, then this may be a Relevant Event entitling the Contractor to an extension of time, by virtue of clause 2.29.7, and also may be a matter materially affecting the regular progress of the Works which may entitle the Contractor to recover loss and expense, by virtue of clause 4.24.5. These obligations of the Employer correspond with the obligation of the Contractor, contained within clause 3.10, to comply with instructions issued by the Architect forthwith, subject to certain exceptions. The obligations of the Contractor are considered in Chapter 5.

In *Neodox Ltd* v. *Swinton and Pendlebury BC* (1958), it was held that what was a reasonable time for the provision of details and instructions necessary for the execution of the work did not depend solely on the convenience and financial interests of the contractor. The employer, through his agents (the engineer in this case), was to have a period of time to provide the information which was reasonable having regard to the point of view of himself and his staff, as well as that of the contractor. It was held in this case that there was an implied term that details and other instructions necessary for the execution of the works should be given by the employer's agent from time to time in the course of the contract and should be given within a time reasonable in all the circumstances.

The employer, through his architect, will therefore be in breach of contract for failure to give details and information in sufficient time to enable the contractors to perform their obligations under the contract. This does not imply an obligation to provide information to contractors such that they can complete ahead of the contractually stipulated date, even if they have indicated that this is their intention, see *Glenlion Construction Ltd* v. *The Guinness Trust* (1987). As regards requests from the contractor for information and instructions required by him, it has been held that a document setting out in diagrammatic form the planned programme for the work and indicating the days by which instructions, drawings, details and levels were required, which was issued by the contractor at the commencement of the work, could amount to a specific application for information. It was held that the date specified for delivery of each set of instructions met the contractual requirement of not being unreasonably distant from nor unreasonably close to the relevant date, see *London Borough of Merton* v. *Stanley Hugh Leach Ltd* (1985).

Failure by the employer to provide the contractor with the drawings and necessary information to enable them to carry out their works may, depending on the importance of the work in question and after a reasonable request for the information by the contractor, constitute a repudiation of the contract, entitling the contractor to rescind. Clause 8.9.2.2 of the SBC specifically provides that if the carrying out of the whole or substantially the whole of the uncompleted Works is suspended for a continuous period of the length specified in the Contract Particulars by reason of any impediment, prevention or default by, among others, the Architect, this will entitle the Contractor to terminate their employment under the contract. That clause would appear wide enough to cover suspension caused by failure to provide necessary information and/or instructions. Termination is considered in Section 9.4.

By its nature, a design and build contract imposes the responsibility for design drawings and information on the contractor and so there is no equivalent in the SBC/DB of the obligation in clause 2.12.1 on the architect/contract administrator to provide further drawings, details and instructions. However, that does not mean that the Employer has no similar obligations at all. The SBC/DB requires the Employer should provide the Employer's Requirements and to satisfy itself that the Contractor's Proposals meet those requirements, and the Contractor has no responsibility for the contents of the Employer's Requirements, save in the case of a divergence between the Employer's Requirements and Statutory Requirements (clause 2.11). This means that if the Contractor becomes aware of any inadequacy in the Employer's Requirements he gives notice to the Employer and the Employer shall issue instructions in that regard (clauses 2.13 and 2.14). Where there is a divergence between the Statutory Requirements and either the Employer's Requirements or the Contractor's Proposals, clause 2.15 requires the Contractor to notify the Employer of its proposed amendment for removing it at his own cost, and the Employer's consent shall not be unreasonably withheld. We would suggest that a failure or delay on the part of the Employer to issue instructions under clause 2.13 or to provide consent under clause 2.15 (where the delay or withholding of consent is unreasonable) could give rise to a claim on the grounds of impediment, prevention or default for extension of time under the SBC/DB clause 2.26.6 and for loss and expense under clause 4.21.5. Clause 8.9.2 of the SBC/DB provides a similar ground for termination by the Contractor for prolonged suspension by reason of impediment, prevention or default as described above in relation to the SBC.

The NEC3 addresses the Employer's obligation to provide information by requiring the Contractor, pursuant to clause 31.2, to show on each programme which he submits for acceptance the dates when, in order to provide the works in accordance with the programme, the Contractor will need plant and material 'and other things' to be provided by the Employer. Failure by the Employer to provide something which he is to provide by the date for doing so in the Accepted Programme is a compensation event under clause 60.1(3). Other compensation events triggered by a failure on the part of the Employer or those acting for him to provide information include clause 60.1(6) (failure by the project manager or supervisor to reply to a communication from the Contractor within the period stated in the contract); and clause 60.1(9) (withholding by the Project Manager of an acceptance for a reason not stated in the contract). This last-mentioned event could include a situation where the Project

Manager has withheld acceptance of the Contractor's design submission under clause 21.2, notwithstanding that the design complies with the Works Information and applicable law.

4.5 Variations

4.5.1 General

The instruction of variations might be described as a right on the part of the employer which imposes a corresponding obligation on the contractor to implement the variation so instructed and an obligation on the employer to pay for the work instructed under the variation (though in some cases the variation can be an omission of work, resulting in a reduction in the contract price). In general, neither of the parties to a building contract has an implied right to vary the works on the basis that, having entered into a contract to carry out certain works for a specific sum of money, the parties are entitled, and obliged, to do no more than they have originally contracted to do. In reality, however, the work which was originally specified may have to be modified for a variety of reasons such as unexpected ground conditions or other circumstances which parties were not able to identify with any degree of certainty at the outset. This is particularly so on a major building project. In addition, the employer may wish to instruct the contractor to carry out certain extra works or, having discovered a quicker or easier way of doing something, to omit certain works that were originally included within the contract.

For these reasons, the contract, if in written form, will almost invariably entitle the employer to vary the works and a contractor will be under an obligation to carry out or omit works in accordance with a variation instruction and in accordance with the variation procedure specified in the contract. This procedure will normally specify the contractor's right to be paid for the work and the basis for valuing such additional payment and the contractor's entitlement to an extension of time if completion of the works is delayed as a result of the additional work instructed.

The contract will also usually define what may be covered by a variation. At the very least, this will extend to a change in the scope of the works, but depending on the terms of the contract, a variation may also extend to the imposition of additional working restrictions on the contract, which may not involve additional work but which may increase the cost of carrying out the original scope of work by changing the assumptions made by the contractor in its pricing. This could include such matters as reducing access routes to and from the site, limiting hours of working, or imposing noise restrictions. If the definition of variation is sufficiently widely expressed in the contract, it could even go so far as to permit the employer to change the conditions of contract. In short, the scope of the variation which the employer may instruct is determined by the terms of the contract. In its most commonly used sense, a variation is a change in the scope of the contract works and will lead to an adjustment of the contract price. However, it is not always immediately obvious whether there has indeed been a change in the original scope. Any work which the contractor is either expressly obliged to do in terms of the contract, or which is necessary by implication, falls within the contractual scope of the works and does not amount to a variation. It will therefore depend upon the terms of the contract as to whether or not an instruction to carry out certain works will amount to a variation of the contract works for which the contractor should be entitled to additional payment. For example, in a contract which can truly be said to be lump sum, there is no obligation on the employer to pay for work by way of a variation even if the work is not described or shown on drawings or if the contractor incurs additional costs due to the impracticable nature of the design. If the contract obliges the contractor to achieve a particular result, the contractor cannot claim as a variation a requirement to use more expensive materials if it becomes obvious that cheaper materials will not be appropriate. This is a risk the contractor takes in tendering a sum to achieve that end result. Of course, the position would be different if the employer changed its mind as to what end result it required. Likewise, if something is missing from the bills of quantities but, nevertheless, is necessary to achieve the end result, the requirement for the contractor to do that work will not amount to a variation. For example, in *Williams* v. *Fitzmaurice* (1858) a contractor was obliged to build a house which was to be ready by a specific date. The specification for the works did not include for any type of flooring and the contractor tried to state that he was entitled to extra payment for having to fix floorboards. However, as the contract was to achieve a particular result, namely, the completed house, the flooring was deemed to be included in the contract and did not amount to a variation.

Where the contract confers power on the employer to vary the contract works (either by itself or through its representative such as the architect), the contractor will be obliged to comply with instructions conforming to the requirements of the contract. Where an employer has varied the contract works, their obligation to provide detailed drawings in respect of the varied works will be the same as their obligation to do so in relation to the contract works; that is to provide such drawings and information within a reasonable time to allow the contractor to perform their obligations in terms of the contract.

In a contract using bills of quantities, a key issue in determining whether or not an increase in quantities is a variation will be the status of the bills, i.e. are they part of the contract documents? In the past, bills of quantities were considered to be only an estimate of the works to be used as a guide for the contractor. They were not contract documents and were not to be taken as having contractual effect. If more materials were required than were stated in the bills, this did not amount to a variation of the contract works. However, bills of quantities are now usually incorporated as a contract document, so that it is the contractor's obligation to carry out the work stated in the bills of quantities, contract drawings and specification, and an increase in the quantities will amount to a variation. This can be the case even if the contract is said to be lump sum, see *Patman & Fotheringham Ltd* v. *Pilditch* (1904).

In a contract which provides for re-measurement, the employer provides drawings and an approximate bill of quantities. The contractor provides a tender sum which is an indication of the likely price of the works. The works are then re-measured and valued as they progress so that the indicative tender figure can be converted into a final sum. This type of contract can be used when there is not sufficient time to prepare detailed drawings in order to produce a final bill of quantities. The contract will detail how items included in the as-built quantities which are different from the approximate quantities should be valued. The valuation of as-built quantities is distinct from the valuation of variations (though the same valuation rules may apply).

4.5.2 Variations under the SBC, the SBC/DB and the NEC3

In the SBC, the term 'Variation' is defined in detail in clause 5.1 and there are two distinct parts to the definition:

First, it means the alteration or modification of the design, quality or quantity of the works. This includes the addition, omission or substitution of any work; the alteration of the kind or standard of any of the materials or goods to be used in the works; or the removal from the site of any work executed, or Site Materials, other than work, materials or goods which are not in accordance with the Contract.

Second, it means the imposition by the Employer of any obligations or restrictions in regard to four specified matters or the addition to, or alteration or omission of, any such obligations or restrictions imposed by the Employer in the Contract Bills or Employer's Requirements (for any Contractor Designed Portion) in relation to these matters. The four specified matters are: (1) access to the site or use of any specific parts of the site; (2) limitations of working space; (3) limitations of working hours; and (4) the execution or completion of the work in any specific order. In the SBC/DB clause 5.1, the term 'Change' is used in place of 'Variation'. The meaning is essentially the same as in the SBC, but the alteration or modification in the design, quality or quantity of the works can only be made by a change to the Employer's Requirements. There is no provision for making a change to the Contractor's Proposals.

The Architect's powers in respect of Variations under the SBC are set out in clause 3.14. The Architect is expressly given the power to issue instructions requiring a Variation and any instruction issued by him in this regard is subject to the Contractor's right of reasonable objection under clause 3.10.1, where the instruction relates to a Variation under clause 5.1.2, i.e. the imposition of any obligations or restrictions in regard to any of the specified matters. Similar powers in respect of a Change are given to the Employer under clause 3.9 of the SBC/DB. In addition to the Contractor's right of reasonable objection under clause 3.9.2 where the instruction relates to a Change under clause 5.1.2, clause 3.9.1 of the SBC/DB provides that the Employer may not effect a Change which is, or which makes necessary, an alteration or modification in the design of the Works without the consent of the Contractor, which is not to be unreasonably withheld or delayed. There is an equivalent provision of sorts in the SBC clause 3.10.3 insofar as the instruction 'injuriously affects the efficacy of the design of the Contractor's Designed Portion', which allows the Contractor to give notice of such injurious effect, in which event the instruction shall not take effect unless confirmed by the Architect.

Under the SBC the Architect also has the power to issue instructions to expend provisional sums included in the Contract Bills or in the Employer's Requirements (clause 3.16). A similar power is given to the Employer under clause 3.11 of the SBC/DB. It should be noted that the NEC3 does not contain any provisions for provisional sums. The Architect is empowered to sanction in writing any Variations made by the Contractor otherwise than pursuant to an instruction of the Architect.

Variations should be instructed or confirmed in writing by virtue of clause 3.12.1. If the Architect purports to issue an instruction otherwise than in writing, it will have no immediate effect. The Contractor is required to confirm the instruction in writing to the Architect within seven days of receipt of that instruction and if not dissented to in writing by the Architect, it will take effect seven days after receipt of the Contractor's confirmation, see clause 3.12.1. Alternatively, the Architect can confirm the instruction in writing within seven days of giving the instruction in which case it takes effect from the date of that confirmation, see clause 3.12.2. If neither party confirms the instruction but the Contractor nevertheless complies with it, the Architect may, at any time prior to the issue of the Final Certificate, confirm the instruction in writing with retrospective effect, see clause 3.12.3. Similar provisions in respect of Changes are contained in the SBC/DB clause 3.7.

The issuing by the Architect under the SBC of an instruction requiring a Variation is a Relevant Event under clause 2.29.1 entitling the Contractor to an extension of time, but only insofar as the works have in fact been delayed by the issue of the instruction and so long as the Contractor has followed the requirements of clause 2.28.6. The Contractor is also entitled to payment for the work carried out in complying with the Architect's instruction and the SBC provides detailed provisions as to how such work should be valued. Those provisions are to be found in clauses 5.2–5.10. These valuation rules expressly exclude any allowance for the effect of the Variation upon the regular progress of the works or for any other loss and/or expense which would be reimbursed under any other provision. This avoids double counting where loss and/or expense resulting from the Variation is separately dealt with as a Relevant Matter under clause 4.24 (though see the case of *WW Gear Construction Ltd* v. *McGee Group Ltd* (2012) discussed in Section 5.2.2). The position will be different if the Architect chooses to use the Variation Quotation procedure set out in clause 5.3 and Schedule Part 2. In that event, the Contractor requires to provide a quotation in respect of the proposed Variation to include amounts for the adjustment to the Contract Sum, the period of extension of time required, and the amount of loss and/or expense and, if such quotation is accepted, the Quantity Surveyor will include in the valuation both the value of the varied works and the loss and/or expense. In the SBC/DB similar valuation rules for Changes are contained in clauses 5.4–5.7. There are also alternative provisions for the Contractor and the Employer agreeing, prior to compliance with the instruction, binding estimates of the valuation of Changes and any extension of time and loss and expense. These are contained in the Supplemental Provisions contained in Schedule Part 2 to the SBC/DB and will apply only if so selected in the Contract Particulars. The value of a Variation will be added to or deducted from the Contract Sum as applicable. See clause 4.3 of the SBC and clause 4.2 of the SBC/DB.

In both the SBC and the SBC/DB an instruction to expend a provisional sum may give rise to both an extension of time and loss and/or expense if the relevant criteria are met. See clauses 2.29.2.1 and 4.24.2.1 of the SBC and clauses 2.26.2.2 and 4.21.2.1 of the SBC/DB. The entitlement to an extension of time and/or loss and expense in relation to the expenditure of provisional sums is excluded in the SBC where the instruction to expend a provisional sum relates to defined work as defined in General Rule 10 of the SMM7. The SMM7 provides that the contractor will be deemed to have made allowance in programming, planning and pricing in relation to defined work. For further details in respect of the procedures in respect of Variations under the SBC and the SBC/DB, see Sections 5.2.2, 8.2.3 and 8.2.4.

The SBC recognizes that a quantity in the bills of quantities may be identified as an approximate quantity, and where such a quantity is so identified and turns out not to be a reasonably accurate forecast of the quantity of work required, clause 2.29.5 provides a right to extension of time and clause 4.24.4 to loss and expense.

The NEC3 deals with changes in a typically concise way. Clause 60.1(1) provides that it is a compensation event (giving rise to both extension of time and costs) if the Project Manager gives an instruction changing the Works Information, except where the change is to accept a defect or is a change to the Works Information provided by the Contractor for his design which is made either at his request or to comply with other Works Information provided by the Employer. Clause 61.2 permits the Project Manager to instruct the Contractor to submit quotations for a proposed instruction and in that event the position will be regulated by the procedure for quotations for compensation events set out in clause 62. The rules for assessing the financial and time consequences are set out in clauses 63 and 64.

For more detail on the procedures for assessment of compensation events and for such quotations, see Sections 5.2.4 and 8.2.5.

There are also circumstances which may give rise to works being deemed a Variation in the SBC. Clause 2.14.3 of the SBC provides that the correction of an error in the bills of quantities or an inadequacy in the design shall be treated as a Variation. Clause 2.14.2 of the SBC/DB provides that where there is a discrepancy in the Employer's Requirements which is not dealt with in the Contractor's Proposals, and as such requires to be amended or otherwise dealt with, such amendment or decision shall be treated as a Change.

The employer cannot vary the contract to the extent that it alters the fundamental nature of the contract works, see *McAlpine Humberoak Ltd* v. *McDermott International Inc* (1990). Nor can he vary the contract by omitting large aspects of it and then employing another contractor to carry out the work. That will amount to a repudiation which, in turn, entitles the contractor to rescind and seek damages, see *Commissioner for Main Roads* v. *Reed & Stuart Pty Ltd* (1974). Contracts so far varied as to make them fundamentally different from that contracted for may give rise to a claim for payment *quantum meruit*, see *ERDC Construction Ltd* v. *H M Love & Co* (1995). *Quantum meruit* is considered in Section 8.4.

Although under the SBC there is a requirement for variation instructions to be in writing, this may be waived in certain circumstances. Examples of such circumstances are: where the work is of such a different character or nature that it is said to be outside the terms of the contract and forms a separate contract; where the main contract is no longer operative; where the final certificate has been issued including a sum for the variations and there is no opportunity for review of this; where an arbitrator is given the power to consider whether the work is or is not a variation and he decides that the work done was a variation; or where, for any other reason, the employer is personally barred from insisting, or has waived his right to dispute, that something was properly a variation where the instruction was not in writing.

4.6 Other obligations

4.6.1 Payment

The other main obligation owed by the employer to the contractor under a building contract, as outlined at the start of this chapter, is to make payment for the works executed under the contract. This obligation is deserving of a chapter of its own and is considered separately in Chapter 8. In the context of the SBC and the NEC3, payment is (subject to the provisions of the 1996 Act) triggered by the issue of certificates and certification is considered in Chapter 7.

Under the SBC/DB there are two alternative mechanisms for interim payments. The appropriate alternative must be selected by an entry in the Contract Particulars. Payment may be made on the basis of stage payments (Alternative A) or on the basis of the value of work executed (Alternative B). In each case, the Contractor must make an application for interim payment. In the case of Alternative A, the application will be made following the completion of a relevant stage, while under Alternative B the applications are made at monthly intervals. The provisions in section 4 of the SBC and the SBC/DB have been amended in the 2011 edition to take account of the changes to the 1996 Act by the 2009 Act.

The payment provisions in section 5 of the April 2013 edition of the NEC3 incorporate the changes required by the 2009 Act.

4.6.2 Insurance and indemnity

In considering the obligations incumbent upon employers under building contracts, the issues of insurance and indemnity are also worthy of mention. At common law, there is no implied obligation incumbent upon an employer, to insure; however, frequently the form of contract used by parties, such as the SBC and the NEC3, will include such obligations. These are considered separately in Chapter 14.

4.6.3 Health and safety

As with all other employers (in the employer/employee sense, as opposed to the building contract sense), employers owe a number of duties in respect of health and safety. Similar duties are incumbent upon contractors. In the area of building contracts such duties are, generally, more pertinent to contractors. These matters are considered in Chapter 20.

Chapter 5
Contractors' Obligations

5.1 Introduction

As with the obligations of the employer, a building contract will ordinarily set out, in express terms, the obligations owed by the contractor to the employer. Similarly, where these are not expressed, certain terms will be implied in the parties' contract. The majority of the obligations considered in this chapter relate to the execution of the works but certain other types of obligation are routinely imposed upon contractors, for example, the obligation to take out and maintain insurance under the SBC.

5.2 Completing the works

5.2.1 Common law

Where a contractor is engaged to carry out specified work, they have an obligation to carry out and complete that work. This carries with it the obligation to execute the work in a good and workmanlike manner using the skill and care to be expected of a builder of ordinary competence. This involves adopting methods which are in accordance with the regular practice in the building trade at the time, see *Morrison's Associated Companies Ltd* v. *James Rome & Sons Ltd* (1964). The exception to this would be if, in particular circumstances, there was an indication of an unusual or extraordinary risk in doing the work in the normal manner. In these circumstances, the contractor would be required to carry out the works in a different way or else would run the risk of being found negligent, see *Morrison's Associated Companies Ltd*.

5.2.2 The SBC provisions

The SBC provisions expressly include the obligation to carry out and complete the works. This is to be found in Article 1 of the Articles of Agreement and clause 2.1 of the Conditions. The obligation is to carry out and complete the works in a proper and workmanlike manner and in compliance with the Contract Documents (namely the Contract Drawings, the Schedule, the Contract Bills, the Agreement

MacRoberts on Scottish Construction Contracts, Third Edition. MacRoberts.

and the Conditions together with, where applicable, the Employer's Requirements, the Contractor's Proposals, the CDP Analysis and any other Contract Documents listed in Part 8 of the Schedule), the Construction Phase Plan and other Statutory Requirements. The Public Sector Supplement 2011 also added to the definition of Contract Documents any agreed Building Information Modelling (BIM) protocol. See Section 1.7 for more on BIM. In relation to the particular provisions of the SBC/DB, see Sections 5.2.3 and 5.3.4.

As the work proceeds, the contract allows the Architect to issue instructions and the Contractor has an obligation, under clause 3.10, to comply with all instructions issued to him in regard to any matter over which the Architect has power under the contract to issue instructions. The instruction may be one which requires a Variation as defined in clause 5.1. Variations are considered above in Section 4.5 and the procedures to be followed in respect of the valuation of Variations are set out in more detail in Section 8.2.3.

The Architect may, in terms of clause 5.3.1, in his instruction for a Variation, state that the Contractor is to provide a 'Variation Quotation'. This is a reference to the procedure set out in Schedule Part 2 for submission by the Contractor of a quotation setting out the amount of the adjustment to the Contract Sum, including the effect of the instruction on other work supported by all necessary calculations, any adjustment to the time required for completion of the Works and/or Section to the extent that this is not covered by any revision to the Completion Date already made, the amount to be paid in lieu of any ascertainment under clause 4.23 or 5.3.3 of direct loss and/or expense not already included elsewhere, a fair and reasonable amount in respect of the cost of preparing the Quotation, indicative information on any additional resources required to carry out the Variation and the method of carrying it out and, where applicable, the base date for the application of Fluctuations Provisions. The Quotation is to be sufficiently detailed to allow it to be evaluated by or for the Employer.

This procedure is time-consuming and requires significant input from both the Architect, in terms of detailing the proposed Variation, and the Contractor in preparing the Quotation. It is therefore likely to be used only for the most significant proposed Variations.

Schedule Part 2 paragraph 2 also provides for an Acceleration Quotation to be requested by the Employer if the Employer wishes to investigate the possibility of achieving practical completion before the Completion Date. On being invited to do so, the Contractor is either to provide an Acceleration Quotation identifying the time that can be saved, the amount of adjustment to the Contract Sum and any other conditions attached or explain why it would be impracticable to achieve practical completion earlier than the Completion Date. The Contractor has no obligation to accelerate until he receives a Confirmed Acceptance of his Acceleration Quotation. This Acceleration Quotation procedure is not referenced within any of the clauses of the contract and is only found in the Schedule. This is because the Contractor cannot be forced to accelerate. This can only be done by agreement.

Schedule Part 2 requires the Contractor to prepare any Variation Quotation or Acceleration Quotation within 21 days of being instructed to do so or of receipt of sufficient information to allow him to do so, whichever is later. The Quotation is to be open for acceptance by the Employer for 7 days from its receipt by the Quantity

Surveyor or Architect. These time limits may be extended by agreement. The Schedule then sets out the procedure for acceptance or rejection of the Quotation.

There are sanctions available against the Contractor, in terms of clause 3.11, if he fails to comply with an instruction. These allow the Employer, if there is non-compliance within seven days after receipt of a notice requiring compliance, to employ and pay others to carry out the work required and provides that the Contractor is responsible for all additional costs incurred as a result. These are deducted from the Contract Sum.

A further matter arising from variations is the Contractor's entitlement to additional time. Under clause 2.27.1 the Contractor is obliged, if and when it becomes reasonably apparent that progress of the work or any section is being or is likely to be delayed, to give written notice to the Architect of the circumstances of this including the cause or causes of the delay, and to identify any event which is a Relevant Event. The contract contains a list of circumstances which are termed Relevant Events, which includes Variations (clause 2.29.1) and instructions of the Architect under certain clauses (clause 2.29.2). The Contractor is required in terms of clause 2.27.2, in respect of each event identified in the notice, either in the notice or in writing thereafter, to give particulars of its expected effects, including an estimate of any expected delay in the completion of the works or any section. The Contractor is in terms of clause 2.27.3 to notify the Architect of any material change in his estimated delay or in other particulars and to supply such further information as the Architect may reasonably require. For further detail in respect of notice requirements, see Section 6.5.4.

Upon an application by the Contractor, if the Architect considers that any of the events stated to be a cause of delay is a Relevant Event and that completion of the work or any section is likely to be delayed because of it beyond the relevant completion date, the Architect shall give an extension of time to the Contractor and fix a later date as the completion date for the works or the sections that he considers to be fair and reasonable all in terms of clause 2.28.1. Extensions of time are considered in more detail in Section 6.5.

Finally, Variations can also give rise to loss and expense being payable to the Contractor. As soon as it has become, or should reasonably have become, apparent to the Contractor that regular progress of the work has been or is likely to be affected by one of the Relevant Matters listed in the contract, the Contractor can, in accordance with clause 4.23, make written application to the Architect stating that he has incurred or is likely to incur direct loss and/or expense for which he would not be reimbursed by a payment under any other provision of the contract. One of the Relevant Matters to which this applies is Variations, see clause 4.24.1. If the Architect considers that the regular progress of the works has been or is likely to be materially affected as stated in the Contractor's application, or that direct loss and/or expense has been or is likely to be incurred, he ascertains (or instructs the quantity surveyor to ascertain) the amount of direct loss and/or expense sustained by the Contractor. Any amount so ascertained falls to be added to the Contract Sum in accordance with both clauses 4.3.3.4 and 4.25. The decision in *WW Gear Construction Lt* v. *McGee Group Ltd* (2012) suggests that claims for loss and/or expense arising from a Variation can (notwithstanding clause 5.10.2) be made under either 5.6 or 4.24.1. This could mean that if the Contractor was prevented from recovering loss and/or expense under clause 4.24.1

because, for example, he had failed to comply with the notice requirements under clause 4.23, then he might still be entitled to recover under clause 5.6. The equivalent provisions under the SBC/DB are clauses 4.21.1, 5.2 and 5.7.2.

There may be work which does not form part of the Contractor's contract which the Employer wishes to carry out himself or which he wishes others to carry out. In such cases, clause 2.7.1 provides that if the Contract Bills provide the information necessary to allow the Contractor to carry out and complete the Works in accordance with the Contract, the Contractor shall permit the execution of such work. Where the Bills do not do so, clause 2.7.2 provides that the Employer may arrange for the work to be executed, though this requires the Contractor's consent.

The Contractor has obligations in relation to the availability of Contract Documents. The Contractor is obliged, under clause 2.8.3 to keep on site and available to the Architect or his representative at all reasonable times a copy of the Contract Drawings, the unpriced bills of quantities, the Contractor's Designed Portion documents (where applicable), the descriptive Schedule or similar documents necessary for use in carrying out the Works as referred to in clause 2.9.1.1, the master programme referred to in clause 2.9.1.2, and the drawings and details referred to in clauses 2.10 and 2.12.

The obligation in relation to the master programme in clause 2.9.1.2 is that the Contractor is required, without charge, to provide to the Architect his master programme for the execution of the Works identifying, where required in the Contract Particulars, the critical paths. Within 14 days of any extension of time awarded under clause 2.28.1 or of agreement of any Pre-agreed Adjustment fixing a revised Completion Date due to acceptance of a Variation Quotation or Acceleration Quotation, an amendment or revision to the master programme to take account of that is to be provided.

In relation to Contractor's Designed Portion Works, the Contractor is, under clause 2.9.4.1, to provide without charge to the Architect, copies of such Contractor's Design Documents and (if requested) related calculations and information, as are reasonably necessary to explain or amplify the Contractor's Proposals and, under clause 2.9.4.2, all levels and setting out dimensions which the Contractor prepares or uses for the purposes of carrying out and completing the Contractor's Designed Portion.

The Contractor's Design Documents and other information referred to in clause 2.9.4.1 are to be provided as and when necessary and in accordance with the Contractor's Design Submission Procedure set out in Schedule Part 1 or as stated elsewhere in the Contract Documents, and the Contractor is not to commence any work to which such a document relates before that procedure has been complied with, all in terms of clause 2.9.5.

The Contractor's Design Submission Procedure in Schedule Part 1 sets out in detail what documents are to be prepared and submitted by the Contractor, the format of these and their timing. It also sets out the procedure for comment by the Architect who is required to respond within 14 days of receiving the documents or, if later, 14 days from expiry of the date or the period for submission stated in the Contract Documents. The Architect can provide an 'A', 'B' or 'C' rating; with 'A' meaning the Contractor is to carry out the work in strict accordance with the document; 'B' meaning the Contractor is to carry out the work in accordance with the document but taking on board the Architect's comments; and 'C' meaning the Architect's comments are to be taken into account and the document resubmitted. Work is not to be carried out in accordance with a document marked 'C'. It is worth noting that if the Architect does not respond within the 14-day period, the document is regarded as marked 'A'.

The Employer has no liability to pay for any work within the Contractor's Designed Portion Works executed otherwise than in accordance with Contractor's Design Documents marked 'A' or 'B'. In cases of disagreement by the Contractor with the Architect's comments, the Schedule sets out a procedure for the contractor to challenge them, in which case the Architect is required to reconsider and either confirm or withdraw the comment. All of this is subject to the overriding obligation of the Contractor to ensure that the Contractor's Design Documents and the Contractor's Designed Portion Works are in accordance with the contract.

In relation to levels required for the execution of the Works, the Architect is, in terms of clause 2.10, to provide these in the form of accurately dimensioned drawings containing the information required to allow the Contractor to set out the Works. This does not apply to Contractor's Designed Portion Works. The Contractor is responsible under clause 2.10 for amending any errors arising from his own inaccurate setting out at no cost to the Employer, unless the Architect instructs that such errors are not to be amended, in which case a deduction is made from the Contract Sum.

The Architect has obligations in terms of clause 2.12.1 to provide further drawings or details as are reasonably necessary to explain and amplify the Contract Drawings. These are, under clause 2.12.2, to be provided at the time it is reasonably necessary for the Contractor to receive them having regard to the progress of the Works. The Contractor's obligation in relation to this is, in terms of clause 2.12.3, where the Contractor has reason to believe the Architect is not aware of the time by which the Contractor needs the information, to advise the Architect sufficiently in advance to allow him to comply with his clause 2.12 obligations. In this way, while the primary obligation is on the Architect, there is still a secondary obligation on the Contractor who, therefore, cannot simply sit back and wait for information to arrive but must proactively seek it. This is consistent with the obligation in clause 2.28.6.1 to constantly use best endeavours to prevent delay in the progress of the Works or any Section, however caused, and to prevent the completion of the Works or any Section being delayed or further delayed beyond the relevant Completion Date.

Related to the design obligations, there are provisions in section 7 of the SBC requiring the Contractor to provide Collateral Warranties or for Third Party Rights to purchaser, tenants and funders as well as to obtain sub-contractor warranties (see Chapter 13).

The Contractor has further obligations in relation to access to the site. In terms of clause 3.1, the Architect is to have access at all reasonable times to the Works but also to any workshop or other premises of the Contractor where work is being prepared for the contract. If work is being prepared in workshops or premises of a sub-contractor, then the Contractor is to include a provision in the sub-contract to secure a similar right of access to the sub-contractor's premises and is to do 'all things reasonably necessary' to make that right effective. Such rights of access are subject to reasonable restrictions which are necessary to protect proprietary rights in the property.

In terms of the Contractor's representatives on site, the Contractor's obligation is, under clause 3.2, to ensure that at all times he has on site a competent person-in-charge. Any instructions or directions issued to that person are deemed to be issued to the Contractor. The Employer can also appoint representatives who can include a clerk of works. The Contractor is required, under clause 3.4, to allow any clerk of works to carry out his duty of acting as inspector on behalf of the Employer. The Contractor's obligation to carry out and complete the Works in accordance with the contract Conditions does not alter, as provided by clause 3.6, regardless of any such inspections by the clerk of works or the Architect.

Sub-contracting of parts of the Works is commonplace but contractors are, in terms of clause 3.7.1, not to sub-let the whole or any part of the works without the consent of the Architect. The same applies under clause 3.7.2, where there is a Contractor's Designed Portion, to sub-letting of the design work. The Contractor maintains his obligation to carry out and complete the works and other contractual obligations regardless of any subletting. The SBCC issued a Named Specialist Update in May 2012 allowing the Employer to name individual specialists as domestic sub-contractors for specified parts of the Works. The position in relation to sub-contractors, suppliers and named specialists is dealt with in more detail in Chapter 11.

As the work progresses, the Contractor may find fossils, antiquities and other objects of interest or value on the site. On making such a discovery the Contractor is obliged in terms of clause 3.22.1 to use best endeavours not to disturb the discovered item and to cease work insofar as continuing with the work would endanger the item or prevent or impede its excavation or removal, to take steps to preserve the item in the exact position and condition in which it was found, and to inform the Architect of its discovery and location. The Architect may issue instructions in relation to the item in terms of clause 3.22.2. The Contractor does have a corresponding right in terms of clause 4.24.3, which makes compliance with clause 3.22.1 or with instructions under clause 3.22.2 a Relevant Matter entitling the Contractor to claim for loss and/or expense for which he would not be reimbursed under any other provision of the contract. Compliance with clause 3.22.1 or with instructions issued under clause 3.22.2 is also a Relevant Event (clause 2.29.4) for the purposes of the extension of time provisions in clause 2.28. The SBC defines completion of the works or a section of work as being when in the opinion of the Architect practical completion of the Works or a Section is achieved and the Contractor has complied sufficiently with the requirements of clauses 2.40 and 3.23.4. These relate to the provision of the Contractor's Design Documents showing or describing the Contractor's Designed Portion as built, where the contract includes a Contractor's Designed Portion, and the provision of information for the Health and Safety file that is required by the CDM Regulations. At this date the Architect issues the Practical Completion Certificate under clause 2.30.

The Courts have made various attempts to define practical completion. In the House of Lords case of *City of Westminster* v. *J Jarvis & Sons Ltd* (1970), the relevant clause provided that when in the opinion of the architect the works were practically complete, he was to issue a certificate to that effect. Viscount Dilhorne set a high standard namely that:

> One would normally say that a task was practically completed when it was almost but not entirely finished but 'practical completion' suggests that that is not the intended meaning and that what is meant is the completion of all the construction work that has to be done.

It was said that there should not be defects apparent at this date and that a practical completion certificate cannot be issued where these exist though it could be issued where there were latent defects which only become apparent later.

In *H. W. Nevill (Sunblest) Ltd* v. *William Press & Son Ltd* (1981), Judge Newey QC's view was:

> I think that the word 'practically' in Clause 15(1) gave the architect a discretion to certify that William Press had fulfilled its obligation under Clause 21(1) where very minor de minimis works had not been carried out, but if there were any patent defects in what William Press had done the architect could not have given a certificate of practical completion.

In *Emson Eastern Ltd (in receivership)* v. *EME Developments Ltd* (1991), Judge Newey QC said that the matrix of facts against which the building contract should be construed is what happened on building sites generally. There was a recognition that a construction project is not like the manufacture of goods in a factory. Factors such as size of the project, site conditions, use of many materials and employment of various types of operative make it impossible to achieve the same degree of perfection as a manufacturer. He took into account the overall contract scheme. After practical completion the employer took occupation and the contractor was required not to do more work but to remedy defects. It was said:

> In my opinion there is no room for 'completion' as distinct from 'practical completion'. Because a building can seldom if ever be built precisely as required by drawings and specification, the contract realistically refers to 'practical completion', and not 'completion' but they mean the same.

In *Borders Regional Council* v. *J Smart & Co (Contractors) Ltd* (1983), practical completion was taken as meaning that the works have been completed for all practical purposes and the employer could take them over and use them for their intended purpose.

Mr Justice Bokhary P.J. in the Hong Kong Court of Final Appeal in the case of *Mariner International Hotels Ltd* v. *Atlas Ltd* (2007) provided a definition of what 'practical completion' is understood to mean in building contracts in general. Preferring the arguments put forward by Counsel for Mariner in which he had referred to practical completion as a 'well known legal term of art with an established meaning in building contracts', it was said by Mr Justice Bokhary:

> In my view, what clause 2.01(b) means by 'practical completion' is a state of affairs in which the Hotel has been completed free from any patent defects other than ones to be ignored as trifling ... True it is that the standard of freedom from non-trifling patent defects is an exacting one. But it does not, after all, demand more than the avoidance of what is apparently defective and, moreover, apparently so to a degree exceeding what can be ignored as trifling.

There does seem to be a theme throughout these cases of practical completion requiring a stage of completeness with an absence of anything other than minor patent defects.

The issue of the Practical Completion Certificate triggers the release of the first half of the retention fund under clauses 4.20.2 and 4.20.3.

The Employer may take possession of the works once practical completion has been achieved. This does not put an end to the Contractors' obligations. They remain responsible for remedying any defects, shrinkages or other faults which may appear during the Rectification Period and which are due to materials or workmanship not in accordance with the contract or any failure of the Contractor to comply with his obligations in respect of the Contractor's Designed Portion, see clause 2.38. This is more fully dealt with in Section 5.3.

Once the defects, shrinkages and other faults as specified in the schedule of defects have been made good, the Architect issues a Certificate of Making Good under clause 2.39. This triggers the release of the second half of the retention fund under clause 4.20.3.

The SBC also includes Supplemental Provisions in Schedule Part 8. These were introduced for the first time in 2009. These apply unless the option in the Contract Particulars not to apply them is selected. These include provisions related to:

- Collaborative working – requiring the parties to work together and with other project team members in a co-operative and collaborative manner, in good faith and in a spirit of trust and respect.
- Health and safety – requiring the parties to endeavour to establish and maintain a culture and working environment in which health and safety are of paramount concern. There are also provisions concerning compliance with codes of practice, training of personnel, advice for personnel and consultation with personnel.
- Cost savings and value improvements – the Contractor is encouraged to propose changes to designs and specifications and/or the programme for the works which may benefit the Employer in terms of reduced construction cost, reduced life-cycle cost or earlier completion. Where the Employer wishes to make a change proposed, the clause envisages a negotiation between the Contractor and Employer to agree the value, financial benefit and any change to the Completion Date. The change, amount of adjustment to the Contract Sum, the share of the financial benefit to be paid to the Contractor and any adjustment to the Completion Date are confirmed in an Architect's instruction. It will be important for the Contractor to ensure that in doing so he does not take on design responsibility which he does not intend. In relation to the value and financial benefit, the Contract is deliberately vague on how these are to be ascertained. This may be particularly difficult to quantify where the Contractor identifies changes which will benefit life cycle cost. The Contract leaves it up to negotiation between the Employer and Contractor to agree on this.
- Sustainable development and environmental considerations – the Contractor is encouraged to suggest economically viable amendments to the works which could result in an improvement in environmental performance either in the carrying out of the works or of the completed works. The Contractor is to provide all information

requested by the Employer regarding the environmental impact of any supply and use of materials and goods which the Contractor selects.

- Performance indicators and monitoring – the Employer monitors and assesses the Contractor's performance by reference to any performance indicators included in the Contract Documents. The Contractor is obliged to provide the Employer with all information reasonably required to assess the Contractor's performance against the targets. If the Employer considers a target may not be met, the Employer may inform the Contractor who is to submit proposals for improvement. This is most likely to be of relevance in the context of a Framework arrangement but there is no reason it cannot be applied in individual contracts.
- Notification and negotiation of disputes – each party is to notify the other promptly of any matter that appears likely to give rise to a dispute or difference. There is then to be a meeting of the senior executives nominated in the Contract Particulars, to be held as soon as practicable, for direct, good faith negotiations to resolve the matter.

These Supplemental Provisions are also included in the SBCC Framework Agreement. Their purpose is to reflect within the contract the principles adopted by the office of Government Commerce's Achieving Excellence in Construction initiative.

5.2.3 The SBC Design and Build Contract (SBC/DB)

The SBC/DB is a stand-alone document and adopts the same format as the other SBC 2011 forms (though the clause numbering is not identical).

The SBC/DB generally follows the principles of the SBC where there is a Contractor's Designed Portion, with the design requirements being set out in the Employer's Requirements and the Contractor's Proposals. The Contractor's Proposals are intended to constitute the Contractor's response to the Employer's design requirements. As stated in the SBCC Guide to the SBC/DB:

> Depending upon the procurement approach adopted, the level of detail within the Employer's Requirements may vary between a performance orientated statement of objectives and a detailed and prescriptive statement of what the Contractor is to provide. Similarly, and responding to the degree of detail within the Employer's Requirements, the Contractor's Proposals may contain the result of substantial design development work or may simply be a reiteration of the Employer's Requirements.

The breakdown of the lump sum contract price is contained in the Contract Sum Analysis and this should be in sufficient detail to allow for monthly interim valuations where Alternative B of the SBC/DB clause 4.7 has been selected. The contract is administered on behalf of the Employer by the Employer's Agent rather than by an architect or contract administrator. Broadly, the duties carried out by the Architect/Contract Administrator under the other SBC forms are carried out under the SBC/DB by the Employer and, in practice, these duties will be delegated to the Employer's Agent. The Articles of Agreement provide that, save to the extent otherwise specified by written notice by the Employer, the Employer's Agent shall have full authority to receive and issue applications, consents, instructions, notices, requests or statements and otherwise to act for the Employer under any of the conditions. Thus it will be the Employer's Agent who will, for example, issue valuations for payment, issue the Practical Completion Statement, issue instructions, and determine applications for extensions of time and loss and expense. Other significant features of the SBC/DB are:

- The inclusion of a Contractor's Design Submission Procedure similar to that in the SBC where there is a Contractor's Designed Portion (see description in Section 5.2.2).
- Collateral Warranties or Third Party Rights from the Contractor to purchasers, tenants and funders and collateral warranties from relevant Sub-Contractors, including warranties to the Employer in the same way as in the SBC (see Chapter 13).
- A requirement on the Contractor to maintain professional indemnity insurance.
- The retention of the term 'change' as opposed to 'variation' (the latter being used in the other SBC forms).
- The use of the terms 'Practical Completion Statement' and 'Notice of Completion of Making Good' instead of the 'certificate' terminology used in the other SBC forms.
- Express limitations on the Contractor's responsibility for the Employer's Requirements (see Section 5.3.4a).
- Clause 3.9 provides that the Employer may not instruct a Change which is, or makes necessary, an alteration or modification in the design of the Works, without the consent of the Contractor, which is not to be unreasonably withheld or delayed.

In relation to the procedures regulating Changes under the DBC/DB, see Section 8.2.4.

5.2.4 The NEC3 contract

The obligation on the contractor in clause 20.1 of the NEC3 is to Provide the Works (which is defined as doing the work necessary to complete the works (as identified in the Contract Data) in accordance with the contract and all incidental works, services and actions which the contract requires) in accordance with the Works Information.

In relation to design, clause 21.1 provides for the Contractor to design the parts of the works which the Works Information states he is to design.

Where Option C, D, E, or F apply, clause 20.3 provides for the Contractor to advise the project manager on the practical implications of the design of the works and on sub-contracting arrangements. Clause 20.4 provides that the Contractor prepares forecasts of the total Defined Cost for the works, in consultation with the Project Manager and for submission to the Project Manager. These are prepared at the intervals specified in the Contract Data along with an explanation of any changes since the previous forecast. Where Option F applies, clause 20.2 provides for the Contractor to manage the Contractor's design, the provision of Site services and the construction and installation of the works. The Contractor in Option F sub-contracts these aspects of the work, except work which the Contract Data states the Contractor will do himself.

Clearly the Works Information is crucial to defining the Contractor's obligation. This is defined as being information which either specifies and describes the works or states any constraints on how the Contractor Provides the Works, and is either in the documents which the Contract Data states it is in or in an instruction given in accordance with the contract.

It would usually be expected that the Works Information would be in the form of a specification and drawings.

The NEC3 has obligations in relation to how parties are to work together. Clause 10.1 provides for the Employer, Contractor, Project Manager and Supervisor to act as stated in the contract and in the spirit of mutual trust and co-operation. Clause 25.1 provides that the Contractor co-operates with Others (i.e. people or organizations who are not the Employer, Project Manager, Adjudicator, Contractor or any employee, Sub-contractor or supplier of the Contractor) in obtaining and providing information which they need in connection with the works. The Contractor is to share the working areas with Others as stated in the Works Information.

As the work proceeds, clause 27.3 requires the Contractor to obey any instruction which is issued by the Project Manager or the Supervisor.

The instruction may be one which changes the Works Information. This will constitute a compensation event in terms of clause 60.1, unless the instruction was to instruct a change to accept a Defect or a change to Works Information provided by the Contractor for his design. Compensation events give rise to both time and cost relief and are considered in more detail in Section 6.11.11. In relation to the procedure for valuing compensation events, see also Section 8.2.5.

Where a compensation event arises from the Project Manager or Supervisor giving an instruction, issuing a certificate, changing an earlier decision or correcting an assumption, clause 61.1 provides that the Project Manager is to notify the Contractor of the compensation event at the time of the communication. He also instructs the Contractor to submit quotations unless the event arises from a fault of the Contractor or quotations have already been submitted. Clause 61.1 also requires the Contractor to put the instruction or changed decision into effect.

The Project Manager may, in terms of clause 61.2, instruct the Contractor to submit quotations for a proposed instruction. The Contractor does not put any proposed instruction into effect. This allows the Project Manager to check on the effect of any proposed instruction before taking a decision on whether or not to implement it.

In terms of clause 61.3, the Contractor is required to notify the Project Manager of any event which has happened or which he expects to happen as a compensation event if the Contractor believes that the event is a compensation event and the Project Manager has not notified the event to the Contractor. This clause contains an important time limit in that the Contractor is required to make the notification within eight weeks of becoming aware of the event. If the Contractor fails to do so, he is not entitled to a change in the Prices, the Completion Date or a Key Date unless the event arises from the Project Manager or Supervisor giving an instruction, issuing a certificate, changing an earlier decision or correcting an assumption. This represents a tightening up in the 2013 edition of the NEC3 from the previous edition of the position in relation to the Contractor's notification.

Where the Project Manager decides that the effects of a compensation event are too uncertain to be forecast reasonably, clause 61.6 requires that he states assumptions about the event in the instruction to the Contractor to submit quotations.

Alternative quotations can be requested, in terms of clause 62.1, where the Project Manager has discussed different practicable ways of dealing with the compensation event with the Contractor or where the Contractor has thought of other ways of dealing with the compensation event which he considers practicable.

In terms of the content of quotations, clause 62.2 provides for these to include proposed changes to the Prices and any delay to the Completion Date and Key Dates assessed by the Contractor. Details of the assessment are to be submitted with the quotation. If the programme for remaining work is altered, the alterations to the Accepted Programme are to be included in the quotation.

Clause 62.3 sets the timescale for submission of quotations by the Contractor as within 3 weeks of the instruction. The Project Manager then has 2 weeks to reply with an instruction to submit a revised quotation, an acceptance of a quotation, a notification that a proposed instruction will not be given or a notification that he will make his own assessment. Clause 62.4 requires the Project Manager to explain his reasons for any instruction to issue a revised quotation. Any revised quotation is to be submitted within 3 weeks of instruction.

Clause 62.5 allows the Project Manager to extend the time allowed for quotations or his reply to quotations if the Project Manager and Contractor agree to this before the deadline.

Where there is a failure by the Project Manager to respond within the time allowed, clause 62.6 allows the Contractor to notify the Project Manager of this failure. If the Project Manager still does not reply within 2 weeks, following the Contractor's notification, and unless the quotation is for a proposed instruction, the Contractor's notification is treated as acceptance of the quotation by the Project Manager.

In terms of clause 65.1, the Contractor is required to implement a compensation event (including an instruction of the Project Manager) when the Project Manager notifies acceptance of the Contractor's quotation, when the Project Manager notifies the Contractor of his own assessment, or when a Contractor's quotation is treated as having been accepted by the Project Manager.

In terms of clause 65.4 of Options A, B, C and D, the changes to the Prices, the Completion Date and the Key Dates are included in the notification implementing a compensation event. In clause 65.3 of Options E and F, the changes to the forecast amount of the Prices, the Completion Date and the Key Dates are included in the notification implementing a compensation event.

In contrast to the usual SBC provisions in relation to variations (with the exception of those related to the Variation Quotation procedure), the NEC3 is frontloaded in terms of requiring quotations and prior agreement to the time and cost consequences of changes to Works Information. It therefore allows assessment of the impact of any changes in advance, allowing informed decisions to be made as to whether or not to proceed with changes.

If an event occurs which stops the Contractor completing the works at all or from completing by the date shown on the Accepted Programme which neither party could prevent and which an experienced Contractor would have judged at the Contract Date to have such a small chance of occurring that it would have been unreasonable for him to have allowed for it, then the Project Manager is to give an instruction to the Contractor stating how he is to deal with the event (clause 19.1).

Although this clause is headed 'Prevention', it is not dealing with prevention in its normal meaning of acts or defaults of the Employer preventing the Contractor from performing the contract. It deals more with '*force majeure*' type events or other matters outside the control of the parties. In considering this, clause 60.1(19) is also relevant. It makes a prevention event a compensation event. The wording of clause 60.1(19) is in almost identical terms to clause 19.1. This means the Employer carries the time and cost risk.

Further, clause 91.7 again repeats the wording and lists this as an event allowing the Employer to terminate if such an event is forecast to delay Completion by more than 13 weeks.

These prevention provisions give the Contractor the potential to argue a wide range of events could not have been prevented by either party and have such a small chance of occurring that it would have been unreasonable to have allowed for them. Examples would include insolvency of suppliers or sub-contractors, supply of defective materials, defective work by sub-contractors, defective design by designers and strikes.

The issue is with the words 'small chance of occurring' which means it is not necessary to say they were not foreseeable. The same applies to 'unreasonable to have allowed for'. This means events which were foreseeable but unlikely to happen provide a potential remedy.

The House of Lords in *Scott Lithgow* v. *Secretary of State for Defence* (a shipbuilding case) considered a term in a contract which allowed Scott Lithgow (the contractor) to be paid for 'exceptional dislocation and delay arising during the construction of the vessel due to alterations, suspensions of work or any other cause beyond the contractor's control'. Delays were caused as a result of defective cables being supplied by one of Scott Lithgow's suppliers. These were found not to have been a matter within the contractor's control and they gave rise to additional entitlements as a consequence.

This is a useful clause for contractors to bear in mind but often employers will delete it from the list of compensation events, as it is regarded as too wide.

The NEC3 can be a Contractor design/Contractor designed portion form of contract, depending on what the Works Information states the Contractor is to design (see clause 21.1).

The Contractor is required by clause 21.2 to submit the particulars of his design, as the Works Information requires, to the Project Manager for acceptance.

The Project Manager can refuse to accept if the design does not comply with either the Works Information or the applicable law. Also in terms of clause 21.2, the Contractor does not proceed with the relevant work until the Project Manager has accepted his design. Clause 21.3 allows the Contractor to submit the design for acceptance in parts if the design of each part can be assessed fully. It should be noted that under clause 13.8 the Project Manager can withhold acceptance of a submission (for design or otherwise) by the Contractor, but if acceptance is withheld for a reason not stated in the contract, this constitutes a compensation event under clause 60.1(9) except where the withholding of acceptance relates to a quotation for acceleration or not correcting a Defect.

The Contractor is also required in terms of clause 23.1 to submit particulars of the design of an item of Equipment to the Project Manager for acceptance if the Project Manager instructs him to. The Project Manager may refuse to accept the design of the item if its design will not allow the Contractor to Provide the Works in accordance with the Works Information, the Contractor's design which the Project Manager has accepted, or the applicable law.

Where required, the Contractor is to obtain approval of his design from Others in terms of clause 27.1.

Secondary Option X15 provides for the Contractor's liability for his design to be limited to reasonable skill and care.

Clause X15.1 provides that the Contractor is not liable for Defects in the works due to his design so far as he proves that he used reasonable skill and care to ensure the design complied with the Works Information.

The Contractor has further obligations in relation to access to the Site. In terms of clause 27.2, the Contractor is to provide access to work being done and to Plant and Materials being stored for the contract for the Project Manager, the Supervisor and Others notified to him by the Project Manager.

In terms of the Contractor's representatives on site, the Contractor's obligations are set out in clause 24.1. The Contractor either employs each key person named to do the job stated in the Contract Data or employs a replacement person who has been accepted by the Project Manager. To have an alternative person accepted, the Contractor needs to submit the name, relevant qualifications and experience of the person to the Project Manager. A reason for the Project Manager not to accept the replacement would be if the relevant qualifications and experience of the replacements were not as good as those of the person who is to be replaced.

Clause 24.2 allows the Project Manager to instruct the Contractor to remove an employee, in which case the Contractor arranges that, after one day, the employee has no further connection with the work.

In relation to sub-contracting, there is specific provision in clause 26.1 to the effect that the Contractor remains responsible for Providing the Works as if he had not sub-contracted. The Sub-contractor's employees and equipment are treated as if they were the Contractor's.

Clause 26.2 requires the Contractor to submit the name of each proposed sub-contractor to the Project Manager for acceptance. It is a reason not to accept if the Sub-contractor's appointment will not allow the Contractor to Provide the Works. The Contractor is not to appoint a proposed Sub-contractor until the Project Manager has accepted him.

In addition to acceptance of the Sub-contractors, clause 26.3 requires the Contractor to submit the proposed conditions of contract for each sub-contractor to the Project Manager for acceptance unless an NEC contract is proposed or the Project Manager has agreed that no submission is required. The Contractor is not to appoint a Sub-contractor on the proposed sub-contract conditions until the Project Manager has accepted them. Reasons for not accepting would be that they will not allow the

Contractor to Provide the Works or they do not include a statement that the parties to the sub-contract shall act in the spirit of mutual trust and co-operation.

In Options C, D, E and F, clause 26.4, the Contractor is to submit the proposed contract date for each sub-contract for acceptance to the Project Manager if an NEC contract is proposed and if the Project Manager instructs the submission. A reason for the Project Manager not to accept the proposed contract date is that its use would not allow the Contractor to Provide the Works.

These provisions related to sub-contracting are significantly more detailed than those in the SBC and the SBC/DB forms of contract and give the Project Manager in the NEC3 a much more hands-on role. See also Section 11.9 in relation to sub-contracting under the NEC3.

As the works progress, the Contractor may find objects of value or historical or other interest within the Site. In terms of clause 73.1, the Contractor is to notify the Project Manager when such objects are found and the Project Manager instructs the Contractor how to deal with it. The Contractor is not to move the object without instructions.

Any Project Manager's instructions for dealing with such objects are a compensation event in terms of clause 60.1(7).

The NEC3, unlike the SBC and the SBC/DB, contains the definition of Completion in clause 11.2(2). It states that Completion is when the Contractor has done all the work which the Works Information states he is to do by the Completion Date and corrected notified Defects which would have prevented the Employer from using the works and Others from doing their work. If the work which the Contractor is to do by the Completion Date is not stated in the Works Information, Completion is when the Contractor has done all the work necessary for the Employer to use the works and for Others to do their work.

Completion is not to be confused with the Completion Date (which is defined in clause 11.2(3) with reference to the completion date inserted in the Contract Data Part One if the Employer is to decide it or Contract Data Part Two if the Contractor is to decide).

Optional Clause X5 provides for Sectional Completion so that, unless references are stated as being to the whole of the works, each reference and clause relevant to the works, Completion and Completion Date applies to either the whole of the works or any section of the works.

In some cases, the Works Information or the applicable law require tests and inspections to be carried out. In that case, the provisions of clause 40 apply to these tests and inspections.

Clause 40.2 provides for the Contractor and Employer to provide materials, facilities and samples for tests and inspections as stated in the Works Information.

Clause 40.3 requires the Contractor and Supervisor each to notify the other of his tests and inspections before it starts and afterwards notify the other of its results. If any subsequent work would obstruct the test or inspection, Clause 40.3 requires the Contractor to notify the Supervisor in time for it to be arranged before doing that work. The Supervisor may watch any test done by the Contractor.

In the event of Defects being identified by a test or inspection, clause 40.4 requires the Contractor to correct the Defect and repeat the test or inspection. In that case, in terms of clause 40.6, the Project Manager assesses the cost incurred by the Employer in repeating a test or inspection after a defect has been found and the Contractor pays the amount assessed.

The Contractor is, in terms of clause 72.1, to remove Equipment from the Site when it is no longer needed unless the Project Manager allows it to be left in the works.

Equipment is defined in clause 11.2(7) as items provided by the Contractor and used by him to Provide the Works and which the Works Information does not require him to include in the works. This can be contrasted with the definition of Plant and Materials in clause 11.2(12) which are items intended to be included in the works.

This removal of Equipment could be on Completion or could be during the progress of the works.

The provisions regarding take-over of the works by the Employer are in clause 35. In the event that the Contractor finishes early, before the Completion Date, clause 35.1 provides that the Employer need not take over the works before the Completion Date if it is stated in the Contract Data that he is not willing to do so. Otherwise, the Employer takes over the works not later than two weeks after Completion.

Clause 35.2 contains the NEC3 equivalent to the SBC and the SBC/DB partial possession provisions. It allows the Employer to use any part of the works before Completion has been certified. If he does so, he takes over the part of the works when he begins to use it except if the use is for a reason stated in the Works Information or to suit the Contractor's method of working.

Clause 35.3 provides that the Project Manager certifies the date upon which the Employer takes over any part of the works and its extent within one week of the date.

In relation to take-over and partial possession under the NEC3, see also Section 6.11.

Optional Clause X6 provides for the Contractor to be paid a bonus for early completion.

Optional Clause X12 provides for partnering and contains detailed provisions as to how that is to operate.

Optional Clause X20 provides for Key Performance Indicators, namely aspects of performance by the Contractor for which a target is stated in the Incentive Schedule.

5.3 *The quality of the work*

5.3.1 Workmanship and design at common law

As far as workmanship is concerned, at common law, the contractor has an obligation to execute the work in a good and workmanlike manner using the skill and care to be expected of a builder of ordinary competence. This obligation subsists while the works are being carried out; it does not arise only at completion, see *Surrey Heath Borough Council* v. *Lovell Construction Ltd and Another* (1998).

The test of what constitutes such skill and care is the standard of the ordinary skilled contractor exercising and professing to have that particular skill. It is not necessary to possess the highest expert skill, but is sufficient to exercise the ordinary skill of an ordinary competent contractor exercising the particular skill, see *Bolam* v. *Friern Hospital Management Committee* (1957).

It may not be sufficient to show the contractor did the same as other contractors if it can be shown the generally accepted practice is not correct, see *Sidaway* v. *Board of Governors of the Bethem Royal Hospital and the Maudsley Hospital* (1985).

The obligation may go further than simply carrying out the work as it is contained on drawings or in other design information provided by the design team.

In a case where the contractor was to provide all materials and perform all work shown on an architect's drawings, but where they were not supervised by an architect or engineer, it was found that the contractor had accepted that the employer was relying on their skill as contractors. There were defects in the plans provided to the contractor. The contractor ought to have recognized the defects and, since they were being relied on by the employer, had a duty to warn the employer of the defects in the plans and the difficulties which would arise if the plans were followed, see *Brunswick Construction Ltd* v. *Nowlan & Others* (1974).

The courts have been willing to imply a term into a contract requiring a contractor to warn of design defects as soon as the contractor came to be aware that they existed. In a case where the court considered it must have become apparent during construction work that the design of curtain walling was unbuildable, it was found this should have been reported. This was on the basis that if, on examining the drawings or as a result of experience on site, a contractor formed the opinion that in some respect the design would not work, or would not work satisfactorily, it would be absurd for them to carry on building in accordance with it. The contractor was found to have a duty of care to warn the employer and architect of design defects known to the contractor, see *Equitable Debenture Assets Corporation Ltd* v. *William Moss Group Ltd and Others* (1984).

The duty to warn goes further. A term has been implied into a contract requiring the contractor to warn the architect of defects in the design which they believe to exist. This does not oblige the contractor to carry out a critical examination of the drawings, bills and specifications looking for mistakes. The contractor's primary duty is to build, not to scrutinize the design. The obligation to warn arises when, in the light of their general knowledge and practical experience, the contractor believes or comes to believe that an aspect of the design is wrong, see *Victoria University of Manchester* v. *Hugh Wilson & Others* (1984).

Even though the contractor has to be satisfied that the design is satisfactory or suitable, the work is still required to be carried out in accordance with good building practice. A contractor cannot rely on drawings or designs produced by a surveyor to relieve them of this duty, see *Mackay* v. *Stitt* (1988). The contractor, if not satisfied with a design that is produced to them, has a duty to raise any queries with the design team. Even if the contractors are given such assurances they may, if still not satisfied, need to take precautions against failure of the design. If they do not do so, then they may be found to have acted with less care than is to be expected of an ordinary competent builder, see *Edward Lindenberg* v. *Joe Canning and Others* (1992).

Where there are health and safety considerations, the duty to warn can go further. In a case where temporary support was to be provided to roof trusses, the sub-contractor considered the propping as designed to be inadequate. They advised the designer but did not follow up their objections. The propping failed, causing the roof to collapse. It was found that the sub-contractor ought to have pressed its objections on safety grounds and made these progressively more formal and insistent, including putting them into writing, approaching higher levels of management, threatening to or actually reporting the problem to the regulatory authorities and, ultimately, suspending work, see *Plant Construction plc* v. *Clive Adams Associates and Others* (2000).

The courts have, however, refused to extend the duty to warn to cover a situation where a design and build contractor who was being advised by structural engineers claimed that specialist underpinning contractors ought to have advised them of the need for temporary lateral support to be provided while the basement was being excavated. This claim was made on the basis that it was said the underpinning contractor knew and/or it was obvious that there was a significant danger that the design and build contractor might excavate the basement without providing such support. Given the danger of the excavation proceeding without the support, it was said it could not be assumed by the underpinning contractor that the excavation would be carried out safely. The court accepted that if the underpinning contractor had been instructed to carry out work it knew to be unsuitable and dangerous, then it would have a duty to warn the contractor, even though the contractor was itself receiving advice from a structural engineer. However, in this case, the work was to be carried out by another party in the future. The underpinning contractor did not know how the contractor intended to carry out the work. It was accepted that the method chosen by the contractor was negligent. It was considered relevant that the contractor was receiving advice from a structural engineer. In these circumstances it was considered unreasonable to impose a duty to warn on the underpinning contractor, see *Aurum Investments Ltd* v. *Avonforce Ltd (In Liquidation) and Others* (2000).

Where the contractor is not responsible for the design of a system or its integration into the works or for the selection of a proprietary system, there is no implied warranty by the contractor that the system will work, see *Greater Glasgow Health Board* v. *Keppie Henderson & Partners* (1989).

The view of what constitutes normal practice may alter depending upon the nature of the development. For example, where plumbing sub-contractors in a multi-storey flat development took the normal steps which would be taken to drain pipes of water in ordinary houses or small developments, they were found to be to blame for damage caused by burst pipes where the burst was caused by water which had not been drained from the pipes. The plumbing sub-contractors had failed to warn anyone that the pipes could not be completely drained. The normal practice of draining pipes, which resulted in some water remaining, did not apply in the face of the extent of the damage which might be expected in a multi-storey development were this to be followed, see *Holland Hannen & Cubitts (Scotland) Ltd* v. *Alexander Macdougall & Co. (Engineers) Ltd* (1968).

Where employers make known the purpose of the building and circumstances indicate that they are relying upon the contractor's skill and judgement to provide it, there is an implied term that the works will be fit for the purpose for which they were intended. Where a contractor expressly undertakes to carry out work which will perform a certain duty or function in conformity with plans or specifications, a 'fitness for purpose' obligation, and it turns out that the works would not perform that function if the plans were followed, the contractor will be liable for the failure to perform even if the work was carried out in accordance with the plans and specifications.

In a case where contractors were employed on a design and build contract to build a new factory, warehouse and offices, it was made known to the contractor that oil drums stored in the warehouse would be moved around by forklift trucks. When constructed, the movement of the forklifts caused vibrations which led to cracking of the floor. It was found that since the employer had made known to the contractor the purpose for which the building was required, they had relied on the contractor's skill and judgement. The contractor was therefore required to ensure the building was reasonably fit for its purpose, see *Greaves & Co. (Contractors) Ltd* v. *Baynham Meikle & Partners* (1975).

General observations about fitness for purpose obligations were made in a case related to the standard of duty to be implied into a town development agreement and whether an obligation was to be implied that the houses to be built were to be fit for human habitation. It was observed that where a contractor has design responsibility, their duties are to carry out the work in a good and workmanlike manner, to supply good and proper materials and to provide a building reasonably fit for human habitation, see *Test Valley Borough Council* v. *Greater London Council* (1979).

In a case concerning the collapse of an aerial mast, the court said it saw no reason why, if contractors contract in the course of their business to design, supply and erect a television aerial mast, they should not be under an obligation to ensure that it is reasonably fit for the purpose for which they knew it was intended. The question to be asked is whether the person for whom the mast was designed, relied on the skill of the supplier to supply and design a mast fit for the purpose for which it was known to be required, see *Independent Broadcasting Authority* v. *EMI Electronics Ltd and BICC Construction Ltd* (1980).

In a case where a contractor was to design and build houses for a local authority and the authority had made known the purpose for which the work was required so as to show reliance on the contractor by the employer, a term that the buildings designed by the contractor as dwellings should be fit for habitation on completion was implied into the contract, see *Basildon District Council* v. *J E Lesser Properties* (1984).

In assessing whether there is an implied fitness for purpose obligation, there is no distinction to be made between reliance by the employer in relation to the quality of materials and their design, the design and specifications of the functional parts of the installation as a whole and the condition of the ground, as all of these are integral parts of the whole and all are interdependent on each other. The term of reasonable fitness for purpose will be implied irrespective of whether there has been any negligence or fault or whether the unfitness results from quality of work, quality of materials or defects in the design, see *Viking Grain Storage* v. *T H White Installations Ltd* (1985).

The duty can be limited to certain aspects of the work only. In a case where an employer advised the contractor that boilers were to be installed in flats which were thereafter to be sold and the contractors recommended boilers, the employer's claim that though the boilers worked satisfactorily they resulted in lower SAP (Standard Assessment Procedure) ratings for the flats, making them more difficult to sell, failed. It was found that as the employer had only given partial information to the contractor, which did not include information upon which they could form a view as to the effect of the boilers on the SAP ratings, the employer had not relied on the contractor to advise in relation to this. The employer had relied on the contractor only to the extent of advice as to the qualities of the boilers generally. The contractor's duty had been fulfilled in relation to the issue on which the employer had relied on them, see *Jewsons Ltd* v. *Leanne Boykan* (2004).

The courts will go to some lengths to establish the factual position when considering the extent of the duty. In *Trebor Bassett Holdings Ltd* v. *ADT Fire and Security plc* (2012), ADT had contracted to design, supply, install and commission a fire suppression system intended to extinguish fires in two parts of a popcorn production factory. The issue in the case was whether ADT's obligation was restricted to reasonable skill and care in the design of the system or whether it went further, so that the system supplied should be reasonably fit for its purpose. The Court of Appeal's analysis focused on whether the service provided by ADT was design (requiring reasonable skill and care) or supply of goods (the system) which required to be reasonably fit for their purpose. ADT were not told and did not know that there is a history of fires caused by burning popcorn. On the evidence, it was found that Trebor had not shown that it made known to ADT a particular purpose for which the system was required. It had not explained the process or its hazards. On this basis, the obligation was restricted to reasonable skill and care. This case would suggest an employer does need to go to some lengths to explain the purpose of work it is acquiring in order to benefit from a fitness for purpose obligation – general statements may not suffice.

The contractor in a contract including an element of design responsibility may have an obligation to complete a design which has been commenced by others. In a case where the contractor was to design piled walls, they relied on the content of a report on ground conditions prepared by the employer's engineer. It was found that the design obligation was to complete the design of the piled walls. This meant developing the conceptual design of the engineers into a complete design capable of being constructed. Understanding the principles underlying the work and forming a view as to their sufficiency was required so that the completed design as a whole was prepared with reasonable skill and care, see *Co-operative Insurance Society Ltd* v. *Henry Boot (Scotland) Ltd* (2002). This has implications for the Contractor's responsibility for inadequacies in the Employer's Requirements under the SBC where there is a Contractor's Designed Portion and under the SBC/DB, see Section 5.3.4.

The implied fitness for purpose obligations described above can, of course, be (and, more often than not, are) displaced by the express terms of the contract imposing an alternative standard, see the SBC clause 2.19.1 (where there is a Contractor's Designed Portion) and the SBC/DB clause 2.17.1, which, as far as any inadequacy in design is concerned, provide that the Contractor shall have the same liability to the Employer as an architect or other appropriate professional designer holding himself out as competent to take on such work, in other words a 'reasonable skill and care' standard. This is referred to in more detail in Section 5.3.3.

5.3.2 Materials: common law

Until 1995 (when the relevant statute was extended to Scotland) it was the case that the common law implied into contracts that required the supply of materials, terms that the materials used would be of good quality and would be reasonably fit for the purpose for which they were used, unless it could be demonstrated that the parties' intention was to exclude the implied terms, or either of them, see *Young & Marten Ltd* v. *McManus Childs Ltd* (1968). Where the purchaser made known to the contractor the particular purpose for which the materials were required, so as to show that reliance was being placed on the contractor's skill and judgement, and where the materials were of a type which it was in the course of the contractor's business to supply, it was implied that the materials would be reasonably fit for that purpose, see *Young & Marten Ltd.*

There was a further obligation in relation to materials, namely that they had to be free from defects which would have meant that they were not of the requisite quality. This also applies to latent defects which could have been detected using due skill and care. This did not, however, go so far as to imply a warranty by the contractor that the materials were suitable for the contract purpose in circumstances where the selection of the materials or their suitability was not a matter for the discretion of the contractor, see *Greater Glasgow Health Board* v. *Keppie Henderson & Partners* (1989).

The matter of implied terms in relation to the supply of materials is now dealt with by the Supply of Goods and Services Act 1982, which was extended to Scotland in relation to contracts made on or after 3 January 1995. Implied terms as to quality and fitness are now to be found in s.11D. S.11D(2) provides that where in a contract for the transfer of goods, the transferor transfers the property in goods in the course of a business, there is an implied term that the goods supplied are of satisfactory quality. S.11D(3) defines satisfactory quality as being where they meet the standard that a reasonable person would regard as satisfactory taking account of any description of the goods, the price (if relevant) and all the other relevant circumstances. In circumstances where the transferee makes known to the transferor any particular purpose for which the goods are being acquired, s.11D(6) provides that there is then an implied term that the goods supplied under the contract are reasonably fit for the purpose. S.11D(7) provides an exception to this where the transferee does not rely, or it is unreasonable for him to rely, on the skill or judgement of the transferor.

5.3.3 Workmanship, design and materials under the SBC

By virtue of clause 2.3.1 of the SBC, the Contractor is obliged to use materials and goods for the Works of the kinds and standards described in the Contract Bills. Materials and goods for any Contractor's Designed Portion Works are to be of the kinds and standards described in the Employer's Requirements, Contractor's Proposals or the Contractor's Design Documents. The Contractor is not permitted to substitute any materials or goods without the consent of the Architect.

There is a further provision in clause 4.1 which provides that the quality and quantity of the work included in the Contract Sum are deemed to be that set out in the

Contract Bills and, where there is a Contractor's Designed Portion, in the Contractor's Designed Portion documents.

Workmanship for the Works, excluding Contractor's Designed Portion Works, is to be of the standards described in the Contract Bills. Workmanship for any Contractor's Designed Portion Works is to be of the standards described in the Employer's Requirements or Contractor's Proposals, all by virtue of clause 2.3.2.

The terms as to quality of materials or goods and standards of workmanship are subject to the proviso in clause 2.3.3 that where and to the extent that approval of these is a matter for the opinion of the Architect, such quality and standards are to be to his reasonable satisfaction. To the extent that the quality of materials or goods or standards of workmanship are not described in the manner set out in clauses 2.3.1 and 2.3.2, nor stated to be a matter for the Architect's opinion or satisfaction, they are required, under clause 2.3.3 in the case of the Contractor's Designed Portion, to be of a standard appropriate to it, and in any other case, of a standard appropriate to the Works.

The Contractor is required by clause 2.3.4, if the Architect requests, to provide reasonable proof that the materials and goods comply with clause 2.3.

The Contractor has an obligation, in terms of clause 2.3.5, to take all reasonable steps to encourage Contractor's Persons to be registered cardholders under the Construction Skills Certification Scheme (CSCS) or qualified under an equivalent recognized qualification scheme. The CSCS was set up in the mid-1990s with the aim of improving workers' competence in the construction industry, reducing accidents by raising health and safety standards and driving up on-site efficiencies. It is managed by CSCS Limited. The owners of the scheme are the Civil Engineering Contractors Association (CECA), the Federation of Master Builders (FMB), the GMB Union, the National Specialist Contractors Council, the UK Contractors Group, UNITE (the Union) and Union of Construction, Allied Trades and Technicians (UCATT).

If the Contractor becomes aware of any departure from the Standard Method of Measurement in the preparation of the Contract Bills, inadequacy in any design in the Employer's Requirements, any error in description or quantity in the Contractor's Proposals or Contractor's Designed Portion Analysis or any omission of items (see clause 2.14), or there is any other discrepancy or divergence in or between the Contract Drawings, the Contract Bills, any Architect's instruction, any drawings or documents issued by the Architect under clauses 2.9 to 2.12 and, if applicable, the Contractor's Designed Portion Documents, clause 2.15 provides that the Contractor is required to give immediate notice to the Architect of this. The Architect is then required, by clause 2.15, to issue instructions to the Contractor as to how this is to be dealt with.

Where any discrepancy or divergence to be notified under clause 2.15 is within or between Contractor's Designed Portion Documents other than the Employer's Requirements, the Contractor is, under clause 2.16.1, to send with his notice, or as soon as reasonably practicable thereafter, a statement setting out his proposed amendments to remove it. Where the discrepancy is within the Employer's Requirements, the Contractor's Proposals prevail. Where the Contractor's Proposals do not deal with such a discrepancy, the Contractor is to inform the Architect of their proposed amendment to deal with it. The Architect either agrees to that or decides how the discrepancy is to be dealt with and that agreement or decision is treated as a Variation, see clause 2.16.2.

If the Contractor becomes aware of any divergence between the Statutory Requirements and the documents referred to in clause 2.15 then, under clause 2.17.1, it is immediately to give notice to the Architect specifying the divergence and, where it is between the Statutory Requirements and any of the Contractor's Designed Portion documents, the Contractor is to inform the Architect of their proposed amendment for removing it. Under clause 2.17.2, the Architect is then to issue instructions in that regard within 7 days of becoming aware of such divergence or within 14 days of receipt of the Contractor's proposed amendment, where applicable.

Clause 2.18.1 makes provision for emergency compliance with Statutory Requirements. If, in any emergency, the Contractor has had to supply materials and/or execute work before receiving instructions, the Contractor is to supply such limited materials and execute such limited work as is reasonably necessary to secure immediate compliance with the Statutory Requirements. The Contractor is then obliged, in terms of clause 2.18.2, to inform the Architect of the emergency and the steps taken under clause 2.18.1.

Where there is any failure to comply with clause 2.1 in regard to the carrying out of the work in a proper and workmanlike manner and/or in accordance with the Construction Phase Plan, clause 3.19 provides that the Architect may issue whatever instructions are reasonably necessary as a result. If this is done, the Contractor receives no addition to the Contract Sum and no extension of time for complying with such an instruction.

Where the Works include a Contractor's Designed Portion, the Contractor is required, in terms of clause 2.2.1, to complete the design, including the selection of any specifications for the kinds and standards of the materials, goods and workmanship so far as these are not described or stated in the Employer's Requirements or Contractor's Proposals, in accordance with the Contract Drawings and Contract Bills.

In terms of clause 2.2.2, the Contractor is to comply with the Architect's directions for the integration of the design of the Contractor's Designed Portion with the design of the Works as a whole. This obligation is subject to the provisions of clause 3.10.3, which allows the Contractor, if in his opinion compliance with a direction under clause 2.2.2 or any Architect's instruction would injuriously affect the efficacy of the design of the Contractor's Designed Portion, to give notice to the Architect specifying the injurious effect. In these circumstances, the direction or instruction shall not take effect unless confirmed by the Architect.

The Contractor has an obligation, in complying with clause 2.2 and in terms of clause 2.2.3, to comply with the CDM Regulations. Health and safety matters are considered in Chapter 20.

The Contractor is not responsible for the content of the Employer's Requirements or for verifying the adequacy of any design contained within them, see clause 2.13.2.

Where there is a Contractor's Designed Portion, clause 2.19.1 provides that, insofar as its design is comprised in the Contractor's Proposals and in what the Contractor is to complete in accordance with the Employer's Requirements and the contract Conditions, the Contractor has, in respect of any inadequacy in the design, the same liability to the Employer as an architect or other appropriate professional designer who held himself out as competent to take on work for such design.

The Contractor's liability for loss of use, loss of profit or other consequential loss arising from the liability under clause 2.19.1 is, under clause 2.19.2, limited to the amount, if any, stated in the Contract Particulars.

5.3.4 Design responsibilities under the SBC/DB

Similar provisions to those referred to in Section 5.3.3 apply to SBC/ DB. The following features should, however, be noted:

- There is no equivalent to clause 2.3.3 of the SBC which relates to approval of the quality of materials or goods or of standards of workmanship by the Architect.
- Clause 2.11 of the SBC/DB replicates clause 2.13.2 of the SBC, i.e. the Contractor shall not be responsible for the contents of the Employer's Requirements or for verifying the adequacy of any design contained within them.
- Clause 2.12 of the SBC/DB provides that if any inadequacy is found in any design in the Employer's Requirements, then to the extent that such inadequacy is not dealt with in the Contractor's Proposals, the Employer's Requirements shall be corrected, altered or modified accordingly and (subject to clause 2.15 in relation to Statutory Requirements), this shall be treated as a Change.

If the Contractor becomes aware of any such inadequacy or any other discrepancy or divergence in or between the Employer's Requirements, the Contractor's Proposals, any instruction by the Employer, or any of the Contractor's Design Documents issued under clause 2.8, the Contractor shall issue a notice to the Employer who shall then issue instructions.

Clause 2.14.1 provides that where the discrepancy or divergence to be notified by the Contractor is within the Contractor's Proposals, the Contractor shall inform the Employer of their proposed amendment to remove the discrepancy and the Employer shall decide between the discrepant items or may accept the Contractor's proposed amendment, in either case without cost to the Employer.

Where the discrepancy is within the Employer's Requirements, the Contractor's Proposals shall prevail without any adjustment to the Contract Sum.

Where the Contractor's Proposals do not deal with such a discrepancy, the Contractor shall inform the Employer of his proposed amendment; and the Employer's decision, either agreeing the amendment or otherwise determining how the discrepancy is to be dealt with, shall be treated as a Change.

Clause 2.15.1 provides that where there is a divergence between the Employer's Requirements or the Contractor's Proposals and a Statutory Requirement, then the Contractor shall immediately give notice specifying the divergence and propose, for the Employer's agreement, an amendment for removing it. The Contractor is then to complete the design and construction of the works in accordance with that amendment at his own cost. The exceptions are (1) where the Statutory Requirements change after the Base Date; (2) an amendment to the Contractor's Proposals which is necessary to comply with Development Control Requirements (i.e. statutory provisions and any decision of a relevant authority controlling the development of the site, such as planning permission) unless the risk is imposed on the Contractor under the Employer's Requirements; or (3) an amendment is necessary to any part of the Employer's Requirements which the Employer has stated is compliant with Statutory Requirements (see clause 2.1.2). In each of these cases the amendment will be treated as a Change.

The equivalent limitation of design liabilities to that contained in clause 2.19 of the SBC (where there is a Contractor's Designed Portion) is to be found in clause 2.17 of the SBC/DB. In effect, this excludes the common law 'fitness for purpose' duty of a design and build contractor (see Section 5.3.1) and replaces it with a liability equivalent to an appropriate professional designer, i.e. one of 'reasonable skill and care'. See also Section 14.4.

The express exclusion of responsibility of the Contractor under the SBC/DB clause 2.11 (and under clause 2.13.2 of the SBC where there is a Contractor's Designed Portion) for the contents of the Employer's Requirements or for verifying the adequacy of any design contained within them is acknowledged by the SBCC Guide to the SBC/DB as being intended to avoid the consequences of the decision in *Co-operative Insurance Society Limited* v. *Henry Boot Scotland Ltd* (2002), see Section 5.3.1. In that case the Contractor was appointed under the Contractor's Designed Portion Supplement to JCT 80. Due to the absence of a clause equivalent to the SBC/DB clause 2.11, the Contractor was held responsible for verifying the adequacy of the design in the Employer's Requirements. In practice, however, developers will often wish the Contractor to undertake exactly the type of risk which clause 2.11 is specifically intended to avoid, and for that reason clause 2.11 will often be deleted.

5.3.5 Workmanship, design and materials under the NEC3

In the NEC3, the obligation in respect of workmanship is, in terms of clause 20.1, that the Contractor Provides the Works in accordance with the Works Information. This is covered in more detail in Section 5.2.4.

There is no derogation from the Contractor's responsibility to Provide the Works or his liability for design as a result of the Project Manager or Supervisor's acceptance of any communication from the Contractor. This is confirmed by clause 14.1. The communications referred to here would include the various design documents which the Contractor is obliged to submit in terms of clauses 21.2 and 23.1. This is dealt with in more detail in Section 5.2.4.

The provisions regarding ambiguities and inconsistencies in or between documents which are part of the contract are in clause 17.1. They simply require the Project Manager or Contractor to notify the other as soon as either becomes aware of any ambiguity or inconsistency. The Project Manager is then to issue an instruction resolving it.

In the event the Contractor considers the Works Information requires him to do anything which is illegal or impossible, clause 18.1 provides for the Contractor to notify the Project Manager. If the Project Manager agrees, he gives an instruction to change the Works Information appropriately.

Secondary Option X15, if selected, allows for limitation of the Contractor's liability for his design to reasonable skill and care. In the absence of this clause the liability would be for the fitness for purpose obligation.

Clause X15.1 provides that the Contractor is not liable for Defects in the works due to his design so far as he proves that he used reasonable skill and care to ensure that his design complies with the Works Information. Clause X15.2 provides that if the Contractor corrects a Defect for which he is not liable under the contract, it is a compensation event.

There is a further opportunity to limit the Contractor's liability in Secondary Option X18. Clause X18.1 provides for the Contractor's liability to the Employer for the Employer's indirect or consequential loss to be limited to the amount stated in the Contract Data. Clause X18.2 provides similarly for loss of or damage to the Employer's property; clause X18.3 for Defects due to the Contractor's design which are not listed on the Defects Certificate; and clause X18.4 for all matters arising under or in connection with the contract, other than excluded matters. The excluded matters are amounts payable by the Contractor as stated in the contract for loss of or damage to the Employer's property, delay damages if Option X7 applies, low performance damages if Option X17 applies, and the Contractor's share if Option C or Option D applies.

Clause X18.5 provides that the Contractor is not liable to the Employer for a matter unless it is notified to the Contractor before the 'end of liability date'. The end of liability date is defined in the Contract Data Part One as a specified number of years after the Completion of the whole of the works.

5.4 *Defective work*

5.4.1 Common law

The contractor, in addition to the obligations already considered, also remains liable for latent defects for the duration of the prescriptive period. This is on the basis that the latent defect is due to an act, neglect or default of the contractor. The contractor will have no liability if it is not possible to show a link between any breach of duty by the contractor and the loss, injury or damage incurred, if the breach and the damage caused are too remote from each other or if a term of the contract excludes the contractor's liability. The subject of prescription (or time bar) is considered in more detail in Section 9.9.

5.4.2 The SBC provisions

During the currency of the contract works, the Contractor can, under clause 3.18.1 of the SBC, be instructed to remove from the site any work, materials or goods that are not in accordance with the contract. As an alternative to this, and following consultation with the Contractor and with the agreement of the Employer, the Architect may allow such work, materials or goods to remain. In these circumstances, a deduction is made from the Contract Sum under the provisions of clause 3.18.2. It should be noted that this provision does not apply to work, materials or goods which are part of the Contractor's Designed Portion and there is no equivalent clause to 3.18.2 in the SBC/DB. Where, as a consequence of any such instruction being issued in terms of clauses 3.18.1 or 3.18.2, instructions requiring a variation are necessary, the Architect may issue these but no addition is made to the Contract Sum as a result and no extension of time is given under the provisions of clause 3.18.3.

Under clause 3.18.4, if work, materials or goods are not in accordance with the contract and having due regard to the related Code of Practice in Schedule Part 4, the Architect may also issue such instructions to open up for inspection or to test, as are reasonable to establish to the Architect's reasonable satisfaction the likelihood or extent of any further similar non-compliance. To the extent such instructions are reasonable, whatever the results of the opening up, no addition shall be made to the Contract Sum. However, the opening up may be a Relevant Event under clause 2.29.2.2, entitling the Contractor to an extension of time under clause 2.28, unless the inspection or test shows that the work, materials or goods are not in accordance with the contract (see clause 3.18.4).

Following practical completion, and if any defects, shrinkages or other faults in the Works appear within the relevant Rectification Period due to materials, goods or workmanship not being in accordance with the contract or any failure of the contractor to comply with his obligations in respect of the Contractor's Designed Portion, these defects, shrinkages or other faults are, in terms of clause 2.38.1, to be specified by the Architect in a schedule of defects which is to be delivered to the Contractor as an instruction not later than 14 days after expiry of the Rectification Period.

The Architect may also, in terms of clause 2.38.2, whenever he considers it necessary, issue instructions requiring any defects, shrinkages or other faults to be made good. No such instruction can be issued after delivery of the schedule of defects under clause 2.38.1 or more than 14 days after the expiry of the relevant Rectification Period.

The Contractor has an obligation, under clause 2.38, and unless the architect otherwise instructs, to make good within a reasonable time after receipt of the schedule of defects the defects, shrinkages or other faults listed in the schedule prepared by the Architect.

The Contractor is obliged, unless the Architect otherwise instructs, to comply with such instructions in terms of clause 2.38. Where the Contractor makes good defects, shrinkages or other faults, this is at no cost to the Employer. Where the Architect instructs the Contractor not to make good, an appropriate deduction is made from the Contract Sum in respect of defects, shrinkages or other faults not made good.

5.4.3 The SBC/DB provisions

Similar provisions to those contained in clause 2.38 of the SBC are contained in clause 2.35 of the SBC/DB, and in the latter case the duties of the Architect are undertaken by the Employer or, in practice, the Employer's Agent.

5.4.4 The NEC3 provisions

As with SBC and the SBC/DB, the Contractor has an ongoing liability in relation to Defects. Defect is defined in clause 11.2(5) as a part of the works which is not in accordance with the Works Information or a part of the works designed by the Contractor which is not in accordance with the applicable law or the Contractor's design which the Project Manager has accepted.

Clause 42.1 provides that until the 'defects date' (which is specified in the Contract Data Part One as a number of weeks after Completion of the whole of the works) the Supervisor may instruct the Contractor to search for a Defect. A reason for the search is given within the instruction. Searching may include uncovering, dismantling, recovering and re-erecting work; providing facilities, materials and samples for tests and inspections done by the Supervisor; and doing tests and inspections which the Works Information does not require.

In addition, clause 42.2 provides that until the defects date the Supervisor notifies the Contractor of each Defect as soon as he finds it and the Contractor does the same.

Correction of Defects is governed by clause 43.1, which states that the Contractor corrects the Defect whether or not the Supervisor notifies him of it.

Where Defects have been notified to the Contractor, clause 43.2 provides for the Contractor to correct them before the end of the 'defect correction period' (which is defined in the Contract Data Part One as a number of weeks). The defect correction period begins at Completion for Defects notified before Completion and when the Defect is notified for other Defects.

In terms of clause 43.3, the Supervisor issues the Defects Certificate at the later of the defects date and the end of the last defect correction period. This does not impact on the Employer's rights in respect of any Defect which has not been found or notified so that liability for latent defects would remain, regardless of the Certificate.

Clause 82.1 provides that until the Defects Certificate has been issued and unless otherwise instructed by the Project Manager, the Contractor promptly replaces loss of and repairs damage to the works, Plant and Materials.

In certain cases, the Contractor and Project Manager may, in terms of clause 44.1, propose to the other that the Works Information should be changed so that a Defect does not have to be corrected. If that is agreed, clause 44.2 provides for the Contractor to submit a quotation for reduced Prices or an earlier Completion Date or both to the Project Manager for acceptance. If the quotation is accepted, the Project Manager gives an instruction to change the Works Information, the Prices and the Completion Date accordingly.

In the event that the Contractor is given access to correct a notified Defect but does not correct it within the defect correction period, clause 45.1 provides for the Project Manager to assess the cost to the Employer for having the Defect corrected by other people and the Contractor pays this. The Works Information is, in that case, treated as being changed to accept the Defect.

If the Contractor is not given access before the defects date, the Project Manager assesses the cost to the Contractor for correcting it and the Contractor pays this amount, in terms of clause 45.2. Again, the Works Information is treated as having been changed to accept the Defect.

Secondary Option X17 provides that if a Defect included in the Defects Certificate shows low performance with respect to a performance level stated in the Contract Data, the Contractor pays the amount of low performance damages stated in the Contract Data.

5.5 *Progress of the works*

5.5.1 Common law

Unless it is specified in the contract that the contractor must complete by a specified date or within a specified time, the contractor is obliged to complete the works within a reasonable time. The reasonableness of the time taken is considered in the light of the circumstances at the time of performance of the contract, see *H & E Taylor* v. *P & W Maclellan* (1891).

The obligation to complete in a reasonable time may also come into play where there are contractual time limits set down but there has been an act or omission of the employer putting the employer in breach of the contract. For example, if the employer failed to give the contractor access to the site timeously, the contractor would not be bound by the contractual time limit.

In these circumstances, the contractor's obligation is to complete within a reasonable time, see *T & R Duncanson* v. *The Scottish County Investment Co. Ltd* (1915).

Further, if the employer is responsible in any way for the failure to achieve the completion date they are not able to recover any contractual liquidated damages from the contractor for the period of delay for which they were responsible, see *Peak Construction (Liverpool) Ltd* v. *McKinney Foundations Ltd* (1970) and *Percy Bilton Ltd* v. *Greater London Council* (1982).

These rules apply even if the failure by the employer is not due to fault on their part. In a case where there were squatters on a site resulting in the employer being unable to give the contractor possession of the site, the court still held that the employer was in breach which meant the contractor's obligation to complete by the completion date changed to an obligation to complete in a reasonable time, see *Rapid Building Group Ltd* v. *Ealing Family Housing Association Ltd* (1984).

Similarly, if the employer orders extra work beyond that specified in the original contract which, as a consequence, increases the time required to complete the work, in the absence of any provisions in the contract that allow an extension of time, the employer is no longer able to claim any penalties as a result of the late completion, see *Dodd* v. *Churton* (1897). See also Section 6.5 in this regard.

Where the contractor's obligation is to complete within a reasonable time, it is still possible for an employer to claim unliquidated damages (i.e. their actual losses which they would be required to calculate and prove) if the contractor takes longer than the time which would be considered reasonable in the circumstances.

It should be noted that in each of the cases referred to, there was no mechanism in the contract to extend the completion date as a result of the matters causing delay. Where there is such a mechanism, this would operate, thereby extending the completion date so that the contractor's obligation would then be to complete by the new completion date rather than within a reasonable time.

In relation to interim progress, the Technology and Construction Court in England considered whether a term requiring a sub-contractor to proceed regularly and diligently with their contract works could be implied into a construction contract (*Leander Construction Limited v. Mulalley & Company Limited* (2011)). The contract identified a commencement date, a sub-contract period and a completion date. There were no contractual terms regarding interim performance, milestones or sectional completion dates.

There was an Activity Schedule which set out programme dates and periods. This was a sub-contract document, but the dates within it were said to be indicative only and subject to change. However, the main contractor argued that the sub-contractor had an implied obligation to proceed regularly and diligently with the works and the Activity Schedule was the best way to measure whether or not they had complied with that term.

The Court refused to imply the term into the contract. It was said that the contract worked perfectly well without the term, so it was not required to give business efficacy to the contract.

The Court recognized that failure to proceed regularly and diligently was a ground for termination in the contract. There was no separate and stand-alone obligation in this contract to proceed in that manner which could result in damages for breach of contract in the event of failure.

Further, the delay and extension of time provisions were all linked to the contract completion date and not to any interim dates or progress. If the term was implied as suggested, it would introduce damages for delay linked to numerous completion dates for individual activities and cut across the express terms, totally changing the sub-contractor's contractual obligations.

If terms as to interim progress are to form part of a contract, they should be expressed as a positive obligation either in the form of sectional completion or as contractually binding milestone dates.

5.5.2 The SBC provisions

The SBC provides, in clause 2.4, that on the Date of Possession the Contractor shall begin the works, regularly and diligently proceed with them and complete them on or before the Date for Completion. This obligation does not sit in isolation. There is a corresponding obligation on the Employer to give the Contractor possession of the site. This is also contained within clause 2.4. The Date of Possession and Date for Completion are specified in the Contract Particulars.

In addition to specified commencement and completion dates, the Contractor tends to work to a programme. Under clause 2.9.1.2 the Contractor is to provide his master programme for the works to the Architect identifying, where required in the Contract Particulars, the critical path(s) and any other details required by the Contract Documents. This is to be done without charge. If the Architect awards an extension of time under clause 2.28.1 or there is a Pre-agreed Adjustment fixing a revised Completion Date due to the acceptance of a Variation or Acceleration Quotation, the Contractor is to provide, within 14 days, an amendment or revision to the programme to take account of that, see clause 2.9.2.

There are consequences in the form of liquidated and ascertained damages if the Contractor fails to meet the Completion Date. These are considered in Sections 6.8 and 6.9 below. Under clause 2.28.6.1, the Contractor is subject to the overriding obligation to constantly use its best endeavours to prevent delay in the progress of the Works, or any Section, and to prevent the completion of the Works or Section being delayed or further delayed. Under clause 2.28.6.2, in the event of any delay, the Contractor is to do all that may reasonably be required to the satisfaction of the Architect to proceed with the Works or Section.

These references give rise to the question of what is required by an obligation to use best endeavours or to do all that is reasonably required. There is some case law which gives guidance on this.

In the English case of *CPC Group Limited* v. *Qatari Diar Real Estate Investment Company* (2010), one of the issues was the interpretation of 'all reasonable but commercially prudent endeavours' and what this required. It was found in this case that this did not mean a party had to act to its commercial detriment. That was perhaps clearer in this case than it might otherwise have been because of the inclusion of the reference to 'commercially prudent'.

In *Phillips Petroleum Co UK Ltd* v. *Enron Europe Ltd* (1997), it was said that use of reasonable endeavours to reach agreement on matters fundamental to a contract did still entitle a party to have regard to its own interests.

The Scottish Courts have agreed with that approach, stating in *R & D Construction Group Limited* v. *Hallam Land Management Limited* (2009) that in using 'all reasonable endeavours to agree the amount of the purchase price', a party was entitled to take into account their own commercial interests.

In *Yewbelle Ltd* v. *London Green Developments Ltd* (2007), it was said, in relation to an 'all reasonable endeavours' obligation, that it was correct to consider whether there was anything else Yewbelle could reasonably have done with any real or significant prospect of overcoming the problem.

In *Mactaggart & Mickel Homes Limited* v. *Charles Hunter and Mrs Sandra Hunter* (2010), the question which arose was whether Mactaggart & Mickel had used 'reasonable endeavours' to obtain planning permission for a site. There the judge said:

> In my opinion the phrase 'reasonable endeavours' in its context imposes obligations on MML which are not as onerous as the phrase 'all reasonable endeavours' … which required the court to consider whether there were reasonable steps which could have been taken but were not. The phrase is also less burdensome on the obligant than the phrase 'best endeavours', which appears to me … to require something more than 'all reasonable endeavours'.

He considered that a reasonable endeavours obligation did not require the obligant to disregard its own commercial interests. The test to be applied was what would a reasonable and prudent Board of Directors, acting properly in the interests of their company and applying their minds to its contractual obligations, have done.

In terms of 'best endeavours', this is regarded as requiring a higher standard than 'reasonable endeavours'. In *Rhodia International Holdings Ltd* v. *Huntsman International LLC* (2007), the judge stated:

> There may be a number of reasonable courses which could be taken in a given situation to achieve a particular aim. An obligation to use reasonable endeavours to achieve the aim probably only requires a party to take one reasonable course, not all of them, whereas an obligation to use best endeavours probably requires a party to take all the reasonable courses he can.

The term 'all reasonable endeavours' was considered in the case of *Jet2.com v Blackpool Airport Ltd* (2011) and, in the particular circumstances of that case, was held to mean the same as 'best endeavours'.

According to *MacBryde, the Law of Contract in Scotland*, it may be that this could include incurring a loss in performing a contract. However, MacBryde suggests that it 'does not normally mean that the limits of reason must be overstepped with regard to cost and effort' (paragraph 8.57).

What is required will depend on the facts and circumstances of each case.

5.5.3 The SBC/DB provisions

Similar provisions to clauses 2.4 and 2.28 of the SBC are contained in clauses 2.3 and 2.25 of the SBC/DB.

5.5.4 The NEC3 provisions

Programme

In relation to the programme, the NEC3 has detailed requirements which are set out in clause 31. If a programme is not identified in the Contract Data, the Contractor submits a first programme to the Project Manager for acceptance within the period stated in the Contract Data (clause 31.1).

Clause 31.2 provides that the Contractor is to show certain specific information on each programme submitted for acceptance. See Section 6.11.2 for details.

Clause 31.3 requires the Project Manager, within 2 weeks of the Contractor submitting a programme, to accept it or to notify reasons for not accepting it. See Section 6.11.2 for details.

Where Option A or C applies, clause 31.4 requires the Contractor to provide information which shows how each activity on the Activity Schedule relates to the operations on each programme which he submits for acceptance.

The programme is then to be revised and a revised programme submitted to the Project Manager. See Section 6.11.3 for the procedure regulating such submission and the required content of such revised programme.

In addition to this is the early warning procedure. The Contractor is to give early warning by notifying the Project Manager as soon as he becomes aware of any matter which could increase the total of the Prices, delay Completion, delay meeting a Key Date or impair the performance of the works in use (clause 16.1).

The Contractor may also give an early warning by notifying the Project Manager of any other matter which could increase his total cost. The Project Manager enters early warning matters in the Risk Register. In terms of clause 16.2, the Project Manager or the Contractor may instruct the other to attend a risk reduction meeting.

Clause 16.3 describes the purpose of a risk reduction meeting as, for those who attend, to co-operate in making and considering proposals for avoiding or reducing the effect of the registered risks, seeking solutions which will bring advantages to those affected, deciding on the actions to be taken and who will take them, and deciding what risks have been avoided or have passed and can be removed from the Risk Register.

The Risk Register

In terms of clause 16.4, the Project Manager revises the Risk Register to record decisions made at each risk reduction meeting and issues the revised Risk Register to the Contractor. In the event that a decision needs a change to the Works Information, the Project Manager instructs this change.

It is worth noting that the Risk Register is simply a management tool for the project, allowing risks to be identified and, through the early warning procedure, managed. The Risk Register does not allocate risk for the purpose of compensation events. This means that even if an item is noted as the Employer having ownership of the risk in the Risk Register, that does not suffice for there to be a compensation event as defined in clause 60.1(14) – an event which is an Employer's risk stated in the contract – if that risk then manifests itself. The only exception to that is if the outcome of the risk reduction meeting is that a change to the Works Information is instructed. If it is wished to truly pass the risk to the Employer, the way to do this is to list additional Employer's risks in the Contract Data Part One. That will bring them within the definition of Employer's risks in clause 80.1. Where this is not done, clause 81.1 is clear in stating that from the starting date until the Defects Certificate has been issued, the risks which are not carried by the Employer are carried by the Contractor.

Acceleration

There is also provision within clause 36.1 for the Project Manager to instruct the Contractor to submit a quotation for acceleration to achieve Completion before the Completion Date. The quotation is to include proposed changes to the Prices and a revised programme showing the earlier Completion Date and changed Key Dates. Clause 36.2 requires the Contractor either to submit a quotation or to give reasons for not doing so within the 'period for reply' which is defined in the Contract Data Part One.

In terms of Options A, B, C and D, clause 36.3, when the Project Manager accepts a quotation for an acceleration, he changes the Prices, Completion Date and the Key

Dates accordingly and accepts the revised programme. In Options E and F, clause 36.4, when the Project Manager accepts a quotation for an acceleration, he changes the Completion Date, the Key Dates and the forecast of the total Defined Cost of the whole of the *works* accordingly and accepts the revised programme.

Float

Where there has been a compensation event, the Project Manager does not change the Completion Date or Key Dates if the event has no effect on Completion or meeting a Key Date (clause 61.4). This means that if a compensation event affects a non-critical path activity (i.e. one with float) and does not push it onto the critical path, the Employer gets the benefit of the float on that activity since there is no impact on the Completion Date.

In relation to any critical path activities affected by a compensation event, a delay to the Completion Date is assessed as the length of time that, due to the compensation event, planned Completion is going to be late as compared to planned Completion in the programme (clause 63.3). Note that this is not the length of time the Completion Date is going to be late.

If there is a gap between planned Completion and the Completion Date this wording means that the Employer does not get the benefit of this float (sometimes referred to as terminal float).

As an example:

- A contract where planned Completion is week 20.
- The Completion Date is week 24.
- That means 4 weeks of terminal float is built in.
- A compensation event takes place with a 6-week duration.
- That causes a 6 week delay to planned Completion.
- In terms of Clause 63.3 the Contactor is entitled to an extension of time for the length of time the planned Completion is going to be late as compared to planned Completion in the Programme.
- The extension of time would be 6 weeks from week 24 to week 30. Not 2 weeks from week 24 to week 26.
- This preserves the terminal float for the Contractor in case it is required for future Contractor delay.

This is illustrated by Figure 5.1.

Float as between the contractor and employer was dealt with in *Henry Boot Construction (UK) Ltd v. Malmaison Hotel (Manchester Ltd* (1999)). In that case, the Engineer argued that the variations and late information relied on by the contractor did not cause any delay because these were not on the critical path (due to there being float). This argument was accepted.

In the more usual situation of a mixture of contractor and employer delays, it can be a 'first come first served' approach. If an activity on the programme has float, it is not a critical path activity. If the employer then instructs a variation to that activity

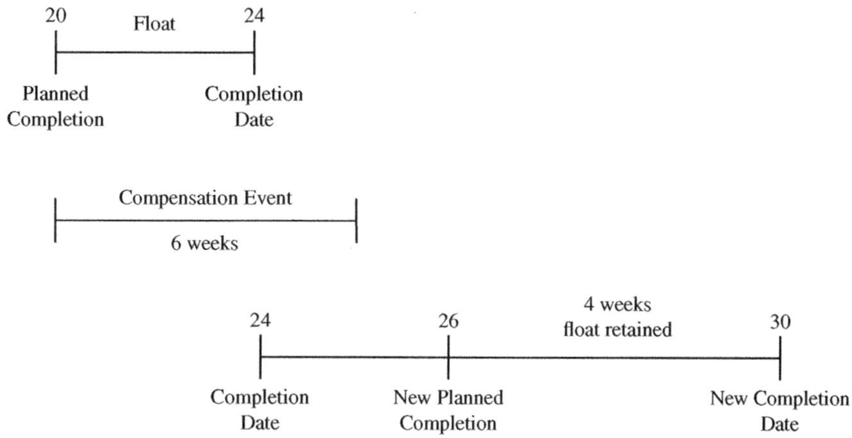

Figure 5.1 Illustration of the float.

which uses up the float, it becomes critical. If the contractor then causes a delay to that activity, the contractor is likely to move into being in culpable delay. In that example, the employer acquires the benefit of the float by getting to it first.

Alternatively, if the contractor's delay happens first, using up float then an instruction is issued by the employer once that activity has become critical, that instruction will have caused a delay to the planned completion date and therefore will trigger an extension of time.

The issue in both of these scenarios will be how to prove the sequence of events and what state the programme was in at the relevant times.

In the NEC3, the Contractor is to submit revised programmes and issue early warning notices of any matters which could cause delay to completion. This prospective analysis of the delay might make it easier to prove the sequence of events than in a contract where this is not done and the analysis is left to be carried out retrospectively at the end. Issuing the revised programmes and giving the required notices could in such a contract protect the Contractor's entitlement to extension of time. These will show if the Contractor has been eating into the float and if activities have become critical as a result.

It might be possible to make an argument in relation to float, even if the contract does not have provisions like those in the NEC3. The argument would be that if the contractor gives notice that a delay is likely which will use up float rendering certain activities critical, that might be sufficient to put the employer on notice that any subsequent change by him will have an effect on the completion date, so that it is that change which is the cause of delay (the float having effectively been pre-booked by the contractor giving notice of this).

The Contractor is to show on each programme submitted for acceptance any provisions for time risk allowances (clause 31.2). This is listed separately from the requirement to show float so each is meant to be identifiable on the programme.

According to the NEC3 Guidance Notes, the Contractor's time risk allowances are to be shown as allowances attached to the duration of each activity or to the duration of parts of the works. They have been described as being the difference between the quickest time an activity can be completed and the time allowed on the programme. It is said that these allowances are owned by the Contractor as part of the planning/programming exercise to cover Contractor risks. This suggests these would be ignored in assessing any delay to Completion due to a compensation event so the Contractor retains the benefit of them.

It is worth noting the terms of clause 31.3. The Project Manager can notify the Contractor that the programme is not being accepted due to it not representing the Contractor's plans realistically.

Assessment of the effect of a compensation event includes risk allowance for cost and time for matters which have a significant chance of occurring and are at the Contractor's risk under the Contract (clause 63.6). According to the Guidance Notes, allowance for time risk should be included in forecasts of Completion in the same way as risks are allowed for in pricing the tender. Clearly more time is allowed where there is a high chance of the Contractor's risk occurring.

5.6 *Insurance and indemnity*

5.6.1 The SBC and the SBC/DB provisions

The SBC contains a number of provisions requiring the Contractor to indemnify the Employer in certain circumstances.

Clause 2.21 requires the Contractor to pay all fees or charges (including rates or taxes) legally demandable under any of the Statutory Requirements and to indemnify the Employer against any liability resulting from any failure to do so. The Contractor is able to recover this through payment for provisional sums, where applicable, or by addition to the Contract Sum. However, that does not apply to fees or charges which are priced in the Contract Bills or which relate solely to the Contractor's Designed Portion, in which event they are deemed to be included in the Contract Sum. The same applies under clause 2.18 of the SBC/DB unless the fees or charges are stated by way of a Provisional Sum in the Employer's Requirements.

Royalties and any other sums payable in respect of the supply and use in the carrying out of the Works of any patented articles, processes or inventions are deemed to be included in the Contract Sum and the Contractor has an obligation in terms of clause 2.22 to indemnify the Employer from and against all claims and proceedings which may be brought or made against the Employer and all damages, costs and expense to which he may be put by reason of the Contractor infringing or being found to have infringed any patent rights. This is not the case where the Contractor has been instructed to use any patented articles, processes or inventions, in which case, under clause 2.23, he is not so liable and any royalties, damages or other monies which the Contractor may be liable to pay shall be added to the Contract Sum. Similar provisions are contained in clauses 2.19 and 2.20 of the SBC/DB.

The Contractor is responsible for loss or damage to Site Materials even though, following payment, title in these materials and goods passes to the Employer, all in terms of clause 2.24. The Contractor maintains a similar responsibility for any materials and/or goods purchased prior to their delivery to site under a separate contract for their purchase in terms of clause 4.17. A similar provision to clause 2.24 is contained in clause 2.21 of the SBC/DB.

The Contractor is, in terms of clause 6.1, liable for and obliged to indemnify the Employer against any 'expense, liability, loss, claim or proceedings whatsoever' in respect of personal injury to or the death of any person arising out of or in the course of or caused by the carrying out of the Works. This is except to the extent this is due to any act or neglect of the Employer or any of the Employer's Persons or any Statutory Undertaker.

Under clause 6.2, the Contractor has a similar liability and similar indemnity obligations in respect of any loss, injury or damage to any property to the extent this is due to any negligence, breach of statutory duty, omission or default of the Contractor or of any of the Contractor's Persons. This liability and indemnity in respect of property does not, in terms of clause 6.3.1, include the Works, work executed and/or Site Materials up to and including whichever is the earlier of the date of issue of the Practical Completion Certificate or the date of termination of the Contractor's employment. There is also an exclusion of this liability and indemnity where Insurance Option C applies and where loss or damage to any property required to be insured under that Insurance Option is caused by a Specified Peril. Similar provisions are contained in clauses 6.1, 6.2 and 6.3 of the SBC/DB.

Notwithstanding these indemnity provisions, the Contractor has obligations related to insurance for personal injury and property damage, the Works and, where there is a Contractor's Designed Portion, professional indemnity insurance. There are also obligations related to obtaining Terrorism Cover. The insurance obligations are considered separately in Chapter 14.

5.6.2 The NEC3 provisions

The NEC3 clause 83.1 requires each party to indemnify the other against claims, proceedings, compensation and costs due to an event which is at his risk. This is a reference back to the Employer's risks provisions in clause 80.1 referred to in Section 5.5.4.

Where there is more than one cause contributing to the claims, proceedings, compensation and costs, clause 83.2 provides that the liability of each party to indemnify the other is reduced. This reduction is in proportion to the extent of the contribution of the other party's risk event and takes into account each party's responsibilities under the contract.

5.7 The Joint Fire Code

5.7.1 The SBC and the SBC/DB provisions

In terms of the SBC, where the Contract Particulars state that the Joint Fire Code applies, clause 6.15 gives all parties an obligation to comply with it. The Contractor is obliged to ensure compliance by all Contractor's Persons. The Joint Fire Code is defined as being the Joint Code of Practice on the Protection from Fire of Construction Sites and Buildings Undergoing Renovation published by the Construction Confederation and the Fire Protection Association, current at the Base Date.

If there is a breach of the Joint Fire Code and the insurers under the Joint Names Policy in respect of the Works specify the Remedial Measures they require, clause 6.16.1.1 requires the Contractor, where these Remedial Measures relate to the obligation of the Contractor to carry out and complete the Works, to ensure that they are carried out by such date as the insurers specify.

Under clause 6.16.1.2, if the Remedial Measures require a Variation to the Works, the Architect is to issue instructions as necessary to ensure compliance. In an emergency, where compliance with the Remedial Measures requires the Contractor to supply materials or execute work before receipt of such instructions, the Contractor is to supply only such limited materials and execute only such limited work as are reasonably necessary to secure immediate compliance. In these circumstances, the Contractor is required to inform the Architect of the emergency and the steps being taken. Clause 6.16.1.2 provides that, save to the extent such emergency work relates to the Contractor's Designed Portion, it is treated as if carried out under an instruction requiring a Variation.

If the Contractor, within 7 days of a notice specifying Remedial Measures not requiring an instruction under clause 6.16.1.2, does not begin to carry out or thereafter fails, without reasonable cause, regularly and diligently to proceed with the Remedial Measures, the Employer may employ and pay others to carry these out and the Contractor shall be liable for all additional costs incurred by the Employer. These costs are deducted from the Contract Sum. This is all in terms of clause 6.16.2.

In the SBC/DB, the provisions regarding the Joint Fire Code are contained in clauses 6.14 to 6.17. Under clause 6.16, where any breach of the Joint Fire Code occurs, any Remedial Measures required by insurers to achieve compliance with the Joint Fire Code are to be carried out by the Contractor, at their own expense.

5.7.2 The NEC3 provisions

There is no equivalent in the NEC3 to the Joint Fire Code provisions in SBC and SBC/DB.

5.8 *Health and safety*

There are detailed provisions under statute, at common law and under the SBC and the NEC3 in relation to health and safety. These are considered in Chapter 20.

Chapter 6
Time

6.1 Introduction

For those with a direct interest in a building project, time is an important subject. The employer will be anxious to fix when and over what period of time the works will be carried out so that they can budget and plan ahead. The contractor will be anxious to plan the commencement and carrying out of the works in order to meet their contractual obligations, express or implied, in relation to the period for completion of the works or perhaps sections or phases of them.

A contractor's tender will normally proceed on the basis that certain operations will cost them a particular amount of money to carry out over a certain period of time. Generally, the longer work takes, the more expensive it is to carry out. Tendering at an appropriate level to take account of time-related costs, forward planning and subsequent on-site control is an essential element of a contractor's consideration of time.

In the traditional manner in which building contracts are let in Scotland, namely, where the employer engages consultants to prepare the design and other requirements, the employer should have specified what they want before the tender stage or, at least, before the formation of the contract. This places a heavy onus on the professional team of architects, engineers, services specialists and quantity surveyors. Changes after the formation of the contract should be kept to a minimum because of the effect that these are likely to have on time and cost. In the absence of agreement about the effect of such changes, they may give rise to claims and disputes.

6.2 Commencement of the works

It is usual for express provision to be made for the date upon which the contractor will be given access to the site for the purpose of carrying out the works. Normally the contract will require the contractor to complete the works either by a specified date, or within a specified period from the agreed date for commencement or the date when access is given to the site. It is very important that the date of commencement of the period for completion is ascertainable and that it is specified what holidays, if any, are to be ignored in computing the period for completion of the works. While these

MacRoberts on Scottish Construction Contracts, Third Edition. MacRoberts.

are matters which common sense dictates should be made clear, experience shows that this is not always done.

6.3 *Time of the essence*

The phrase 'time of the essence' is one that is much used but often without much understanding. The need to do something by a specified date or time does not of itself make time of the essence. However, if the contract specifically makes time of the essence, the failure to complete the works by the stipulated date amounts to a material breach of contract that entitles the innocent party to rescind the contract and claim damages. If there is no express stipulation in the contract that time is of the essence, it can be made so by serving a notice fixing a specified time for completion. This, however, must be a reasonable one.

It is unusual for time to be of the essence in building contracts. Usually the failure of the contractor to complete is to be regarded as a breach of contract that will form the basis of a claim for damages for late completion. Even where the phrase is used, the contract must be considered as a whole. In one case the contractual clause was as follows: 'Time shall be considered as of the essence of the contract ... and in case the contractor shall fail in the due performance ... [the contractor] shall be liable to pay the [employer] ... liquidated damages. In the circumstances it was nevertheless held that time was not of the essence as the contract included other terms, such as an extension of time clause, which were inconsistent with time being of the essence, see *Peak Construction (Liverpool) Ltd* v. *McKinney Foundations Ltd* (1970).

6.4 *Progress of the works*

6.4.1 Common law

As discussed in Section 6.2, the contractor's obligation is usually to complete by a particular date or within a particular period. In the absence of a relevant express contractual obligation, delay in progress prior to that date or before the end of the period probably does not in itself give the employer any rights, see *Greater London Council* v. *Cleveland Bridge and Engineering Co. Ltd and Another* (1986). In such a case a claim for damages will only arise if there is actual delay in completion. In extreme cases, however, particularly if combined with other failings, it may amount to anticipatory breach of contract, see *Sutcliffe* v. *Chippendale & Edmondson* (1982) and *Carr* v. *J A Berriman Pty Ltd* (1953).

Some commentators suggest that there should be an implied term that the contractor will proceed with reasonable diligence and maintain reasonable progress while others state that to imply such a term would be going too far. The courts remain reluctant to imply such a term (see *Leander Construction Limited* v. *Mulalley and Company Limited* (2011) where, after an extensive consideration of the relevant authorities, the court refused to imply such a term). This topic is discussed in more detail in Sections 3.4.4 and 5.5.1.

6.4.2 The SBC provisions

Clause 2.4 of the SBC expressly requires the Contractor to proceed 'regularly and diligently' with the works. It has been held that proceeding 'regularly and diligently' indicates a sense of activity, of orderly progress and of industry and perseverance probably such as will ensure completion according to the contract, see *London Borough of Hounslow* v. *Twickenham Garden Developments Ltd* (1971) and *West Faulkner Associates* v. *London Borough of Newham* (1994).

Moreover, clause 8.4 of the SBC gives the Employer the right to determine the Contractor's employment if they without reasonable cause wholly or substantially suspend the carrying out of the Works or if they fail to proceed regularly and diligently. Accordingly, if this right is to be exercised, it is important that the Architect is satisfied that there is sufficient and reliable evidence of the Contractor's failure.

6.5 *Adjustment of the Completion Date and extension of time for completion*

6.5.1 General

It is a basic principle applicable in all contracts that one party cannot seek to enforce a contractual obligation of the other party where they have prevented the other from performing that obligation. As Lord Denning put it in *Trollope & Colls Ltd* v. *North West Metropolitan Regional Hospital Board* (1973):

> It is well settled that in building contracts … where there is a stipulation for work to be done in a limited time, if one party by his conduct – it may be quite legitimate conduct, such as ordering extra work – renders it impossible or impracticable for the other party to do the work within the stipulated time, then the one whose conduct caused the trouble can no longer insist upon strict adherence to the time stated. He cannot claim any penalties or liquidated damages for non-completion in that time … The time becomes at large … The work done must be done within a reasonable time.

Given the complex nature of most building contracts, the need to instruct variations and to take account of unforeseen matters, it is almost inevitable that the employer will fall foul of this doctrine. To accommodate such matters, building contracts usually set out an express mechanism by which the original completion date can be changed and specify the circumstances in which an extension of time for completion can be obtained. It is less common to find a contractual term which allows the contractor to make up time by way of some form of acceleration.

It is a common but misguided view that extensions of time benefit only the contractor. Clearly, they give the contractor more time to complete the works and reduce or extinguish their liability for liquidated damages. However, were it not for extension of time provisions an employer would not be entitled to claim liquidated damages where they have been the cause of some delay.

Dealing with an employer's default is not usually the sole reason for having extension of time clauses. Most include an entitlement to an extension of time for what might be described as neutral events. These arise through the fault of neither party, for example, war, riots and bad weather. In this sense, such clauses do benefit the contractor by giving them an extension of time for some matters that might otherwise be at their risk.

Indeed, the terms of a particular contract may allow an extension in circumstances which some may think go beyond what might be regarded as neutral.

6.5.2 Adjustment of the Completion Date under the SBC

Clauses 2.26–2.29 of the SBC contain very complex provisions detailing the circumstances in which the Contractor is entitled to an extension of time. The entitlement to an extension only arises as a result of delay due to specified 'Relevant Events', which are considered below. It is noteworthy that the SBC contains substantially revised Relevant Events. The actual number of Relevant Events has decreased but the nature of some of the Relevant Events is more generally expressed as if to give rise to a less prescriptive set of carefully defined Relevant Events. Accordingly, matters which were previously the subject of a particular, specific Relevant Event may be included within a new, more generally expressed, Relevant Event.

6.5.3 Relevant Events under the SBC

Compliance with Variations

Variations comprise Variations (as defined in clause 5.1) and any other matters or instructions which under the Conditions are to be treated as, or as requiring, a Variation.

Compliance with certain specified instructions

There are two categories of Architect's instruction that can entitle the Contractor to an extension of time. First, there are the matters listed in clause 2.29.2.1, which cover a wide range of circumstances where the Architect is obliged to issue instructions under the Contract. Second, an instruction requiring the opening up of the Works for inspection or testing may give rise to such an entitlement, unless the inspection or test shows that work, materials or goods are not in accordance with the Contract.

Deferment by the employer of the giving of possession of the site

The Employer must give the Contractor the possession of the site anticipated in the contract to allow him to carry out the works. This is of critical importance. If he fails to do so, the Employer is in breach of the contract. To address such circumstances, clause 2.5 makes specific provision for extension of time due to the deferment of possession.

Approximate quantity which is not a reasonably accurate forecast of the quantity of work required

This Relevant Event is provided to address any significant understatement in the Contract Bills of a quantity of work required. In such circumstances, the Contractor can be confronted with a requirement to execute far more work than envisaged at the time of entering into the Contract.

Suspension by the Contractor under clause 4.14

Assuming the Contractor meets the requirements of the contract in respect of suspension for non-payment (which requirements are considered in Section 10.9.3), and thereafter suspends performance, such a suspension will be a Relevant Event which will entitle the Contractor to an extension of time.

Impediment, prevention or default

This event extends to any impediment, prevention or default whether by act or omission, by the Employer, the Architect, the Quantity Surveyor or any of the Employer's Persons, except to the extent caused or contributed to by any default, whether by act or omission of the Contractor or of any of the Contractor's Persons. This is a general catch-all provision which replaces previous wording which sought to set out specific acts or omissions for which the Employer was held responsible. Employer's Persons are defined as all persons employed, engaged or authorized by the Employer excluding the Contractor, Contractor's Persons, the Architect, the Quantity Surveyor and any Statutory Undertaker but including any such third party as is referred to in clause 3.23.

Matters falling within this heading will include, but are not limited to, matters for which previously there was express provision such as:

- delay in receipt of instructions from the Architect;
- delay on the part of persons employed on behalf of the Employer to do other work associated with the Works;
- delay caused by the late supply of materials and goods which the Employer has undertaken to supply;
- failure by the Employer to give access to the site in accordance with the contract;
- compliance or non-compliance by the Employer with contractual provisions in relation to the CDM Regulations.

Work by a Statutory Undertaker in pursuance of its statutory obligations or failure to carry out such work

This includes such matters as electricity, gas, water and other services which need to be installed in most, if not all, buildings.

This falls to be contrasted with a quite separate situation where delay to the works is caused by the need to work close to, or in physical contact with, pipes and other apparatus of local authorities or statutory undertakers. These matters will be governed by the terms of the particular contract in question. If there are no express provisions that deal with the Contractor's rights in such a situation, implied terms will need to be relied upon, see *Henry Boot Construction Ltd* v. *Central Lancashire New Town Development Corporation* (1981).

It should be noted that this Relevant Event applies only to circumstances in which the Statutory Undertaker is independently carrying out work pursuant to its statutory obligations. It will not apply where a Statutory Undertaker carries out other work under contract to the Contractor.

Exceptionally adverse weather conditions

Bad weather is not, in itself, a good reason for not completing on time. That means that, save where the contract contains provisions which recognize the need for an extension of time due to bad weather, the contractor will be held to have accepted the risk of completing on time notwithstanding bad weather as they are the party best able to deal with it. It is open to the parties to agree where the risk falls.

The SBC has taken a particular route, but the words used require careful consideration. There is a requirement that there is not just exceptionally adverse weather conditions but also delay to the progress of the works as a result.

Proof that the weather has been exceptionally adverse is usually provided by examining local weather records and comparing the actual weather experienced at a particular time of year against that of previous years at that time.

Under JCT 63 conditions it was held to be important to note that the test was whether the weather itself was 'exceptionally inclement' so as to give rise to delay, and not whether the amount of time lost by the inclement weather was exceptional, see *Walter Lawrence and Son Ltd* v. *Commercial Union Properties (UK) Ltd* (1984).

Further, any delay due to weather is to be determined at the time the work is carried out, not when it was programmed to be carried out.

Loss or damage occasioned by any of the Specified Perils

The Specified Perils are those listed in clause 6.8, namely, fire, lightning, explosion, storm, flood, escape of water from any water tank, apparatus or pipes, earthquake, aircraft and other aerial devices or articles dropped therefrom, riot and civil commotion but excluding the Excepted Risks which are also listed in that clause.

Civil commotion or the use or threat of terrorism and/or the activities of the relevant authorities dealing with such an event or threat

A civil commotion appears to be some form of insurrection of the people that is different from a riot or a civil war. The potentially wide-reaching consequences of terrorist activity, whether actual or threatened, are obvious.

In recognition of the particular problems that could be occasioned by this, a relevant event dealing with terrorism was first introduced by JCT in July 1993, by way of Amendment 12 to JCT 80.

Strikes, lock-out or local combination of workmen

Where there is no express provision for such matters in a contract they may be covered by general provisions relating to *force majeure* or special circumstances. The SBC allows the Architect to take into account the possible far-ranging effects of strike action. In other building contracts it may be difficult to know if an extension is to be granted only where the strike relates to on-site work or whether it extends to strikes which have an impact upon the performance of sub-contractors and suppliers. This should be made clear in the contract. Difficulties can sometimes arise, for example, in determining whether the extension should be for strikes or delay due to work to be carried out by statutory undertakers, see *Boskalis Westminster Construction Ltd* v. *Liverpool City Council* (1983). A 'local combination of workmen' is not defined. However, it is thought that it might cover, for example, a go-slow.

Government intervention

The Contractor may be entitled to an extension of time by reason of the United Kingdom Government and/or the Scottish Government exercising a statutory power after the base date which directly affects the execution of the Works.

Force majeure

The term *force majeure* is thought to have been taken from the Code Napoleon. In a contract governed by Scots law, it does not have any particular technical meaning. *Force majeure* is considered at Section 9.3.

6.5.4 Requirements for the adjustment of completion date under the SBC

The building contract should set out the procedure which is to be followed when dealing with extensions of time and it is important that those involved in that process adhere to the requirements of the contract. For example, clause 2.27.1 of the SBC provides that if and whenever it becomes reasonably apparent that progress of the Works or any Section is being or is likely to be delayed, the Contractor should forthwith give written notice to the Architect of the material circumstances including the cause or causes of delay and identifying any event which in his opinion is a Relevant Event.

It is important to note that the requirement is to give notice irrespective of whether the Contractor is seeking an extension and irrespective of whether an event is a Relevant Event, provided it becomes reasonably apparent that progress of the Works is being or is likely to be delayed. While the Contractor will always be reluctant to advise the Architect of matters for which they are responsible, for example, defective planning, poor supervision or inefficient working, the logic of this appears to be that, as the Architect is only obliged to grant such extension as is fair and reasonable, it is important that he is aware of all the facts which are relevant in determining what is fair and reasonable.

It is only in respect of Relevant Events that the contract requires the Contractor to give, in the notice or as soon as possible thereafter, particulars of the expected effects including an estimate of the extent of any expected delay beyond the Completion Date, see clause 2.27.2. Under clause 2.27.3 the Contractor must forthwith notify the Architect in writing of any material change in the estimated delay or in any other particulars and supply such further information as the Architect may at any time reasonably require.

Provisions like this often give rise to arguments about whether proper and timeous notice by the contractor is a condition precedent to an award of an extension of time. Architects and employers often argue that that is the case, but the contractual provisions in each case require careful consideration, see *London Borough of Merton* v. *Stanley Hugh Leach Ltd* (1985) and other cases in this area such as *Education 4 Ayrshire Ltd* v. *South Ayrshire Council* (2009). However, where it is expressly stated in the contract that the contractor shall not be entitled to an extension of time where they have failed to give proper notice under the contract, such a provision will be upheld, see *City Inn Limited* v. *Shepherd Construction* (2003).

A contractor should always consider the terms of an extension of time clause very carefully. Should they give notice of those matters or events that they consider at the time are likely to be non-critical? Should they refrain from giving notice where they believe that they have an adequate float in terms of time to allow them to complete within the required period? Much will depend on the particular wording of the extension of time clause, but it is important to remember that things can change over the course of a contract through no fault of the contractor. In most, if not all, cases it will be prudent to give notice. A failure to give written notice of delay may, in certain circumstances, constitute a breach of contract.

It has been suggested that if the architect, because of a failure on the part of the contractor to give notice, has been unable to avoid or reduce a delay to completion, the contractor should not be awarded an extension greater than that which they would have received had they given notice, see *London Borough of Merton*.

The SBC provides that no extension is to be granted unless the Contractor has constantly used their best endeavours to prevent delay, however caused, and they have done all that may reasonably be required to the satisfaction of the Architect to proceed with the Works. Unfortunately there is little guidance on what is meant by 'best endeavours' and 'all that may reasonably be required' in this context. In other commercial contexts the courts have interpreted 'best endeavours' as importing a high standard (see, e.g. *Sandhu* v. *Sandhu* (2010)). Some take the view that the Contractor must, if necessary, re-programme, increase resources and work overtime. Others take the view that, strictly, they are not obliged to take steps which would result in them incurring any material additional costs. See also the discussion on this topic, and the cases cited, in Section 5.5.2.

6.5.5 Fixing a new Completion Date under the SBC

Under the SBC clause 2.28 the Architect must consider if the events notified are Relevant Events and whether, as a result, completion is likely to be delayed beyond the Completion Date. As soon as is reasonably practicable, and in any event within 12 weeks of receiving the required particulars, the Architect must notify the Contractor in writing of his decision in respect of any notice under clause 2.27. This applies whether or not an extension of time is given. In his decision the Architect must state the extension of time he has attributed to each Relevant Event and (in the case of a decision under clause 2.28.4 or 2.28.5) the reduction in the time he has attributed to each Relevant Omission.

Under clause 2.28.4, if work is omitted after an extension has been granted, the Architect may by notice in writing to the Contractor fix an earlier Completion Date.

Although clause 2.28 states that the Architect 'shall' act appropriately within 12 weeks, it is generally regarded that, taken in the context of the other terms of the contract, the timescale is directory only and not mandatory.

After Practical Completion has been achieved, the Architect is obliged to review and reach a final view on the fair and reasonable extension of time to which the Contractor is entitled having regard to any Relevant Events, and that not later than 12 weeks after Practical Completion. He can confirm the Completion Date previously fixed. In fixing a later Completion Date he can review a previous decision and have regard to all Relevant Events, whether notified to him or not. He can only fix a Completion Date earlier than a previously revised Completion Date if that is fair and reasonable taking into account Relevant Omissions. In no circumstances can the Architect fix a date earlier than the relevant Date for Completion (clause 2.28.6.3), nor is any alteration allowed to the length of any Pre-agreed Adjustment in terms of a Schedule Part 2 Quotation except in the case of a Variation Quotation where the relevant Variation is itself the subject of a Relevant Omission (clause 2.28.6.4).

Refusal by the Architect to consider an application for extension of time may, in certain circumstances, constitute a breach of contract by the Employer.

6.5.6 Calculation of extension of time and proof of entitlement

These matters are the subject of much controversy and a detailed examination of them is beyond the scope of this book. Indeed they appear to have generated a whole industry of consultants who profess an expertise in this area, using critical path analysis, computer technology and other techniques to provide delay analysis which is said to be as accurate as is capable of being achieved. In October 2002, the Society of Construction Law published a Delay Protocol 'to provide guidance to all parties to the construction process when dealing with time/delay matters'. Although it is said that the Protocol 'recognises that transparency of information and methodology is central to both dispute prevention and dispute resolution', the terms of the Protocol have been controversial given the stance it adopts and the divergence of opinion among experts on how delay analysis should be conducted. The particular method used can vary and experts in this field can often disagree as to which method is the correct one to use in the circumstances of the particular case.

6.5.7 Contractor's programmes

Under most forms of building contract the contractor's programme is not part of the contract. In the unusual event that the parties have agreed that the contractor's programme does form part of their contract and is binding upon them, it is of considerable significance. In such an event the contractor is obliged to work to that programme and, perhaps more significantly, the employer is obliged to allow the contractor to work to that programme. Conversely, a contractor is not obliged to work to a programme where that has not been made a requirement of the contract, see *Pigott Foundations Ltd* v. *Shepherd Construction Ltd* (1993).

Contractors are entitled to programme the works in order to complete in less time than that allowed in the contract. That does not alter the obligation of the employer, which is not to impede the contractor in completing the works in the time allowed by the contract, see *Glenlion Construction Ltd* v. *The Guinness Trust* (1987) and *J F Finnegan Ltd* v. *Sheffield City Council* (1988).

However, most analyses of delay use the contractor's original programme as part of the process. Whether this can be used as a basis for any proper analysis depends upon whether the original programme was put together properly. Usually the analysis will also involve an 'as built' programme showing the actual start and finish dates for each activity specified on the programme. This allows the actual start and finish dates to be compared with those set out in the contractor's original programme. If the 'as built' programme also highlights the nature and timing of the matters upon which the contractor bases as the source of alleged delay, for example, instructions and variations, suspensions and the like, such an 'as built' programme can give, at the very least, a useful picture of the factual background to the carrying out of the works.

6.5.8 Causation

The contractor must prove that a relevant event, and not their own inefficiencies or other matters for which they must accept responsibility under the contract, caused delay. Difficulty is caused by the fact that some contractors do not keep sufficiently detailed records of events and their impact upon the works. Rarely is it the case that there is one clear event that can be shown to be the only cause of delay to the works. If it can, there is usually little scope for real dispute. More usual is the situation where there are different events that are productive of delay, some of which are the responsibility of the contractor and others of which are the responsibility of the employer. They may all have an impact upon the works at the same time or they may have an impact upon the works at different times. This makes the calculation of extensions of time a very difficult area. In such circumstances the strict application of certain legal rules relating to causation may not be possible or may give rise to very unsatisfactory results.

Until fairly recently there has been very little authority from the Courts in this area. However, the decision of the Scottish Appeal Court in *John Doyle Construction Ltd* v. *Laing Management (Scotland) Ltd* (2004) set down general guidance in relation to claims for loss and expense caused by delay and disruption which has had a significant impact far outside Scotland. In that case it was stated that the question of causation must be addressed by 'the application of common sense to the logical principles of causation'. A very recent and, perhaps, most controversial decision from the Scottish Courts in this area is *City Inn Limited* v. *Shepherd Construction* (2010) in relation to the thorny issue of whether and to what extent a contractor is entitled to an extension of time where there are concurrent causes of delay, one of which is a Relevant Event under JCT conditions and the other is not (being a matter for which the contractor is responsible). In that case, which related to a contract governed by JCT80 conditions (the provisions are not materially different in the SBC 2011 conditions), it was held by the majority of the judges that:

> where a situation exists in which two causes are operative, one being a relevant event and the other some event for which the contractor is to be taken to be responsible, and neither of which could be described as the dominant cause, the claim for extension of time will not necessarily fail. In such a situation, which could as a matter of language be described as one of concurrent causes, in a broad sense … it will be open to the decision maker, whether the architect, or other tribunal, approaching the issue in a fair and reasonable way, to apportion the delay in completion of the works occasioned thereby as between the relevant event and the other event.
>
> (Lord Osborne at paragraph 42 with whom Lord Kingarth concurred)

Lord Carloway, in the dissenting judgment, took quite a different view. In such circumstances, he considered that:

> [T]he architect does not engage in an apportionment exercise. Where the contractor can show that an operative cause of delay was a Relevant Event, he is entitled to an extension to such new date as would have allowed him to complete the works in terms of the contract. The words 'fair and reasonable' in the clause are not related to the determination of whether a Relevant Event has caused the delay in the Completion Date, but to the exercise of fixing a new date once causation is already determined.

It is the majority's view on such apportionment which has caused significant controversy. Indeed, it has led to reference to what has been called 'the Scottish approach' and 'the English approach'. The English Courts have to date not followed the approach of the majority in *City Inn*. They have essentially taken the route of Lord Carloway. In the leading case of *Walter Lilly & Company Limited* v. *Giles Patrick Cyril MacKay DMW Developments Limited and* (2012) and after reviewing earlier English authority, Mr Justice Akenhead held that in such circumstances:

> [T]he Contractor is entitled to an extension of time for the whole period of delay caused by the Relevant Event in question. There is nothing in the wording ... which expressly suggests that there is any sort of proviso to the effect that an extension should be reduced if the causation criterion is established. The fact that the Architect has to award a 'fair and reasonable' extension does not imply that there should be some apportionment in the case of concurrent delays. The test is primarily a causation one. It therefore follows that, although of persuasive weight, the *City Inn* case is inapplicable within this jurisdiction.

It is clearly very unfortunate that the courts in Scotland and those in England have taken quite a different approach on the construction of the same contractual terms which are regularly in use on both sides of the border. Until the decision of the majority in *City Inn* is reconsidered and possibly overturned, it remains the law of Scotland. Some have argued that the position adopted by Lord Osborne, i.e. 'it will be open to the decision maker .... to apportion', does not require the decision-maker to apportion and that in appropriate circumstances it would still be open to the decision-maker to allow the contractor a full period of extension of time. It is submitted, however, that on the question of construction of the particular provisions, there are strong arguments that the approach of Lord Carloway and 'the English approach' is to be preferred.

Given the large number, complex nature and interaction of events on most building sites, it is submitted that the extension of time to which the contractor is entitled will always be very much a matter of opinion.

6.6 *Partial possession, sectional completion and acceleration*

There may be good reason why the parties to a building contract wish to make provision for partial possession or sectional completion of the works, or for acceleration. The employer may need the building desperately. The contractor may wish to be relieved of obligations such as those regarding insurance and site security.

6.6.1 Partial possession

Partial possession refers to the situation where the employer takes possession of part or parts of the works before completion of the whole. If the contract does not make express provision allowing the employer to take partial possession, he will normally be unable to do so without the consent of the contractor. Clause 2.33 of the SBC makes such provision and alters the Contractor's obligations in respect of liquidated damages (clause 2.37), insurance (clause 2.36) and defects liability (clause 2.35).

There is deemed practical completion of the part taken over by the Employer for certain purposes.

6.6.2 Sectional completion

Sectional completion refers to the situation where the works are defined in advance in separate sections and a different date is given for the completion of each section. This requires the precise definition of each section making sure that the whole of the works are covered, and a date of possession, date or period for completion and liquidated damages for each section. Problems arise where parties do not use a tried and tested standard form. In such circumstances they run the risk that unless great care is taken with, for example, the liquidated damages provisions, they may be inoperable, see, for example, *Taylor Woodrow Holdings Ltd and Another* v. *Barnes & Elliott Ltd* (2004).

6.6.3 Acceleration

In some circumstances it may be possible for the contractor to make up delay by way of acceleration of the works. For example, Schedule Part 2 of the SBC contains provisions that regulate the procedure where the Employer wishes to investigate the possibility of achieving practical completion before the Completion Date for the Works or a Section. In such circumstances, the Architect is to invite proposals from the Contractor in that regard (an 'Acceleration Quotation'). Upon receipt of such an invitation the Contractor shall either (a) provide an Acceleration Quotation identifying the time that can be saved, the amount of the adjustment to the Contract Sum and any other conditions attached or (b) explain why it would be impracticable to achieve practical completion earlier than the Completion Date. The Employer may on or before receipt of the quotation seek revised proposals. The Contractor is under no obligation to accelerate or take any steps for that purpose until he receives a Confirmed Acceptance of his Acceleration Quotation. For more details on acceleration, see Section 5.2.2.

In relation to partial possession, acceleration and sectional completion under the NEC3, see respectively Sections 6.11.6, 6.11.7 and 6.11.8.

6.7 *Completion of the works*

6.7.1 Timescale for completion

If no timescale has been specified, the contractor is obliged to complete the works within a reasonable time, see *H & E Taylor* v. *P & W Maclellan* (1891). The implication of a contractual term requiring that the works should be completed within a reasonable time is most common in building operations of small value where more importance is placed on the price and specification of the work than the period within which the work is to be carried out.

Where a completion date has been agreed, it may become unenforceable by virtue of some later agreement, or by waiver on the part of the employer, or where the contractor has been prevented from completing on time by acts or omissions of the employer or those for whom he is responsible. In such circumstances, unless a new completion date is agreed or the contract provides a mechanism for an extension of time, time is said to be at large. Contractors like to argue that time is at large in the sense that there is then no date by which the works must be completed. However, that is an erroneous view of what is meant by 'time at large'. In such circumstances the contractor is still obliged to complete the works within a reasonable time.

What is a reasonable time is a question of fact, to be determined in the light of all the surrounding circumstances of each particular case. If time has come to be at large, the employer is unable to recover liquidated damages under the contract because there is no fixed date for completion that can be used in the calculation of the damages. However, it is still possible for the employer to claim unliquidated damages for breach of the contractor's implied obligation to complete within a reasonable time. These matters are discussed in more detail in Sections 6.8 and 6.9.

6.7.2 'Practical completion'

It is perhaps trite to say that the works are complete when the contractor has executed all the work that he has contracted to perform. In building contracts, however, things are not always as simple as they might be.

Completion of the works is an important event. Most building contracts will require completion to the satisfaction of the employer, the architect or some other specified third party. It is the date of completion that is used to determine whether the contractor has completed timeously. Accordingly, where the contractor is in culpable delay, it marks the end of the period for which damages for late completion are payable. Given the importance of this, most building contracts require some kind of formal certification that the works are complete.

Whether the works are complete is a matter that is ripe for dispute. If the contractor is in culpable delay and liable to pay damages for late completion, they will be seeking certification of completion as soon as possible. The employer will be more concerned that the works are truly complete.

Clause 2.30 of the SBC provides for certification of 'practical completion'. The term is not defined but is broadly accepted as meaning a stage of completeness with an absence of anything other than minor patent defects. For a more detailed discussion on this topic, see Section 5.2.2.

6.8 *Damages for late completion*

The failure of the contractor to complete the works on time as required by the contract is a breach of contract. Like any other breach of contract, it gives rise at common law to the possibility of a claim for damages for that breach. The damages are determined after the breach has occurred and require proof of loss by the employer. Such a claim must also meet the requirements of the general law of damages. If the loss is too remote, it will not be recoverable, see *Liesbosch Dredger* v. *Edison Steamship* (1933), *Hadley* v. *Baxendale* (1854), and *Victoria Laundry (Windsor) Ltd* v. *Newman Industries Ltd* (1949). The general law of damages is considered in Section 10.4.

It is usually possible for employers to estimate, with a fair degree of certainty, the loss that they will sustain if the contractor does not complete on time. This can be done in a number of ways, for example, by estimating additional financing costs, loss of rental and the like. For an interesting discussion of this area, see *Multiplex Constructions Pty Ltd* v. *Abgarus Pty Ltd* (1992). As a result of this, and the desire of contractors to fix the level of their liability to the employer for damages in the event of late completion, most building contracts are drafted in such a way that the parties fix in advance the damages that will be payable for late completion. If these damages are a genuine pre-estimate of the loss likely to be suffered by the employer, they are called 'liquidated damages'. This subject is considered below in the next section.

What is a genuine pre-estimate of loss in the context of liquidated damages is an issue which has prompted much debate, not least where the project in question is said not to be commercial in nature, see *Clydebank Engineering and Shipbuilding Co. Ltd* v. *Don Jose Ramos Yzquierdo y Castaneda* (1904). That will rarely, if ever, be the case in a building contract.

There is much to be said for the view that, if the contractor does not like the liquidated damages, he should negotiate them down before entering into the contract. It is submitted that the view that liquidated damages provisions, where operable, provide an exhaustive remedy to the employer for late completion is to be preferred to the view, sometimes expressed, that it is not. The contrary view gives insufficient weight to the considerable benefits of the agreed nature of such damages.

6.9 Liquidated damages

6.9.1 General

It is normal in modern building contracts to find a liquidated damages provision to the effect that the contractor will pay or allow the employer a sum for each specified period, for example, per day or per week that the works remain incomplete after the contractual date for completion.

It is less common to find a liquidated damages provision in a sub-contract. A sub-contract may contain a provision putting the sub-contractor on notice that in the event of the main contractor's failure to complete the works timeously the employer may impose liquidated damages upon the main contractor. If the sub-contractor is on notice of this, the damages which they may become liable to pay to the main contractor in the event that a breach of contract on their part causes delay to the completion of the main contract may include the amount of liquidated damages payable by the main contractor to the employer as a result of the sub-contractor's breach.

Clauses that specify liquidated and ascertained damages for delay apply where the works are completed in natural course, but not to contract time. They do not apply where the original contractor does not complete the works. They are ineffective if time has become at large since there is no fixed date from which damages can be calculated, see *British Glanzstoff Manufacturing Co. Ltd* v. *General Accident, Fire and Life Assurance Corporation Ltd* (1912).

A valid liquidated damages clause removes the need for proof of actual loss, which may be difficult and costly. It should be recognized that if the clause is valid and applicable, employers are entitled to the agreed liquidated damages even if they have in fact sustained no loss. In circumstances where the liquidated damages clause is inapplicable, the employer must prove the loss which has been caused by the contractor's breach of contract. Sometimes arguments arise about whether the provisions for liquidated damages apply to the particular circumstances or whether damages can still be sought at common law, see *Scottish Coal Company Ltd* v. *Kier Construction Ltd* (2005).

A typical example of a provision for payment or allowance of liquidated damages is to be found in clause 2.32 of the SBC. This provides that, subject to the issue of a Non-Completion Certificate under clause 2.32.1 and provided that the Employer has informed the Contractor in writing before the date of the Final Certificate that he may require payment of, or may withhold or deduct liquidated damages, then the Employer may, not later than five days before the final date for payment of the debt due under the Final Certificate, give notice in the terms set out in clause 2.32.2. That clause requires that the notice under clause 2.32.1 shall state that the Employer requires that the Contractor pay or allow the Employer liquidated damages at the rate stated in the Contract Particulars (or at such lesser rate as may be specified in writing by the Employer) for the period between the Completion Date and the date of practical completion.

The certificate issued under clause 2.32.1.1 is a prerequisite to the Employer's right to deduct liquidated damages under the SBC. Such a certificate may not be issued before the expiry of the period for completion or after the issue of the Final Certificate, see *H Fairweather Ltd* v. *Asden Securities Ltd* (1979). Notwithstanding the issue of such a certificate, liquidated damages may not be payable if the liquidated damages clause does not apply, is invalid or is inoperable. If there is a dispute as to whether the Employer is entitled to deduct liquidated damages from the sum certified the Employer may do so, at his own risk, pending resolution of the dispute. The interaction of contractual provisions dealing with the final date for payment, non-timeous completion and deduction or allowance of liquidated damages can produce very complex situations, see *Reinwood Ltd* v. *L Brown & Sons Ltd* (2008), a case decided in the House of Lords.

If a new, later, Completion Date is fixed after a Non-Completion Certificate has been issued under clause 2.32.1.1, the certificate is superseded and a new one needs to be issued if the Contractor fails to meet the revised Completion Date. No new notice of deduction is required to be given by the Employer, see clause 2.32.4. The Employer is obliged to repay any liquidated damages already recovered for the period up to the new Completion Date, see clause 2.32.3. If the Contractor should fail to complete by the new Completion Date, the Employer may not deduct liquidated damages unless a new valid Non-Completion Certificate has been issued, see *A Bell & Son (Paddington) Ltd* v. *CBF Residential Care and Housing Association* (1989),

Jarvis Brent Ltd v. *Rowlinson Construction Ltd* (1990) and *J F Finnegan Ltd* v. *Community Housing Association Ltd* (1996).

If the employer's losses arising from the breach for which liquidated damages have been stipulated are greater than the stipulated amount, they are not entitled to ignore the liquidated damages clause and claim for such losses as they can prove. In effect, the clause operates as a limitation of the contractor's liability. Doubt remains as to whether, in the event that the liquidated damages provisions of a contract become inoperable, the employer can recover more by way of unliquidated damages than the amount stated in the contract as liquidated damages.

Occasionally the rate of 'nil' has been inserted as the rate of liquidated damages. Although it has been argued that this simply means that the parties have not agreed the sum payable by way of liquidated damages and that unliquidated damages may still be payable, the better view is probably that it is to be treated as an agreement that no damages are to be paid to the employer for delay, see *Temloc Ltd* v. *Errill Properties Ltd* (1988). It should be noted that the reasoning in *Temloc Ltd* was not followed in *Baese Pty Ltd* v. *R A Bracken Building Pty Ltd* (1990). Although the contract in *Temloc Ltd* was one under JCT 80 conditions, it may have a wider application in other contracts where the terms are similar. It does not assist in determining what the position is where there is a dash (–) inserted in the contract or the rate of damages is left blank.

It has been said that liquidated damages clauses are to be construed *contra proferentem*, (i.e. against the party putting it forward) but this requires detailed consideration of issues such as whether the parties had equal bargaining power to negotiate the contract and whether the contract is in a standard form drawn by a body on which employers, contractors and sub-contractors are represented. In some cases, insofar as the clause has a limiting effect on the contractor's liability for damages for late completion, it may be capable of being challenged by the employer under the Unfair Contract Terms Act 1977.

In relation to delay damages under the NEC3, see Section 6.11.10.

6.9.2 Where liquidated damages provisions are not enforceable

A liquidated damages clause is unenforceable if the amount specified is a penalty. The classic discussion of the differences between a penalty clause and a valid liquidated damages clause is contained in the speech of Lord Dunedin in *Dunlop Pneumatic Tyre Co. Ltd* v. *New Garage & Motor Co. Ltd* (1915) where he said that:

> (1) Though the parties to a contract who used the words 'penalty' or 'liquidated damages' may *prima facie* be supposed to mean what they say, yet the expression used is not conclusive. The court must find out whether the payment stipulated is in truth a penalty or liquidated damages ... (2) The essence of a penalty is a payment of money stipulated as *in terrorem* of the offending party; the essence of liquidated damages is a genuine covenanted pre-estimate of damage ... (3) The question of whether a sum stipulated is a penalty or liquidated damages is a question of construction to be decided upon the terms and inherent circumstances of each

> particular contract, judged of as at the time of the making of the contract, not as at the time of the breach … (4) To assist this task of construction various tests have been suggested which, if applicable to the case under consideration, may prove helpful, or even conclusive. Such are (a) it will be held to be a penalty if the sum stipulated for is extravagant and unconscionable in amount in comparison with the greatest loss that could conceivably be proved to have followed from the breach; (b) it will be held to be a penalty if the breach consists only in not paying a sum of money and the money stipulated is a sum greater than the sum which ought to have been paid … (c) There is a presumption (and no more) that it is a penalty when a simple lump sum is made payable by way of compensation, on the occurrence of one or more or all of the events, some of which may occasion serious and others but trifling damage … On the other hand (d) it is not an obstacle to the sum stipulated being a genuine pre-estimate of damage that the consequences of the breach are such as to make precise pre-estimation almost an impossibility. On the contrary that is just the situation when it is probable that pre-estimated damage was the true bargain between the parties.

The authority of these rules was reaffirmed by the Judicial Committee of the Privy Council in *Philips Hong Kong Ltd* v. *The Attorney General of Hong Kong* (1993). The rules may appear clear but their application is not always easy. For a more recent discussion of the general rules and their application, see *Alfred McAlpine Capital Projects Ltd* v. *Tilebox* (2005).

Doubt remains in Scots law about the enforceability of a liquidated damages clause where the works could not have been completed in the time specified, see *Robertson* v. *Driver's Trustees* (1881).

An employer is not entitled to enforce a liquidated damages clause if he has agreed not to enforce the clause or he has waived his right to do so. In the absence of contractual provision to the contrary, payment of the contract price does not constitute waiver of the right to claim liquidated damages, see *Clydebank Engineering and Shipbuilding Co. Ltd* v. *Don Jose Ramos Yzquierdo y Castaneda* (1905).

6.10 The SBC/DB

In the SBC/DB the provisions relating to adjustment of the Completion Date are contained in clauses 2.23–2.26. They are substantially in the same terms as those in the SBC except that appropriate changes are made to reflect that there is no Architect/Contract Administrator. Accordingly, in relation to notices to be given by the Contractor of delay to the progress of the Works (clause 2.24), such notice is to be given to the Employer. It is the Employer who is to give such extension of time as is fair and reasonable (clause 2.25). There are certain changes to the Relevant Events reflecting the need to take account of Changes by the Employer and Employer's Instructions (clauses 2.26.1 and 2.26.2) as well as delay in receipt of any necessary permission or approval of any statutory body which the Contractor has taken all practicable steps to avoid or reduce (clause 2.26.13).

When, under the SBC/DB, practical completion is achieved in relation to the Works or a Section, the Employer is to issue a 'statement' (not a certificate) to that effect, the Practical Completion Statement (clause 2.27). If the Contractor fails to complete the Works or a Section by the relevant Completion Date, the Employer is to issue a notice (not a certificate) to that effect, a Non-Completion Notice. Payment or allowance of liquidated damages follows the same scheme as in the SBC based on the requirement of a valid Non-Completion Notice (clause 2.29). In relation to Partial Possession, the notice identifying the part or parts taken into possession and giving the date when the Employer took possession (the Relevant Part and the Relevant Date) is to be given by the Contractor (clause 2.30). Practical completion of the Relevant Part is deemed to have occurred and the Rectification Period in relation to defects is deemed to have commenced on the Relevant Date (clause 2.31). When any defects, shrinkages or other faults in the Relevant Part which the Employer has required to be made good have been made good, the Employer is to issue a notice to that effect (clause 2.32). The effects of partial possession upon insurance and liquidated damages are dealt with in clauses 2.33 and 2.34 respectively.

6.11 The NEC3

The following is a brief synopsis of the relevant terms of the NEC3 contract in relation to time.

Section 3 of the NEC3 deals with certain requirements as to time and certain secondary option clauses may be relevant in this area; Option X5 Sectional Completion, Option X6 Bonus for Early Completion, and Option X7 Delay Damages. Compensation Events are dealt with in Section 6 of the NEC3.

6.11.1 Starting, Completion and Key Dates

Clause 30 deals with Starting, Completion and Key Dates. Clause 30.1 contains two important but quite separate provisions. First, it provides that the Contractor is not able to start work on the Site until the first access date. It is important to note that it does not prevent the start of any work that can be done off-site including procurement and design. Access dates are to be identified by the Employer in part one of the Contract Data for parts of the Site. Second, it is provided that the work is to be completed on or before the Completion Date. It is noteworthy that there are no express provisions which require that the Contractor should (a) proceed regularly and diligently with the works or (b) use a particular level of endeavour to prevent or reduced delay. For the reasons given above the courts will be very slow to imply such terms. However, some argue that the obligation in clause 30.1 to 'do the work so that Completion is on or before the Completion Date' imposes 'an obligation to progress so that Completion by the Completion Date is always achievable'.

Clause 30.2 also contains two important but separate provisions. First, it states that the Project Manager decides the date of Completion. There are no express provisions entitling the Contractor to apply for a completion certificate or obliging the Contractor to give notice to the Project Manager when the Contractor believes that the works are nearing or have reached Completion. If there is a dispute between the Contractor and the Project Manager as to whether Completion has been achieved, there is no easy way to discern what procedure is to be adopted. Second, clause 30.2 states that the Project Manager is to certify Completion within one week of Completion (see clauses 13.1 and 13.6. as to form and intimation). If he does not do so, this may be the Project Manager 'not having taken action' for the purposes of the adjudication table in option W1. Accordingly, care should be taken to give notice of dispute within the four-week period so as not to lose the right to dispute the certified Completion date. A failure to certify Completion would give rise to a dispute under option W2 dispute procedure capable of being referred to adjudication.

Clause 30.3 deals with Key Dates. This requires the Contractor to do the work so that the condition stated for each Key Date is met by the Key Date. Unlike the sectional completion option, there is no provision for delay damages for failure to achieve a Key Date but clause 25.3 sets out the Employer's rights to obtain certain costs from the Contractor in the event of failure to achieve this and the clause makes it clear that this 'is his only right in these circumstances'.

For further details in relation to assessing the consequences of a compensation event on the Completion Date and Key Dates, see Section 5.5.4.

6.11.2 Programmes

Clause 31.1 deals with the programme. The operation of the NEC3 relies upon there being an Accepted Programme. A programme may have been identified by the Contractor in the Contract Data. In that case, such a programme will usually become the first 'Accepted Programme'. If there is no programme so identified, the Contractor is to submit a first programme for acceptance within the period stated in part one of the Contract Data. Clause 50.3 provides that a failure of the Contractor to submit a first programme for acceptance where there is none identified in the Contact Data, entitles the Employer to retain one quarter of the amounts due as interim payments until the first programme is submitted.

Clause 31.2 sets out very detailed provisions about what each programme is to contain:

- the starting date, access dates, Key Dates and Completion Date;
- planned Completion;
- the order and timing of the operations which the Contractor plans to do in order to Provide the Works;
- the order and timing of the work of the Employer and Others as last agreed with them by the Contractor or, if not so agreed, as stated in the Works Information;
- the dates when the Contractor plans to meet each Condition stated for the Key Dates and to complete other work needed to allow the Employer and Others to do their work;

- provisions for:
  - float;
  - time risk allowances (see clause 63.6);
  - health and safety requirements;
  - the procedures set out in the contract;
- the dates when, in order to Provide the Works in accordance with his programme, the Contractor will need:
  - access to a part of the Site if later than its access date,
  - acceptances;
  - Plant and Materials and other things to be provided by the Employer;
  - information from Others.
- for each operation, a statement of how the Contractor plans to do the work identifying the principal Equipment and other resources which he plans to use; and
- other information which the Works Information requires the Contractor to show on a programme submitted for acceptance.

It is quite clear from the nature of what is required to be included in the programme that a simple bar chart type programme simply will not do. To meet the requirements of clause 31 a number of different documents will be required.

Clause 31.3 requires the Project Manager to respond to the Contractor's programme within two weeks of its submission by either (a) accepting it or (b) notifying the Contractor of his reasons for not accepting it. Reasons for non-acceptance are:

- the Contractor's plans which it shows are not practicable;
- it does not show the information which the contract requires;
- it does not represent the Contractor's plans realistically; or
- it does not comply with the Works Information.

Non-acceptance for a reason other than one of the reasons stated above would be a compensation event under clause 60.1(9). See Section 6.11.11.

Clause 31.4 applies only if Option A (Priced Contract with Activity Schedule) or Option C (Target Cost with Activity Schedule) is used, i.e. the contracts with Activity Schedules. The clause requires the Contractor to provide information which shows that each activity on the Activity Schedule relates to the operations on each programme which the Contractor submits for acceptance.

6.11.3 Revising the programme

Clause 32.1 deals with the submission of revised programmes. While the contract does not expressly require the Contractor to comply with his programme, it is important to note that the contract requires the Contractor to submit revised programmes.

The Contractor is to submit a revised programme to the Project Manager for acceptance (a) within the period for reply after the Project Manager has instructed the Contractor to do so; (b) when the Contractor chooses to do so; and, in any case (c) at no longer an interval than that stated in the Contract Data from the starting date until Completion of the whole of the works.

The Contractor is to show on each revised programme (a) progress achieved and its effect upon the timing of remaining work; (b) effects on compensation events and notified early warning matters; (c) how the Contractor plans to deal with any delays and to correct notified Defects; and (d) any other changes which the Contractor proposes to make to the Accepted Programme.

In addition to the programme requirements, clause 16.1 of the NEC3 also requires the Contractor to give early warning by notifying the Project Manager as soon as he becomes aware of any matter which could increase the total of the Prices, delay Completion, delay meeting a Key Date, or impair the performance of the works in use.

6.11.4 Access to and use of the Site

Clause 33.1 provides that the Employer is to allow access to and use of each part of the Site to the Contractor which is necessary for the work included in the contract. Access and use are allowed on or before the later of its access date and the date for access shown on the Accepted Programme.

6.11.5 Instructions to stop or not stop work

Clause 34.1 provides that the Project Manager may instruct the Contractor to stop or not to start any work and may later instruct him that he may re-start or start it.

6.11.6 Take over

Clause 35.1 provides that the Employer need not take over the works before the Completion Date if it is stated in the Contract Data that he is not willing to do so. Otherwise the Employer is to take over the works not later than two weeks after Completion.

Clause 35.2 provides that the Employer may use any part of the works before Completion has been certified. If he does so, the Employer takes over the part of the works when he begins to use it except if the use is (a) for reasons stated in the Works Information or (b) to suit the Contractor's method of working.

Clause 35.3 provides that the Project Manager is to certify the date upon which the Employer takes over any part of the works and its extent within one week of the date.

6.11.7 Acceleration

Clause 36.1 provides that the Project Manager may instruct the Contractor to submit a quotation for an acceleration to achieve Completion before the Completion Date.

The Project Manager is to state changes to the Key Dates to be included in the quotation. Such a quotation is to include changes to Prices and a revised programme showing the earlier Completion Date and the changed Key Dates. The Contractor is to submit details of his assessment with each quotation. The Contractor is to submit a quotation or give reasons for not doing so within the period for reply (clause 36.2). Where the Project Manager accepts a quotation for an acceleration, clause 36.3 applies if any of Options A, B, C and D (priced and target contracts) are used and clause 36.4 if Options E and F (the cost reimbursable and management contracts) are used. The Project Manager changes the Prices, the Completion Date and the Key Dates accordingly and accepts the revised programme. In the case of clause 36.4, the forecast of the total Defined Cost of the whole of the works is changed rather than the Prices.

6.11.8 Sectional completion

Secondary Option clause X5 contains provisions in relation to Sectional completion. It states that in the conditions of contract, unless stated as the whole of the works, each reference and clause relevant to (a) the works; (b) Completion; and (c) Completion Date, applies as the case may be, to either the whole of the works or any section of the works. This clause allows the Employer to specify sectional completion dates for identified sections of the works.

6.11.9 Bonus for early completion

Secondary Option clause X6 provides that the Contractor is to be paid a bonus calculated at the rate stated in the Contract Data for each day from the earlier of (a) Completion and (b) the date on which the Employer takes over the works until the Completion Date. This provision, if used, can operate as an incentive to the Contractor to finish before the Completion Date.

6.11.10 Delay damages

Secondary Option clause X7 contains provisions in relation to delay damages. The Contractor is to pay delay damages at the rate stated in the Contract Data from the Completion Date for each day until the earlier of (a) Completion and (b) the date on which the Employer takes over the works (or for each section if Option X7 is used with Option X5, Sectional completion). The damages should be a genuine pre-estimate of loss and not a penalty. There is no provision for a certificate or notice of non-completion, nor is there any provision for notice of deduction or allowance of delay damages by the Employer. If the Completion Date is changed to a later date after delay damages have been paid, the Employer is to repay the overpayment of damages with interest. If the Employer takes over a part of the works before Completion, the delay damages are reduced from the date on which the part is taken over. Unlike other forms of contract where the damages are reduced in proportion to the value of the works taken into possession by the Employer, the Project Manager is to assess 'the benefit to the Employer of taking over the part of the works' as a proportion of the benefit to the Employer of taking over the whole of the works not previously taken over. The delay damages are reduced in this proportion. This unusual provision is one which could give rise to significant disputes. There may be circumstances in which reduction in the delay damages could be assessed at nil.

6.11.11 Compensation events

It should be noted that, in contrast to the SBC and the SBC/DB, all compensation events under the NEC3 carry both time and cost relief. See also Sections 6.5.3 and 6.10 and Section 8.3. In relation to the procedures to be followed in respect of a compensation event under the NEC3, see Section 5.2.4.

Clause 60.1 of NEC3 sets out 19 compensation events as follows:

1. The Project Manager gives an instruction changing the Works Information (as he is entitled to do under clause 14.3); except where the change (a) is made in order to accept a Defect or (b) is to Works Information provided by the Contractor for his design and is made either at his request or to comply with other Works Information provided by the Employer. For a more detailed description of the provisions in the NEC3 in respect of changes, see Section 8.2.5.

2. The Employer does not allow access to and use of a part of the Site by the later of its access date specified in the Contract Data and the date shown in the Accepted Programme. This is consistent with the Employer's obligation under clause 33.1. It is not unusual for an Employer to amend this provision to exclude rights of access which are expressly restricted under the Works Information.

3. The Employer does not provide, by the date for so doing shown in the Accepted Programme, something which he is required to provide.

4. The Project Manager gives an instruction to stop or not to start any work or to change a Key Date. The Project Manager is entitled to instruct a change to a Key Date under clause 14.3 and under clause 34.1 may instruct the Contractor to stop or not start any work. There will be time and cost consequences of issuing such instructions by virtue of this compensation event.

5. The Employer or others not engaged by the Contractor do not work within the times shown on the Accepted Programme or the conditions stated in the Works Information, or carry out work on the Site that is not stated in the Works Information. The Contractor's obligation to co-operate and share working areas with such others under clause 25.1 is therefore subject to this compensation event, where the others do not perform as the Contractor is entitled to assume.

6. The Project Manager or the Supervisor does not reply to a communication from the Contractor within the period required by the contract. Unless otherwise stated in the contract this period is the 'period for reply' to be completed in the Contract Data, or as may be extended by agreement (clauses 13.3 – 13.5). There are separate time periods in clause 31 for responding to the Contractor's programmes.

7. The Project Manager gives an instruction for dealing with an object of value or of historical or other interest found within the Site. The provisions in respect of such objects are set out in clause 73.1.

8. The Project Manager or the Supervisor changes a decision which he has previously communicated to the Contractor.

9. The Project Manager withholds an acceptance (other than an acceptance of a quotation for acceleration or for not correcting a Defect) for a reason not stated in the contract. This could, for example, apply to the situation where the Project Manager withholds consent to a sub-contractor or sub-contract conditions for reasons not stated in clauses 26.2 or 26.3 (see Section 11.9), where the Contractor's design is not accepted for reasons not stated in clause 21.2, or the Contractor's programme is not accepted for reasons not stated in clause 31.3 (see Section 6.11.2). The underlying principle in respect of the Project Manager's acceptance of a Contractor's submission is that the Project Manger is entitled to withhold acceptance, but if he does so for a reason not stated in the contract, then this gives rise to a compensation event. See clause 13.8. This is an alternative and possibly preferable approach to the common 'consent not to be unreasonably withheld or delayed' drafting seen in other contracts.

10. The Supervisor instructs the Contractor to search for a Defect and no Defect is found except where the search is needed only because the Contractor gave insufficient notice of doing work obstructing a required test or inspection.

11. A test or inspection done by the Supervisor causes unnecessary delay. This ties in with the Supervisor's obligation under clause 40.5 to do his tests and inspection without causing unnecessary delay to the work or to a payment which is conditional on a successful test or inspection. It is not clear at what stage a delay caused by a test or inspection becomes 'unnecessary'. This is sometimes clarified by an amendment to the effect that this will only be a compensation event if the test or inspection is not provided for in the contract.

12. The Contractor encounters physical conditions which are within the Site, are not weather conditions, and an experienced contractor would have judged at the date the contract came into effect to have such a small chance of occurring that it would have been unreasonable for him to have allowed for them; and in assessing the compensation event only the difference between the physical conditions encountered and those for which it would have been reasonable

to have allowed is taken into account. Clause 60(2) states that in judging the physical conditions, the Contractor is assumed to have taken into account the Site Information (and publicly available information referred to therein) referred to in the Contract Data, information obtainable from a visual inspection of the site, and other information which an experienced contractor could reasonably be expected to have or obtain. In *Atkins Limited* v. *The Secretary of State for Transport* (2013) the contractor under an NEC3 routine maintenance contract claimed the existence of an excessive number of potholes as a compensation event. This was rejected by the court, the judge considering that, in the particular circumstances of that case, any such excess volume was not an occurrence with such a small chance of being present that it would have been unreasonable to have allowed for it.

13. A weather measurement is recorded at the place stated in the Contract Data, the value of which, by comparison with the weather data referred to in the Contract Data, is shown to occur on average less frequently than once in 10 years. The Contract Data specifies the weather measurements to be recorded each calendar month, with provision for additional measurements on an individual contract basis. The source of the records to be used as past weather data for comparison against the monthly weather measurement is also to be specified in the Contract Data, or where no recorded data are available, assumed values are to be specified. In assessing the compensation event, only the difference between the recorded weather measurement and the weather which the weather data show to occur on average less frequently than once in 10 years is taken into account. This is a much more objective, precise and, arguably, stricter test than the equivalent 'exceptionally adverse weather conditions' under the SBC and the SBC/DB (and as noted above the latter contracts, unlike the NEC3, do not confer cost relief for weather events).

14. An event which is stated in the contract to be an Employer's risk. These are the risks listed in clause 80.1 (see Section 14.1.6) and any additional Employer's risks stated in the Contract Data.

15. The Project Manager certifies take over of a part of the works before both actual Completion of the works and the contractual Completion Date. See Section 6.11.6.

16. The Employer does not provide materials, facilities and samples for tests and inspections as stated in the Works Information. This is consistent with the Employer's obligation to do so under clause 40.2.

17. The Project Manager notifies a correction to an assumption which he has stated about a compensation event. This relates to clause 61.6 which provides that if the Project Manager decides that the effects of a compensation event are too uncertain to be forecast reasonably, he states assumptions about the event in his instruction to the Contractor to submit quotations, and the

assessment of the event is based on these assumptions. If any of them is later found to have been wrong, the Project Manager is required to notify a correction.

18. A breach of contract by the Employer which is not otherwise a compensation event.

19. An event which stops the Contractor completing the works or completing the works shown on the Accepted Programme, which neither Party could prevent, which an experienced contractor would have judged at the effective date of the contract to have such a small chance of occurring that it would have been unreasonable for him to have allowed for it, and which is not otherwise a compensation event. This 'prevention' category of compensation event should be read along with clause 19.1 which requires the Project Manager to issue an instruction if such a 'prevention' event occurs and also clause 91.7 which allows the Employer to terminate the contract (without penalty to the Contractor) if the forecast delay caused by the prevention event exceeds 13 weeks. This compensation event is commonly amended by the Employer, for example by restricting it to an act of prevention by the Employer. See also Sections 5.2.4 and 9.3.3.

In addition to the above compensation events, if Secondary Option Y(UK) (Housing Grants Construction and Regeneration Act 1996) is used, as it invariably will be where the law of any part of the UK applies to the contract, then clause Y2.4 also provides that if the Contractor exercises his rights under the 1996 Act to suspend performance, this is a compensation event. This complies with section 112(3A) added by the amendments to the 1996 Act (see Sections 10.9.2 and 10.9.3).

Chapter 7
Certification

7.1 Introduction

The use of certificates in building contracts is both common and, it is submitted, essential for the proper administration of the contract. Although certificates fulfil a number of wide-ranging functions, their central use is to provide triggers or mechanisms that regulate the rights and obligations of the parties to a contract during its currency and on completion. In particular, certificates often play an important role in the contractual mechanisms which regulate payment, both interim and final, progress and completion of the works and the rectification of defects in the works. The issue of a certificate is regularly a condition precedent to one of the parties obtaining rights in terms of the building contract.

In order to ascertain whether a building contract has any requirements for the issue of certificates, it is necessary to look at the express terms of the contract. If the express terms of the contract do require certification, then the terms have to be carefully considered in order to ascertain what certificates need to be issued, the contractual pre-conditions which must be satisfied before a certificate can be issued, and the rights and obligations which flow from or are extinguished by the issue of a certificate, see *Ata Ul Haq* v. *The City Council of Nairobi* (1985). In the absence of any express terms dealing with certification, then the question of certification does not arise. The requirement for certification is not implied by operation of law.

7.2 Formal requirements of certificates

As the requirement for the use of certificates has to be expressed in the building contract, similarly the requirements as to the form of a valid certificate can also be stipulated. In the commonly used standard forms of building contract, it is unusual to find the form of certificates specified in detail and even more uncommon for style or specimen certificates to be provided. However, any requirements as to form which are stipulated should be strictly followed, failing which there is a danger that the certificate will be open to challenge and ultimately held to be invalid. See, for example, *B R Cantrell, E P Cantrell* v. *Wright & Fuller Ltd* (2003), in which it was held that a certificate was not a valid certificate in form, substance or intent.

MacRoberts on Scottish Construction Contracts, Third Edition. MacRoberts.

If the contract is silent as to the form a certificate is to take, then no particular form is required. If the certifier only has to pronounce himself satisfied in respect of certain matters, then the certification may be given orally. Notwithstanding this, it is clearly preferable for certificates to be in writing, if for no other reason than to avoid evidential difficulties in subsequently proving whether certification has or has not been given. In this connection, clause 13.1 of the NEC3 expressly provides that certificates require to be communicated in a form which can be read, copied and recorded. Although it does not expressly mention certificates, clause 1.7 of the SBC and the SBC/DB similarly provides that notices and other communications referred to in the contract shall be in writing.

Disputes do regularly arise in the course of building contracts as to whether certification has been given and whether or not a written document amounts to a certificate in terms of the contract. In *Halliday Construction Ltd and Others* v. *Gowrie Housing Association Ltd* (1995), a dispute arose as to whether letters written by the architect to the contractor amounted to non-completion certificates in terms of the contract. The letters simply advised the contractor that the architect had notified the employer that the contract had overrun and that liquidated and ascertained damages might be deducted. The contractor argued that such letters did not constitute certificates as they lacked the necessary form, substance and intent. The court held that the letters did amount to certificates, but only after some considerable hesitation. Similarly in *Norwest Holst Ltd* v. *Carfin Developments Limited* (2008), the Scottish courts held that a letter from the employer's project manager to the contractor was clearly a valid payment certificate in form, substance and intent particularly as it was in the same form as previous certificates which had been paid by the employer.

In contrast, a letter written by the architect and relied on by the employer to deduct liquidated and ascertained damages was held insufficient to constitute a certificate in the case of *Token Construction Co. Ltd* v. *Charlton Estates Ltd* (1976). The reasoning of the court was that it was unclear and ambiguous whether the architect had intended the letter to constitute a certificate.

In order to avoid such difficulties, when purporting to issue a certificate, the certifier should make it clear that certification is being given. In this connection the use of the word 'certificate' is not essential though it is submitted that its use is prudent, see *Minster Trust Ltd* v. *Traps Tractor Ltd and Others* (1954). See also *H Fairweather Ltd* v. *Asden Securities Ltd* (1979), in which the court attached weight to the fact that a letter relied on as a certificate did not contain the word 'certify'. The use of words such as 'checking', 'approving' and 'satisfies' may not in itself be sufficient and can give rise to ambiguity.

The certificate should leave the parties in no doubt as to its intention and effect. The rights and obligations which flow from the issue of the certificate should be clear. Where possible, a certificate should refer to the relevant clause of the contract under which it is being issued and, insofar as possible, follow the wording contained in the clause.

In the event that a dispute does arise as to whether a document is or is not a certificate, then the use of extrinsic evidence may be permissible. For example, the terms of a covering letter sent with a purported certificate may provide assistance in ascertaining whether a document is truly a certificate. In *H Fairweather Ltd*, the court considered a whole course of correspondence to ascertain whether certain letters were intended to constitute certificates.

Unless otherwise required by the terms of the contract, a certificate does not have to include any reasons in support of the matters decided by or the opinions expressed in the certificate, and a lack of reasons can make the challenge of the certificate by an aggrieved party more difficult. Whether this is a perceived advantage or disadvantage is a matter for the parties to decide and take account of when drafting the contract.

In addition to any requirements regarding the form of certificates, there are other matters that should be borne in mind when preparing and issuing certificates. In particular, one should ensure that any express pre-conditions to the issue of a certificate have been complied with. Such pre-conditions may include when the certificate needs to be issued; by what mechanism the certificate should be issued; by whom the certificate should be issued; and to whom the certificate should be issued.

For example, clause 4.15 of the SBC contains a number of pre-conditions to the issue of a Final Certificate, namely the end of the Rectification Period in respect of the Works or where there are Sections the last such period to expire, the issue of the Certificate of Making Good, and the sending by the architect to the contractor of an ascertainment of any loss and expense and a statement of all adjustments to be made to the Contract Sum. In addition, the Final Certificate must be issued no later than two months after whichever of the foregoing is last to occur. Clause 4.15 goes on to stipulate what the Final Certificate should include, being the adjusted Contract Sum; the sum of the amounts already stated as due in Interim Certificates plus the amount of any advance payment; the difference between the two sums expressed as a balance due to the contractor from the employer or vice versa; and the basis on which that balance has been calculated. As an example of a case where pre-conditions were not followed, see *G A Group Ltd* v. *Scottish Metropolitan Property plc* (1992) where a certificate of non-completion was held to be invalid due to the fact that it was issued prior to the expiry of the period for completion. See also *Crestar Ltd* v. *Michael John Carr and Joy Carr* (1987).

It is common for standard forms of building contract to stipulate that the certificate must actually be delivered to the parties to the contract. The requirements for delivery may also be expressed including the method of delivery and the address to which delivery has to be made, for example, to a limited company at its registered office. The SBC provides, in clause 1.8, that each certificate issued by the Architect shall be issued to the Employer and the Contractor at the same time and clause 1.7 contains detailed provisions in relation to how all notices and communications under the contract are to be transmitted. Identical provisions on the transmission of notices and communications are contained in clause 1.7 of the SBC/DB. In the event that the contract does not stipulate that the certificate needs to be delivered to the parties, then it is probably implied in any event. See, for example, the comments of Lord Justice Edmund Davies in the case of *Token Construction Co. Ltd* v. *Charlton Estates Ltd* (1973).

Minor errors in complying with any of the formal requirements of a certificate may not result in the certificate being held to be invalid, provided that the substance and effect of the certificate are correct and provided none of the parties to the contract have been misled or prejudiced. Nevertheless, such comfort should not be relied upon and the prudent course is to ensure that the certificate complies entirely with all the contractual requirements. If the certifier issues a certificate which is invalid, it may be open to him to reissue the certificate in a form which is valid provided he is not *functus officio* (disempowered because his role is concluded) and provided he does not alter the substance of the certificate unless the contract permits him to do so, see *Kiu May Construction Co. Ltd* v. *Wai Cheong Co. Ltd and Another* (1983).

7.3 Interim certificates

One of the most widely used types of certificate found in building contracts is the interim certificate, sometimes known as a progress certificate. Such a certificate is issued during the course of the contract works and is commonly designed to fulfil the dual function of monitoring the progress of the works and at the same time regulating instalment or interim payments to the contractor.

At common law, unless the contract provides otherwise, a contractor has no implied right to interim or instalment payments. This position has been altered in respect of most building contracts by the 1996 Act, see Chapter 8 for a fuller discussion of this point. Nevertheless, most building contracts do expressly provide for interim or instalment payments to the contractor during the currency of the works.

In the commonly used standard forms of building contract, interim or progress certificates are the mechanism most often used as a means of regulating the timing and amount of such interim or instalment payments. See, for example, clause 4.10 of the SBC which makes provision for the issue of Interim Certificates by the Architect no later than 5 days after the due dates for interim payments. Similarly clause 51.1 of NEC3 makes provision for the Project Manager to certify payments within one week of each assessment date. In contrast, under the SBC/DB it is the Employer who issues a Payment Notice no later than 5 days after the due date in terms of clause 4.9.2, as there is no independent architect or project manager to issue certificates (although in practice the Employer's Agent often does so on the Employer's behalf).

When issued, an interim certificate commonly operates in one of two ways. The interim certificate either certifies the value of work carried out at the date of the certificate and triggers payment of that amount to account of the final contract sum or, alternatively, the interim certificate certifies that the works have been completed to a particular stage triggering the release of an agreed instalment payment for that stage.

The SBC recognizes both these alternatives. It provides a mechanism in clauses 4.9 and 4.16 for ascertaining the amount to be included in an Interim Certificate using the former alternative but also stipulates, in the opening lines of clause 4.9.2, that this is subject to any agreement between the parties as to stage payments. Under clauses 50.1 and 50.2 of the NEC3, the Project Manager assesses the amount due to the Contractor at the assessment date, which amount is the price for the work done to date plus any other amounts to be paid to the Contractor. How the price is calculated depends on which of Options A–F has been selected as the main pricing option.

As a consequence of the fact that interim certificates are issued during the currency of the contract, the valuation of the work carried out, or any assessment of the quality of the work carried out, at the date of the certificate is not an exact science.

Accordingly, interim certificates are not normally stipulated to be conclusive in respect of either the amount to be paid to the contractor or to the extent that they provide that the works and materials are of satisfactory quality. In relation to the last point, see *Clark Contracts Ltd* v. *The Burrell Co. (Construction Management) Ltd* (2002). Interim certificates simply have provisional validity, see *Beaufort Developments (NI) Ltd* v. *Gilbert-Ash NI Ltd and Another* (1998). The amount certified in an interim certificate can usually be challenged during the currency of the works. The procedure for challenging interim certificates is considered in more detail below.

In any event, the effect of interim certificates may be superseded by subsequent developments and subsequent interim certificates. In relation to the valuation of the work carried out, this can normally be revised at the time of issue of the next interim certificate, see *Scottish Equitable plc* v. *Miller Construction Ltd* (2001). Most standard forms of building contract allow for a revaluation of all work carried out in terms of the contract at the date of issue of an interim certificate and not simply a valuation of the work carried out since the issue of the previous certificate, see, for example, clause 4.16 of the SBC. Confirmation that each interim certificate is intended to be based on a revaluation of all work carried out can be found in *William Verry Ltd* v. *North West London Communal Mikvah* (2004). In this case the court indicated that each month the works have to be revalued so as to ensure that the total value of work properly executed is ascertained. In addition, if work previously valued is discovered to be defective, then there will have to be a downward adjustment of the gross value previously certified. The NEC3 puts the position beyond doubt by providing expressly in clause 50.5 that the Project Manager corrects any wrongly assessed amount due in a later payment certificate and this was confirmed in the case of *RBG Ltd* v. *SGL Carbon Fibers Ltd* (2010) and *SGL Carbon Fibres Ltd* v. *RBG Ltd* (2012).

The fact that each interim certificate supersedes its predecessor has important implications with regard to the prescription of claims. It was held in *Scottish Equitable plc* v. *Miller Construction Ltd* (2001) that the whole structure of the contract in question allowed challenges to be made against certificates, notwithstanding the fact that a challenge on the same basis could have been made against an earlier certificate. In short, it appears that a failure to challenge one interim certificate in respect of a particular issue does not automatically trigger the start of the prescriptive period in respect of that issue as it also falls to be valued in subsequent interim certificates. Support for such a position is also to be found in the English case of *Henry Boot Construction Ltd* v. *Alstom Combined Cycles Ltd* (2005).

With regard to the quality of work and materials, the issue of an interim certificate does not normally prevent the issue of instructions or directions in relation to remedying defective or unsatisfactory work. Indeed many defects may not be apparent at the time of issue of an interim certificate. Clause 1.10 of the SBC specifically provides that a certificate is not conclusive evidence that any work, materials or goods to which it relates are in accordance with the contract. Similarly, clause 3.6 of the SBC stipulates that the Contractor remains wholly responsible for carrying out the Works in accordance with the contract notwithstanding the fact that the value of that work has been included in a certificate for payment.

In many standard forms of building contract, the issue of an interim certificate is a condition precedent to payment of interim amounts to the contractor. In such cases if the contractor does not receive a certificate, then it will have no right to payment under the contract. Similarly, an employer is only obliged to pay to a contractor the amount contained in an interim certificate, see *Nicol Homeworld Contracts Ltd* v. *Charles Gray Builders Ltd* (1986), *Costain Building & Civil Engineering Ltd* v. *Scottish Rugby Union plc* (1993) and *Karl Construction Ltd* v. *Palisade Properties plc* (2002). See also the English case of *Lubenham Fidelities and Investments Co. Ltd* v. *South Pembrokeshire DC and Another* (1986).

Until the coming into force of the 1996 Act, if the employer challenged an interim certificate (and provided they could aver a genuine dispute regarding the issue of the certificate), then they could attempt to avoid making payment on the certificate, see *W & J R Watson Ltd* v. *Lothian Health Board* (1985). Employers could also attempt to avoid making payment on an interim certificate if they could rely on their common law rights of retention and set-off, or if there were any contractual rights to make deductions from amounts certified. A fuller discussion of these matters is found in Chapter 10. The position regarding the withholding of payment is now regulated by section 111 of the 1996 Act and generally an employer cannot withhold payment of a certified sum without first serving a valid notice of intention to pay less, see *Rupert Morgan Building Services (LLC) Ltd* v. *Jervis* (2003).

For many years following upon the decision in *Northern Regional Health Authority* v. *Derek Crouch Construction Co. Ltd* (1984), it was the position that an aggrieved party could only challenge the amount of or the lack of an interim certificate by means of arbitration proceedings and not through the courts. In that case, the parties to a building contract conferred on an arbitrator the power to open up, review and revise certificates. The English Court of Appeal held that this special power had been expressly conferred on the arbitrator, that the courts did not have a similar power and, accordingly, could not open up, review or revise certificates.

The decision in *Northern Regional Health Authority* was followed in England and also in Scotland for 14 years, see *D & J McDougall Ltd* v. *Argyll & Bute DC* (1986) and *Stanley Miller Ltd* v. *Ladhope Developments Ltd* (1988). As a result, great care was required when drafting arbitration clauses to ensure that an arbitrator was given sufficiently wide powers to alter certificates. Similarly, great care had to be taken when deleting arbitration clauses, and it became increasingly common for parties to insert provisions in building contracts providing that the courts could open up, review and revise certificates where there was no arbitration clause. Whether such provisions competently gave the courts power to review certificates was never clear.

In 1998, however, the position altered dramatically with the decision of the House of Lords in the case of *Beaufort Developments (NI) Ltd* v. *Gilbert-Ash NI Ltd and Another* (1998) which overruled *Northern Regional Health Authority*. In *Beaufort Developments (NI) Ltd*, the court held that merely because an arbitration clause gave an arbitrator power to open up, review and revise certificates that did not mean that the courts could not consider the matter. The wording of such an arbitration clause did not confer on an arbitrator wider powers than those enjoyed by the courts whose normal powers to enforce contracts were sufficiently wide to achieve the same result. Accordingly, interim certificates do not have binding and conclusive effect before a court and a party to a building contract can sue for payment in the courts of sums not yet certified in an interim certificate.

The reasoning behind the House of Lords' decision was that the parties to the contract had conferred on the arbitrator the power to open up, review and revise certificates. This illustrated that interim certificates were not intended to be binding and conclusive at all and accordingly, could not be binding and conclusive before the courts, see also *Robins* v. *Goddard* (1905). The House of Lords did stress that the position would be different in respect of certificates which are expressly stipulated to be binding and conclusive.

Unfortunately, it is not clear from the court's decision in *Beaufort Developments (NI) Ltd* how the contractor's right to raise proceedings against the employer in the absence of a certificate is to be analysed. In *Beaufort Developments (NI) Ltd*, the most helpful discussion of available remedies is found in the speech of Lord Hope of Craighead who stated:

> On this approach the court will be able to exercise all its ordinary powers to decide the issues of fact and law which may be brought before it and to give effect to the rights and obligations of the parties in the usual way. It will have all the powers which it needs to determine the extent to which, if at all, either party was in breach of the contract and to determine what sums, if any, are due to be paid by one party to the other whether by way of set-off or in addition to those sums which have been certified by the architect. It will not be necessary for it to exercise the powers which the parties have conferred upon the architect in order to provide the machinery for working out that contract. This is because the court does not need to make use of the machinery under the contract to provide the parties with the appropriate remedies. The ordinary powers of the court in regard to the examination of the facts and the awarding of sums found due to or by either party are all that is required.

It was unclear, however, what effect the decision had on the authorities referred to above in which it had already been held that a certificate was a condition precedent to payment of the contractor, see *Nicol Homeworld Contracts Ltd* v. *Charles Gray Builders Ltd* (1986) and *Costain Building & Civil Engineering Ltd* v. *Scottish Rugby Union plc* (1994).

This issue has since been further considered by the English Court of Appeal in the case of *Henry Boot Construction Ltd* v. *Alstom Combined Cycles Ltd* (2005). In that case the court confirmed that certificates were a condition precedent to the contractor's entitlement to payment under the contract in question but that it did not follow that the absence of a certificate was a bar to the right to payment. In particular, Lord Justice Dyson stated that:

> By 'condition precedent' I mean that the right to payment arises when a certificate is issued or ought to be issued, and not earlier. It does not, however, follow from the fact that a certificate is a condition precedent that the absence of a certificate is a bar to the right to payment. This is because the decision of the engineer in relation to certification is not conclusive of the rights of the parties, unless they have clearly so provided. If the engineer's decision is not binding, it can be reviewed by an arbitrator (if there is an arbitration clause which permits such a review) or by the court. If the arbitrator or the court decides that the engineer ought to have issued a

> certificate which he refused to issue, or to have included a larger sum in a certificate which he did issue, they can, and ordinarily will, hold that the contractor is entitled to payment as if such certificate had been issued and award or give judgment for the appropriate sum.

He went on to confirm that he did not consider that the decision in *Beaufort* compelled the conclusion that certificates were not a condition precedent to the right to payment.

The issue was also considered in the Scottish case of *Karl Construction Ltd* v. *Palisade Properties plc* (2002). In this case Lord Drummond Young held that payment under the standard JCT forms was conditional on the issue of a certificate by the certifier or on the decree of an arbitrator or a court on the basis that the decree of an arbitrator or court was equivalent to a certificate. He stated that:

> The equivalence of a court decree to an architect or engineer's certificate follows from *Beaufort*, on the basis that a decree of the court can achieve the same result as a decree of an arbiter. In every case, however, until an appropriate certificate or decree has been obtained, the debt due by the employer to the contractor is contingent.

Consideration also needs to be given to whether an adjudicator has the power to open up, review and revise interim certificates. It is submitted that he must have such a power as the 1996 Act stipulates that a party has the right to refer 'any difference' to an adjudicator for his decision, see s.108(1) – 108(4) of the 1996 Act. Accordingly, it is arguable that this must imply a right to open up, review and revise interim certificates. If no such right is implied and if the adjudication provisions in a building contract do not allow an adjudicator to open up, review and revise certificates, then it is possible that the contract will not meet the requirements of s.108(1) – 108(4) and the Statutory Scheme for Construction Contracts will apply, see s.108(5). The Scheme specifically provides that an adjudicator may open up, review and revise any decision taken or any certificate given, see Part I of the Schedule to the Scheme for Construction Contracts (Scotland) Regulations 1998, paragraph 20(2)(a).

This potential issue is avoided by clause 9.2 of the SBC and the SBC/DB which states that if a dispute or difference arises which either party wishes to refer to adjudication, then (subject to some minor qualifications) the Scheme will govern the process and, accordingly, an adjudicator has in respect of disputes under that form of contract an express power to open up, review and revise interim certificates. The position is slightly less clear under the NEC3, although Options W1 and W2 dealing with dispute resolution both provide that the adjudicator may review and revise any action or inaction of the Project Manager which is likely to cover the position. It is submitted that if parties drafting a building contract wish to ensure that their adjudication provisions comply with the requirements of the 1996 Act, and thus avoid any risk of the statutory scheme applying, it is prudent to expressly confer such a power upon the adjudicator. It appears that when challenging interim certificates the burden of proof will fall on the party mounting the challenge to establish that the interim certificate was incorrect. If the contractor seeks further payment, then he will have the burden and the employer will carry it if he contends that the certification is too high. This was determined in the context of an NEC3 contract in the case of *SGL Carbon Fibres Ltd* v. *RBG Ltd* (2012).

7.4 Final certificates

7.4.1 General

A second type of certificate regularly encountered in building contracts is the final certificate. The issue of a final certificate usually signals the end of the contract and can deal with a number of matters, including the final amount payable in terms of the contract which often includes any amount payable for additional or extra work. It can also include, but more normally excludes, any amounts payable in respect of damages for delay or other breaches of contract. It can mean that the contract works have been completed to the satisfaction of the certifier; that additional or extra work has been completed to the satisfaction of the certifier; and that the rectification of patent defects has been carried out to the satisfaction of the certifier. Neither the SBC/DB nor the NEC3 contain final certificate provisions. The SBC/DB does contain provisions regarding a Final Statement which regulates many of the same issues as a final certificate but that is a document that is issued by the Contractor (whom failing the Employer) and then agreed (or disputed) by the parties. The provisions regulating the Final Statement and the effect thereof are to be found in clauses 4.12 and 1.8 of the SBC/DB. The SBC, on the other hand, does contain very detailed provisions that mirror the JCT provisions on final certificates and which merit closer consideration.

7.4.2 The Final Certificate under the SBC

In order to ascertain the matters that are covered by the Final Certificate, it is necessary to consider the express terms of the contract. It is also necessary to consider the express terms of the contract to ascertain the effect of the issue of the Final Certificate, see *Ata Ul Haq* v. *The City Council of Nairobi* (1962).

Clause 1.9 of the SBC sets out the effect of the Final Certificate. It provides that the Final Certificate shall have effect in any proceedings, whether by adjudication, arbitration or legal proceedings, as conclusive evidence that where and to the extent that any of the particular qualities of any materials or goods or any particular standard of an item of workmanship was described expressly in the Contract Drawings, the Contract Bills, an instruction of the Architect or in any drawing or document issued by the Architect to be for the approval of the Architect, the particular quality or standard was to the reasonable satisfaction of the Architect. The Final Certificate is not conclusive evidence that materials or goods or workmanship comply with any other requirement or term of the contract.

It is conclusive evidence that necessary effect has been given to all the terms of the contract which require that an amount be added to or deducted from the Contract Sum or that an adjustment is to be made to the Contract Sum, save where there has been any accidental inclusion or exclusion of any work, materials, goods or figure in any computation or any arithmetical error in any computation. In such circumstances, the Final Certificate is conclusive evidence as to all other computations.

It is conclusive evidence that all and only such extensions of time, if any, as are due under clause 2.28 have been given.

Finally, it is conclusive evidence that the reimbursement of direct loss and/or expense, if any, to the Contractor pursuant to clause 4.23 is in final settlement of all and any claims which the Contractor has or may have arising out of the occurrence of any of the relevant matters referred to in clause 4.24 whether such claims are for breach of contract, duty of care, statutory duty or otherwise.

The issue of a final certificate is unlikely to cover claims in respect of damages for breach of contract, as it is unusual to find such matters within the certifier's remit. The issue of the final certificate can preclude the employer's ability to deduct liquidated damages. Clause 2.32 of the SBC provides that the Employer may require the Contractor to pay liquidated damages provided they give notice in writing prior to the issue of the Final Certificate. See also *Robert Paterson & Sons Ltd* v. *Household Supplies Co. Ltd* (1974).

A final certificate will not have conclusive effect if the certifier has exceeded his jurisdiction or the issue of the certificate is challengeable on other grounds, for example, where the certifier has not acted independently, has acted in bad faith or has acted fraudulently. These issues are more fully considered in Section 7.6. Clause 1.9 of the SBC specifically provides that the Final Certificate will not be conclusive in the event of fraud. The Final Certificate is conclusive unless and until it is successfully challenged.

7.4.3 The final certificate as conclusive evidence

It will be noted that clause 1.9 of the SBC states on a number of occasions that the Final Certificate has the effect of being 'conclusive evidence' on a matter. In many standard forms of building contract, it is common to find provisions that the final certificate is to some extent conclusive and binding upon the parties. Where a final certificate is stated to be conclusive and binding and has been properly issued, then its effect is final in respect of the matters covered by the certificate. Accordingly, the parties to a contract cannot challenge the certificate by adjudication, arbitration or in the courts, or ask the adjudicator, arbiter or courts to review the certificate, simply on the grounds that they are aggrieved by it or disagree with its terms. The Statutory Scheme for Construction Contracts in Scotland stipulates that an adjudicator can open up, review and revise any certificate unless the contract states the certificate is final and conclusive, see Part I of the Schedule to the Scheme for Construction Contracts (Scotland) Regulations 1998, paragraph 20(2)(a).

The certificate is the final expression of the certifier's decision and cannot be interfered with simply on the basis that the certificate is wrong or negligently issued, see, for example, *Rush & Tompkins Ltd* v. *Deaner* (1989) where, due to an error on the part of the certifier, the balance due to the contractor in terms of the final certificate was mistakenly based on sums certified rather than sums certified and paid. It is submitted that the reason for this is that the certifier has been selected from a professional discipline because he possesses and can exercise the requisite skills and knowledge when issuing certificates. It is further submitted that the certifier should have an intimate knowledge of the contract as a result of his involvement, which knowledge would not be available to an independent third party such as an adjudicator, arbiter or court. Accordingly, the certifier is often the person best placed to decide any issues between the parties which fall within his remit.

7.4.4 The English and Scottish approaches

The matters in respect of which the final certificate is conclusive and binding differ from contract to contract and it is necessary to consider every contract on its own terms. This may not be an easy exercise particularly if the contract is not clearly drafted. Furthermore, the authorities which exist in this area are often inconsistent.

The English courts have tended to interpret the conclusive effect of final certificates very broadly. In *Crown Estates Commissioners* v. *John Mowlem & Co. Ltd* (1994), the English Court of Appeal considered the then wording of clause 30.9.1.1 of JCT 80. The clause under consideration by the court differs from the wording of clause 1.9.1.1 of the SBC (the current equivalent of clause 30.9.1.1 of JCT 80) in that it provided that the final certificate was to have effect as conclusive evidence that, where the quality of materials or the standard of workmanship were to be to the reasonable satisfaction of the architect, the same were to such satisfaction. The court held that, on a true construction of this clause, all matters of standards and quality of work and materials were for the reasonable opinion of the architect and so were concluded by the issue of a final certificate. Accordingly, if a final certificate was issued and not challenged timeously, then all claims for defects arising from the standard or quality of work or materials would be defeated by the conclusive effect of the final certificate. Following this reasoning, claims for latent defects not apparent at the date of issue of the final certificate would also be excluded. Similarly, claims would also be excluded for defects arising from work or materials failing to meet prescribed criteria found, for example, in the bill of quantities or specification. The case of *Colbart Ltd* v. *Kumar* (1992) is a further example of the broad interpretation favoured by the English courts.

It appears that this broad interpretation did not reflect the intention of the Joint Contracts Tribunal when it originally drafted clause 30.9.1.1 of JCT 80. The intention of the Joint Contracts Tribunal was that the final certificate should only be conclusive evidence that the architect was satisfied that certain requirements of the work had been complied with where both the contractor and employer had agreed to abide by the architect's decision in respect of those requirements. It was not intended to have conclusive effect where the contractor had failed to comply with prescribed requirements of the contract documents. Following the decisions in *Colbart Ltd* and *Crown Estates Commissioners*, the Joint Contracts Tribunal revised clause 30.9.1.1 and that revision is carried through to clause 1.9.1.1 of the SBC to try and reflect its original intention. Accordingly, the decisions in *Colbart Ltd* and *Crown Estates Commissioners* may now be of limited application.

In any event, in Scotland the courts appear to have taken a narrower view of the conclusive effect of final certificates. Such a view reflects the original intention of JCT.

In *Firholm Builders Ltd* v. *McAuley* (1982), the court considered a clause with wording similar to that considered by the English court in *Crown Estates Commissioners*. The court held, however, that the existence of the final certificate did not necessarily defeat a claim for defective workmanship or materials. The existence of the certificate simply allowed the contractor to rely on it as conclusive evidence that, where materials or workmanship were to be to the architect's reasonable satisfaction, then the final certificate demonstrated that they were to his reasonable satisfaction. It was still open to the employer to argue that the architect should not reasonably have been satisfied.

The decision of the Court of Session in *Belcher Food Products Ltd* v. *Miller & Black and Others* (1998) also appears to support a narrower interpretation of the conclusive effect of final certificates. In *Belcher Food Products Ltd*, the court attempted to distinguish *Crown Estates Commissioners* by holding that even where a final certificate was conclusive evidence of the architect's reasonable satisfaction, it did not necessarily follow that it had the further effect of being conclusive evidence that the relevant standard and quality of workmanship or materials had been achieved in a question between the employer and the contractor. In other words, the final certificate would not be conclusive evidence of the standard and quality of workmanship or materials but only conclusive evidence that the architect was satisfied with the standard and quality. It followed that the court could only decide whether the quality and standard of workmanship was satisfactory once it had heard evidence. The final certificate would, however, have strong evidential value in this connection. This was not a distinction which the court had been asked to consider in *Crown Estates Commissioners*. The court in *Belcher Food Products Ltd* further attempted to distinguish *Crown Estates Commissioners* and *Colbart Ltd* on the basis that they both concerned issues which were inherently matters for the subjective opinion of the architect. In contrast, many of the issues in *Belcher Food Products Ltd* related to whether there had been actual compliance with express contractual requirements as to the quality of materials, which could be assessed objectively.

Accordingly, the Scottish courts appear to have taken a slightly different approach to that of the English courts. The authors respectfully suggest that the approach of the Scottish courts is to be preferred. The court in *Belcher Food Products Ltd*, however, clearly felt that *Crown Estates Commissioners* and *Colbart Ltd* were sufficiently authoritative that it was necessary to distinguish them rather than openly disagree with them or refuse to follow them. The result of all this is that it is difficult to draw any general principles from the cases dealing with the conclusive effect of final certificates. Each contract has to be considered on its own terms and against the background of the particular facts which have arisen. The subsequent amendment to the relevant clause is also of significance, the decisions in each of the cases referred to on this point being in relation to the effect of the old, now superseded, wording.

7.4.5 Challenging the final certificate

A number of the standard forms of building contract contain provisions providing that final certificates may be challenged by arbitration or other proceedings following their issue. Accordingly an arbitrator is usually expressly empowered to open up, review and revise certificates. The SBC specifically provides for this in clause 9.5.3.

Following the decision in *Beaufort Developments (NI) Ltd* v. *Gilbert-Ash NI Ltd and Another* (1998), the courts also enjoy such an inherent power without the requirement for an express power to be given to them. Where a final certificate can be challenged by arbitration or other proceedings, then it should also be challengeable by means of adjudication under the 1996 Act and the Scheme for Construction Contracts expressly provides for this, as discussed at paragraph 7.3 in the context of interim certificates.

Clearly such challenges to final certificates are inconsistent with the concept of final certificates having, in some circumstances, binding and conclusive effect.

The means by which this inconsistency is often dealt with in the standard forms of building contract is to provide that a final certificate does not have conclusive effect immediately on being issued; rather its conclusive effect is suspended for a stipulated period of time. During this period of time, an aggrieved party is given the opportunity to challenge the final certificate by raising the appropriate proceedings. At the end of the stipulated period of time, the final certificate will have conclusive effect to the extent that it has not been challenged. In respect that the final certificate is challenged, then it will still have conclusive effect subject to any award of the adjudicator, arbiter or court. In this connection see, for example, clause 1.9.3 of the SBC which provides that the Final Certificate will have conclusive effect 60 days after issue save in relation to any matters in respect of which adjudication, arbitration or other proceedings have been raised. If the challenge to the Final Certificate is by adjudication, then parties are given a further 28 days to raise arbitration or legal proceedings following conclusion of the adjudication in terms of clause 1.9.4. In the English case of *Tracy Bennett* v. *FMK Construction Limited* (2005), the court refused to declare that the final certificate was conclusive despite the fact that an adjudication commenced within the 28-day period was brought to end by the adjudicator and only re-started after the expiry of the 28-day period. It is submitted, however, that the facts of this case were peculiar and in all cases parties should ensure that time limits are strictly adhered to. For the relationship between adjudicator's decisions and final certificates, see *Castle Inns (Stirling) Ltd* v. *Clarke Contracts Ltd* (2005).

7.5 *Other certificates*

In addition to interim and final certificates, a number of other types of certificate are commonly found in the standard forms of building contract. These have a wide variety of different functions. The following are some examples.

7.5.1 Completion certificates

These certificates are commonly used to record when the works or sections of the works have been substantially or practically completed. This can have important consequences for a number of matters, including the start of the rectification, defects liability or maintenance period; the release of retention monies; and the end of the period for which the employer is entitled to deduct liquidated damages. Clause 2.30 of the SBC expressly provides for the issue by the Architect of a Practical Completion Certificate which should specify the day on which practical completion of the Works has taken place. In terms of the Contract Particulars, the date specified in the Practical Completion Certificate signals the beginning of the Rectification Period referred to in clause 2.38, which runs for a period of six months unless the parties have agreed otherwise. It also triggers the release of one half of the Retention in terms of clause 4.20. It is, however, specifically provided in clause 3.6 of the SBC that the Contractor remains wholly responsible for carrying out the Works in accordance with the contract whether or not a Practical Completion Certificate has been issued. The NEC3 also contains similar provisions requiring the Project Manager to certify Completion in clause 30.2, which again can trigger the release of half of the retention in terms of Secondary Option clause X16. Under the SBC/DB the position is slightly different as there is no independent architect or project manager. Accordingly in terms of clause 2.27 it is the Employer who issues a Practical Completion Statement once the works have reached practical completion.

7.5.2 Non-completion certificates

In some standard forms of building contract, as a prerequisite to the deduction of liquidated damages by the employer, the certifier needs to issue a certificate of non-completion indicating that the works are not substantially or practically complete by the date for completion agreed between the parties or any extended date thereof. The SBC provides for the issue of such a Non-Completion Certificate in clause 2.31. Such a certificate is a precondition to the deduction of liquidated damages using the mechanism set out in clause 2.32, see, for example, *Halliday Construction Ltd and Others* v. *Gowrie Housing Association Ltd* (1995). Under the SBC/DB a similar certificate is required although this is called a Non-Completion Notice and is issued by the Employer under clause 2.28 in the absence of an independent architect or project manager. In contrast there are no provisions in relation to non-completion certificates in the NEC3.

7.5.3 Partial possession certificates

Often building contracts contain provisions whereby the employer can take possession of part of the works despite the fact that the whole of the works is not yet substantially or practically complete. This often requires the issue by the certifier of a certificate or statement identifying what part or parts of the works is/are being taken into possession by the employer. This can have important consequences for liquidated damages and protection of the works, including the question of which party is responsible for insuring the works. See, for example, clause 2.33 of the SBC which provides for the issue by the Architect of a written statement identifying the part or parts taken into possession and the date when the Employer took possession. The effect of partial possession being taken upon practical completion, defects, insurance and liquidated damages in respect of the Relevant Part are set out in clauses 2.34–2.37. Almost identical provisions are to be found in the SBC/DB at clauses 2.30–2.34, though in the absence of an architect or project manager it is the Contractor who issues a notice identifying what part or parts of the Works have been taken into possession by the Employer. The NEC3 has similar provisions in clause 35 in terms of which the Employer can take over part of the works before Completion and this requires to be certified by the Project Manager.

7.5.4 Certificates of making good defects

Such certificates are regularly found in the standard forms of building contract. They are normally issued at the end of the rectification, defects liability or maintenance period when all the defects have been rectified to the satisfaction of the certifier. The issue of such a certificate often triggers the release of any remaining retention and is usually a precondition to the issue of a final certificate, see, for example, clause 4.20.3 of the SBC which provides that the Employer need release only half the Retention Percentage where work has reached practical completion but where no Certificate of Making Good has been issued. Similarly, under clause 4.15.1, the issue of the Certificate of Making Good is a precondition to the issue of the Final Certificate. The provision permitting the Certificate of Making Good to be issued is found in clause 2.39. Clause 3.6 nevertheless provides that the issue of a Certificate of Making Good does not relieve the Contractor of his responsibility for carrying out the Works in accordance with the contract. Similarly under the SBC/DB there is provision for the Employer to issue a Notice of Completion of Making Good under clause 2.36 when all defects appearing during the Rectification Period have been made good. This again triggers release of any remaining Retention in terms of clause 4.18.

The NEC3 has slightly different provisions regarding the certification of the correction of defects. Under clause 43.3 the Supervisor issues a Defects Certificate once the contractual period for correcting Defects has ended. That Defects Certificate either confirms that there are no Defects or provides a list of any uncorrected Defects. In the latter case this triggers the Employer's remedies in terms of clause 45.1 and could result in low performance damages if Option X17 has been selected. In terms of clause 43.3 the issue of the Defects Certificate does not affect the Employer's rights in relation to Defects that have not been found or notified. It does, however, appear to bring to an end the Contractor's risks in terms of clause 81 and removes the Contractor's obligation to repair loss or damage to the works in terms of clause 82.

7.5.5 Challenging such certificates

The effect of the foregoing types of certificates and any other certificates which may be provided for in a building contract depends on the express terms of the building contract in question. Such certificates may or may not have conclusive effect, though the position under most standard forms of building contract is that they do not normally have conclusive effect and are susceptible to challenge either by way of adjudication, arbitration or court proceedings.

7.6 Roles and duties of certifiers

7.6.1 Who is the certifier?

In the same way that the certification process requires to be expressly stipulated in a building contract, so should the identity of the certifier. In most standard forms of building contract the certifier is a person with relevant skill and knowledge selected from an appropriate professional discipline, for example, an architect, quantity surveyor or engineer. Rather than naming an individual, it is possible and, indeed, more common to nominate a firm or partnership as the certifier. Naming an individual can give rise to issues of *delectus personae* in the event of the illness, death or incapacity of such person, considered in Section 9.6. It is also possible to specify the certifier by making reference to the holder of a particular post.

The certifier named in the contract must be the person who issues the relevant certificates. Although he can delegate some of the detailed work, for example, the carrying out of measurements and calculations or the inspection of work, it is the certifier himself who must issue the certificates. He cannot delegate this function. Similarly, if a firm or partnership fulfils the role of certifier, then it must be a partner in the firm who issues the certificate, though, again, work of a detailed nature may be delegated to others within the partnership, see *London Borough of Hounslow* v. *Twickenham Garden Developments Ltd* (1970). Obviously the contract can provide otherwise and clause 14.2 of the NEC3, for example, makes express provision for the Project Manager and Supervisor to delegate any of their actions after notification to the Contractor.

In some cases, contracts provide that the employer is also the certifier. Such provisions are infrequent in that the contractor is unlikely to agree to them. Under the SBC/DB we do, however, see examples of the Employer exercising a certification-like function, for example, in relation to Payment Notices under clause 4.9.2, Practical Completion Statements under clause 2.27, and Notices of Completion of Making Good under clause 2.36. Such provisions are generally not to be recommended to either party to a contract in that there is an obvious potential for abuse and disputes. An employer who does fulfil the role of certifier clearly has to act honestly and fairly, failing which, his actions will be open to challenge. In addition, the courts are likely to be unsympathetic towards an employer acting as certifier, particularly where certification is a precondition of payments to the contractor or where decisions require a strong element of subjectivity. The decision of the court in *Scheldebouw BV* v. *St James Homes (Grosvenor Dock) Ltd* (2006) is a good example of the unsympathetic view taken by the courts when the employer assumes the role as certifier.

Where a certifier resigns, dies or becomes incapable of fulfilling the role of certifier, then normally the contract will stipulate the procedure for making a new appointment. The SBC, for example, provides in clause 3.5.1 that, in the event of the Architect ceasing to hold that post for the purposes of the contract, then the Employer has to nominate a replacement within 21 days. This is subject to the Contractor's right of objection within seven days of the nomination. In contrast, the NEC3 simply provides in clause 14.4 that the Employer may replace the Project Manager or Supervisor after he has given the Contractor notification of the name of the replacement. Where the contract is silent, then it is submitted that there is, nevertheless, an implied term that the employer has a right to appoint a new certifier, subject to reasonable objections from the contractor. If the employer fails to appoint a new certifier, then this may amount to a breach of contract on their part particularly in the event that certification is a precondition to rights on the part of the contractor, for example, to payment, see *Croudace Ltd* v. *London Borough of Lambeth* (1986). Similarly, where an employer fails to appoint a new certifier, then he may be prohibited from relying on the lack of certification to defend claims by the contractor.

The certifier is normally engaged by the employer and has no direct contractual link with the contractor. The certifier's rights and obligations are set out in his terms and conditions of appointment or implied by law. Usually the certifier is a member of a professional body and his conditions of appointment may be based on standard forms issued by his professional body. As the certifier is normally engaged by the employer, then he must act on the instructions of his employer and any acts or omissions on the part of the certifier can amount to a breach of his contract with the employer giving rise to a claim for damages against him. Furthermore, the certifier is normally also acting as the agent of the employer in respect of the building contract and any acts or omissions in this connection can place the employer in breach of contract with the contractor. In this respect, however, a difficult distinction requires to be drawn between the certifier's duties as agent of the employer and his duties as certifier.

7.6.2 Jurisdiction of the certifier

The jurisdiction of the certifier is defined by the express terms of the building contract. If the certifier goes beyond his jurisdiction, then any certificates issued will be invalid and open to challenge by either party, see, for example, *Hall & Tawse Construction Ltd* v. *Strathclyde Regional Council* (1989), where the certifier made deductions in a certificate which he had no authority to make in terms of the contract. This will apply even to certificates stated to have final and conclusive effect. See, for example, the English case of *Menolly Investments 3 SARL* v. *Cerep SARL* (2009). The certifier also needs to ensure that he has complied with any formal requirements for issuing certificates, particularly those which are essential preconditions and that he has complied with any time limits set out in the contract.

Once the certifier has completed his functions, then his jurisdiction falls and he is *functus officio*, having discharged his function. In some contracts this occurs when he issues the final certificate. Thereafter, he is precluded from issuing any further valid certificates, see *H Fairweather Ltd* v. *Asden Securities Ltd* (1979).

The contract can also indicate a number of other events or contingencies which will bring a certifier's jurisdiction to an end. Raising arbitration proceedings, however, will not normally make the certifier *functus officio*, see *GA Group Ltd* v. *Scottish Metropolitan Property plc* (1992). In certain instances the certifier may be disqualified. This can be a difficult area because of the dual function which certifiers are regularly required to fulfil as both agent of the employer and as independent certifier.

7.6.3 General duties in certification

As already indicated, the certifier is normally engaged by the employer and acts as his agent in respect of the building contract. Accordingly, he must have regard for his employer's interests and, in respect of many matters, act upon the instructions of his employer. When acting as certifier, however, he is under a duty to act fairly, honestly and independently. See *London Borough of Hounslow* v. *Twickenham Garden Developments Ltd* (1970), *Sutcliffe* v. *Thackrah and Others* (1974), and *Costain Ltd and Others* v. *Bechtel Ltd* (2005).

The comments of Lord Hoffman in the case of *Beaufort Developments (NI) Ltd* v. *Gilbert-Ash NI Ltd and Another* (1998) are instructive in this regard. He stated that: 'The architect is the agent of the employer. He is a professional man but can hardly be called independent.'

A certifier must apply the terms of the contract exercising his own skill and judgement, failing which, his certificates will be invalid and he will be open to disqualification. He should not be unduly influenced by or act on his client's instructions to the extent that he jeopardizes his independence and impartiality, see *Nash Dredging (UK) Ltd* v. *Kestrel Marine Ltd* (1986). Similarly, the parties are under a duty to ensure that they do not interfere with the certifier exercising his certification duties, see *Perini Corporation* v. *Commonwealth of Australia* (1969) and *Nash Dredging Ltd* v. *Kestrel Marine Ltd* (1986). A certifier does not, however, have to apply the strict rules of natural justice and he has a discretion as to how to gather information and whether to allow parties a hearing, see *The North British Railway Company* v. *William Wilson* (1911) and *London Borough of Hounslow* v. *Twickenham Garden Developments Ltd* (1970).

The application of the rules of natural justice was considered further in the English case of *Amec Civil Engineering Ltd* v. *Secretary of State for Transport* (2005). The contract between the parties in that case incorporated the Institution of Civil Engineers' Conditions of Contract (Fifth Edition). It was the role of the engineer acting under the dispute provisions in clause 66(1) of that contract which came under review. The court held that the engineer in that role was acting as a certifier and that though he was not obliged to comply with the rules of natural justice applicable to judges, he was required to act independently, honestly and fairly. Lord Justice May puts the point simply:

> The rules of natural justice are formalised requirements for those who act judicially. Compliance with them is required of judges and arbitrators and those in equivalent position, but not of an engineer giving a decision under clause 66 of the ICE Conditions ... Under clause 66, the engineer is required to act independently and honestly.

In the light of the dual role of a certifier, correspondence or discussion of matters with the employer regarding certification will not necessarily invalidate a certificate nor will the giving of advice by the certifier to the employer. The certifier can consult the employer and the employer's advisers on legal matters that arise out of or in connection with the building contract provided any response is treated solely as advice and not as a direction or instruction. Many certifiers will take independent legal advice and it is submitted that this is the most appropriate course of action as the employer's legal advisers have their own clients' interests at heart and any advice given by them may be tainted or may unduly influence the certifier.

If the certifier does lose his independence, then he will become disqualified and any certificates issued by him thereafter will be invalid. Fraud or dishonesty or taking account of unduly influential matters and advice would disqualify a certifier. Similarly, fraudulent misrepresentation on the part of either party to the contract can also invalidate certificates issued in reliance upon such a misrepresentation, see *Gray and Others (The Special Trustees of the London Hospital)* v. *T P Bennett & Son and Others* (1987) and *Ayr Road Trustees* v. *W & T Adams* (1883).

Concealed interests on the part of the certifier can also lead to disqualification. It is implied that certain interests are already known about and accepted by both parties. For example, it is known that the certifier will be paid by the employer for the work carried out by him, is liable to the employer for breach of contract and that, to an extent, he has to look after the employer's interests. Interests in the welfare of either party may, however, disqualify: for example, if the certifier is a shareholder in either of the parties or has a financial interest in the outcome of the certification process. This can lead to particular difficulties where, for example, the certifier has given advice to the employer in respect of the cost of the contract. If the certifier has guaranteed a price to the employer or will receive incentive payments for savings made, then this could lead to disqualification. Where, however, the certifier has simply estimated the cost of the job for the employer, then it is unlikely that this would amount to an interest sufficient to disqualify.

A certifier may also be disqualified if his dual roles as employer's agent and certifier become incompatible. This may occur where the certifier becomes a witness to fact or a key witness in support of one party's position.

A certifier is not disqualified simply because of an error in judgement or an error in the exercise of his discretion. Where such errors occur, then it may be possible for one party to the contract to challenge the certificate by means of adjudication, arbitration or through the courts. There is authority which suggests that there is no obligation on an employer who becomes aware of errors on the part of the certifier to bring them to his attention and ensure that he adequately performs his duties, see *Lubenham Fidelities and Investments Co. Ltd* v. *South Pembrokeshire District Council and Another* (1986). This decision should be contrasted with the decisions in *Perini Corporation* and *Panamena Europea Navigación Compañía Limitada* v. *Frederick Leyland & Co. Ltd* (1947).

Failures by the certifier can place him in breach of contract with his employer and may result in the certifier being liable to the employer in damages, see, for example, *Jameson* v. *Simon* (1899) where the certifier was found liable in damages to his employer for certifying work which did not conform to the contract as a result of failing to exercise reasonable supervision. See also *Sutcliffe* v. *Thackrah and Others* (1974) and *Atwal Enterprises Ltd* v. *Donal Toner Associates* (2006).

Such failures can also place the employer in breach of contract with the contractor though the case of *Karl Construction Ltd* v. *Palisade Properties plc* (2002) suggests that a failure to certify by the certifier will not provide the contractor with a right to damages from the employer but simply the right to arbitrate or sue for payment of any sums that would have been due had a certificate been issued. The certifier himself will generally have no liability to the contractor as there is no direct contractual link between certifier and contractor. The certifier could only be liable to the contractor in circumstances where he has procured the employer to breach his contract with the contractor, see *John Mowlem & Co. plc* v. *Eagle Star Insurance Co. Ltd and Others* (1992), or in delict. A simple under-certification may not generally be sufficient to give rise to delictual liability, see, for example, *Pacific Associates and Others* v. *Baxter* (1988), *Lubenham Fidelities and Investments Co. Ltd*, and *Leon Engineering & Construction Co. Ltd* v. *Ka Duk Investment Co. Ltd* (1989), though a warning is to be found in the case of *William McLaughlan and Another* v. *Keith Edwards* (2004). Although having no direct contractual link with the ultimate purchasers of a house, an architect was found to owe the purchasers a duty to take reasonable care in the provision of architectural services in connection with the design and construction of the house and a duty to take reasonable care to ensure that the statements contained in the certificates which he issued were true and accurate. A similar warning can be found in the more recent English case of *Hunt & Others* v. *Optima (Cambridge) Ltd & Others* (2013) in which an architect was held to owe a duty of care to a landlord with whom they had no direct contractual link in relation to inspections carried out and certificates issued confirming that work had been carried out to a satisfactory standard. In light of such authorities, certifiers should be aware that they may, depending on the particular facts and circumstances, owe duties to those other than their employer.

Chapter 8
Payment

8.1 *Contractual payment*

8.1.1 Introduction

One of the main obligations owed by the employer under a building contract is to make payment to the contractor for the work carried out by them. The building contract will ordinarily contain express provisions relating to payment and the parties are, subject to the provisions of the 1996 Act (as amended), free to agree between them the sum that is to be paid for the works, whether instalment payments are to be made, when payments are to be made, and the mechanism or procedure to facilitate payment.

The commonly used standard forms of building contract contain detailed provisions in respect of payment, see, for example, clause 4 of the SBC and core clause 5 of the NEC3 (as supplemented by Option Y(UK)2).

8.1.2 When are payments due?

Traditional Scots law position

In any building contract one of the contractor's prime concerns is the timing of payments for work carried out. Traditionally under Scots law contractors have no implied right to payment until they have completed all the work they have contracted to carry out, see *Muldoon* v. *Pringle* (1882) and *Readdie* v. *Mailler* (1841). It is a general principle of Scots law that, in the absence of any provisions in a contract providing for interim or instalment payments, there is no obligation to make payment until the entire contract has been fulfilled. Whether or not this applies to building contracts was considered in the case of *Charles Gray & Son Ltd* v. *Stern* (1953). In this case, the contractor demanded payments to account while building a house, maintaining that it was normal building trade practice to pay contractors for nine-tenths of the work completed. The contract did not expressly provide for interim payments to be made. The court held that the contract was a lump sum contract and that, as the contractors had failed to carry out a material part of the works, they could not sue for payment under the contract.

MacRoberts on Scottish Construction Contracts, Third Edition. MacRoberts.

Exceptions to the basic principle

However, while it may be the general principle that contractors are not entitled to payment from an employer before they have carried out all their obligations under the contract, there are situations in which the contractor may be entitled to payment notwithstanding the fact that they have failed to complete.

In certain instances, if the work can be said to be 'substantially complete', the contractor may be entitled to payment of the whole contract price, less payment for that part not completed or, alternatively, the cost of remedying any defects in the works. In *Ramsay & Son* v. *Brand* (1898), Lord President Robertson stated that:

> A building contract by specification necessarily includes minute particulars, and the law is not so pedantic as to deny action for the contract price on account of any and every omission or deviation. It gives effect to the principle [that if builders choose to depart from the contract, they lose their right to sue for the contract price] by deducting from the contract price whatever sum is required to complete the work in exact compliance with the contract.

Similar authority can be found in the cases of *Speirs Ltd* v. *Petersen* (1924), *Hoenig* v. *Isaacs* (1952) and *Stewart Roofing Co. Ltd* v. *Shanlin* (1958).

Following the case of *Forrest* v. *The Scottish County Investment Company Ltd* (1915), it appears that the contractor's right to payment where there are defects may be stronger in a measurement contract than a lump sum contract (see Section 8.1.8 for an explanation of these different types of contract). This approach also receives some support in *Speirs Ltd*.

If, however, the contractor's deviation from the contract is material, then he will be prevented from claiming payment under the contract, see *D Ramsay & Son* and also *Dakin & Co. Ltd* v. *Lee* (1916). Unfortunately, it is not always clear whether a deviation is material and this will require to be considered in the light of the facts of each individual case, see, for example, *McMorran* v. *Morrison & Co.* (1906). This can include consideration of the dimension, complexity and value of the contract and consideration of the cost of rectifying the deviation relative to the whole contract price, see *Speirs Ltd*. On the other hand, if the employer has benefited from the work carried out by the contractor, then the contractor may still be able to claim compensation under the principle of *quantum lucratus* even where there are material deviations. A fuller discussion of this principle is contained in Section 8.5.

Where the contractor is prevented from completing the contract works by a matter outwith their control, then he may still be able to claim payment for the work carried out. For example, where the employer denies the contractor access to the site to complete the contract works, then the contractor will be able to claim payment for the work he has carried out. Similarly, the contractor will be entitled to claim payment for all work executed in accordance with a contract in the situation where the works are destroyed and cannot be completed, see, for example, the cases of *Andrew McIntyre and Company* v. *David Clow and Company* (1875) and *Richardson* v. *County Road Trustees of Dumfriesshire* (1890). This may, however, depend upon which party has assumed the risk of damage to the works during their construction and careful consideration may need to be given to any insurance provisions found in the contract. Insurance is considered in Chapter 14.

Finally, and most importantly, the contractor may be entitled to claim payment prior to completing the contract works where the express terms of the building contract specifically provide for the making of instalment or interim payments. In modern construction contracts such provisions are almost always included, as many contractors would be unable to fund the on-going construction costs pending payment upon completion. Procuring external funding would significantly increase construction costs.

8.1.3 Payment under the 1996 Act

In respect of building contracts entered into after 1 May 1998, the position in relation to payment altered dramatically following the coming into force of the 1996 Act. The 1996 Act materially improved the rights of a party to be paid under a construction contract as compared to the common law position. Part 8 of the Local Democracy Economic Development and Construction Act 2009 ('the 2009 Act') made amendments to the 1996 Act that apply to construction contracts entered into in Scotland from 1 November 2011 and in England from 1 October 2011.

In the following sections of this chapter, where reference is made to the 1996 Act, it is a reference to the 1996 Act as amended by the 2009 Act.

The amendments to the 1996 Act have made sweeping changes to the payment provisions in the 1996 Act. The amendments have created a notice-driven interim payment mechanism for construction contracts, seeking to create certainty in respect of interim payments by prescribing that the payer must pay the payee the 'notified sum' on or before the final date for payment.

Sections 109 and 110

Section 109 of the 1996 Act deals with payment and provides that a party to a relevant construction contract is entitled to payment by instalments, stage payments or other periodic payments in respect of work carried out under the contract. Where the parties are entitled to interim payments under section 109, then the 1996 Act provides that the parties are free to agree the amounts of such payments and the intervals or circumstances in which they will become due.

Section 110(1) of the 1996 Act stipulates that every construction contract must provide an adequate mechanism for determining what payments become due and when they become due and provide a final date for payment in relation to any sum which becomes due. The parties can, however, agree the interval between the date on which a sum becomes due and the final date for payment.

Subsections (1A)–(1D) of section 110 of the 1996 Act provide details of the means by which parties can, and cannot, satisfy the requirement that every construction contract must provide an adequate mechanism for determining what payment becomes due and when.

Subsection (1A) provides that the underlying requirement to provide an adequate mechanism for payment under section 110(1) is not satisfied where a construction contract makes payment conditional on (i) the performance of obligations under another contract, or (ii) a decision by any person as to whether obligations under another contract have been performed (though subsection (1B) provides that the reference to 'obligations' in subsection (1A) does not apply to obligations to make payment).

Subsection (1C) provides that subsection (1A) does not apply where (i) the construction contract is an agreement between the parties for the carrying out of construction operations by another person, whether under sub-contract or otherwise; and (ii) the obligations referred to in subsection (1A) are obligations on that other person to carry out those operations (see Section 8.1.6).

Subsection (1D) provides that the requirement to provide an adequate mechanism for payment under section 110(1) is not satisfied where a construction contract provides that the date on which a payment is to become due is to be determined by reference to the date on which a notice is given to the person to whom the payment is due, identifying what payments the payer considers are due under the contract. This means that provisions which state that the 'due' date is identified by the date upon which a payment notice is issued to the payee (by the payer), do not amount to an adequate payment mechanism and are ineffective.

The Scheme for Construction Contracts (Scotland) Regulations 1998 as amended by the Scheme for Construction Contracts (Scotland) Amendment Regulations 2011

Should the parties fail to provide for all or any of the matters which are required in terms of section 109 and/or section 110 (including section 110A and/or section 110B – see Section 8.1.7), then the payment provisions found in Part II of the Schedule to the Scheme for Construction Contracts (Scotland) Regulations 1998 as amended by the Scheme for Construction Contracts (Scotland) Amendment Regulations 2011 will apply to the extent required to cover any matter not otherwise agreed, see *Hills Electrical & Mechanical plc* v. *Dawn Construction Ltd* (2004).

The Schedule deals first of all with 'relevant construction contracts', being any construction contract other than one which specifies that the duration of the work is to be less than 45 days or in respect of which the parties agree that the duration of the work is estimated to be less than 45 days. For the definition of a 'construction contract' for the purposes of the 1996 Act, see Section 1.2.2.

Paragraph 4 of the Scheme provides that interim payments under a relevant construction contract will become due when the payee has made a claim and seven days have expired following the relevant period. Paragraph 12 provides that the 'relevant period' is as specified in or calculated by reference to the contract, failing which, it is a period of 28 days.

Paragraph 5 of the Scheme provides that the final payment under a relevant construction contract, namely the difference between the contract price and the aggregate of any interim payments, will be due when the payee has made a claim and 30 days have expired following completion of the work.

In respect of contracts where the duration of the work is less than 45 days or where the parties agree that the duration of the work is estimated to be less than 45 days, then paragraph 6 of the Schedule provides that payment of the contract price shall become due when the payee has made a claim and 30 days have expired following completion of the work. Paragraph 7 provides that any other payment under a construction contract will be due when the payee has made a claim and seven days have expired following the completion of the work to which the payment relates.

Paragraph 8 provides that the final date for any payment under a construction contract will be 17 days from the date when the payment becomes due.

8.1.4 Payment under the SBC

The payment provisions in the SBC are to be found in clause 4. The payment provisions have been amended in the 2011 editions in order to comply with the 1996 Act, as amended. The previous editions are no longer 1996 Act compliant.

Clause 4.9 provides that the due dates for interim payments are to be the monthly dates specified in the Contract Particulars up to Practical Completion and bi-monthly thereafter. Interim valuations are made by the Quantity Surveyor whenever the Architect considers them necessary for the purpose of ascertaining the amount to be stated as due in an Interim Certificate (clause 4.10.2). Clause 4.11.1 provides that the Contractor may issue an Interim Application to the Quantity Surveyor not less than seven days before the due date, stating the sum the Contractor considers will be due at the due date and the basis on which that sum has been calculated. Clause 4.10 provides that Interim Certificates are to be issued by the Architect/Contract Administrator not later than 5 days after the due date, stating the sum that the Architect/Contract Administrator considers to be due at the due date and the basis on which that sum has been calculated.

In terms of clause 4.11.2, if an Interim Certificate is not issued in accordance with clause 4.10, then in circumstances where the Contractor has issued an Interim Application in accordance with clause 4.11.1 the Contractor's Interim Application will be treated as an Interim Payment Notice. Where the Contractor has not issued an Interim Application, the Contractor can issue an Interim Payment Notice stating the sum the Contractor considers to be due at the due date and the basis on which that sum has been calculated (clause 4.11.2.2).

Clause 4.12 provides that the final date for payment is 14 days from the due date. Where an Interim Payment Notice is issued by the Contractor due to a failure by the Architect/Contract Administrator to issue an Interim Certificate, the final date for payment is postponed by the same number of days as the number of days after expiry of the 5-day period for the Interim Certificate that the Interim Payment Notice is given (clause 4.12.4).

Clause 4.12.5 provides that if the Employer intends to pay less than the sum in the Interim Certificate or Interim Payment Notice, a Pay Less Notice is required to be issued no later than 5 days before the final date for payment. Clause 4.13 stipulates what is required of a Pay Less Notice and provides that it must state the sum the Employer considers to be due at the date the notice is given and the basis on which that sum has been calculated. The Pay Less Notice can be issued by the Employer or on its behalf by the Architect, Quantity Surveyor, the Employer's representative or other person notified by the Employer as being authorized to do so (clause 4.13.1.1).

Clause 4.12 provides that the sum payable by the final date for payment is that stated as due in the Interim Certificate or, where there is none, the sum stated in the Contractor's Interim Payment Notice, or where a Pay Less Notice is issued, the sum stated as due in that Pay Less Notice. Figures 8.1, 8.2 and 8.3 depict how the interim payment provisions under the SBC operate in the different circumstances described above.

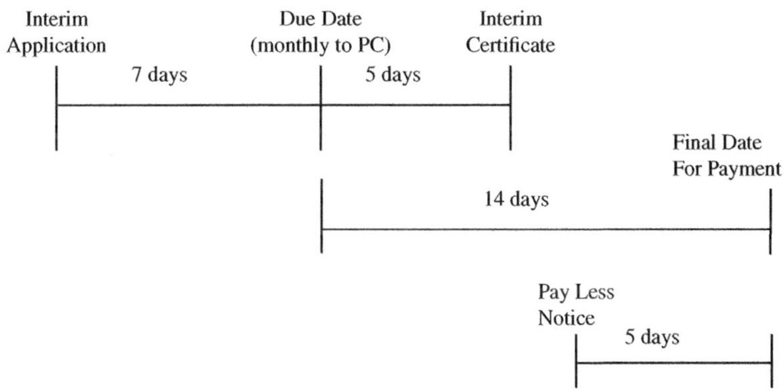

Figure 8.1 Payment timeline: Interim Certificate issued by Architect/Contract Administrator.

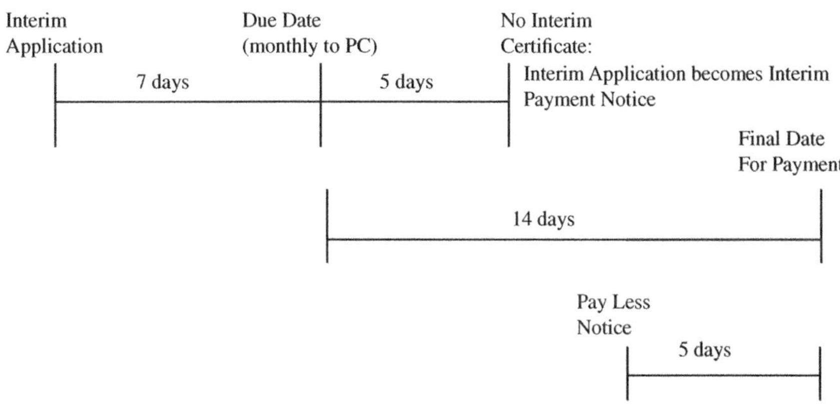

Figure 8.2 Payment timeline: no Interim Certificate issued, Contractor has issued Interim Application.

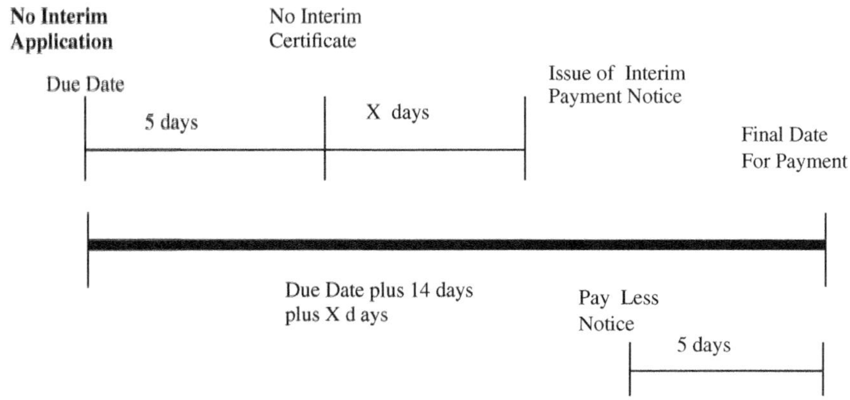

Figure 8.3 Payment timeline: no Interim Certificate issued, Contractor has not issued Interim Application but subsequently issues Contractor Payment Notice.

Failure by the Employer to make payment by the final date for payment entitles the Contractor to simple interest at the rate of 5% above the Bank of England base rate on the amount not paid in terms of clause 4.12.6.

The timing of the final payment to be made to the Contractor under the SBC is also governed by the provisions of clause 4. Clause 4.5 stipulates that, not later than six months after the issue of the Practical Completion Certificate or last Section Completion Certificate, the Contractor shall provide the Architect/Contract Administrator or (if so instructed) the Quantity Surveyor, with all documents necessary for the purposes of adjustment of the Contract Sum. The Architect/Contract Administrator, or, if he instructs, the Quantity Surveyor, then has three months to prepare a statement of all adjustments to be made to the Contract Sum and to ascertain any loss and/or expense due to the Contractor. This statement and ascertainment have to be sent to the Contractor forthwith upon preparation. In terms of clause 4.15, once the statement and ascertainment have been sent to the Contractor under clause 4.5, the Rectification Period has ended and a Certificate of Making Good has been issued, then the Architect shall not later than two months thereafter issue a Final Certificate. The Final Certificate should state the final adjusted Contract Sum and the sum of the amounts already stated as due in Interim Certificates together with any advance payment. The difference between the two is expressed as a balance due to the Contractor by the Employer or vice versa. Clause 4.15.3 provides that the due date of the final payment is the date of issue of the Final Certificate, or if the Final Certificate is issued outwith the two-month period referred to in clause 4.15.1, the last day of that two-month period. Clause 4.15.3 further provides that the final date for payment of this balance is 28 days from the due date. As with the provisions in respect of interim payments, any Pay Less Notice in respect of the final payment requires to be issued no later than five days before the final date for payment (clause 4.15.4).

If no Final Certificate is issued by the Employer in accordance with clauses 4.15.1 and 4.15.2, the Contractor may issue a Final Payment Notice stating what sum the Contractor considers to be the amount of the final payment due to him and the basis on which this has been calculated (clause 4.15.6.1). If the Contractor gives a Final Payment Notice, the final date for payment of the sum specified in it shall be postponed by the same number of days as the number of days after expiry of the two-month period that the Final Payment Notice is given (clause 4.15.6.2). Subject to any Pay Less Notice issued by the Employer, that Final Payment Notice shall set the amount of the final payment. If the Employer intends to pay less than the amount specified in the Final Certificate, or Final Payment Notice, a Pay Less Notice is required, and it must be given no later than five days before the final date for payment (clause 4.15.6.3).

8.1.5 Payment under the SBC/DB

The payment provisions in the SBC/DB are also to be found in clause 4. The payment provisions are broadly similar to those under the SBC, but there are differences in terminology, particularly as there is no provision for certification by an Architect/Contract Administrator. SBC/DB provides the parties with two alternatives for interim payments, either by way of specified 'Stage Payments' or 'Periodic Payments', which are monthly from a specified date.

Clause 4.8.1 of the SBC/DB *obliges* the Contractor to make an Interim Application to the Employer stating the sum the Contractor considers to be due to it and the basis on which that sum has been calculated. Where Alternative A (stage payments) applies, an Interim Application is required on completion of each stage and, following completion of the last stage, at two-monthly intervals. The due date would be the later of the date for completion of the stage and the date of receipt by the Employer of the Interim Application.

Where Alternative B (Periodic Payments) applies, Interim Applications are to be made monthly at the dates specified in the Contract Particulars until practical completion and two-monthly thereafter. The due date would be the later of the specified date or receipt of the Interim Application by the Employer.

Clause 4.9.1 provides that the final date for payment of an Interim Payment shall be 14 days from the due date. Clause 4.9.2 provides that not later than five days after the due date, the Employer is to give a Payment Notice to the Contractor specifying the sum it considers to be due at the due date and the basis upon which that sum has been calculated. In terms of clause 4.9.3, if an Interim Payment Notice is not issued by the Employer in accordance with clause 4.10, then the amount of the Interim Payment to be made by the Employer will be the sum stated as due in the Interim Application (but subject to any Pay Less Notice).

Clause 4.9.1 provides that the final date for payment is 14 days from the due date. Clause 4.9.4 provides that if the Employer intends to pay less than the sum in the Payment Notice or Interim Application, as the case may be, a Pay Less Notice requires to be issued no later than five days before the final date for payment. Clause 4.10 stipulates what is required of a Pay Less Notice and, as with the terms of the SBC, provides that it must state the sum the Employer considers to be due at the date the notice is given and the basis on which that sum has been calculated. The Pay Less Notice can be issued by the Employer or any person whom the Employer notifies the

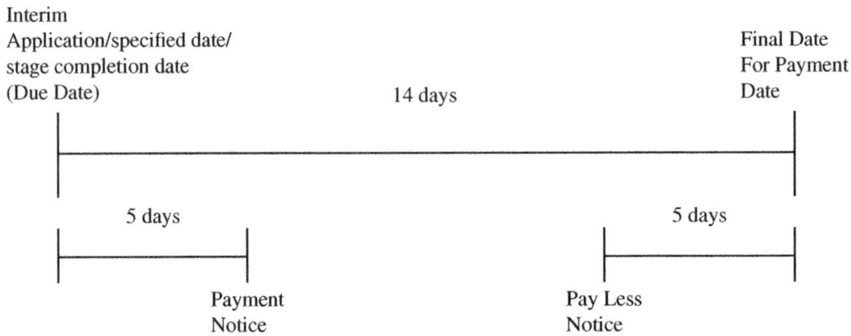

Figure 8.4 Payment timeline: Periodic Payments; Interim Application submitted after specified date.

Contractor as being authorized to do so (clause 4.10.3). Figure 8.4 depicts how the interim payment provisions under SBC/DB operate.

The timing of the final payment to be made to the Contractor under the SBC/DB is also governed by the provisions of clause 4. Clause 4.12.1 stipulates that, following practical completion of the Works the Contractor shall submit the Final Statement to the Employer and supply him with such supporting documents as he may reasonably require. If the Contractor does not submit the Final Statement within three months of practical completion of the Works, the Employer may give notice that unless that statement is submitted within two months from the date of the notice the Employer may himself issue a an Employer's Final Statement.

Clause 4.12.5 provides that the due date of the final payment is the date one month after whichever of the following occurs last:

(a) the end of the Rectification Period in respect of the Works;
(b) the date stated in the Notice of Completion of Making Good; or
(c) the date of submission to the other Party of the Final Statement or, if issued first, the Employer's Final Statement.

Clause 4.12.7 further provides that the final date for payment is 28 days from the due date. Not later than five days after the due date, the party, by whom the Final Statement shows a payment is to be made (the paying party) is obliged to give a Payment Notice to the other party. Subject to any Pay Less Notice, the payment to be made on or before the final date for payment is the sum stated in the Payment Notice. If the payer intends to pay less than the sum stated in the Payment Notice or, in default of such notice, less than the amount stated in the relevant statement, the paying party must issue a Pay Less Notice. As with the provisions in respect of interim payments, any Pay Less Notice in respect of the final payment requires to be issued no later than five days before the final date for payment (clause 4.12.8).

8.1.6 Payment under the NEC3

The six main Options under the NEC3 are based on different mechanisms for assessing payments due to the Contractor, and provide different bases of allocation of price risk between the Employer and Contractor. An analysis of the various options and methods of assessing the payments due to the Contractor is beyond the scope of this chapter; however, these main options are as follows:

Option A: Priced contract with activity schedule
Option B: Priced contract with bill of quantities
Option C: Target contract with activity schedule
Option D: Target contract with bill of quantities
Option E: Cost reimbursement contract
Option F: Management contract.

Notwithstanding the different options available, the core payment provisions in the NEC3 are to be found in core clause 5. Where the contract is a construction contract in terms of the 1996 Act, Option Y(UK)2 of the NEC3 applies. Core clause 5 requires to be read in conjunction with Option Y(UK)2 as the latter seeks to make the payment provisions compliant with the 1996 Act. This section will consider the payment provisions of the NEC3 where Option Y(UK)2 is applicable.

Clause 51.1 provides that the Project Manager certifies a payment within one week of each assessment date. Part one of the Contract Data should identify the starting date, at section 3, and the assessment interval at section 5. The first assessment date must fall within this period, though the exact date is within the discretion of the Project Manager. Further assessment dates occur at the end of each assessment interval thereafter.

Clause 51.2 provides that each certified payment is made within three weeks of the assessment date. Clause Y2.2 of Option Y(UK)2 provides that the date on which payment becomes due is seven days after the assessment date, while the final date for payment is 14 days (or such other period as stated in the Contract Data) after the date on which payment becomes due. Clause Y2.2 of Option Y(UK)2 also provides that the Project Manager's certificate is the notice of payment to the Contractor specifying the amount due at the payment due date and stating the basis on which the amount was calculated. No obligation is imposed upon the Contractor to make an application for payment to the Project Manager, but clause 50.4 permits the Contractor to submit an application for payment, and accordingly, if the Project Manager fails to certify within the relevant time period, the amount stipulated in the Contractor's application for payment would become the notified sum and be payable (subject to the issue of a Pay Less Notice). Clearly it would be good practice for a Contractor to submit an application for payment in accordance with clause 50.4.

Clause Y2.3 of Option Y(UK)2 provides that if either party intends to pay less than the notified sum, he must notify the other party not later than seven days (the prescribed period) before the final date for payment by stating the amount considered to be due and the basis on which that sum is calculated. Parties are free to revise the seven-day notice period, so long as a prescribed period is retained.

Unlike the SBC and the SBC/DB, the NEC3 does not provide for a final payment process; rather the Project Manager can assess changes to the amount due at completion of the works, as and when required.

8.1.7 Pay when paid and pay when certified

Prior to the coming into force of the 1996 Act, the timing of payments under building contracts was often stipulated to be dependent upon the receipt of funds by the paying party from another source. The most common example was for main contractors to make it a condition of a sub-contractor's entitlement to payment that the main contractor had in turn received payment from the employer under the main contract, see *Taymech Ltd* v. *Trafalgar House Construction (Regions) Ltd* (1995). Such provisions, commonly known as 'pay when paid' clauses, were outlawed by the 1996 Act, except in very limited circumstances.

Section 113 provides that any contractual provision which makes payment under a construction contract conditional on the payer receiving payment from a third person is ineffective unless that third person is insolvent. Perhaps unsurprisingly, many contractors include provisions in their construction contracts that seek to rely upon the insolvency exception contained in section 113 of the 1996 Act. The case of *William Hare Ltd* v. *Shepherd Construction Ltd* (2009) is, however, a salutary lesson that where a party wishes to rely upon the insolvency exclusion contained in section 113 of the 1996 Act, care should be taken to specify in the contract the precise situations or events that are to define that third party's insolvency. If, as in this case, a particular insolvency event or situation arises that is not covered by the contractual provisions, the court is unlikely to confer the benefit of the carve-out from section 113 of the 1996 Act. Where a payment provision is ineffective, then the relevant provisions found in Part II of the Schedule to the Scheme will apply.

Prior to the amendments to the 1996 Act, it was open to parties to a building sub-contract to agree to a provision that made payments to a sub-contractor conditional upon the value of work carried out by the sub-contractor being valued and included in a certificate issued under the main contract, though the English case of *Midland Expressway Ltd* v. *Carillion Construction Ltd (No. 2)* (2005) cast some doubt on this. The amendments to the 1996 Act introduced by the 2009 Act have put the issue beyond doubt. As mentioned earlier, subject to limited exceptions, section 110 (1A) of the 1996 Act provides that a construction contract will not provide an adequate mechanism for payment, as required by section 110, where payment is conditional on (i) the performance of obligations under another contract or (ii) decisions by a third party that obligations under another contract have been performed (e.g.. a certificate of making good defects under a main contract).

However, the Construction Contracts (Scotland) Exclusion Order 2011 expressly excluded contracts pursuant to which a party to a relevant contract has sub-contracted to a third party some or all of its obligations under that contract to carry out, or arrange that others carry out, construction operations. In effect, the Exclusion Order excludes section 110 (1A) from operating in the PFI sphere, in relation to contracts between a project company and construction sub-contractor. The exclusion does not, however, apply to the tier of contracts beneath the project company and construction contractor.

While section 110 (1A) was intended to improve cash flow for sub-contractors, main contractors have sought to dilute the impact of the section. Rather than agreeing that payment is to be conditional on payment or some other event up the contractual chain, contractors are seeking to include contractual provisions which provide that payment is due on a specific date a considerable period into the future, thus reducing the risk for the contractor of paying the sub-contractor, while not being in receipt of funds from the employer.

8.1.8 The amount to be paid

The final contract price, or a mechanism for ascertaining the final contract price, is a fundamental and essential part of the contract and should be agreed at the time the parties enter into the contract, see *Uniroyal Ltd* v. *Miller & Co. Ltd* (1985). If no price or mechanism has been agreed, then it may be possible for a contractor to obtain payment on the basis of *quantum meruit*. This is discussed in Section 8.4. The price to be paid by the employer to the contractor for carrying out the contract works may be ascertained by a number of different methods. In many building contracts it may not be possible to calculate the contract price until after completion of the works, see, for example, *Arcos Industries Pty Ltd* v. *The Electricity Commission of New South Wales* (1980).

Lump sum contracts

In what are commonly known as lump sum contracts the employer and the contractor agree the price for the contract works at the time of entering into the contract. Assuming the contractors complete the contract works, then they will be entitled to be paid the agreed price regardless of what the works have actually cost to construct. See, for example, *Mitchell* v. *Magistrates of Dalkeith* (1930). Even in lump sum contracts, however, the price can alter as a result of a number of matters including additions to or omissions from the contract works, events giving rise to loss and expense, and fluctuations in cost.

An error on the part of the contractor in calculating the price will not, however, result in an alteration to the price and the contractor is bound by the price even if the work costs more than they allowed for, see *Seaton Brick and Tile Company Ltd* v. *Mitchell* (1900). A contractor will only be entitled to additional payment if he has a contractual entitlement thereto.

Under the SBC and the SBC/DB, the Employer and Contractor agree a Contract Sum at the time of entering into the contract and Article 2 provides that the Employer shall pay to the Contractor the Contract Sum or such other sum as shall become payable in accordance with the conditions of contract. Clause 4.2 further provides that the Contract Sum shall not be adjusted other than in accordance with the express provisions of the conditions and, subject to clause 2.14, any error (whether arithmetic or not) in the computation of the Contract Sum shall be deemed to have been accepted by the parties. Under the SBC, clause 2.14.1 provides for the correction of errors in the preparation of the Contract Bills (which may be errors of quantity or description) and such correction is treated as a Variation.

Under the SBC/DB, clause 2.14.1 provides that where the error or discrepancy is in the Contractor's Proposals, its correction is without cost to the Employer. However, in terms of clause 2.14.2, where the discrepancy or error is in the Employer's Requirements and the Contractor's Proposals do not deal with the error/discrepancy, its correction is treated as a Change. Under the SBC, the provisions in the conditions governing final adjustment of the Contract Sum are found in clause 4.3 (clause 4.2 under the SBC/DB). A fuller discussion of the matters which can give rise to adjustment of the Contract Sum is contained in Section 8.2.

Measurement contracts

In what are commonly known as measurement contracts, no agreed price is ascertainable prior to the carrying out of the works. The price is calculated by measurement of the work actually carried out during the currency of and on completion of the contract and this work is then valued by applying a schedule of rates agreed at the time of entering into the contract. The schedule of rates will often take the form of a bill of quantities. Very basic forms of measurement contracts can be seen in the cases of *Jamieson* v. *McInnes* (1887) and *Wilkie* v. *Hamilton Lodging-House Company Ltd* (1902). Measurement contracts are often used where it is impossible to ascertain the full extent of the contract works at the time of contracting, or where there is insufficient information available to do so. For example, in a contract to construct a road, it may not be possible, at the time of contracting, to ascertain the exact ground conditions that the contractor will encounter.

Reimbursement contracts

A third method often employed to ascertain the price to be paid is that used in what are commonly known as reimbursement contracts. In such contracts contractors are paid for the cost of the work carried out by them, normally with an additional allowance for overheads and profit or, alternatively, a fee for managing the contract. The additional allowance to be paid to the contractor is normally agreed at the time of entering into the contract and is usually a specified sum or, alternatively, an agreed percentage of the total contract price. It is important when drafting reimbursement contracts to ensure that the contractor only recovers costs properly and reasonably incurred in order to avoid extravagance or inefficient working on their part.

Interim payments

Where the contractor has an entitlement to receive interim payments the contract will normally contain a mechanism for ascertaining the amount of such payments. This often involves the issue of interim certificates, discussed in Section 7.3. Interim payments are payments to account of the final contract sum and normally represent either an agreed instalment payment due on a particular date, or an agreed instalment due at completion of a particular stage of the works (see, for example, *The Government of Newfoundland* v. *The Newfoundland Railway Company and Others* (1888)), or a valuation of the work carried out at a particular date (see, for example, *F R Absalom Ltd* v. *Great Western (London) Garden Village Society Ltd* (1933)).

Interim payments are often subject to review by later payments and on completion of the contract, see, for example, *The Tharsis Sulphur and Copper Company Ltd* v. *McElroy & Sons* (1878) in which Lord Chancellor Cairns stated that 'payments made under [interim certificates] are altogether provisional, and subject to adjustment or readjustment at the end of the contract'. See also *Beaufort Developments (NI) Ltd* v. *Gilbert-Ash NI Ltd and Another* (1998), *RBG Ltd v SGL Carbon Fibers Ltd* (2010) and *SGL Carbon Fibres Ltd v RBG Ltd* (2012). Thus interim certificates are not usually conclusive as to either the value of work carried out at the date of payment or the quality of work carried out.

It has, however, been suggested that it is for the employer to prove that an interim valuation is inaccurate. In *Johnston* v. *Greenock Corporation* (1951) Lord Sorn stated:

> [T]he proper time for an employer to challenge any item in the contractor's monthly account is at the time he receives it, and when he is checking it with a view to payment. That is the time when the facts are fresh in the contractor's mind and when he can best give explanations, or make the necessary inquiries or investigation into any matter which requires explanation. At that stage the onus is clearly on the contractor to justify and explain every item in his account if called upon to do so. But if the accounts are checked, and all explanations asked for having been satisfactorily given, payment is made on them, so that the contractor naturally thinks that the business is over and done with, and then, after the lapse of a year or two, the employer seeks to reopen particular items in the account I think the situation is different ... [W]here objections can be so infinitely varied in character, it may not be advisable to attempt to lay down any general rule about onus, but it is at least clear that great care must be taken to see that the contractor is not prejudiced by the delay in bringing forward the challenge. Perhaps it would not be going too far to say that, instead of it being for the contractor to justify his charge, it is, at least initially, for the employer to show why the item, which he had already passed and paid for, should not stand.

This view appears to be consistent with the later decision in *SGL Carbon Fibres Ltd v RBG Ltd* (2012) though if it is the contractor seeking further payment, then he will have the burden of proving that the valuation is too low.

Section 110(1)

Section 110(1) of the 1996 Act stipulates that every construction contract must provide an adequate mechanism for determining what payments become due under the contract. If the contract fails to do so, then the relevant provisions of Part II of the Schedule to the Scheme will apply. Paragraph 2(1) of the Scheme provides that the amount of any interim payments shall be the difference between the amount determined in accordance with paragraph 2(2) and the amount determined in accordance with paragraph 2(3).

The amount determined in accordance with paragraph 2(2) is the aggregate of:

- an amount equal to the value of any work performed in accordance with the contract from commencement of the contract to the end of the relevant period;
- where the contract provides for payment for materials, an amount equal to the value of any materials manufactured on site or brought onto the site from commencement of the contract to the end of the relevant period; and
- any other amount which the contract specifies shall be payable from commencement of the contract to the end of the relevant period.

The relevant period is defined in paragraph 12 as the period specified in, or calculated by reference to, the construction contract or (where no period is specified or so calculable) a period of 28 days. The amount determined in accordance with paragraph 2(3) is the aggregate of any sums which have been paid or are due for payment by way of interim payments during the period from commencement of the contract to the end of the relevant period. It is further provided in paragraph 2(4) that the amount of any interim payment shall not exceed the difference between the contract price and the aggregate of the interim payments which have become due.

Section 110A

Section 110A of the 1996 Act provides that for every payment provided for by the construction contract, a payer notice or a payee notice must be given not later than five days after the payment due date. In order for a notice to comply with section 110A it must specify the sum that the payer/specified person (payer notice) or payee (payee notice) considers to be or to have been due at the payment due date and the basis on which that sum is calculated.

A payer or payee notice must be issued even if the sum that is considered due is zero. To the extent that the construction contract does not comply with section 110A, the relevant provisions of the Scheme apply.

Section 110B

Section 110B applies in a case where the construction contract provides for the payer, or a specified person to give a payer notice not later than five days after the payment due date, but fails to do so. In those circumstances, section 110B allows the payee to give the payer a payee notice at any time after the date on which the payer notice should have been issued (a 'payee notice in default'). Where a payee issues a payee notice in default, the final date for payment of the sum specified in the notice will be postponed by the same number of days after the date upon which the payer notice should have been issued that the payee notice in default was issued. Figure 8.5 shows the timing of the various notices referred to above.

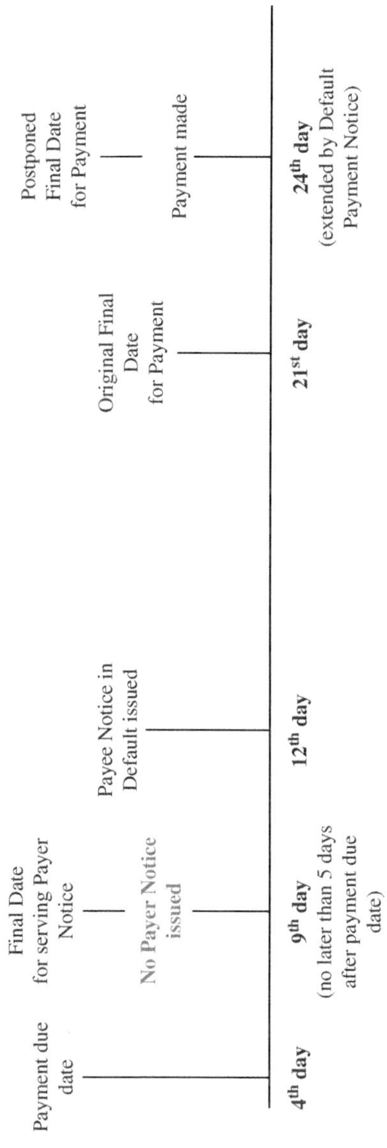

Figure 8.5 Payee notice in default.

Section 110B (4) provides that where the construction contract requires the payee (before the date on which the payer notice is to be given) to notify the payer or a specified person of the sum that the payee considers will become due on the payment due date and the basis upon which that sum is calculated, then if that notification is given and the payer or specified person fail to issue a payer notice, that notification is to be treated as a payee notice in default and no further payee notice in default is required. In those circumstances the final date for payment will not be postponed.

Section 111

Section 111 provides that the payer must pay the 'notified sum'. The notified sum is defined in section 111(2) as the amount specified in (i) a payer notice; (ii) a payee notice; or (iii) a payee notice in default. The payer is obliged to pay the notified sum by the final date for payment, subject only to the issue of a notice of the payer's intention to pay less than the notified sum (a 'Pay Less Notice') as provided for in section 111(3) of the 1996 Act. A Pay Less Notice under section 111(3) must specify:

- the sum that the payer considers to be due on the date the notice is served; and
- the basis on which that sum is calculated.

The Pay Less Notice must be given no later than the prescribed period before the final date for payment and must not be given before the relevant notice which determines the notified sum. That prescribed period can be agreed between the parties, failing which, paragraph 10 of the Scheme for Construction Contracts prescribes a period of seven days.

Section 111(10) provides that a Pay Less Notice is not required where (i) the contract provides that, if the payee becomes insolvent, the payer need not pay any sum due in respect of the payment; and (ii) the payee has become insolvent after the prescribed period. This provision reflects the decision of the House of Lords in *Melville Dundas Ltd (in receivership)* v. *George Wimpey UK Ltd* (2007), that no notice of intention to withhold payment will be required to withhold payment following termination of a contract where the contract expressly provides that no further payment requires to be made until after a final accounting has been completed.

The SBC provisions

The SBC contains complex provisions for ascertaining the amount of interim payments. The procedure for ascertaining amounts to be included in Interim Certificates is to be found in clause 4.9. This stipulates that the amount due as an interim payment shall be the gross valuation of the work carried out by the Contractor pursuant to clause 4.16, less (i) the aggregate of the total amount stated as due in Interim Certificates previously issued; (ii) any amount which may be deducted and retained by the Employer by way of retention under clauses 4.18–4.20; (iii) the total amount of any advance payment due to be reimbursed to the Employer in terms of the Contract

Particulars and clause 4.8; and (iv) any sums paid in respect of an Interim Payment Notice given after the issue of the latest Interim Certificate (whether as adjusted by a Pay Less Notice or otherwise). Section 8.6 contains a fuller discussion of retention. The provisions for ascertaining the gross valuation of the work carried out by the Contractor are to be found in clauses 4.16 and 4.17.

The SBC/DB provisions

The SBC/DB provisions for ascertaining the amount of interim payments are similar to those contained in the SBC.

The procedure for ascertaining amounts to be included in Interim Certificates is to be found in clause 4.7. This stipulates that the amount due as an interim payment shall be an amount equal to the Gross Valuation under clause 4.13, where Alternative A applies (Stage Payments) or clause 4.14, where Alternative B applies (Periodic Payments), less (i) the aggregate of any amount which may be deducted and retained by the Employer by way of retention under clauses 4.16 to 4.18; (ii) the total amount of any advance payment due to be reimbursed to the Employer in terms of the Contract Particulars for clause 4.6; and (iii) the amounts paid in previous Interim Payments. The provisions for ascertaining the gross valuation of the work carried out by the Contractor are to be found in clauses 4.13–4.15.

The NEC3 provisions

As previously discussed, the six main options in the NEC3 are essentially different payment structures. Options A and B are priced contracts; options C and D are target contracts; and options E and F are cost reimbursable contracts.

A detailed review of each of the NEC3 options is beyond the scope of this chapter. However, in terms of all six options, clause 50.2 provides that the Contractor is entitled to be paid the 'Price for Work Done to Date' (PWDD), plus other amounts to be paid to the Contractor, less amounts to be paid or retained from the Contractor, following each assessment date. PWDD is, however, defined differently in respect of each option:

- Option A – the total of the 'Prices' for the completed activities;
- Option B – the quantity of completed work for each item in the Bill of Quantities multiplied by the relevant rate;
- Options C to F – the 'Defined Cost' that the Project Manager forecasts the Contractor will have incurred before the next assessment date plus the Fee.

The other amounts that could be paid to the Contractor could arise from such things as interest in terms of clause 51.2, compensation events or secondary options (e.g. Option X6 - Bonus for early Completion).

Amounts to be paid or retained by or retained from the Contractor depend on the operation of clause 50.3 and the particular Secondary Options that are selected (e.g. Option X7 - Delay Damages).

Section 5 of the core clauses sets out a number of terms of general application in connection with the assessment and timing of payments. Clause 50 governs what is due, and clause 51 governs when it is due.

Amounts to be paid by the Contractor to the Employer form part of the calculation of PWDD, and accordingly do not require to be included as part of a separate Pay Less Notice, unless those sums due by the Contractor arose after the assessment date.

As previously discussed, clause 51 governs the process for payment of sums due to the Contractor. For construction contracts in the UK, section 5 of the core clauses requires to be read in conjunction with Option Y(UK)2 in order to comply with the 1996 Act.

The terms of Option Y(UK) 2 do not revise the terms of clauses 51.1 or 51.2 and in terms of those clauses the Contractor's entitlement to payment is to the 'amount due' as provided for in clause 50.2, rather than the sum 'certified' by the Project Manager. However, it appears that despite this anomaly, the intention of Option Y(UK)2 is that the sum due to the Contractor and paid on the final date for payment is to be the sum notified in the Project Manager's certificate, which failing, the Contractor's application for payment (subject to any Pay Less Notice).

8.2 Adjustment of the contract price

8.2.1 Introduction

As indicated above, the final contract price or a mechanism for ascertaining the final contract price should be agreed at the time of entering into the contract. Even where such an agreement has been reached, matters can still arise during the carrying out of the works which will result in either an increase or a decrease in the price. The most common matters which give rise to an adjustment in the price are contractual variations; fluctuations in cost, and claims for direct loss and expense. There are, however, many other matters which can potentially give rise to an adjustment in the price.

Most building contracts provide either expressly or by implication that the contractor is obliged to carry out, and the employer is obliged to pay for, the work which the parties have agreed will be carried out. Failure by either of the parties will amount to a breach of contract giving rise to the possibility of the contract being brought to an end (as discussed in Chapter 9) and also to other remedies becoming available to the innocent party (as discussed in Chapter 10).

If the works cost more than the contractor priced for, they are still bound by the price or the mechanism for ascertaining the price which was agreed between the parties, see *Seaton Brick and Tile Company Ltd* v. *Mitchell* (1900). Similarly, the employer is so bound even if the works cost the contractor less than the price agreed or the price ascertained using the agreed mechanism, see *Mitchell* v. *Magistrates of Dalkeith* (1930). In many instances, however, the parties will attempt to claim an increase or a decrease in the contract price as a result of work they claim was added, omitted or varied from the original agreed scope of the works.

If a contractor claims additional payment for work which actually formed part of the original contract works, then clearly they will have no entitlement to such additional payment. If claims are made in respect of work which was added, omitted or varied in terms of the express provisions of the contract, then there may be an entitlement to an adjustment of the contract price, see Section 8.2.2. The most difficult situation, however, is where there are no express provisions dealing with such matters.

In such cases, the removal by the employer of any work from the contractor may amount to a breach of contract on the part of the employer unless the work is paid for. If the contractor does not agree to the omission and price reduction, then he may have a valid claim against the employer for his loss of profit on the portion of the work which has been omitted.

Similarly, an unauthorized variation by the contractor may amount to a breach of contract on the part of the contractor. Where the deviation is material, the contractor may have difficulty claiming payment not only for the cost of the varied work but also for the work he has executed in accordance with the contract, see *Ramsay & Son* v. *Brand* (1898) and the related cases referred to in Section 8.1.2. Obviously if the contractor can show that the variation was agreed to by the employer, then it will not amount to a breach of contract and the contractor will be entitled to be paid for the varied work. It appears that such agreement may be established by words or conduct on the part of the employer or his agent, see *Holland Hannen & Cubitts (Northern) Ltd* v. *Welsh Health Technical Services Organisation and Others* (1981). (Contrast, however, the decision in *Burrell & Son* v. *Russell & Company* (1900).) In a situation where a variation is agreed to by the employer because it assists or is convenient to the contractor, then the contractor will not be entitled to any additional payment, see *The Tharsis Sulphur and Copper Company Ltd* v. *McElroy & Sons* (1878).

Where the contractor executes additional work, he will have no entitlement to payment unless he can show that the employer agreed to pay for the work. This is a general principle of Scots law which does not only apply to building contracts. For example, in *Walter Wright & Co. Ltd* v. *Cowdray* (1973) an electrical contractor was instructed to dry out and test two motors. In addition, the contractor carried out certain repairs to the motors which had not been instructed. The court held that the contractor had no entitlement to payment for the performing of the repairs. See also *Wilson* v. *Wallace and Connell* (1859). In such situations the only means of obtaining payment may be by using the principle of *quantum lucratus*, which is discussed in Section 8.5.

It is relatively common in building contracts to find provision for other payments to be made to the contractor in addition to the contract price, and sometimes the parties will agree after the contract has been concluded and the works have commenced that additional payments are to be made.

In many of the standard forms of building contract, provision is made for the opening up for inspection or testing of work carried out by the contractor. It is usually provided that the contractor will be paid for all such work, unless the opening up discloses that work carried out is not in accordance with the contract, see, for example, clause 3.17 of the SBC, clause 3.12 of the SBC/DB and core clause 42 of the NEC3. Similar provisions are to be found in ICE contracts, see, for example, *Hall & Tawse Construction Ltd* v. *Strathclyde Regional Council* (1990).

Many of the standard forms of building contracts, for example, clause 2.21 of the SBC, provide that the contractor is required to pay any fees or charges in respect of the contract works demandable under any Act of Parliament, byelaw or similar regulation.

Under certain contracts, including the SBC, these payments may be reimbursed to the contractor in addition to the contract price. Similarly, clauses 2.22 and 2.23 of the SBC provide that where a Contractor incurs liability by way of royalties, damages or other monies as a result of any infringement or alleged infringement of any patent right arising from compliance with an Architect's instruction, then any such liability will be added to the Contract Sum.

Where the employer wishes completion earlier than contracted for, or where the works have been delayed, then the employer may agree a bonus payment to the contractor if they achieve early or timeous completion of the works, see, for example, *Williams* v. *Roffey Bros & Nicholls (Contractors) Ltd* (1990). In such cases, the circumstances in which the payment is to be made should be clearly set out. In particular, it should be made clear whether the payment is truly a bonus in the sense that it is in addition to any other right to payment which the contractor may have in terms of the contract between the parties, for example, direct loss and/or expense. It should also be made clear what is to happen to such payment if the works are delayed by subsequent variations instructed by the employer.

In addition to the contract price, the contractor will also be entitled to payment of any value added tax which is applicable. Normally, the agreed contract price is exclusive of value added tax, see, for example, Article 2 and clause 4.6.1 of the SBC. (But see Section 19.1.4 for the implications of the contract being silent on VAT.) In certain circumstances the contractor may also be entitled to interest on the contract price where the employer has failed to make payments timeously in terms of the contract. Clause 4.12.6 of the SBC and clause 4.9.5 of the SBC/DB expressly stipulate that where the Employer fails to pay any amount due to the Contractor by the final date for payment, then the Employer shall also pay simple interest to the Contractor at the rate of 5 per cent over the official dealing rate of the Bank of England which is current at the date the payment becomes overdue. Clause 51.2 of NEC3 stipulates that if a certified payment is late or the certificate is not issued by the Project Manager, interest (at the rate stated in part one of the Contract Data) is paid on the late payment, and is assessed from the date the payment should have been made until the date payment is made. See Section 10.6 for a more detailed discussion of interest.

In certain circumstances the employer may be entitled to deduct sums from the contract price before making payment. Where the contractor has delayed completion of the works, then the employer may be entitled to deduct liquidated damages, see Section 6.9. The employer may also have an obligation to deduct tax from payments to the contractor. Employers may also have rights under the contract to make deductions in respect of costs they have incurred. For example, clause 6.4.3 of the SBC and the SBC/DB provides that if the Contractor fails to take out insurance to cover death or personal injury to third parties, then the Employer can take out such insurance and deduct the premiums from any payments otherwise due to the Contractor.

8.2.2 Payment for contractual variations

As a result of the magnitude and complexity of most building contracts, there will inevitably arise during the carrying out of the works a need to add, omit or vary some of the work to be executed. For this reason most building contracts, and certainly all the common standard forms of building contract, contain detailed provisions to govern the instruction, execution and payment of variations. See Section 4.5 for a discussion of variations.

A building contract can provide that the contractor will be entitled to no additional payment in respect of such variations. As the contractor is obviously unlikely to agree to such provisions, they are uncommon. It is more normal for provision to be made that the contractor will be entitled to additional payment for variations. Clauses 4.16.1.1 and 5.5 of the SBC and clauses 4.13.1.2 and 5.3 of the SBC/DB expressly provide that the Contract Sum should be adjusted by the value of any Variations/Changes. As discussed earlier in Sections 4.5.2 and 5.2.4, the NEC3 deals with changes by giving the Project Manager the power to give an instruction which changes the Works Information (see clause 14.3), and obliging the Contractor to obey such an instruction (see clause 27.3). Clause 60.1(1) provides that such an instruction amounts to a compensation event (giving rise to both extension of time and costs), except where the change is to accept a Defect, or is a change to the Works Information provided by the Contractor for his design which is made either at his request or to comply with other Works Information provided by the Employer. Clause 61.2 permits the Project Manager to instruct the Contractor to submit quotations for a proposed instruction and in that event the position will be regulated by the procedure for quotations for compensation events set out in clause 62. The rules for assessing the financial and time consequences are set out in clauses 63 and 64.

Nevertheless, in order to ensure that the contractor is entitled to recover additional payment, it is important that any variations are instructed in accordance with the terms of the contract. If the variations are not so instructed, or if the contractor cannot prove that they were so instructed, then they may have difficulty recovering payment, see, for example, *Robertson* v. *Jarvie* (1907).

In order to try and avoid disputes as to whether variations were or were not instructed, many building contracts provide that all variations should be instructed in writing. In such cases, the contractor may, in the absence of a written instruction, be unable to recover payment under the terms of the contract, see, for example, *Brown* v. *Lord Rollo and Others* (1832) and *Holland Hannen & Cubitts (Northern) Ltd* v. *Welsh Health Technical Services Organisation and Others* (1981). Accordingly, where the contractor is instructed to execute a variation by a mechanism other than that provided for in the contract, the advisable course of action for the contractor is to refuse to carry out the work until the correct procedure has been followed. Otherwise the contractor runs the risk that they will be unable to recover payment.

Disputes often arise as to whether an instruction to the contractor is or is not a variation under the contract. For example, a dispute may arise as to whether the contractor is obliged to execute a particular item of work as part of the original contract works or whether that particular item of work is a variation. In such a case, by refusing to carry out the work because of the absence of an instructed variation, the contractor runs the risk of being in breach of contract if it is ultimately determined that the work did form part of the original contract works. In practice, the contractor often simply carries out the work and makes a claim for payment despite the lack of an appropriate variation order.

Similar difficulties can arise where the contractor is instructed to rectify or alter work which it is alleged is not in accordance with the contract. If the work was in accordance with the contract, then the contractor will have incurred the cost of carrying out further work for which they have no appropriate variation order. In such cases the courts have, in certain circumstances, held that the contractor is entitled to payment where a variation order should have been issued, but was improperly withheld, see *Brodie* v. *Corporation of Cardiff* (1919). In other cases, the courts have construed instructions not purporting to be variations to be in fact, variations, see *Shanks & McEwan (Contractors) Ltd* v. *Strathclyde Regional Council* (1994).

Where the variation is made for the contractor's benefit, however, the courts have not been willing to construe instructions as variations entitling the contractor to additional payment, see *The Tharsis Sulphur and Copper Company Ltd* v. *McElroy & Sons* (1878).

8.2.3 Payment for variations under the SBC

Where a building contract provides a mechanism to regulate variations to the works, it is common and, it is submitted, prudent for the contract also to contain provisions to govern how any such variations are to be valued. The method of valuation can take a number of different forms and indeed many of the standard forms of building contract contain more than one method of valuation. The relevant provisions in the SBC are to be found in clause 5, which provides a number of possible alternatives for valuation.

In the first place, it is provided that the Employer and Contractor can agree the value of a Variation. Unfortunately, this rarely happens in practice as many Variations require immediate compliance, leaving little time for agreement of a price in advance between the Contractor and Employer. It is, however, a useful mechanism, particularly in relation to major Variations. To facilitate the agreement of a price in advance, when instructing a Variation, the Architect/Contract Administrator can in terms of clause 5.3.1 stipulate in the instruction that the Contractor is to provide a quotation in accordance with the provisions of Schedule Part 2. In that case, those provisions will regulate the valuation of the Variation unless the Contractor disagrees in writing to the procedure envisaged by the clause within seven days of receipt of the instruction. In terms of Schedule Part 2, the Contractor is required, not later than 21 days from the later of (i) the date of receipt of the instruction, or (ii) the date of receipt by the Contractor of sufficient information to enable the Contractor to quote, to submit a Schedule Part 2 quotation to the Quantity Surveyor. Paragraph 2 of Schedule Part 2 specifies in detail what the quotation should contain. The quotation can then be accepted by the Employer and the acceptance must be confirmed in writing by the Architect. Schedule Part 2 sets out the timescales within which this should happen. Until the quotation is accepted, the Variation is not executed by the Contractor. If the quotation is not accepted by the Employer, then the Architect should either instruct that the Variation is not to be carried out or that it is to be carried out but valued in accordance with the provisions of the Valuation Rules in clauses 5.6–5.10.

The purpose of clause 5.3 and Schedule Part 2 is to deal with Variations which the Employer may require but for which they wish to know the price prior to confirming that the work is to be carried out. Clauses 5.6–5.10, on the other hand, provide a mechanism for valuing Variations which the Contractor is contractually bound to carry out when instructed and for which the Employer is contractually bound to pay. Accordingly, clauses 5.6–5.10 are used to value all instructed Variations other than those to which clause 5.3 and Schedule Part 2 apply or those where the Employer and Contractor have agreed a price in advance, outwith the Schedule Part 2 procedure.

Clauses 5.6 and 5.7 provide two distinct methods for valuing Variations, namely, 'Measurable Work' and 'Daywork'.

Under clauses 5.6 to 5.10, the Quantity Surveyor must value the Variation in accordance with the valuation rules contained within those clauses depending on the nature of the work to be carried out. Clause 5.6 provides that where the Variation requires the execution of additional or substituted work which can properly be valued by measurement, then it shall be valued in accordance with specific rules.

Where the additional or substituted work is of similar character to, is executed under similar conditions as, and does not significantly change the quantity of, work set out in the Contract Bills, then the rates and prices for the work set out in the Contract Bills shall determine the valuation, see clause 5.6.1.1.

Where the additional or substituted work is of similar character to work set out in the Contract Bills but is not executed under similar conditions and/or significantly changes the quantity thereof, then the rates and prices for the work so set out shall be the basis for determining the valuation and the valuation shall include a fair allowance for such difference in conditions and/or quantity, see clause 5.6.1.2.

Where the additional or substituted work is not of similar character to work set out in the Contract Bills, then the work shall be valued at fair rates and prices, see clause 5.6.1.3.

To the extent that the Variation relates to the omission of work set out in the Contract Bills, then the rates and prices for such work therein set out shall determine the valuation of the work omitted, see clause 5.6.2.

To the extent that the Variation requires the execution of additional or substituted work which cannot properly be valued by measurement, then clause 5.7 provides for valuation on a cost/daywork basis.

If the valuation of a Variation cannot reasonably be effected using any of the foregoing methods, then clause 5.10.1 provides that a fair valuation shall be made.

Regardless of whether the valuation is calculated under the Valuation Rules, agreement between Employer and Contractor or Schedule Part 2, clause 5.5 provides that the valuation will be effected by addition to or deduction from the Contract Sum. In a change from previous JCT and SBCC standard forms of contract, the SBC provides, at clause 5.10.2, that no allowance shall be made under the Valuation Rules for any effect upon the regular progress of the Works or of any part of them or for any other direct loss and/or expense for which the Contractor would be reimbursed under any other provision. In those circumstances, the loss and expense provisions in clauses 4.23–4.26 should be operated. However, see the case of *Gear Construction Ltd v McGee Group Ltd* (2012) discussed in Section 5.2.2.

This is not the case in relation to Schedule Part 2 quotations and the Employer's acceptances of those quotations. Paragraph 2 of Schedule Part 2 provides that the

Contractor's quotation should contain an amount to be paid in lieu of direct loss and/or expense ascertained under clause 4.23.

See also Sections 4.5.2 and 5.2.2 above in relation to the valuation of Variations under the SBC.

8.2.4 Payment for changes under the SBC/DB

Clause 4.13.1.2 (Alternative A – Stage Payments), clause 4.14.1.1 (Alternative B – Periodic Payments) and clause 5.3 of the SBC/DB expressly provide that the Contract Sum should be adjusted by the value of any variations (which are defined in clause 5.1.1 as 'Changes'). The provisions in the SBC/DB which govern the valuation of these Changes are also to be found in clause 5 which provides a number of possible alternatives for valuation in the same way as under the SBC.

As with the SBC, it is provided that the Employer and Contractor can agree the value of a Change, which failing, the change is valued by reference to clauses 5.4–5.7 ('the Valuation Rules'). The two methods for valuing a change are identical to the methods under the SBC - 'Measurable Work' and 'Daywork'.

Clause 5.6 provides that, if as a result of compliance with either an instruction requiring a Change or compliance with an instruction as to the expenditure of a Provisional Sum, there is a substantial change in the conditions under which any other work is executed, then that other work is to be treated as if it had been the subject of an instruction requiring a Change, and valued accordingly. Clause 5.4 provides that valuations of Changes are to be made in accordance with clause 5.4, and insofar as is relevant, clauses 5.5–5.7.

The valuation of additional or substituted work is to be consistent with the values of work of similar character set out in the Contract Sum Analysis, but making due allowance for the conditions under which the work is carried out and any significant change in the quantity of work set out, see clause 5.4.2.

The valuation of the omission of work which is set out in the Contract Sum Analysis is to be in accordance with the values that are contained in that Analysis, see clause 5.4.3. Valuation of work under clauses 5.4.2 and 5.4.3 must include an allowance for any necessary addition to or reduction of the provision of site administration, site facilities and temporary works.

Where the execution of additional or substituted work cannot be valued in accordance with clause 5.4 (Measurable Work), the valuation will comprise of, either:

1. the prime cost of the work (calculated in accordance with the 'Definition of Prime Cost of Daywork carried out under a Building Contract' issued by the RICS and the Construction Confederation) together with Percentage Additions to each section of the prime cost at the rates provided for in the Contract Particulars; or

2. where the work is within the province of a specialist trade, the prime cost of the work as calculated by reference to any definition of prime cost agreed and issued by that specialist trade body, together with Percentage Additions to each section of the prime cost at the rates provided for in the Contract Particulars.

If the valuation of a Change cannot reasonably be effected using any of these methods, then clause 5.7.1 provides that a fair valuation shall be made.

As with the SBC, clause 5.7.2 provides that no allowance shall be made under the Valuation Rules for any effect upon the regular progress of the Works or of any part of them or for any other direct loss and/or expense for which the Contractor would be reimbursed under any other provision. In those circumstances, the loss and expense provisions in clauses 4.20–4.23 should be operated. However, see the case of *Gear Construction Lt* v. *McGee Group Ltd* (2012) discussed in Section 5.2.2.

8.2.5 Payment for changes under the NEC3

Under the NEC3, changes instructed by the Project Manager are treated as compensation events in terms of clause 60. Clause 60.1 provides that an instruction from the Project Manager which varies the works by changing the Works Information, qualifies as a compensation event, except where the change is made in order to accept a defect, or the change is made by the contractor for his design at his request or to comply with other Works Information provided by the employer.

Importantly, clause 60.1(1) provides that it is only an instruction from the *Project Manager* which can change the Works Information. Accordingly a contractor should take care to ensure that instructions are received from the project manager, or else run the risk that changes to the Works Information instructed by the employer will not amount to compensation events for which the contractor is entitled to payment.

Core clauses 61–65, as supplemented by the relevant Options, govern the notification, assessment and implementation of compensation events. Clause 61.1 provides that where the compensation event arises from an instruction of the Project Manager, the Project Manager instructs the Contractor to submit quotations at the same time as the relevant instruction. Clause 61.2 permits the Project Manager to instruct the Contractor to submit quotations for a proposed instruction and in both events the position will be regulated by the detailed procedure for quotations for compensation events set out in clause 62. The rules for assessing the financial and time consequences are contained in clauses 63 and 64.

Clause 63 governs the assessment of compensation events and clause 63.1 provides that the changes to the Prices are assessed by the effect of the compensation event on (i) the actual Defined Cost of the work already done; (ii) the forecast Defined Cost for the work not yet done; and (iii) the resulting fee. The core clause is supplemented for each Main Option having regard to the definition of Prices applicable to that Main Option and clause 63 of each Main Option sets out the basis under that particular Option for assessments for changed Prices for compensation events.

Core clause 63.1 seeks to divide work done and work that is not yet done by reference to the date when the Project Manager gave an instruction, issued a certificate, changed an earlier decision or corrected an assumption which gave rise to the compensation event. The date which divides the work already done from the work not yet done is the date of the communication mentioned above. In all other cases the date is the date of the notification of the compensation event. In effect, the assessment is an assessment of the impact of the variation on the work already done, and an assessment of the impact of the variation on the work that is still to be done. Crucially, clause 63.4 provides that the only rights the Employer and Contractor have in respect of compensation events are changes to the Prices, the Completion Date and the Key Dates. It is beyond the scope of this chapter to discuss all of the various provisions based on the different Options; however, the aim of the procedure is to ensure that compensation events are assessed and agreed as early as possible, to avoid the traditional final account dispute which commonly includes a large number of disputed variations. It should be noted that certain amendments were made to clause 63 (both in the core clauses and in the Main Options) in the 2013 edition of the NEC3.

See also Sections 4.5.2 and 5.2.4 in relation to the valuation of changes as compensation events under the NEC3.

As can be seen, the provisions for valuing variations in the SBC, the SBC/DB and the NEC3 are extremely detailed. They have developed over a number of years. Similar provisions are to be found in a number of the other standard forms of building and engineering contracts, though it is open to the parties at the time of contracting to agree any method they choose for valuing variations. Where the contract does not provide a mechanism, then variations will require to be valued on the basis of *quantum meruit*, see Section 8.4.

8.2.6 Fluctuation in cost

In concluding a price for any building contract, there is a risk to both the contractor and the employer that the costs involved in constructing the building can fluctuate dramatically due to changes in economic factors entirely outwith the control of either of them. Such economic factors can include inflation, which can affect the price of both labour and materials, and also changes in tax legislation.

The risk to the employer is that costs decrease and as a consequence they end up paying the contractor far more than the building actually cost to construct. On the other hand, the employer will have budgeted for the contract price prior to concluding the contract, so this is probably more a case of a disappointment than a risk.

The risk to the contractor is that the costs increase, resulting in a diminution in profit or, more seriously, the constructions costs exceeding the contracted price. Such an eventuality is not in the interest of either the contractor or the employer as it can result in the contractor having difficulties completing the building and, in extreme cases, may give rise to the contractor trying to cut corners to minimize costs.

Fluctuations in cost are a particularly serious risk where the contract price is substantial or the contract period particularly lengthy. Without any provision in the contract to deal with such fluctuations in cost, both parties would be bound by the price agreed. Assessing the potential effect of fluctuations can be extremely speculative and factoring the risk into the price may not provide value for money to the employer.

In an effort to minimize the risk of fluctuation in cost, many of the standard forms of building contract provide a mechanism for adjusting the contract price to take account of increases or decreases in cost during the contract period. Such provisions can be found in clauses 4.21, 4.22 and Schedule Part 7 of the SBC, and clauses 4.19 and

Schedule Part 7 of the SBC/DB. Schedule Part 7 provides three alternatives for dealing with fluctuations. The parties should choose which alternative is to be employed at the time of entering into the contract and the choice should be inserted in Part 1 of the Contract Particulars.

If no choice is made, then Part 1 of the Contract Particulars stipulates that Option A is to apply. It should be noted that, in terms of clause 4.22 of the SBC, the fluctuation in cost provisions do not apply to Variations where a price has been agreed in advance under Schedule Part 2, see Section 8.2.3.

Of the three options, Option A provides for adjustment in the contract price as a result of changes in contributions, levies and taxes; Option B provides for adjustment in the contract price as a result of fluctuations in the cost of labour and materials and as a result of tax changes; and Option C provides for adjustment using a price adjustment formula.

The NEC3 addresses potential price adjustment for inflation in Secondary Option X1 (to be used only with Main Options A, B, C or D). Option X1 allows parties to factor in a mechanism for increasing the contractor's right to payment in line with inflation. The Option places the risk of inflation on the Employer and the risk, if any, of deflation on the Contractor. If the Option is selected, the parties must select a base index in the Contract Data (B), which will be used as the measure of inflation as against the latest index available (L) before the date of assessment of an amount due. A price adjustment factor is determined by calculating the total of the products of each of the proportions stated in the Contract Data multiplied by (L-B)/B for the index linked to it (see clause X1.1). Clause X1.5 provides that each time the amount due is assessed, an amount for price adjustment is added to the total of the Prices, which is the sum of the change in Price for Work Done to date since the last assessment of the amount due multiplied by the Price Adjustment Factor/Price Adjustment Factor +1 and correcting amounts not included elsewhere, which arise from changes to indices used for assessing previous amounts for price adjustment.

If either party is to rely on the fluctuation provisions, then it is important that they fully comply with any prerequisites contained within the relevant option, see *John Laing Construction Ltd* v. *County and District Properties Ltd* (1983).

8.3 Loss and expense

In building contracts, it is not uncommon for the contractor's progress of the works to be affected by events that are within the control of the employer or the architect. This can significantly increase the cost to the contractor in carrying out the works. Clauses 4.23–4.26 of the SBC contain a mechanism that, in certain circumstances, entitles the Contractor to recover direct loss and/or expense where the regular progress of the Works has been materially affected by certain specified Relevant Matters or where there has been a deferment of giving possession of the site to the Contractor, under clause 2.5.

As soon as the Contractor becomes aware, or should reasonably have become aware, that the regular progress has been affected, they need to make written application to the Architect. A list of the Relevant Matters in respect of which direct loss and/or expense can be claimed is contained in clause 4.24. The matters are:

- Variations (excluding any loss and/or expense relating to an acceptance of a Schedule Part 2 quotation but including any other matters or instructions which under the contract are to be treated as, or requiring a Variation). In relation to loss and/or expense arising from Variations under the SBC, the SBC/DB and the NEC3, see also Sections 4.5.2 and 5.2.2.
- Instructions of the Architect under clause 3.15 (postponement of work) or clause 3.16 (expenditure of provisional sums).
- Instructions of the Architect for the opening up for inspection or testing of any work, materials or goods under clause 3.17, unless the cost is provided for in the Contract Bills or the inspection shows that the work was not in accordance with the Contract.
- Any discrepancy in or divergence between the Contract Drawings, the Contract Bills and/or the documents referred to in clause 2.15.
- Compliance by the Contractor with clause 3.22.1 (Antiquities) or with the Architect's instructions under clause 3.22.2.
- The execution of work for which an approximate quantity is not a reasonable accurate forecast of the quantity of work required.
- Any impediment, prevention or default, whether by act or omission, by the Employer, the Architect, the Quantity Surveyor or any of the Employer's Persons, except to the extent caused or contributed to by any default, whether by act or omission of the Contractor or of any of the Contractor's Persons.

The number of Relevant Matters stipulated within the SBC has been greatly reduced from previous JCT and SBCC standard form contracts, although the final Relevant Matter mentioned above is very much a catch-all matter, consolidating a number of previously discrete matters.

Having received the Contractor's written application, the Architect is to form an opinion as to whether the regular progress has been or is likely to be materially affected as stated in the application or whether direct loss and/or expense has been or is likely to be incurred due to deferment. He can request from the Contractor such further information as is reasonable to enable him to form such an opinion. Assuming he forms such an opinion, then the Architect from time to time shall ascertain, or shall instruct the Quantity Surveyor to ascertain, the amount of such loss and/or expense which has been or is being incurred by the Contractor. Only the Architect can decide upon the validity of an application, see *John Laing Construction Ltd* v. *County and District Properties Ltd* (1982). Either the Architect or the Quantity Surveyor can request further details as may be reasonably necessary to enable ascertainment of the loss and/or expense due to the Contractor. In terms of clause 4.25, any loss and expense so ascertained is added to the Contract Sum.

Clauses 4.23–4.25 are not exhaustive of the Contractor's rights in such circumstances. Clause 4.26 provides that the provisions of clauses 4.23–4.25 are without prejudice to any other rights and remedies which the Contractor may possess. It should be noted, however, that under clause 1.10.1.4, a Final Certificate is conclusive evidence that the reimbursement of direct loss and/ or expense, if any, to the Contractor pursuant to clause 4.23 is in final settlement of all and any claims which the Contractor has or may have arising out of the occurrence of any of the Relevant Matters referred to in clause 4.24, whether such claim be for breach of contract, duty of care, statutory duty or otherwise. Accordingly, despite the terms of clause 4.26, the Final Certificate is conclusive as to loss and expense in respect of any of the matters described in clause 4.24, notwithstanding the legal basis upon which payment arising out of such matters was sought.

The provisions contained in clauses 4.20–4.23 of the SBC/DB are broadly similar to those contained in clauses 4.23–4.26 of the SBC, except there is no reference to the Architect/Contract Administrator. The Contractor must make written application to the Employer, in the same manner as application to the Architect is required under the SBC. A list of the Relevant Matters in respect of which direct loss and/or expense can be claimed under the SBC/DB is contained in clause 4.21. The matters are:

- Changes and any other matters or instructions which under the conditions are to be treated as, or requiring, a Change. In relation to loss and/or expense arising from changes under the SBC/DB, see also Sections 4.5.2 and 5.2.3.
- Instructions of the Employer under clauses 3.10 (postponement of work) or 3.11 (expenditure of provisional sums).
- Instructions of the Employer for the opening up for inspection or testing of any work, materials or goods under clause 3.12, unless the inspection or test shows that the work, materials or goods are not in accordance with the contract.
- Compliance by the Contractor with clause 3.15.1 (Antiquities) or with the Employer's instructions under clause 3.15.2.
- Delay in receipt of any permission or approval for the purposes of Development Control Requirements necessary for the Works to be carried out or proceed, which delay the Contractor has taken all practicable steps to avoid or reduce.
- Any impediment, prevention or default, whether by act or omission, by the Employer, the Architect, the Quantity Surveyor or any of the Employer's Persons, except to the extent caused or contributed to by any default, whether by act or omission of the Contractor or of any of the Contractor's Persons.

Having received the Contractor's written application, the Employer is to form an opinion as to whether the regular progress has been or is likely to be materially affected as stated in the application or whether direct loss and/or expense has been or is likely to be incurred due to deferment. As with the SBC, the Employer can request from the Contractor such further information as is reasonable to enable him to form such an opinion. Again, as with the SBC, clauses 4.20–4.22 are not exhaustive of the Contractor's rights in such circumstances and clause 4.23 provides that they are without prejudice to any other rights and remedies which the Contractor may possess; though it should be noted that the Final Statement under clause 1.8.1.3 has the same conclusive effect in respect of the reimbursement of direct loss and/ or expense, if any, as the Final Certificate under the SBC.

The NEC3's approach to loss and expense is somewhat different from the SBC and the SBC/DB and the most striking difference is that the NEC3 does not distinguish between events which give rise to extension of time only and those which confer a right to money. Core clause 60.1 sets out a number of compensation events for which the Contractor is entitled to claim both time and money. These are described in more detail in Section 6.11.11. There are detailed notification provisions contained in clauses 61 and 62 of the NEC3, while clause 63 governs the assessment of the compensation events. As discussed earlier in Section 8.2.5, clause 63 provides that the changes to the Prices are assessed as the effect of the compensation event on the actual Defined Cost of the work already done, the forecast Defined Cost of the work not yet done and the resulting Fee. The effect of the compensation event on the cost of carrying out the works, whether by way of delay or disruption, will form part of that assessment. In contrast to the SBC and the SBC/DB, clause 63.4 provides that the rights of the Employer and the Contractor to changes to the Prices, the Completion Date and Key Dates are their only rights in respect of a compensation event. In those circumstances, care should be taken by a Contractor to ensure that if a compensation event arises it strictly adheres to the terms of core clause 6, as otherwise it will not be entitled to additional payment in respect of the compensation event. In relation to compensation events in respect of changes under the NEC3, see also Sections 4.5.2 and 5.2.4.

As one might imagine, the loss and expense provisions to be found in the standard form building contracts have been considered by the courts on numerous occasions. It is not possible to give detailed consideration to all the available case law in this book. However, Scotland does have the benefit of an Inner House decision reviewing the law in this area. In the case of *John Doyle Construction Ltd* v. *Laing Management (Scotland) Ltd* (2004) the court held that for a loss and expense claim under a building contract to succeed, the contractor is required to prove three matters:

- the existence of one or more events for which the employer is responsible, for example, a Relevant Matter under clause 4.24 of SBC;
- the existence of loss and expense suffered by the contractor; and
- a causal link between the event or events and the loss and expense.

This statement is consistent with earlier cases in this area in which it had been held that direct loss and expense was that which flowed naturally in the usual course of things. See, for example, *F G Minter Ltd* v. *Welsh Health Technical Services Organisation* (1980) following *Saint Line Ltd* v. *Richardsons Westgarth & Co. Ltd* (1940).

However, it is not always straightforward to establish the causal link between an event and a particular item of loss and expense. Indeed, in many instances loss and expense to the contractor are caused by a number of events which may be the responsibility of the employer, the responsibility of the contractor or the responsibility of neither party. In such instances it may not be possible to identify the causal link between each individual event and the items of loss and expense incurred by the contractor. In such cases attempts have been made by contractors to pursue global claims against employers, where little or no attempt is made to establish any causal link, with varying degrees of success. See, for example, *London Borough of Merton* v. *Stanley Hugh Leach Ltd* (1985) and *Wharf Properties Ltd* v. *Eric Cumine Associates (No. 2)* (1991).

The area of global claims has been reviewed in *John Doyle Construction Ltd* and guidance produced for the first time by a Scottish court. In delivering the court's opinion Lord Drummond Young indicated that if a contractor can prove that all the events on which he relies are the responsibility of the employer, then there is no need to identify any causal link between the individual events and particular items of direct loss and expense. If, however, an event for which the employer is not responsible plays a significant part in the causation of the loss and expense, then the claim will fail unless it is possible to separate out the effect of the event for which the employer is not responsible. The court considered, however, that this was subject to three mitigating considerations as follows:

> In the first place, it may be possible to identify a causal link between particular events for which the employer is responsible and individual items of loss ... In the second place, the question of causation must be treated by the application of common sense to logical principles of causation ... In this connection, it is frequently possible to say that an item of loss has been caused by a particular event notwithstanding that other events played a part in its occurrence. In such cases, if an event or events for which the employer is responsible can be described as the dominant cause of an item of loss, that will be sufficient to establish liability, notwithstanding the existence of other causes that are to some degree at least concurrent ... In the third place, even if it cannot be said that events for which the employer is responsible are the dominant cause of the loss, it may be possible to apportion the loss between the causes for which the employer is responsible and other causes. In such cases it is obviously necessary that the event or events for which the employer is responsible should be a material cause of the loss. Provided that condition is met, however, we are of the opinion that apportionment of loss between the different causes is possible in an appropriate case.

Although *John Doyle Construction Ltd* provides useful guidance in relation to global claims, it should be borne in mind that it does not provide *carte blanche* to use a global claim and that evidence of the causal links will still be of crucial importance in loss and expense claims. The issue for the court was whether a global claim fell to be rejected at the stage of legal debate and the court's decision was simply that the claim should be allowed to proceed to a full hearing on the evidence. If, however, evidence could not be produced to satisfy any of the three principles quoted above, then the claim would still fail in its entirety. Similarly if evidence did satisfy any of the three principles, this may still result in only a small proportion of the claim succeeding with the remainder of the claim failing. Accordingly a contractor should still endeavour where possible to prove a causal link between the relevant events and the items of loss and expense. A failure to keep satisfactory records will not be a legitimate excuse for failing to do so.

It should be noted that there is no direct connection between the loss and expense provisions in clauses 4.23–4.26 of the SBC (clauses 4.20–4.23 of the SBC/DB) and the provisions dealing with extension of time to be found in clauses 2.26–2.29 (clauses 2.23–2.26 of the SBC/DB). The two sets of clauses have distinct and separate purposes. Support for this proposition can be found in *Methodist Homes Housing Association Ltd* v. *Scott & McIntosh* (1997).

8.4 *Quantum meruit*

It is a general principle of Scots law that the recipient of services in terms of a contract is under an implied obligation to pay for the services. As mentioned previously, the final contract price or a mechanism for ascertaining the final contract price should be agreed at the time of entering into a building contract. Similarly, it is prudent to agree the basis on which any variations or additions will be priced. In practice, this may not happen. In such circumstances, the contractor may still be entitled to payment *quantum meruit*, that is, payment of a reasonable sum for the work carried out by them.

In order for a claim for *quantum meruit* to succeed, there must be a contract between the parties, see, for example, *Alexander Hall & Son (Builders) Ltd* v. *Strathclyde Regional Council* (1989). There is no scope for a *quantum meruit* claim if the contract between the parties contains an agreed contract price or a mechanism for ascertaining the contract price, see, for example, *Interbild Components Ltd* v. *Fife Regional Council* (1988). As has been indicated previously, however, agreement of the price, or a mechanism for ascertaining the price, is an essential term of any building contract and, in the absence of such agreement, there may be no binding contract between the parties.

In certain circumstances, however, where work has been carried out by agreement and only the price has not been agreed, the courts have been prepared to hold that an implied contract exists between the parties. See, for example, *Avintair Ltd* v. *Ryder Airline Services Ltd* (1994) in which services were performed by one party but no price was agreed. The court held that in those circumstances the law would imply, from the parties' conduct, a contract that a reasonable sum be paid and that the appropriate claim in those circumstances was an implied contract on the principle of *quantum meruit*.

Before the law can imply such a contract, it appears that the services must already have been performed by the party seeking payment. See *British Bank for Foreign Trade Ltd* v. *Novinex* (1949), in which Lord Denning stated:

> In the ordinary way, if there is an arrangement to supply goods at a price 'to be agreed', or to perform services on terms 'to be agreed', then although, while the matter is still executory, there may be no binding contract, nevertheless, if it is executed on one side, that is, if the one does his part without having come to an agreement as to the price or the terms, then the law will say that there is necessarily implied, from the conduct of the parties, a contract that, in default of agreement, a reasonable sum is to be paid.

In addition to the contract price, *quantum meruit* can also apply to additional work instructed where no price has been agreed, see *Taylor* v. *Andrews-Weatherfoil Ltd* (1991). If, however, the additional works have not been instructed, then the claim will fail, see *T & R Aitken* v. *Cordner* (1958). *Quantum meruit* can also apply where a price has been agreed but it becomes inapplicable through the passage of time, see *Constable Hart & Co. Ltd* v. *Peter Lind & Co. Ltd* (1978) in which a price was agreed which was fixed until a particular date. The contract was delayed through no fault on the part of the sub-contractor and the court held that work carried out after the agreed date was to be paid for at reasonable rates.

In certain circumstances, even where a price has been agreed, it may be possible for a contractor to put forward a claim that they can ignore the contract price and demand payment on the basis of *quantum meruit*. In certain circumstances, it may be possible for the contractor to show that, as a result of breaches of contract on the part of the employer, the contract works have altered so dramatically from those contracted for that the contract price is no longer applicable and a new contract term should be implied that they be paid on the basis of *quantum meruit*, see *Lodder* v. *Slowey* (1904). Before contractors can put forward a claim on the basis of *quantum meruit*, however, it appears that they must rescind the contract in order to make it clear that they no longer consider themselves bound by the original contract price, see *ERDC Construction Ltd v. H M Love & Co.* (1995), *Boyd & Forrest* v. *Glasgow & South Western Railway Company* (1915) and *Smellie* v. *Caledonian Railway Company* (1916). In order to allow the contractor to rescind, any breach by the employer will have to be material. In the event that the breach is not material, or should the contractor choose not to rescind, then the contractor's remedies will be limited to payment of the contract price for the work executed together with damages for breach of contract. In *Morrison-Knudsen Co. Inc* v. *British Columbia Hydro & Power Authority* (1978) it was stated that:

> It is well established law that a plaintiff's remedies for a defendant's default under a contract between them are limited to those provided in the contract or which may be awarded for breaches of the contract for so long as the contract remains open and available to the parties. To enable the court to award compensation by *quantum meruit* the Respondents must show that [the contract] has been rescinded or discharged and that mutual obligations thereunder have ceased to exist. While it continues to exist the obligation of [the plaintiff] and the rights of the Respondents are limited by its terms.

This statement was cited with approval in *ERDC Construction Ltd* in which the court indicated that a party faced with a breach of contract had to elect between affirming the contract and holding the other party to the performance of its obligations or, alternatively, rescinding the contract and suing at once for damages or *quantum meruit* for performance to the date of rescission. The court stated that the election must be made promptly and communicated to the employer, and once made would be binding on the parties and could not be changed. If the contractor simply continues to carry out the contract works, then they waive their right to claim payment *quantum meruit*, see *Smellie*.

Another situation in which it may be possible to ignore the contract price and seek payment on the basis of *quantum meruit* is where the nature of the work carried out is altered fundamentally from that which the contractor originally contracted to carry out. As a result, the works may become more difficult and more expensive. In such circumstances, it may be open to the contractor to maintain that the original contract has been frustrated by the fundamental alterations and that they are entitled to maintain a claim based on *quantum meruit*, see, for example, *Head Wrightson Aluminium Ltd* v. *Aberdeen Harbour Commissioners* (1958) and *Smail* v. *Potts* (1847).

The alteration to the work may not amount to breach of contract as the employer may, for example, have power to instruct variations in terms of the contract. Similarly, the contractor may suggest variations which are approved by the employer, see *Mercer* v. *Wright* (1953). It appears that if a contractor wishes to claim for payment *quantum meruit*, he may have to advise the employer at the time when the difficulty becomes apparent, see *Mackay* v. *Lord Advocate* (1914). It is submitted that this is correct as it offers the employer an opportunity to choose between proceeding with the contract works on the basis of payment *quantum meruit* or halting the work because of the frustration.

If, however, the works have become more difficult or more expensive because of matters which existed at the time of contracting, but which the contractor did not foresee, then the contract will not be frustrated and the contractor will have no entitlement to payment *quantum meruit*, see *Davis Contractors Ltd* v. *Fareham UDC* (1956).

Where a contractor is entitled to make a claim for payment *quantum meruit*, then they are entitled to be paid at ordinary or market rates, or where no such rates are available, they are entitled to be paid a reasonable rate, see *Avintair Ltd*. The party seeking payment *quantum meruit* is entitled to lead evidence to prove what would be a reasonable rate in the circumstances, see *Wilson* v. *Gordon* (1828). It appears that a building contractor can include in their rate elements for work carried out, material supplied, overhead costs and reasonable profit, see *Monk Construction Ltd* v. *Norwich Union Life Assurance Society* (1992). In addition to proving that the rate charged is reasonable, the party claiming payment will also have to prove to the court that the amount of time spent carrying out the work was reasonable, see *Scottish Motor Traction Co.* v. *Murphy* (1949).

8.5 Quantum lucratus

Disputes sometimes arise in a situation where a person has in good faith carried out works to another's land or property where there is no contract between the parties. In such a situation the party carrying out the work cannot claim payment in terms of the contract or payment *quantum meruit*. There may, however, be an entitlement to payment on a *quantum lucratus* basis, that is, seeking to recover the value of the land or property owner's enrichment, see, for example, *Newton* v. *Newton* (1925). *Quantum lucratus* is a branch of the law of recompense and is an equitable remedy requiring an owner of land to pay for works carried out by another on their land as a result of which they are enriched. *Quantum lucratus* does not apply to the situation where a third party is enriched by work carried out by the owner of land or property to that land or property, see *Edinburgh and District Tramways Company Ltd* v. *Courtenay* (1909). It may, however, apply in the situation where work is carried out to common property, see *Stark's Trustees* v. *Cooper's Trustees* (1900).

Where work is carried out to land or property without the owner's permission or agreement, then the owner can insist that the works are removed. If the owner does not do so, then the principle of *quantum lucratus* deems him to have accepted the benefit and requires him to pay for that benefit. If, however, the person who has carried out the works has done so in bad faith, then he will have no claim based on *quantum lucratus*, see *Barbour* v. *Halliday* (1840) and *Duke of Hamilton* v. *Johnston* (1877).

If works are carried out in terms of a contract between the parties, there is no scope for a *quantum lucratus* claim, so long as the contract remains applicable, see *Thomson* v. *Pratt* (1962). It appears that where the contract becomes inapplicable because it has been determined by the employer who retains the benefit of any works already carried out, then the contractor may be entitled to a *quantum lucratus* claim, see *Alexander Graham & Co.* v. *United Turkey Red Company Ltd* (1922), *NV Devos Gebroeder* v. *Sunderland Sportswear Ltd* (1990) and *R & J Scott* v. *Gerrard* (1916). A *quantum lucratus* claim cannot be sustained if the contract makes provision for payment on its determination, see, for example, clause 8 of the SBC.

The courts have had difficulty laying down a general definition for *quantum lucratus* and have indicated that each case requires to be looked at on its own particular circumstances, see *Edinburgh and District Tramways Company Ltd, Varney (Scotland) Ltd* v. *Burgh of Lanark* (1976) and *Lawrence Building Co. Ltd* v. *Lanarkshire County Council* (1979). It appears, however, that some essential features must exist in order for a claim to succeed. In both *Varney (Scotland) Ltd* and *Lawrence Building Co. Ltd* it was held that for a claim to succeed, the pursuer must have incurred a loss though the cost of carrying out the works will suffice in this respect; the pursuer must not have intended to make a gift to the defenders; and there must be a quantifiable benefit to the defender who is thereby *lucratus*.

Accordingly, if the pursuers carry out the work for their own benefit, then they will not be entitled to claim *quantum lucratus*, see *Edinburgh and District Tramways Company Ltd* and *Rankin* v. *Wither* (1886). Some incidental benefit will not, however, bar a claim. It also appears that a claim for *quantum lucratus* cannot succeed where the pursuer has any other legal remedy available, see *Stewart* v. *Stewart* (1878). In *Varney (Scotland) Ltd* it was stated that:

> Recompense is an equitable doctrine. That being so, it becomes a sort of court of last resort, recourse to which can only be made when no other legal remedy is or has been available. If a legal remedy is available at the time when the action which gave rise to the claim for recompense has to be taken, then normally that legal remedy should be pursued to the exclusion of a claim for recompense.

Some authorities have indicated that a claim for *quantum lucratus* can only succeed where there has been an error or mistake of fact on the part of the person making the claim, see *Rankin* v. *Wither* (1886), *Buchanan* v. *Stewart* (1874) and *Gray* v. *Johnston* (1928). This can be contrasted, however, with the comments of Lord Justice Clerk Alness in *Gray* who stated that he did not think error was essential in all cases for a claim for *quantum lucratus* to succeed. Similarly in *Varney (Scotland) Ltd*, the court indicated that error may found a claim for *quantum lucratus* but that the absence of an error or mistake of fact will not invalidate a claim if the other circumstances justify its imposition.

8.6 *Contractual retention*

8.6.1 Retention in general

In the standard forms of building contract it is common to find a mechanism whereby the employer can deduct an amount from any payment otherwise due to the contractor by way of contractual retention. The purpose of the retention is to allow the employer to retain a proportion of any payment due in respect of work already carried out as security against the risk of any failure by the contractor to complete their obligations under the contract, including the making good of defects. Once the contractor has completed all his obligations in terms of the contract, then the retention is released.

Accordingly, the employer can use the retention as a lever to ensure the contractor completes the works or, alternatively, as a fund to pay for completion of the works in the event that the contractor does not fulfil his obligations. This can be particularly important in the event of the contractor's insolvency where, without any retention, the employer might simply be left with an unsecured claim against the contractor for breach of contract, see *Asphaltic Limestone Concrete Co. Ltd and Another* v. *Corporation of the City of Glasgow* (1907).

Most standard forms of building and engineering contracts contain detailed rules governing both the deduction of contractual retention and its release.

8.6.2 Retention under the SBC

Clause 4.9.2 of the SBC provides that the sum due as an interim payment shall be the gross valuation of the work carried out by the Contractor, less any Retention which may be deducted and retained by the Employer as provided for in clauses 4.18–4.20.

Clause 4.20 provides that the Retention which the Employer may deduct and retain shall be a percentage of the total amount included under clause 4.16.1 in the gross valuation for any Interim Certificate. The percentage ('Retention Percentage') is stipulated to be 3% unless a different rate is agreed between the parties and inserted in the Contract Particulars.

The Retention Percentage may be deducted from the amount certified in any Interim Certificate insofar as the amount certified relates to work which has not reached practical completion. Where the work has reached practical completion but no Certificate of Making Good has been issued, then the Employer may only deduct half the Retention Percentage. In practice this operates on the basis that the Employer deducts the whole Retention Percentage from amounts included in Interim Certificates issued to the Contractor. Half the Retention is then released on practical completion of the works with the remaining half being released on the issue of a Certificate of Making Good.

Clause 4.18 of the SBC stipulates that the Employer's interest in the Retention is fiduciary as trustee for the Contractor. At the date of each Interim Certificate, the Architect has to prepare or instruct the Quantity Surveyor to prepare a statement specifying the Retention deducted in arriving at the amount stated as due in the Interim

Certificate and this statement is issued to the Contractor. Thereafter, the Contractor can request the Employer, at the date of payment under each Interim Certificate, to place the Retention to be deducted in a separate bank account and certify that this has been so done, see clause 4.18.3. It appears that the Contractor can also make this request at a later date if he has not done so at the date of payment, see *J F Finnegan Ltd* v. *Ford Sellar Morris Developments Ltd* (1991). The Employer is entitled to any interest accruing on the Retention while it remains in this separate account.

8.6.3 Retention under the SBC/DB

The position in respect of retention under SBC/DB is almost identical to the position under the SBC. Clause 4.7.2 of the SBC/DB provides that the sum due as an interim payment shall be the gross valuation of the work carried out by the Contractor, less any Retention which may be deducted and retained by the Employer as provided for in clauses 4.16–4.18.

Clause 4.18 provides that the Retention which the Employer may deduct and retain shall be a percentage of the total amount included under clause 4.13.1 (Stage Payments) or 4.14.1 (Periodic Payments) in the gross valuation for any interim payment. The percentage ('Retention Percentage') is stipulated to be 3% unless a different rate is agreed between the parties and inserted in the Contract Particulars.

The Retention Percentage may be deducted from the amount certified in any Interim Certificate insofar as the amount certified relates to work which has not reached practical completion. Where the work has reached practical completion but no Notice of Completion of Making Good has been issued, then the Employer may only deduct half the Retention Percentage. As under SBC, half the Retention is then released on practical completion of the works with the remaining half being released on the issue of a Notice of Completion of Making Good.

As with the SBC, the SBC/DB stipulates that the Employer's interest in the retention is fiduciary as trustee for the Contractor (see clause 4.16.1), and the Contractor can request the Employer, at the date of each Interim Payment, to place the retention to be deducted in a separate bank account and notify the Contractor that this has been so done (see clause 4.16.2.).

The objective of clause 4.18.3 of the SBC and clause 4.16.2 of the SBC/DB is to provide a mechanism whereby the Retention deducted by the Employer is to be held in trust on behalf of the Contractor, see *Wates Construction (London) Ltd* v. *Franthom Property Ltd* (1991). This is to afford the Contractor a degree of protection in the event of the insolvency of the Employer. If it is not placed in a separate account, then it appears that the Contractor will have no protection, see *Mac-Jordan Construction Ltd* v. *Brookmount Erostin Ltd* (1991). If the Employer fails to put the money in a separate account, then the Contractor's remedy would be an action for specific implement (see Section 10.3). Unless an interim order can be obtained, given the time it may take to conclude such an action, this remedy may be of little practical assistance. This is particularly so in cases where the Employer disputes that the Contractor is entitled to the retention because the Employer has other claims which he wishes to meet out of the contractual retention, for example, liquidated damages, see *Henry Boot*

Building Ltd v. *The Croydon Hotel & Leisure Co. Ltd* (1985) and *GPT Realisations Ltd (in Administrative Receivership and in Liquidation)* v. *Panatown Ltd* (1992). Contrast, however, the decision in *Concorde Construction Co. Ltd* v. *Cogan Co. Ltd* (1984).

Unfortunately, it appears that under Scots law the terms of clause 4.18.3 of the SBC and clause 4.16.2 of the SBC/DB are insufficient to create a trust without other actions on the part of the Employer, see *Clark Taylor & Co. Ltd* v. *Quality Site Development (Edinburgh) Ltd* (1981) and *Balfour Beatty Ltd* v. *Britannia Life* (1997). This appears to mean that in Scotland under the SBC and the SBC/DB, if an Employer becomes insolvent, then, in respect of the payment of retention which has been deducted, the Contractor may find themselves in no better a position than other ordinary creditors. This may differ from the position under English law, where a trust can be established by less formal means. A detailed examination of the law of creation of trusts is beyond the scope of this book.

8.6.4 Retention bonds

A practice has grown up whereby, in order to receive payment of the full amount of interim certificates and the final account, the contractor will often provide the employer with a bond equivalent to the amount which would otherwise have been retained by the employer until full satisfaction of the works by the contractor, including remedying defects. These bonds are either put in place from the commencement of the works or are put in place at the time of practical completion in respect of the remaining half of the retention fund which would otherwise not be payable until after (in the case of the SBC and the SBC/DB) the issue of a Certificate of Making Good/Notice of Completion of Making Good. This practice has now been formalized in clause 4.19 of the SBC and clause 4.17 of the SBC/DB. The parties can choose to apply these clauses by filling in the Contract Particulars appropriately. These clauses set out the mechanics of operating the retention bond. See also Section 23.2.6.

8.6.5 Retention under the NEC3

The core clauses of the NEC3 do not make provision for retention. Secondary Option X16: Retention does, however, contain retention provisions that the parties can agree to include. Secondary Option X16 is, however, not applicable to Option F - Management Contracts.

If Option X16 is used, the Employer requires to enter a retention percentage and a retention free amount in the Contract Data. At each payment assessment date the retention percentage is applied to the Price for Work Done to Date (PWDD), and the relevant sum is retained by the Employer from the amount due to the Contractor. The retention sum is retained by the Employer until the earlier of the Completion of the works or the date on which the Employer takes over the whole of the works, at which point the retention is halved. The final moiety of retention is released to the Contractor when the Defects Certificate has been issued, and no amount is retained in the assessments made after the Defects Certificate has been issued. To all intents and purposes, Option X16 operates as a traditional retention clause.

However, the NEC3 differs from other standard form contracts in the use of a 'retention free amount'. The Employer is not entitled to retain any retention from the Contractor until the PWDD reaches the 'retention free amount' that is stipulated in the Contract Data. The Contract Data can of course stipulate (and very often does) that the 'retention free amount' should be zero. Clearly the higher the retention free amount, the better the cash flow for the Contractor. However, the mechanism may be of limited benefit. If there is a relatively high retention free amount, when the PWDD actually reaches that figure, and the retention percentage applied to it, there may be a significant deduction in the sums dues to be paid to the Contractor, which could have a more significant impact on the Contractor's cash flow than if retention had been deducted from the outset.

The NEC3 makes no provision for retention bonds.

8.7 Project bank accounts (PBA)

A project bank account (PBA) is a form of ring-fenced payment mechanism which is gradually finding more favour as an alternative to the traditional cascade of payments through the contractual chain. Albeit their use is currently not as widespread as some would like, the recent support for PBAs by central government and the introduction of an appropriate option in the most commonly used standard forms are likely to see a growth in their popularity. PBAs incorporate equitable principles associated with trusts into a contractual business arrangement and their principal purpose is to improve both security and speed of payment for contractors and sub-contractors.

With the government now committing to use PBAs 'unless there are compelling reasons not to do so', their total value in public sector contracts is expected to reach £4 billion by 2014. The Scottish Government's Deputy First Minister Nicola Sturgeon has commented that 'using project bank accounts guarantees a diverse and competitive marketplace, meaning that Scotland's many SMEs are given the confidence to compete for Scottish construction contracts', and the Scottish Government committed in 2013 to trial the use of PBAs in public sector projects. The Welsh Government announced in January 2014 that it had decided to trial PBAs in its 21st Century Schools Programme, which means that PBAs are now being used in major government projects throughout the UK.

There are two common ways to create a PBA. First, it can be set up jointly in the names of the employer and contractor, in trust for the key sub-contractors to allow them to benefit from the security of the trust. Alternatively, it can be opened in the name of the employer, with bank mandates put in place authorizing the contractor to act. In either case, to confer protection in the event of insolvency it is important that a separate account is opened, and that it is not used for any other funds (e.g. for different purposes of the employer).

A key benefit of PBAs is security of payment for the beneficiaries of the trust in the event of insolvency of the employer (or in the case of sub-contractor beneficiaries, the insolvency of the main contractor). However, in order to achieve this benefit under Scots law, it is crucial that a trust obligation securing the sums due is properly established. The basic requirements for the creation of trust are:

- the existence of distinct property which can be identified to form the trust;
- named beneficiaries to benefit from the trust; and (most importantly)
- a clear and declared intention to create the trust.

In Scotland the requirements for creation of a trust and are more extensive than under English law. For example, under Scots law, creation of a separate bank account is not normally sufficient on its own to grant protection to contractors against the employer's insolvency, if not accompanied by the other criteria for the creation of a trust. See also the comments in Section 8.6 in relation to a trust in the context of retention.

It is also important to remember that the trust arrangement will protect only the named beneficiaries and not other parties. For example, small sub-contractors may in practice not experience any benefit from the arrangement, as there is every likelihood that they will not be named in the account documentation. Furthermore, the cost of set-up and administration, together with project team training, may make the use of PBAs inefficient in small-scale ventures.

It has become common for parties to use the trust deed precedent published by the Office of Government Commerce in the *Guide to Best 'Fair Payment' Practices* in 2007. That document became the basis for the JCT's PBA standard form and is also used by some of the banks.

The SBCC also published in 2013 PBA documentation (PBA/Scot/13), facilitating the option of using PBAs in Scotland, see Section 1.5. The documentation is divided into three parts. The main part is the Project Bank Account Agreement. This specifies the general rules relating to the operation of the account and rules on new sub-contractors and removal of sub-contractors. It also deals with confidentiality, assignation and termination. The other parts comprise the Additional Party Agreement and the Enabling Provisions. The latter need to be given careful consideration as they are important for the set-up and operation of the PBA.

The April 2013 edition of the NEC3 contained optional PBA provisions in new secondary option Y(UK)1. This allows for a joint deed in terms of which new suppliers can benefit from the arrangement. However, the effectiveness of this in creating a trust under Scots law needs to be considered carefully in view of the above comments. See also Section 1.6.

Chapter 9

Ending a Construction Contract

9.1 Introduction

Having examined what is required to constitute a construction contract in Scotland, and the rights and obligations arising from it, it is important to establish when, and in what circumstances, a construction contract and the obligations arising from it will be brought to an end. The law of Scotland contains a number of general rules which relate to the extinction of contractual obligations, and these apply equally to the extinction of the obligations under a construction contract.

Certain of the methods by which an obligation may be extinguished are of general application rather than being peculiar to construction contracts. For example, an obligation may be extinguished by acceptilation, where the creditor discharges his right without payment or performance. An obligation for the payment of money may also be extinguished by confusion, where the same person becomes creditor and debtor in the obligation. This does not apply where there are continuing rights and obligations beyond the payment of money and, thus, were the employer under a building contract to take over the contractor (or vice versa) during the currency of the contract, the doctrine of confusion would not apply. A detailed examination of these doctrines is beyond the scope of this book.

It must be borne in mind that, at any time during the currency of a contract, it is open to parties to enter into an agreement whereby their respective obligations are extinguished. Construction contracts usually contain detailed mechanisms whereby either or both of the parties are entitled to terminate.

In this chapter we will examine certain methods of extinction of obligations that are of particular significance to construction contracts. Certain others, such as payment (Chapter 8) and novation (Section 12.8) are dealt with elsewhere in this book.

9.2 Frustration and impossibility

Frustration occurs whenever, without fault on the part of either party, intervening circumstances have rendered a contract incapable of being performed, or so altered the conditions that, if there were to be performance, it would, in essence, be performance of a different contract, see *Davis Contractors Ltd* v. *Fareham UDC* (1956) and *National Carriers Ltd* v. *Panalpina (Northern) Ltd* (1981). In judging whether or not a

MacRoberts on Scottish Construction Contracts, Third Edition. MacRoberts.

contract has been frustrated, the contract must be viewed as a whole. The question to be considered is whether the purpose of the contract, as gathered from its terms, has been defeated, see *James B Fraser & Co. Ltd* v. *Denny, Mott & Dickson Ltd* (1944). It follows that if parties had regard to the possibility of a particular event and made provision for it in their contract, the occurrence of such an event cannot have the effect of frustrating the contract, see *Cricklewood Property & Investment Trust Ltd* v. *Leighton's Investments Trust Ltd* (1945). If the contract does not contemplate the intervening circumstances, it will be frustrated. The intervening circumstances must, however, be thoroughly investigated before it can be concluded that the contract has been frustrated where, though not contemplated by the contract, the circumstances still allowed for aspects of the contract to continue, see *Islamic Republic of Iran Shipping Lines* v. *Steamship Mutual Underwriting Association (Bermuda) Ltd* (2010).

Whether or not a contract has been frustrated will, in each case, be a question of fact to be decided upon the true construction of the terms of the contract, read in the light of the nature of the contract and of the relevant surrounding circumstances when the contract was made, see *Head Wrightson Aluminium Ltd* v. *Aberdeen Harbour Commissioners* (1958).

The propositions relevant to the doctrine of frustration were set out by Lord Justice Bingham (as he then was) in *J Lauritzen AS* v. *Wijsmuller BV ('The Super Servant Two')* (1990):

> Certain propositions, established by the highest authority, are not open to question:
>
> 1. The doctrine of frustration was evolved to mitigate the rigour of the common law's insistence on literal performance of absolute promises … The object of the doctrine was to give effect to the demands of justice, to achieve a just and reasonable result, to do what is reasonable and fair, as an expedient to escape from injustice where such would result from enforcement of a contract in its literal terms after a significant change in circumstances.
>
> 2. Since the effect of frustration is to kill the contract and discharge the parties from further liability under it, the doctrine is not to be lightly invoked, must be kept within very narrow limits and ought not to be extended.
>
> 3. Frustration brings the contract to an end forthwith, without more and automatically.
>
> 4. The essence of frustration is that it should not be due to the act or election of the party seeking to rely on it … A frustrating event must be some outside event or extraneous change of situation.
>
> 5. A frustrating event must take place without blame or fault on the side of the party seeking to rely on it.

In the leading case of *Davis Contractors Ltd*, the House of Lords held that a contract which had been scheduled to take eight months, and was said to be subject to there being adequate supplies of labour available as and when required, but which took 22 months to complete due to unanticipated shortages of labour and materials, had not been frustrated. The qualification as to the availability of adequate supplies of labour was contained in a letter which accompanied the contractor's tender. That letter was held not to form part of the contract and the contractor had to bear the additional costs.

Nevertheless, there may be circumstances where modifications, which necessarily and fundamentally alter the whole design of a project, frustrate the original contract and entitle the contractor to a claim based upon *quantum meruit*. This is considered more fully in Section 8.4. The absence of intimation by a contractor that he is proceeding upon a *quantum meruit* basis may be an important element in deciding whether there has, in fact, been frustration.

Another example of frustration is where the performance of a contract is dependent upon a certain thing existing and that thing is either destroyed or is so fundamentally altered that the contract cannot be performed. This is known as *rei interitus*. If this occurs prior to the contractor taking possession of the site, then neither party will have a claim against the other. If it occurs when building works are underway, the contractor has a claim for the work carried out and the materials supplied. By the doctrine of accession, property in the building passes to the owner of the ground upon which it is erected and the contractor's entitlement to payment arises under the principle of *res perit domino* (a thing perishes to its owner). A contractor may not be entitled to payment if the work carried out is so defective that the employer would have a defence to an action raised against him. If payments have been made in advance and the contract is subsequently frustrated, the payments made can be recovered under the doctrine known as *condictio causa data causa non secuta* (a claim that the consideration has failed of its purpose), see *Cantiere San Rocco, SA* v. *Clyde Shipbuilding and Engineering Co. Ltd* (1923).

Where the contract is frustrated, it is more accurately parties' rights and obligations as to future performance under the contract that are frustrated. In the context of building contracts this distinction is important as, even after frustration, certain clauses, most notably arbitration clauses, may continue to be enforceable, see *Heyman and Another* v. *Darwins Ltd* (1942). The same position is likely to prevail with adjudication clauses, see *A & D Maintenance and Construction Ltd* v. *Pagehurst Construction Services Ltd* (2000).

In *Robert Purvis Plant Hire Ltd* v. *Alex Brewster & Sons* (2009), it was held that a planning enforcement notice which prevented the intended use of a site was not a supervening event which frustrated a contract for the lease of that site. The tenant argued that the enforcement notice prevented the use of the land as specified in the user clause of the lease, but the court rejected this argument, holding that there was no supervening event as the planning status of the site had been known at the time the lease was entered into.

9.3 *Force majeure*

9.3.1 General

As frustration cannot apply where the parties to a contract have had regard to the possibility of a particular event and made provision for it in their contract, difficulties of interpretation may arise in determining whether frustration has occurred. To address this problem many contracts expressly provide for events that might ordinarily be sufficient to frustrate the contract. Such clauses are known as *force majeure* clauses.

The term *force majeure* is believed to originate from France and in particular the Code of Napoleon. It has no particular technical meaning in Scotland. The term covers events, such as war, epidemics and strikes, beyond the control of the party to the contract who seeks to rely upon the clause. It is also said to encompass any direct legislative or administrative interference, see *Lebeaupin* v. *Richard Crispin & Co.* (1920).

By its very nature a *force majeure* clause will have to be read carefully in conjunction with the remaining terms of the contract to establish, precisely, its scope.

9.3.2 *Force majeure* under the SBC and the SBC/DB

By virtue of clause 2.29.14 of the SBC and clause 2.26.14 of the SBC/DB, *force majeure* constitutes a relevant event which may give rise to an extension of time. Civil commotion and the use or threat of terrorism and/or the activities of the relevant authorities in dealing with such events or threat are separate relevant events under clause 2.29.11 of the SBC and clause 2.26.10 of the SBC/DB, though were they not specifically provided for, they might otherwise, in any event, fall within the ambit of a *force majeure* clause.

Clause 8.11.1.1 of both the SBC and the SBC/DB provides that if, before practical completion of the works, *force majeure* causes the carrying out of the whole or substantially the whole of the uncompleted works to be suspended for a continuous period of time specified in the Contract Particulars, either the Employer or the Contractor is, in defined circumstances, entitled to terminate the employment of the Contractor. Termination is considered in Section 9.4.

9.3.3 *Force majeure* under the NEC3

Although the words *force majeure* are not used, clause 19 (Prevention) covers situations that would normally be within the scope of, in defined circumstances, a *force majeure* clause. It covers an event which stops the Contractor from completing the works or stops the Contractor completing the works by the date shown on the Accepted Programme and which neither party could prevent and which an experienced contractor would have judged at the point of entering into the contract to have such a small chance of occurring that it would have been unreasonable for him to have allowed for it. If an event occurs which falls within this clause, the Project Manager gives an instruction to the Contractor stating how he is to deal with the event.

An event such as this is a compensation event under clause 60.1(19). Clause 91.7 provides that, if such an event occurs, the Employer may terminate the Contractor's obligation to Provide the Works. Termination is considered in Section 9.4. For a more detailed discussion on the prevention provisions, see Section 5.2.4.

9.4 Termination

9.4.1 Contractual provision

Most construction contracts contain express provisions regulating the rights of either or both of the parties in defined circumstances to terminate the contract, or bring it to an end. Such rights should be used with caution and the party exercising the right should ensure that the termination procedure laid down in the contract is strictly adhered to, see *Muir Construction Ltd* v. *Hambly Ltd* (1990). Where the contract has already been brought to an end, it may not be terminated, see W *Hanson (Harrow) Ltd* v. *Rapid Civil Engineering Ltd and Another* (1987). A party who purportedly operates a termination clause in circumstances where they are not entitled to do so may be treated as having repudiated the contract, see *Architectural Installation Services Ltd* v. *James Gibbons Windows Ltd* (1989), though see also *Lockland Builders Ltd* v. *Rickwood* (1995).

By purporting to terminate the contract a party is clearly indicating that they are not going to perform in the future. Unless they are permitted to do that under the contract, such an intention constitutes a repudiation. In such circumstances the other party to the contract is entitled to accept the repudiation, rescind the contract and seek damages. The concepts of repudiation and rescission are considered at Section 9.5. Such a state of affairs may prove welcome to the recipient of the purported termination notice, particularly if that party was finding it difficult to perform in the first place!

Assuming the contract is properly terminated, what is the effect of that termination? The contract as a whole is not terminated. While many of the obligations under the contract, including what might conveniently be termed the principal obligations (e.g. the obligation of the contractor to execute the contract works), will no longer be enforceable, the remaining obligations are fundamentally altered but continue to have effect, see *Mac-Jordan Construction Ltd* v. *Brookmount Erostin Ltd* (1991). Ordinarily a termination clause will also provide for the respective rights and duties of the parties in the event of such a clause being operated. By their nature, those provisions are intended to operate upon the contract being terminated. In each case, it is the employment of the contractor under the contract that is terminated, not the contract itself.

An arbitration clause will continue to be operative, notwithstanding the fact that the contract has been terminated, see *R & J Scott* v. *Gerrard* (1916). The same position is likely to prevail with adjudication clauses, see *A & D Maintenance and Construction Ltd* v. *Pagehurst Construction Services Ltd* (2000). Where the contract contains provisions that deal with an assessment of the sums due to or by either party and an accounting of such sums on termination, the courts in Scotland have held that a claim based upon an alleged breach of contract is irrelevant. The correct way to proceed is to claim for payment based upon the contractual provisions, see *Muir Construction Ltd.*

In this section we will examine the termination provisions under the SBC, the SBC/DB and the NEC3.

9.4.2 General termination provisions under the SBC and the SBC/DB

The general termination provisions of the SBC and the SBC/DB are to be found in clause 8. Clause 8.1 sets out the meaning of insolvency and is considered in Section 9.8.2.

Clause 8.2 provides that a notice served under clause 8 is not to be given unreasonably or vexatiously. For a consideration of the phrase 'unreasonably or vexatiously', see *J M Hill & Sons Ltd* v. *London Borough of Camden* (1980); *John Jarvis Ltd* v. *Rockdale Housing Association Ltd* (1986); and *Ferrara Quay Ltd* v. *Carillion Construction Ltd* (2009). In *Ferrara Quay Ltd*, the Employer was a special purpose vehicle, the Contractor threatened to terminate to safeguard its financial position, due to the Employer's failure to make timeous payments and the refusal by the guarantor to step into the Employer's shoes to secure funding for the rest of the project. The Employer obtained an interim injunction to prevent the Contractor terminating and then applied for a continuation of the hearing in respect of the injunction on the basis that it would be unreasonable and vexatious of the Contractor to give notice to terminate, particularly given that the project was only a few months from completion. The court discharged the interim injunction and held that, despite the presence of a guarantor (in respect of which the guarantee was limited), the Employer remained 'an uncreditworthy special purpose vehicle' and in the prevailing economic climate, the Contractor would not be acting unreasonably or vexatiously in giving notice to terminate. Whether the giving of a notice of termination is unreasonable or vexatious will be a question of fact in each case.

Termination takes effect upon receipt of the relevant notice, see clause 8.2.2. Each notice referred to in clause 8 requires to be given in writing. Unlike many other notices under the SBC and the SBC/DB (see clause 1.7), notices relevant to termination require to be given by actual, special or recorded delivery, see clause 8.2.3. There is a deeming provision in relation to receipt of notices given by special or recorded delivery post (see clause 1.7.4). Failure to comply with the provisions of the contract as to the giving of notice can be fatal, see, for example, *Muir Construction Ltd* v. *Hambly Ltd* (1990). Compare, however, the decision of the House of Lords in *Mannai Investment Co. Ltd* v. *Eagle Star Life Assurance Co. Ltd* (1997).

The relevant provisions of clause 8 are, in terms of clause 8.3.1, stated to be without prejudice to any other rights and remedies available to the Employer and to the Contractor. The other remedies open to the Employer and to the Contractor are considered in Chapter 10.

It should be borne in mind that irrespective of the grounds of termination, the Contractor's employment may be reinstated at any time and on such terms as the parties agree, see clause 8.3.2.

9.4.3 General termination provisions under the NEC3

The termination provisions in the NEC3 are contained in clauses 90–93. Clause 90.1 states that if either party wishes to terminate the Contractor's obligation to Provide the Works he notifies the Project Manager and the other party, giving details of his reason for terminating. The Project Manager will then issue a termination certificate to both parties promptly, providing that the reason given complies with the contract. Unlike the SBC and the SBC/DB, the termination provisions are not split into separate sections for termination by the Employer, the Contractor or either party. There are four relevant clauses, namely, termination (clause 90); reasons for termination (clause 91); procedures on termination (clause 92); and payment on termination (clause 93). Clause 90.2 contains the 'Termination Table'. This Table sets out which party can terminate; for what reasons; the procedure to be followed; and how the amount due is calculated. There is no specific notification procedure for termination under the NEC3. Parties should follow the requirements of clause 13. The procedures for termination are implemented immediately after the Project Manager has issued the termination certificate and, within 13 weeks of termination, the Project Manager certifies a final payment to or from the Contractor. This is the Project Manager's assessment of the amount due on termination less the total of previous payments. Payment is then made within three weeks of the certificate. Once a termination certificate has been issued, the Contractor is not permitted to do any further work.

9.4.4 Termination by the Employer under the SBC and the SBC/DB

Clause 8.4 of the SBC and the SBC/DB entitles the Employer to terminate the employment of the Contractor if the Contractor continues a 'specified default' for 14 days after receiving a notice specifying that default. The 'specified defaults' relied upon must arise before the date of practical completion of the Works and are set out in clause 8.4.1. These are where:

- The Contractor without reasonable cause, wholly or substantially suspends the carrying out of the Works (or the design of the Contractor's Designed Portion in the case of the SBC);
- fails to proceed regularly and diligently with the Works or the design of the Contractor's Designed Portion in the case of the SBC, and with the performance of his obligations under the contract in the case of the SBC/DB. The inclusion of such a termination event does not necessarily mean a separate implied obligation to proceed regularly and diligently, see *Leander Construction Limited* v. *Mulalley & Company Limited* (2011) discussed in Section 3.4.4. For a recent decision on an employer's right to terminate for the contractor's failure to proceed with 'due diligence', see *SABIC UK Petrochemicals Ltd* v. *Punj Lloyd Ltd* (2013).
- The Contractor refuses or neglects to comply with a written notice or instruction from the Architect under the SBC and from the Employer under the SBC/DB requiring the removal of any work, materials or goods not in accordance with the contract, where the Works are materially affected by such refusal or neglect.

- The Contractor fails to comply with the provisions of either clause 3.7 or clause 7.1 of the SBC or clause 3.3 or clause 7.1 of the SBC/DB (which relate to sub-letting or assignation without written consent).
- The Contractor fails to comply with the provisions of clause 3.23 of the SBC or clause 3.16 of the SBC/DB (which require compliance with the requirements of the Construction (Design and Management) Regulations 2007).

The Employer's right to terminate the contract arises on, or within, 21 days from the expiry of the 14-day period. That right is exercised by serving a further notice upon the Contractor, see clause 8.4.2. The termination takes effect on the date of receipt of that further notice.

If the Contractor remedies the specified default, or the Employer elects not to terminate, and the Contractor repeats the specified default (whether they have previously repeated it or not), then upon, or within a reasonable time after, such repetition the Employer is entitled to serve notice of termination on the Contractor, see clause 8.4.3. In these circumstances, the specified default need not continue for 14 days; repetition alone is sufficient. A reasonable time need not elapse between the repetition of the specified default and the giving of notice, see *Reinwood Ltd* v. *L Brown & Sons Ltd* (2007).

Certain insolvency events will also entitle the Employer to terminate. The effect of insolvency is considered in Section 9.8.

By virtue of clause 8.6 the Employer is entitled to terminate the employment of the Contractor if the Contractor or any person they employ has been involved in any of the corruption-related activities specified in that clause, such as the commission of an offence under the Bribery Act 2010, see Section 22.8.

The consequences of termination by the Employer are set out in clause 8.7. The Employer may employ other persons to carry out and complete the Works and to make good any defects. In addition, under the SBC, the Employer may also employ others where applicable to carry out and complete the design of the Contractor's Designed Portion. The Employer may also enter upon and take possession of the site and the Works and, subject to obtaining any necessary third party consents, may use all temporary buildings, plant, tools, equipment and site materials for those purposes.

When required in writing by the Architect under the SBC and the Employer under the SBC/DB so to do (but not before), the Contractor is required to remove or procure the removal from the Works of any temporary buildings, plant, tools, equipment, goods and materials belonging to the Contractor or Contractor's Persons. Under the SBC where there is a Contractor's Designed Portion, the Contractor is obliged, without charge, to provide to the Employer copies of all the Contractor's Design Documents then prepared, whether or not previously provided by the Contractor to the Employer. There is a similar provision in the SBC/DB, however, the words 'without charge' are omitted. Clause 8.7.2.3 provides that, if required to do so by the Employer (or by the Architect on the Employer's behalf under the SBC) within 14 days of the date of termination, the Contractor is obliged to assign (so far as assignable and so far as the Contractor may lawfully be required to do so) to the Employer, without charge, the benefit of any agreement for the supply of materials or goods and/or for the execution of any work for the purposes of the contract. The footnote to this clause points out that it may not be effective in cases of the Contractor's insolvency.

Perhaps more significantly in the context of termination by the Employer, by virtue of clause 8.7.3 no further sum shall become due to the Contractor other than any amount that may become due to him under clauses 8.7.5 or 8.8.2. The Employer need not pay any sum that has already become due either (a) insofar as the Employer has given or gives a Pay Less Notice under clause 4.12.5 of the SBC or clause 4.9.4 of the SBC/DB or (b) if the Contractor, after the last date upon which such notice could have been given by the Employer in respect of that sum, has become insolvent within the meaning of clauses 8.1.1 – 8.1.3.

The accounting provision upon termination is to be found in clause 8.7.4. Following the completion of the Works and the making good of defects (or of instructions otherwise as referred to in clause 2.38 of the SBC and clause 2.35 of the SBC/DB), an account requires to be set out within three months thereafter in a certificate issued by the Architect under the SBC or a statement prepared by the Employer under the SBC/DB. This contains the amount of expenses properly incurred by the Employer, including any direct loss and/or damage caused to the Employer and for which the Contractor is liable, whether arising as a result of termination or otherwise; the amount of payments made to the Contractor; and the total amount which would have been payable for the Works in accordance with the contract. The difference between the sum of the amount of expenses properly incurred and the amount of payments made to the Contractor, on the one hand, and the total amount which would have been payable for the Works, on the other, is a debt payable by the Contractor to the Employer or, in the rare circumstances of the Works being completed for less money than originally contracted for, by the Employer to the Contractor. Clause 8.8 makes provision in relation to the circumstances where the Employer elects not to complete the Works.

9.4.5 Termination by the Contractor under the SBC and the SBC/DB

The Contractor's right to terminate their own employment is governed by clause 8.9. As with Employer termination, the clause sets out certain specified defaults. In addition, there are also what are termed 'specified suspension events', the occurrence of which can entitle the Contractor to terminate their own employment under the contract.

Specified defaults

Unlike termination by the Employer, the specified defaults which entitle the Contractor to determine their employment under the contract can arise both before and after practical completion. The specified defaults by the Employer are set out in clause 8.9.1. These are:

- The Employer does not pay by the final date for payment an amount properly due in accordance with clause 4.12 of the SBC and clause 4.9 of the SBC/DB and/or any VAT due thereon.

- The Employer fails to comply with the provisions of clause 7.1 (assigning the contract without the written consent of the Contractor); and
- The Employer fails to comply with their undertakings in respect of the Construction (Design and Management) Regulations 2007.

Additionally, in the case of the SBC only, it is a specified default by the Employer if they interfere with or obstruct the issue of any certificate under the contract.

Specified suspension events

The specified suspension events, which must arise prior to the date of practical completion under the SBC and after the Date of Possession or any deferred Date of Possession but before practical completion under the SBC/DB, are set out in clause 8.9.2 of each contract. This provision applies when the carrying out of the whole, or substantially the whole, of the uncompleted Works is suspended for a continuous period of the length specified in the Contract Particulars due to specified events, namely:

- any impediment, prevention or default, whether by act or omission, by the Employer or any of the Employer's Persons and (in the SBC only) the Architect or the Quantity Surveyor, unless caused by negligence or default of the Contractor or any of the Contractor's Persons; and/or
- under the SBC, Architect's instructions issued under clause 2.15 (discrepancies in or divergence between contract documents), clause 3.14 (instructions requiring a Variation) or clause 3.15 (postponement of any work to be executed under the contract) unless caused by the negligence or default of the Contractor, or any of the Contractor's Persons, or any Statutory Undertaker.

Once a notice has been given, the ensuing procedure is to all intents and purposes identical to that which operates in the case of Employer termination, the only difference being the necessary modifications made to accommodate specified suspension events. Similarly, the provisions which deal with the repetition of a specified default or of a specified suspension event are virtually identical to those in Employer termination.

The consequences of termination by the Contractor are specified in clause 8.12. First, the other provisions of the contract which require any further payment or any release of Retention to the Contractor cease to apply. Upon termination, the Contractor shall with all reasonable dispatch remove or procure the removal from the site of any temporary buildings, plant, tools and equipment which belong to the Contractor or to the Contractor's Persons and, unless they have become the property of the Employer, all goods and materials (including Site Materials). In the SBC, where there is a Contractor's Designed Portion, the Contractor is obliged, without charge, to provide to the Employer two copies of the as-built drawings then prepared. In the SBC/DB, the Contractor is obliged to provide the Employer with copies of the documents referred to in clause 2.37 (the Contractor's Design Documents) then prepared. As with the termination provisions of clause 8.4, the words 'without charge' are omitted in the SBC/DB and there appears to be no restriction on the number of copies.

Where the Contractor's employment is terminated by reason of default by the Employer or the insolvency of the Employer, the Contractor shall as soon as reasonably practicable prepare and submit an account or, not later than two months after the date of termination, provide the Employer with all documents necessary for the Employer to prepare the account, which the Employer shall do with reasonable dispatch (and in any event within three months of receipt of such documents).

The account, which is prepared in accordance with clause 8.12.3, sets out the total value of the work properly executed at (and in the SBC/DB, of any design work properly carried out before) the date of termination; any sums ascertained in respect of direct loss and/or expense (whether ascertained before or after the date of termination); the reasonable costs of removal of any temporary buildings, plant, tools and equipment; the cost of materials or goods (including Site Materials) properly ordered for the Works for which the Contractor then has paid or is legally bound to pay; and any direct loss and/or damage caused to the Contractor by the termination.

After taking into account amounts previously paid to the Contractor, the Employer is required to pay to the Contractor the amount properly due in respect of the account within 28 days of its submission by the Employer to the Contractor (or vice versa), without deduction of any Retention. Payment by the Employer for any such materials and goods as are referred to in the account shall be subject to such materials and goods thereupon becoming the property of the Employer.

Certain insolvency events entitle the Contractor to terminate. These are considered in Section 9.8.

9.4.6 Termination by either party under the SBC and the SBC/DB

Clause 8.11 of the SBC provides for certain circumstances which will entitle either the Employer or the Contractor to terminate the employment of the Contractor. Each of the specified circumstances must arise before the date of practical completion and must cause the carrying out of the whole or substantially the whole of the uncompleted Works to be suspended for the relevant continuous period of time set out in the Contract Particulars. The events, provided for by clause 8.11.1, are:

- *force majeure;*
- loss or damage to the Works occasioned by any of the Specified Perils set out in clause 6.8;
- civil commotion (which has been defined as a stage between riot and war, see *Levy* v. *Assicurazioni Generali* (1940)) or the threat of terrorism and/or the activities of the relevant authorities in dealing with such event or threat;
- under the SBC, Architect's instructions under clauses 2.15, 3.14 or 3.15 issued as a result of the negligence or default of any Statutory Undertaker or in the SBC/DB, Employer's instructions under clauses 2.13, 3.9 or 3.10 issued in the same circumstances;
- the exercise by the United Kingdom Government of any statutory power which directly affects the execution of the Works; and

- under the SBC only, delay in receipt of any permission or approval for the purposes of any statutory provisions and any decision of a relevant authority thereunder which controls the right to develop the site ('Development Control Requirements') necessary for the Works to be carried out or proceed, which delay the Contractor who has taken all practicable steps to avoid or reduce.

Upon the occurrence of one or more of these events, and once the period specified in the Contract Particulars has expired, either party may give notice to the other to the effect that unless the suspension ceases within seven days after receipt of that notice the employment of the Contractor may be terminated. This is done by way of further notice, see clause 8.11.

The Contractor is not entitled to give notice where the loss or damage to the Works occasioned by one or more of the Specified Perils is caused by negligence or default on their part or on the part of any of the Contractor's Persons. The consequences of termination under this clause are identical to those where the Contractor terminated, see clause 8.12 and Section 9.4.5.

9.4.7 Termination by the parties under the NEC3

Under the NEC3, if either party wishes to terminate the Contractor's obligation to Provide the Works he notifies the Project Manager and the other Party, giving details of his reasons for terminating, the Project Manager then issues a termination certificate to both Parties promptly if the reason complies with the contract, see clause 90.1. The permitted reasons for termination are set out in clause 91. There are 21 permitted reasons in total, covering:

- insolvency (reasons 1 – 10) for which either party can terminate;
- defaults by the Contractor which he has failed to put right within four weeks of notification by the Project Manager (reasons 11 – 15), in terms of which only the Employer can terminate. The defaults include substantially failing to comply with his obligations, not providing a bond or guarantee which the contract requires, appointing a Sub-contractor for substantial work before the Project Manager has accepted the Sub-contractor, substantially hindering the Employer or others, and substantially breaking a health or safety regulation.
- failure by the Employer to make payment to the Contractor of an amount due under the contract within 11 weeks of the date that it should have been paid (reason 16), in terms of which only the Contractor can terminate. In the 2005 edition of the NEC3, that clause referred to a failure by the Employer to make payment to the Contractor of an amount certified by the Project Manager within 13 weeks of the date of the certificate.
- where the Parties have been released under the law from further performance of the whole of the contract (reason 17) for of which either party can terminate;
- an instruction given by the Project Manager to the Contractor to stop or not to start any substantial work or all work has not been followed within 13 weeks by a further instruction allowing the work to re-start or start. If the instruction was

due to a default by the Contractor, the Employer may terminate (reason 18). If the instruction was due to a default by the Employer, the Contractor may terminate (reason 19). If the instruction was due to any other reason, either party may terminate (reason 20); and

- the occurrence of an event which stops the Contractor completing the works or stops the Contractor completing the works by the date shown on the accepted programme and is forecast to delay completion by more than 13 weeks and which neither party could prevent and an experienced contractor would have judged at the date of entering into the contract to have such a small chance of occurring that it would have been unreasonable for him to have allowed for it (reason 21), in terms of which case only the Employer can terminate.

The procedures to be followed on termination are found in clause 92, which requires to be read along with the Termination Table in clause 90.2 specifying which procedures are applicable to which termination reasons.

Procedure 1: the Employer may complete the works and use any plant and material intended to have been included in the works to which he has title;
Procedure 2: the Employer may instruct the Contractor to leave the site, remove any equipment, plant and materials from the site and assign the benefit of any sub-contract or other contract related to performance of the contract to the Employer;
Procedure 3: the Employer may use any Equipment (as defined in clause 11.2(7)) to which the Contractor has title to complete the works. The Contractor must promptly remove the Equipment from site when the Project Manager notifies him that the Employer no longer requires it to complete the works; and
Procedure 4: the Contractor leaves the Working Areas (as defined in clause 11.2(18)) and removes the Equipment.

The procedures governing payment on termination are found in clause 93. The amount due on termination includes an amount due assessed as for normal payments; the Defined Cost for plant and materials within the working areas or to which the Employer has title and of which the Contractor has accepted delivery; other Defined Cost reasonably incurred in expectation of completing the whole of the works; any amounts retained by the Employer; and a deduction of any un-repaid balance of an advanced payment. In addition, the amount due on termination includes one or more of the following, depending on the reason for termination as set out in the Termination Table:

- the forecast Defined Cost of removing the Equipment;
- a deduction of the forecast of the additional cost to the Employer of completing the whole of the works; and
- the direct fee percentage applied to (i) for Options A, B, C and D, any excess of the total of the Prices at the Contract Date over the Price for Work Done to Date; or (ii) for Options E and F, any excess of the first forecast of the Defined Cost for the works over the Price for Work Done to Date less the Fee.

The termination provisions apply to all of the Options A–F, subject to the following modifications for specific Options:

- Option A (priced contract with activity schedule) inserts a clause 93.3 which states that the amount due on termination is assessed without taking grouping of activities into account.
- Option C (target contract with activity schedule) inserts two sub-clauses. First, clause 93.4 provides that on termination the Project Manager assesses the Contractor's share after he has certified termination. His assessment uses, as the Price for Work Done to Date, the total of the Defined Cost which the Contractor has paid and committed to pay for work done before termination. The assessment uses as the total of the Prices the lump sum price for each activity which has been completed and, for each incomplete activity, the appropriate proportion of the lump sum price for the work that has actually been completed. Second, clause 93.6 provides that the Project Manager's assessment of the Contractor's share is added to the amount due to the Contractor on termination if there has been a saving or deducted if there is an excess.
- Option D (target contract with bill of quantities) inserts two sub-clauses. First, clause 93.5 provides that on termination the Project Manager assesses the Contractor's share after he has certified termination. His assessment uses as the Price for Work Done to Date, the total of the Defined Cost which the Contractor has paid and committed to pay for work done before termination. The second sub-clause is the same clause 93.6 as is used in Option C.

9.5 *Repudiation and rescission*

Notwithstanding the absence in a contract of detailed termination provisions, circumstances may arise in which a party is freed from future performance. The concepts of repudiation and rescission are inextricably linked. Rescission is considered in more detail in Section 10.2. In certain circumstances a material breach of contract by one party may entitle the other party, the 'innocent party', to terminate the contract. Such a material breach of contract is referred to as a repudiation and gives the innocent party a choice. They can accept the repudiation and rescind the contract or, alternatively, they may elect to ignore the repudiation and continue with the performance of the contract. This option exists because, as stated by Lord Keith in *Woodar Investment Development Ltd* v. *Wimpey Construction UK Ltd* (1980), the doctrine of repudiation exists for the benefit of the innocent party. Whether the option is restricted in certain circumstances is considered in Section 10.2.

If the repudiation is accepted, the acceptance should be communicated to the party in breach. The method by which communication is made would appear to be immaterial, see *Monklands DC* v. *Ravenstone Securities* (1980).

The remedy that is open to an innocent party who elects to rescind, namely damages, is considered at Section 10.4. Should the innocent party elect to ignore the repudiation, they may be barred from relying upon the material breach at a later date.

It is difficult to generalize as to what conduct is, and what conduct is not, a repudiation. Not every material breach will constitute a repudiation, see *Blyth* v. *Scottish*

Liberal Club (1982) approved in *Tehrani* v. *Argyll and Clyde Health Board (No.2)* (1989). Should one party refuse to perform their obligations under the contract, that is likely to constitute a repudiation. Should an employer prevent a contractor from carrying out the contract works, for example, by engaging another contractor to carry out all or part of those works, that too is likely to constitute a repudiation, see *Sweatfield Ltd* v. *Hathaway Roofing Ltd* (1997).

In Scotland, the precise effect of the acceptance of a repudiation and resultant rescission of the contract has been examined by the Inner House of the Court of Session. The case of *Lloyds Bank plc* v. *Bamberger* (1993) provides a clear and succinct exposition of the position under Scots law where a contract has been rescinded.

In *Lloyds Bank plc*, Lord Ross stated that, following rescission, both parties are freed from future performance of their primary obligations under the contract. Nevertheless, parties continue to be bound by the primary obligations which are extant at the time of rescission. The contract does not come to an end. The innocent party is entitled to sue the party in default for damages for breach of contract. Ancillary clauses which the parties intended would survive rescission, such as arbitration clauses, may be enforced after rescission. Apart from such ancillary clauses, the contract may also contain clauses which affect damages due for breach of contract, such as a liquidated damages clause. The language of the contract may be such as to demonstrate that the parties intended such clauses to be enforceable after rescission.

Rescission should be distinguished from contractual termination, considered above in Section 9.4. In the latter, the employment of the contractor is terminated, in certain instances upon the occurrence of events that are not the fault of either party. Contractual termination clauses seek to bring some degree of certainty to the circumstances in which the parties' contract comes to end prematurely.

9.6 Death and illness

The effect of the death or incapacity of a party to a building contract will primarily depend upon whether or not the contract involves an element of *delectus personae*. A contract that involves *delectus personae* means that one party to the contract entered into it in reliance upon certain qualities possessed by the other. Where such qualities are a necessary element of the contract, the death or incapacity of the party who is bound to perform clearly prevents the contract being performed and, thus, brings it to an end. For examples of this, see *Hoey* v. *McEwan & Auld and Others* (1867) and *Smith* v. *Riddell* (1886). The existence of *delectus personae* in a contract has a direct bearing upon whether or not that contract is capable of being assigned. This is considered in Section 12.4.

The delegation of building work (through the use of sub-contractors) is an everyday occurrence and, therefore, in the absence of special circumstances or an express contractual provision to the contrary, *delectus personae* will not apply and the obligation to perform will pass to the personal representatives of the deceased party. It will be for the representatives to secure alternative contractors to carry out the works, or to complete them themselves.

As a building contract is ordinarily divisible (unlike a contract for a painting or a sculpture), it would appear that remuneration can be claimed by a deceased party's representatives for work partially carried out up to the date of death. The valuation of that work may be problematic, particularly if the contract does not have a mechanism for valuation, and may require equitable adjustment.

Illness and incapacity need to be treated in a like manner, though the position is, perhaps not surprisingly, not as clear-cut as in the case of death. The effect of illness or incapacity is one of degree and will depend upon the whole circumstances, most notably the likely duration of the illness or incapacity in relation to the length of the contract. Even where illness is not sufficient to bring the contract to an end, it has been enough to entitle the employer to rescind the contract, see *Manson* v. *Downie* (1885), and to constitute a breach of contract, see *McEwan* v. *Malcolm* (1867). However, in *Atwal* v. *Rochester* (2010), the contractor (a sole trader) became seriously ill and was unable to work. The employer claimed that the contractor had breached his contractual obligations and sought delay damages together with the additional expense incurred as a result of engaging alternative contractors. The contractor argued that the contract was discharged by frustration rather than a breach and counterclaimed for payment of the value of the work done to date. The court agreed with the contractor on the basis that the contract was a personal service contract in terms of which the contractor himself undertook to carry out the work and manage any specialist work carried out by sub-contractors.

Since the vast majority of contractors are now limited companies, the problems occasioned by death and illness are unlikely to arise on a regular basis. However, it should be noted that *delectus personae* may arise in employer/contractor relationships that do not involve individuals, as demonstrated by the case of *Scottish Homes* v. *Inverclyde DC* (1997). The issue may also arise in the case of architects or engineers appointed under a construction contract or in the case of an adjudicator named in a construction contract, see *Amec Capital Projects Ltd* v. *Whitefriars City Estates Ltd* (2004).

9.7 Illegality

While a detailed examination of the concept of illegality is beyond the scope of this book, it does merit some consideration.

In general terms, an illegal contract is one which the law will not enforce. However, there is a distinction between illegal contracts and those that are associated with an unenforceable transaction. Perhaps the best example that can be given to illustrate the latter is gambling. While gambling is not illegal, until the coming into force of the Gambling Act 2005, the Scottish courts would not entertain actions to determine wagers. This was for reasons of public policy, not illegality.

What precisely constitutes an illegal contract is open to question. A number of vague and differing concepts such as 'moral turpitude' and 'subversive of the interests of the State' have been used, see *Jamieson* v. *Watt's Trustee* (1950). The matter is far from clear, as is demonstrated by the decision in *Cuthbertson* v. *Lowes* (1870) in which it was held that a contract which contravened a statute was not necessarily illegal.

Whether or not either party questions the legality of the contract the court will have regard to it, see *F W Trevalion & Co.* v. *J J Blanche & Co.* (1919). Where a contract is held to be illegal the court will not interfere as between the rights of the parties to the contract. This is consistent with the general principle that the courts will not assist the party who is in breach of a statute, albeit that the corollary of this is that the other party to the contract is entitled to keep the advantage gained by them. This may be regarded as unfortunate where the parties were equally aware of the illegal nature of the transaction.

If the contract is not itself illegal, but has a connection with some other illegal transaction, the contract is said to be tainted with illegality. If one of the parties was unaware of the illegality, they will be entitled to enforce their rights under the contract. However, should they fail to resile from the contract, after becoming aware of the illegality, they may be held to have acquiesced. As a consequence, they may not be entitled to enforce their rights under the contract or be entitled to an equitable remedy, see *Dowling & Rutter* v. *Abacus Frozen Foods Ltd (No. 2)* (2002).

Where only part of a contract is illegal, that part may be capable of being severed from the remainder of the contract. In a building contract which contains the power to instruct variations it may well be possible to instruct a variation to remove the 'offending' part of the contract. The question of severance is a complex one upon which there is little Scottish authority, though the English authorities on this subject are likely to be regarded as highly persuasive. Whether or not an illegal provision is capable of being severed will depend upon the nature of the illegality.

The contract may be valid, but the works executed under it illegal. For example, in *Townsend (Builders) Ltd* v. *Cinema News and Property Management Ltd* (1959), the works as built, but not as specified, contravened a byelaw. In the rather special circumstances of that case the contractor was held to be entitled to recover payment, though it would appear that but for those special circumstances the contractor would not have succeeded. In *Robert Purvis Plant Hire Ltd* v. *Alex Brewster & Sons Ltd* (2009), the use specified in a lease was not permitted by law as there was no planning permission for such use. However, the court held that this did not affect the binding nature of the lease as the planning status of the site was known at the time the lease was signed.

If a contractor carries out work in the absence of necessary consents they take the risk that the work is illegal and they may be unable to recover payment for that work, see *Designers and Decorators (Scotland) Ltd* v. *Ellis* (1957). The contract will often expressly oblige the contractor to give statutory notices and comply with statutory requirements, and may also provide for the circumstances in which there is a change in the statutory requirements. For example, clause 2.1 of the SBC makes such provision.

9.8 Insolvency

9.8.1 General

Insolvency, in itself, does not affect a contract, but has potentially far-reaching implications that merit some examination within the confines of this book.

The insolvent party may be unable to implement their obligations under the contract, which would entitle the other party to withhold performance of their obligations under the contract, see *Arnott and Others* v. *Forbes* (1881). In a case of personal insolvency (known as bankruptcy), the party contracting with the insolvent debtor can compel the debtor's representative (known as a trustee) to make his position clear in relation to the contract. Section 42 of the Bankruptcy (Scotland) Act 1985 provides that the trustee is deemed to have refused to adopt the contract unless he responds within 28 days from the receipt by him of a request in writing from any party to a contract entered into by the debtor, or within such longer period of that receipt as the court on application by the trustee may allow, to adopt or refuse to adopt the contract.

Insolvency can also impact upon a party's right to refer a dispute to adjudication. In *Enterprise Managed Services Ltd* v. *Tony McFadden Utilities Ltd* (2009), the liquidator of a company which had been party to a number of sub-contracts with Enterprise had assigned rights to outstanding and unresolved accounts from some of those contracts to McFadden. Enterprise claimed money under some contracts and the liquidated company under others. McFadden referred a dispute relative to one of the final accounts. However, it was held that the effect of insolvency was that the contracts between Enterprise and the company had ceased to exist, and so the accounts for both parties in the different contracts had to be balanced and set off against each other, leaving only a net balance payable to the company and, in turn, McFadden. Adjudication allows only for the referral of one dispute under one contract and because the net balance in this case arose out of multiple contracts, the dispute was not capable of adjudication.

9.8.2 Insolvency under the SBC and the SBC/DB

Meaning of insolvency

The SBC and the SBC/DB define a party as 'Insolvent' in the circumstances set out in clause 8.1, depending on the type of entity.

If the party is a company, the circumstances are:

- if it enters administration within the meaning of Schedule B1 to the Insolvency Act 1986;
- if it has appointed to it an administrative receiver or a receiver or a manager of its property under Chapter I of Part III of the Insolvency Act 1986 or a receiver under Chapter II of that Part;
- if it has passed a resolution for voluntary winding up without a declaration of solvency under section 89 of the Insolvency Act 1986; or
- if it has a winding up order made against it under Part IV or V of the Insolvency Act 1986.

If the party is a partnership, the circumstances are:

- if it has a winding up order made against it under any provision of the Insolvency Act 1986 as applied by an order under section 420 of that Act; or

- if it has sequestration awarded on the estate of the partnership under section 12 of the Bankruptcy (Scotland) Act 1985 or the partnership grants a trust deed for its creditors.

If the party is an individual, the circumstances are if:

- he has a bankruptcy order made against him under Part IX of the Insolvency Act 1986; or
- his estate is sequestrated under the Bankruptcy (Scotland) Act 1985 or he grants a trust deed for his creditors.

A party also becomes insolvent if he enters into an arrangement, compromise or composition in satisfaction of his debts (excluding a scheme of arrangement as a solvent company for the purposes of amalgamation or reconstruction); or in the case of a partnership, each partner is the subject of an individual arrangement or any other event or proceedings referred to above. Each of the above circumstances also includes any analogous arrangement, event or proceeding in any other jurisdiction.

Insolvency of the Contractor

Clause 8.5 deals with the insolvency of the Contractor. If the Contractor is Insolvent as defined above, the Employer may at any time by notice to the Contractor terminate the Contractor's employment under the contract.

The Contractor is obliged to immediately inform the Employer in writing if they make any proposal, give notice of any meeting or become the subject of any proceedings or appointment relating to any of the matters referred to in the definition of Insolvency above. As from the date the Contractor becomes Insolvent, whether or not the Employer has given such notice of termination, the provisions of the contract setting out the payment and accounting procedure upon termination apply as opposed to any other payment provisions; the Contractor's obligations to carry out and complete the Works are suspended, which in the SBC includes the design of the Contractor's Designed Portion; and the Employer may take reasonable measures to ensure that the site, the Works and Site Materials are adequately protected and that such Site Materials are retained on site. The Contractor is obliged to allow and not to hinder or delay the taking of those measures.

Insolvency of the Employer

Clause 8.10 deals with the Insolvency of the Employer. If the Employer is Insolvent, the Contractor may by notice to the Employer terminate the Contractor's employment under the contract.

The Employer is obliged to immediately inform the Contractor in writing if they make any proposal, give notice of any meeting or become the subject of any proceedings or appointment relating to any of the matters referred to in the definition of Insolvency above. As from the date the Employer becomes Insolvent, the Contractor's obligations to carry out and complete the Works, and (in the case of the SBC) the design of the Contractor's Designed Portion, are suspended.

9.8.3 Insolvency under the NEC3

The NEC3 defines a party as insolvent in the circumstances set out in clause 91.1. If the party is an individual, the circumstances are if he has done one of the following or its equivalent, namely:

- presented his petition for his bankruptcy;
- had a bankruptcy order made against him; or
- made an arrangement with his creditors.

If the party is a company or a partnership, the circumstances are that it has done one of the following or its equivalent, namely:

- had a winding up order made against it;
- had a provisional liquidator appointed to it;
- passed a resolution for winding up (other than in order to amalgamate or reconstruct);
- had an administration order made against it;
- made an arrangement with its creditors; or
- had a receiver, receiver and manager or administrative receiver appointed over the whole or a substantial part of its undertakings or assets.

If an insolvency event as set out in clause 91.1 occurs, this provides a 'reason' for the purpose of clause 90.1 and the other party may terminate in terms of the procedure described in Section 9.4.7.

9.9 *Prescription*

9.9.1 General

Prescription is the establishment or definition of a right or the extinction of an obligation through the lapse of time. The former is termed positive prescription, the latter negative prescription. Positive prescription applies to interests in land, servitudes and public rights of way. Since it has no direct relevance to building contracts, it is not considered further in this chapter.

Prescription falls to be contrasted with limitation. Limitation does not affect the subsistence of rights and obligations. It is a doctrine that denies certain rights of action after the passage of a certain lapse of time, see *Macdonald* v. *North of Scotland Bank* (1942). Limitation periods may be statutory or conventional. Conventional limitation is where the parties set out in their contract that a particular obligation will be extinguished by the lapse of a stipulated time period without a claim being made. Such provisions can appear in construction contracts.

The law in relation to both prescription and limitation is to be found in the Prescription and Limitation (Scotland) Act 1973 ('the 1973 Act'). While a detailed examination of this subject is beyond the scope of this book, we shall consider those aspects of it which are most pertinent to construction contracts, namely short negative prescription and long negative prescription.

Under the 1973 Act the party under the relevant obligation is known as 'the debtor' and the party to whom the obligation is owed is known as 'the creditor'. In relation to the short negative and the long negative prescriptive periods, the general rule is that the burden of proof in establishing whether or not an obligation has prescribed rests with the party alleging the affirmative. For example, if the assertion is that the obligation had subsisted for the prescriptive period, it would be for the party so affirming to prove, see *Strathclyde Regional Council* v. *W A Fairhurst & Partners* (1997) and *Pelagic Freezing (Scotland) Ltd* v. *Lovie Construction Limited* (2010).

9.9.2 Short negative prescription

The short negative prescriptive period of five years is the one most familiar to those in the construction industry. Section 6(1) of the 1973 Act provides that if, after the 'appropriate date', an obligation which is set out in Schedule 1 to the 1973 Act has subsisted for five years (a) without any relevant claim having been made in relation to it, and (b) without the subsistence of the obligation having been relevantly acknowledged, then as from the expiration of the five-year period the obligation in question is extinguished.

A number of technical expressions are used in s.6(1). As we will see below, many of these are equally relevant to long negative prescription. We shall examine each of these expressions in turn.

9.9.3 The appropriate date

The short negative prescriptive period commences upon what is termed the 'appropriate date'. This date varies from obligation to obligation. Schedule 2 to the 1973 Act sets out various obligations and the appropriate date relative to each of them. None of the Schedule 2 obligations are particularly relevant to construction contracts. With the exception of obligations of the kind specified in Schedule 2, the appropriate date in relation to an obligation is the date upon which that obligation became enforceable.

Section 11 of the 1973 Act defines when certain types of obligation become enforceable. For example, an obligation to make reparation for loss, injury or damage caused by an act, neglect or default is regarded as having become enforceable on the date when the loss, injury or damage occurred or was discovered.

There must be an act, neglect or default and resultant loss, injury or damage. The obligation to make reparation does not arise, and thus does not become enforceable, until the loss, injury or damage occurs, see *Watson* v. *Fram Reinforced Concrete Co.*

(Scotland) Ltd (1960), *Dunlop* v. *McGowans* (1979) and *Strathclyde Regional Council* v. *W A Fairhurst & Partners* (1997).

The loss, injury or damage must arise from the act, neglect or default. For example, in *Sinclair* v. *MacDougall Estates Ltd* (1994), it was held that the defenders' act, neglect or default founded upon by the pursuers was not a breach of the general duty to construct in accordance with the contract, but was constituted by certain specified failures on the defenders' part to design and construct the building in a workmanlike manner in terms of the contract. The minor breaches of the contract which had caused damage discovered at earlier stages (in 1972 or 1977) were not sufficient to constitute *injuria* in relation to major and different failures to design and construct the building properly which had resulted in the damage discovered in 1988. The loss, injury or damage sustained in 1988 did not arise from the act, neglect or default discovered in 1972 or 1977 and the case was held not to be time-barred.

A number of the relevant cases on this subject arise from construction contracts and these usefully illustrate the position. In *George Porteous (Arts) Ltd* v. *Dollar Rae Ltd* (1979), contractors were refused planning permission and the work executed by them had to be demolished. In that case it was held that the prescriptive period ran from the date of service of the enforcement notice, that being the date upon which the pursuers suffered loss. In *Scott Lithgow Ltd* v. *Secretary of State for Defence* (1989), the prescriptive period was held to have commenced as from the date when the materials in question were found to be defective.

In *Scottish Equitable plc* v. *Miller Construction Ltd* (2001), the contractors sought payment of sums in respect of direct loss and expense that they maintained ought to have been included in an interim certificate issued by the architect on 18 June 1992. That was the last interim certificate issued under the parties' contract. No final certificate was issued. The events upon which the claim was based all took place prior to practical completion being certified on 6 August 1990. A notice to concur in the appointment of an arbitrator was issued on 7 May 1996, that being the relevant date for the purposes of prescription. The employer argued unsuccessfully that the rights founded upon by the contractor had prescribed, the Inner House holding that time only began to run on the loss and expense claim from the date of issue of the relevant interim certificate, namely 18 June 1992.

Section 11(2) of the 1973 Act provides that where, as a result of a continuing act, neglect or default, loss has occurred prior to the act, neglect or default ceasing, the loss is deemed to have occurred on the date when the act ceased.

Where the creditor is not aware, and could not with reasonable diligence have become aware that loss, injury or damage has occurred, the prescriptive period does not commence until the date on which the innocent party first became, or could with reasonable diligence have become so aware, see s.11(3). This provision has particular relevance in the case of latent defects; the five-year period will commence from the date of discoverability of the defect, subject to the long-stop of the 20-year-long negative prescriptive period, which is considered in Section 9.9.7.

Section 6(4) of the 1973 Act provides that if there is any period where fraud or error has occurred which resulted in the creditor from refraining from making a relevant claim in relation to the obligation, it shall not be reckoned as, or as part of, the prescriptive period. This point was canvassed in *ANM Group Limited* v. *Gilcomston*

North Limited and Others (2008), in which the question of 'reasonable knowledge' was discussed in depth. Premature cracking was occurring in the roof of the pursuer's building, which some of the defenders had inspected in secret and without notification to the pursuer until much later. The court accepted that the pursuer was not aware of the damage, and could not with reasonable diligence have been aware of the damage.

9.9.4 Schedule 1 obligations

The types of obligations that are affected by the short negative prescriptive period are defined in Schedule 1 to the 1973 Act. Unlike long negative prescription, short negative prescription applies only to a limited number of obligations. Of these, certain are particularly relevant to construction contracts. These are set out in paragraph 1 of Schedule 1 and are any obligation:

- based on unjustified enrichment (including restitution, repetition or recompense);
- arising from liability to make reparation; and
- arising from, or by reason of any breach of, a contract or promise, not being an obligation falling within any other provision of paragraph 1.

An obligation arising under a contract will include an obligation to refer disputes under an engineering contract to the contract engineer, see *Douglas Milne Ltd* v. *Borders Regional Council* (1990). The same will apply in the case of an arbitration clause. A performance bond has been held to be a cautionary obligation subject to the short negative prescriptive period, see *City of Glasgow DC* v. *Excess Insurance Co. Ltd* (1986). The appropriate date in such a case is the date of issue of an architect's certificate ascertaining the extent of the damages due for default, see *McPhail* v. *Cunninghame DC* (1983) and *City of Glasgow DC* v. *Excess Insurance Co. Ltd (No. 2)* (1990).

9.9.5 Relevant claims

An obligation affected by short negative prescription will be extinguished if it has subsisted for a continuous period of five years without either of two events occurring, namely, the making of a relevant claim or the giving of a relevant acknowledgement.

If a relevant claim is made, the prescriptive period is said to have been interrupted and a new five-year period commences as from the date of interruption, see s.9 of the 1973 Act.

A 'relevant claim' is one made by or on behalf of the creditor in an obligation for implement or part implement of the obligation in 'appropriate proceedings' or in certain insolvency-related circumstances. An examination of the latter is beyond the scope of this book. 'Appropriate proceedings' means court proceedings in Scotland; an arbitration in Scotland; or an arbitration outside Scotland in which an award would be enforceable in Scotland.

The date of the relevant claim is the date of service of court proceedings, except in relation to Court of Session proceedings which do not subsequently call. If Court of Session proceedings do not call, a relevant claim is not made. In the case of an arbitration, the date of the relevant claim is the date when the claim is made in the arbitration or the preliminary notice is served, whichever is the earlier. If no preliminary notice is served the relevant claim in an arbitration will be made on the date when the claim is actually made. To be a relevant claim, the preliminary notice must state the nature of the claim, see *Douglas Milne Ltd* v. *Borders Regional Council* (1990).

9.9.6 Relevant acknowledgements

Section 10 of the 1973 Act defines a 'relevant acknowledgement'. The subsistence of an obligation is regarded as having been relevantly acknowledged if, and only if, either of two defined conditions is satisfied. First, there must have been such performance by or on behalf of the debtor towards implement of the obligation as clearly indicates that the obligation still subsists. Second, and alternatively, there has to have been made by or on behalf of the debtor to the creditor or his agent an unequivocal written admission clearly acknowledging that the obligation still subsists. As with a relevant claim, if a relevant acknowledgement is made the prescriptive period is interrupted and a new five-year period commences as from the date of interruption.

9.9.7 Long negative prescription

In Scotland, the long negative prescriptive period is 20 years. Section 7(1) of the 1973 Act provides that if, after the date when an obligation became enforceable, the obligation has subsisted for a continuous period of 20 years without either a relevant claim or a relevant acknowledgement, then as from the expiration of the 20-year period the obligation is extinguished. Long negative prescription does not apply to obligations arising under s.22A of the 1973 Act (liability under the Consumer Protection Act 1987 for a defect in a product) or specified in Schedule 3 to the 1973 Act (imprescriptible rights and obligations) or to obligations under Schedule 1 of the 1973 Act to which the short negative prescriptive period applies.

Other than the length of the period, the main difference between the short negative and long negative prescriptive periods is the point in time at which they commence. As we have seen in Section 9.9.3, the former commences as from the 'appropriate date'. The latter commences as from the date upon which the obligation in question became enforceable.

The practical consequence of this distinction in the context of construction contracts is significant. The concept of 'discoverability' of a latent defect (which applies to the five-year period) does not apply to the 20-year period, and thus an obligation arising from a latent defect will, in the absence of a valid interruption of the prescriptive period, prescribe 20 years after the date upon which the obligation became enforceable which is, broadly, when there has been both an act, neglect or default and loss, injury or damage arising therefrom. Thus a latent defect that is discovered 19 years after the obligation in question became enforceable will prescribe 20 years after the date upon which the obligation became enforceable, i.e. in these circumstances only one year after the discovery. A defect that is discovered less than 15 years after it became enforceable will prescribe five years after it is discovered.

It should be noted that, in long negative prescription, there is no equivalent provision to s.11(3) of the 1973 Act. Accordingly, even if the creditor was not aware, and could not with reasonable diligence have been aware, that loss, injury or damage had been caused, the long negative prescriptive period continues to run. Discoverability is not relevant in long negative prescription.

As with short negative prescription, the long negative prescriptive period can be interrupted by the making of a relevant claim or by the giving of a relevant acknowledgement. In this regard, Sections 9.9.5 and 9.9.6 apply equally to long negative prescription.

Chapter 10
Remedies

10.1 Introduction

Disputes arise under construction contracts as with any other type of contract. While the resolution of such disputes is considered in Chapters 15, 16 and 17, in this chapter we will consider certain of the remedies that are open to parties where a dispute arises.

While certain of the remedies are, perhaps, peculiar to building contracts, the ordinary remedies that are open to the parties to any form of commercial contract are also available. The remedies that are most commonly associated with building contracts are to be found within the provisions of the standard form contracts, such as the SBC, the SBC/DB and the NEC3. Certain of these remedies, such as liquidated and ascertained damages and extensions of time (Section 6.5), and termination (Section 9.4) have been considered previously. However, certain others are considered in this chapter. Separately, we will consider the general common law remedies open to parties, some of which are quite independent of those arising under the terms of a specific contract.

Ordinarily, the general common law remedies and the remedies provided for in a specific contract will exist at the same time, see *Gilbert Ash (Northern) Ltd* v. *Modern Engineering (Bristol) Ltd* (1974). A party's common law rights can only be taken away by clear, unequivocal words, see *Redpath Dorman Long Ltd* v. *Cummins Engine Co. Ltd* (1981). The extent to which that is achieved will depend upon the terms of the contract in question. See, for instance, the Scottish case of *Eurocopy Rentals Ltd* v. *McCann Fordyce* (1994), in which it was held that the contractual termination provision was the exclusive method of termination.

There are both advantages and disadvantages associated with each of the types of remedy. For example, a liquidated damages provision of the nature contained in both clauses 2.32 of the SBC and 2.29 of the SBC/DB is of advantage to the employer in that they are not required to prove the actual loss they have sustained as a result of the contractor failing to complete the works on time. The downside of clauses such as this is that the employer must adhere strictly to the provisions of the clause to entitle them to deduct liquidated damages.

MacRoberts on Scottish Construction Contracts, Third Edition. MacRoberts.

10.2 Rescission

The concepts of repudiation and rescission are considered in Chapter 9. In certain circumstances a material breach of contract by one party may be such as to entitle the other party (the 'innocent party') to terminate the contract. If the breach constitutes a repudiation of the contract, the innocent party has a choice. They can either accept the repudiation and rescind the contract or, alternatively, they may elect to ignore the repudiation and continue with the performance of the contract. The extent to which there is a right to continue with performance may, however, be limited.

The remedy open to the innocent party is to rescind the contract. Where a contract has been rescinded, both parties are freed from future performance of their primary obligations under it. Parties continue to be bound by the primary obligations that were extant at the time of rescission. The contract does not come to an end. The innocent party is entitled to sue the party in default for damages for breach of contract. Ancillary clauses which the parties intended would survive rescission, such as arbitration clauses, may be enforced after rescission. Apart from such ancillary clauses, the contract may also contain clauses which affect the amount of damages due for breach of contract, such as a liquidated damages clause. The language of the contract may be such as to demonstrate that the parties intended such clauses to be enforceable after rescission, see *Lloyds Bank plc* v. *Bamberger* (1994).

It is well-established law that the innocent party's remedies for the other party's breach of contract are limited to those provided for in the contract, and for those breaches committed while the contract subsisted. However, where the contract has been rescinded and the mutual obligations under it have ceased to exist, the court has the power to award compensation on the basis of *quantum meruit*, see *Morrison-Knudsen Co. Inc* v. *British Columbia Hydro and Power Authority* (1991), approved by the Inner House of the Court of Session in *ERDC Construction Ltd v. H M Love & Co.* (1995). The subject of payment *quantum meruit* is considered in Section 8.4.

The ordinary position in contract is that the innocent party is entitled to ignore the repudiation and continue with the performance of the contract. This is supported by the decision of the House of Lords in *White & Carter (Councils) Ltd* v. *McGregor* (1962) and, in the case of building contracts, by the decision of the Inner House of the Court of Session in *ERDC Construction Ltd*. The courts have, however, recognized that there may be limitations upon the right of the innocent party to insist upon performance. In *White & Carter (Councils) Ltd*, Lord Reid made it clear that:

> [H]ad it been necessary for the defender to do or accept anything before the contract could be completed by the pursuers, the pursuers could not and the court would not have compelled the defender to act, the contract would not have been completed, and the pursuers' only remedy would have been damages.

It must be noted that in *White & Carter (Councils) Ltd* the pursuers did not require any co-operation, either active or passive, on the part of the defender. Building contracts patently cannot be performed without co-operation between employer and contractor. The courts in England have made it clear that they will not enforce an agreement for two people to live peaceably under the same roof, see *Thompson* v. *Park* (1944). They have also expressed the opinion that a repudiatory breach cannot be ignored where it can be shown that the innocent party has no legitimate interest in performing the contract, provided that damages would be an adequate remedy and keeping the contract alive would be unreasonable, see *Ocean Marine Navigation Ltd* v. *Koch Carbon Inc ('The Dynamic')* (2003). The rationale behind this is identical to that identified by Lord Reid in *White & Carter (Councils) Ltd.* A multitude of practical problems would arise if the courts compelled performance.

In these circumstances, the authors respectfully suggest that the *obiter* comments of Lord Reid in *White & Carter (Councils) Ltd* should generally apply in relation to building contracts and the innocent party's right to insist upon performance limited. Support for this proposition is to be found in the case of *London Borough of Hounslow* v. *Twickenham Garden Developments Ltd* (1970). In essence, the innocent party may be forced to accept a repudiation and rescind the contract, see *Decro-Wall International SA* v. *Practitioners in Marketing Ltd* (1971) and *Ocean Marine Navigation Ltd.* The nature of building contracts is such that the option to ignore an employer's repudiation may not be one that is open to a contractor.

10.3 Specific implement

When parties enter into a contract, they each undertake certain obligations. In Scotland, it is presumed that contractual obligations will be enforced by the courts, unless there are considerations which make implement impossible or unjust, see *Stewart* v. *Kennedy* (1890) and *Beardmore* v. *Barry* (1928). The remedy open to the innocent party to compel performance of contractual obligations by the party in breach is known as specific implement.

As with rescission, the innocent party, in most circumstances, has a choice. Either they can insist upon their entitlement under the contract, or they can seek damages for the breach, see *Holman & Co.* v. *Union Electric Co.* (1913). Damages are considered in Section 10.4. It will, however, always be at the discretion of the court as to whether the remedy of specific implement or that of damages is the appropriate one, see *Graham* v. *Magistrates and Police Commissioners of Kirkcaldy* (1881). Specific implement is not an appropriate remedy in every case. It is not available in respect of the enforcement of a party's monetary obligations under a contract, see *White & Carter (Councils) Ltd* v. *McGregor* (1962). It is also an inappropriate remedy where performance would require the party in breach to become a partner in a commercial undertaking, see *Pert* v. *Bruce* (1937). It is an inappropriate remedy where performance is impossible, see *McArthur* v. *Lawson* (1877). There are a number of other instances in which specific implement has been held to be inappropriate. A more detailed examination of these is beyond the scope of this book.

The issue of specific implement in the context of building contracts is a difficult one. It raises similar issues to those that arise in relation to rescission. In *London Borough of Hounslow* v. *Twickenham Garden Developments Ltd* (1970), it was argued on behalf of the borough that the contract (which incorporated the RIBA conditions) was not specifically enforceable. While a decision on this point was not necessary to resolve the case, Lord Megarry (as he then was) stated that he could not see why the contract should not be held to be specifically enforceable. In contrast to rescission, whether or not specific implement is an appropriate remedy in relation to a particular contractual obligation will depend upon whether the party in breach is required to do, allow or accept something. Such co-operation may be essential in relation to certain obligations, but unnecessary in respect of others.

In the context of building contracts, the authors respectfully submit that where co-operation by the party in breach is required, specific implement is not an appropriate remedy. Such an approach is consistent with the comments of Lord Reid, and indeed the dissenting opinion of Lord Morton of Henryton, in *White & Carter (Councils) Ltd*. The innocent party must have an interest to insist upon specific implement of the obligation. If the court is not satisfied that they have such an interest, their claim will be for damages, see *Clea Shipping Corp* v. *Bulk Oil International Ltd* (1984). The reality is that, in the majority of cases, the innocent party in a building contract may have no legitimate interest in performing the contract, rather than claiming damages. In such cases, it has been suggested that the innocent party can, in one sense, be said to be forced to claim damages. To insist upon any other remedy would be of little value, see *Decro-Wall International SA* v. *Practitioners in Marketing Ltd* (1971) and *Ocean Marine Navigation Ltd* v. *Koch Carbon Inc ('The Dynamic')* (2003). As stated by Lord Justice Sachs in *Decro-Wall International SA*, 'in such cases it is the range of remedies that is limited, not the right to elect'.

In Scotland, the remedy of specific implement is available both in the Court of Session and in the Sheriff Court. By virtue of section 47(2) of the Court of Session Act 1988, interim orders for specific implement can competently be granted, see *Scottish Power Generation Ltd* v. *British Energy Generation (UK) Ltd* (2002), *Va Tech Wabag UK Ltd* v. *Morgan Est (Scotland) Ltd* (2002) and *Purac Ltd* v. *Byzak Ltd* (2005). These cases are considered in Section 15.2.3. No equivalent remedy is available in the Sheriff Court.

10.4 Damages for breach of contract

10.4.1 General

Perhaps the most commonly used legal remedy is that of damages. Damages are expressed in monetary terms. The purpose of damages for breach of contract is to place the innocent party in the position they would otherwise have been in had the breach not occurred. Damages may also be recoverable where no contract exists but one party owes the other a duty of care and is in breach of that duty. In those circumstances damages will be recoverable under the law of delict rather than for breach of contract. Delictual claims are considered in Section 10.10.

The law of damages is a vast and complex subject and a detailed examination is beyond the scope of this work.

It is open to the parties to a contract to decide in advance what damages, if any, will be payable in the event of a breach by one, or any, of them. In essence, it is open to parties to exclude or limit liability. This is a common occurrence in building contracts, for example, clause 2.32 of the SBC, which provides for the payment of liquidated and ascertained damages for non-completion. Under clause 2.17.2 of the SBC/DB the parties may elect to limit the Contractor's liability for loss of use, loss of profit or other consequential loss arising from an inadequacy in design. Option X7 of the NEC3 can also provide for liquidated damages, as it simply provides for payment by the Contractor of delay damages at the rate stated in the Contract Data. See also *Elvamite Full Circle Limited* v. *AMEC Earth & Environmental (UK) Limited* (2013).

Care must be taken when seeking to exclude liability in respect of a certain type of loss. Sufficient detail must be given as to what precisely is to be excluded in order for it to be of effect, see *McCain Foods GB Limited* v. *Eco-Tec (Europe) Limited* (2011). It has also been held that a financial cap on liability was unreasonable and contrary to the Unfair Contract Terms Act 1977 where an insurance policy had been taken out in excess of the level of the cap, see *The Trustees of Ampleforth Abbey Trust* v. *Turner & Townsend Project Management Limited* (2012). It should also be borne in mind that claims for damages are subject to the law of prescription, see Section 9.9.

10.4.2 Causation, foreseeability and remoteness

The loss which the pursuer is entitled to recover is that which has been caused by the defender's breach, see *Bourhill* v. *Young* (1942). The pursuer must establish not only that there has been a breach of contract and they have suffered loss, but also that there is a causal connection between the breach and the losses sought to be recovered, and, further, that such losses are not too remote. In complex building contract disputes, there can be a multitude of losses and breaches of contract, which can make it extremely difficult to link a specific loss to a specific breach. This has often resulted in the presentation of what is termed a 'global claim'. It is competent, in certain circumstances, to present a claim in this manner, see *John Doyle Construction Ltd* v. *Laing Management (Scotland) Ltd* (2004), discussed in Section 8.3.

Only losses that were foreseeable as being the likely consequences of a breach at the time the contract was entered into are recoverable. If the losses were not foreseeable at that time, they are too remote and cannot be recovered.

These issues were considered in the case of *Hadley* v. *Baxendale* (1854). This case laid down the rules that apply in assessing the measure of damages in a breach of contract case. In delivering the judgment of the court, Baron Alderson stated:

> Where two parties have made a contract which one of them has broken, the damages which the other party ought to receive in respect of such breach of contract should be such as may fairly and reasonably be considered either arising naturally, i.e., according to the usual course of things, from such breach of contract itself, or such as may reasonably be supposed to have been in the contemplation of both parties, at the time they made the contract, as the probable result of the breach of it.

That part of the judgment of the court has been widely repeated and relied upon since 1854. In the context of building contracts, however, the following part of the judgment is also significant. Baron Alderson went on to state:

> [I]f the special circumstances under which the contract was actually made were communicated by the [pursuers] to the [defenders], and thus known to both parties, the damages resulting from the breach of such a contract, which they would reasonably contemplate, would be the amount of injury which would ordinarily follow from a breach of contract under these special circumstances so known and communicated. But on the other hand, if these special circumstances were wholly unknown to the party breaking the contract, he, at the most, could only be supposed to have had in his contemplation the amount of injury which would arise generally ... from such a breach of contract.

An example of such special circumstances is to be found in *Balfour Beatty Construction (Scotland) Ltd* v. *Scottish Power plc* (1994). In that case the pursuers were engaged in the building of a roadway and associated structures, including an aqueduct. They contracted with the defenders' predecessors for the supply of electricity to operate a concrete batching plant. The construction of the aqueduct required a continuous pour operation. In the course of the construction of the aqueduct, the batching plant stopped working. It was established that the electricity supply had been interrupted and that the interruption was a breach of contract by the defenders' predecessors. The pursuers claimed the cost of demolishing and rebuilding a substantial part of their works, this having been rendered necessary by the interruption of the electricity supply and the consequent interruption of the required continuous pour. It was established that the defenders' predecessors had not known of the need for a continuous pour. The Lord Ordinary concluded that the need to rebuild part of the works as a result of the interruption of the continuous pour had not been within the defenders' reasonable contemplation and the action failed. Ultimately, the House of Lords upheld the decision of the Lord Ordinary. Had the special circumstances, namely, the need for a continuous pour, been known to the defenders, it is likely that the pursuers would have succeeded.

10.4.3 Damages recoverable and mitigation

The law of Scotland is clear in respect of the method of assessment of damages, assuming the necessary prerequisites considered in Section 10.4.2 have been met. As was stated by Lord Pearson in *The Govan Rope & Sail Co. Ltd* v. *Andrew Weir & Co.* (1897):

> [I]t appears to me that the criterion of damage now adopted by the pursuers is in accordance with the principle which governs the whole law on the subject, namely, that the party observing the contract is to be put as nearly as possible in the same position as he would have been if the contract had been performed.

As damages is a monetary remedy, the party suffering the loss can only be put in the position it would have been in, but for the breach, insofar as a payment of money to them allows.

Building contracts will often confer on the contractor a right to an increase in the contract sum for direct loss and/or expense incurred by the contractor on the occurrence of certain events, e.g. failure or delay in issuing instructions or information, which otherwise would be treated as damages for breach of contract. Although the nature of the claim in such cases is conceptually different insofar as it is a claim for payment under the contract rather than a claim for damages, the calculation of the direct loss and/or expense is likely to be little different from a claim for damages for breach of contract. See clauses 4.23–4.26 of the SBC, 4.2–4.23 of the SBC/DB and clauses 60–65 of the NEC3 relating to compensation events, as discussed in Section 8.3.

If the contract is rescinded by the contractor due to a material breach by the employer, the contractor is likely to be entitled to recover by way of damages the profit he would have made had the contract been completed in the ordinary course. If the works had been partially carried out, the contractor retains the right to payment for the value of such works. The rescission of a contract in consequence of a repudiation does not affect accrued rights to payment under the contract, unless the contract provides that it was to do so, see *Hyundai Heavy Industries Co. Ltd* v. *Papadopoulos and others* (1980).

In the event of a breach of contract by the contractor, again the contract may provide the remedy in certain circumstances, see, for example, the liquidated and ascertained damages provisions in respect of non-completion under clause 2.32 of the SBC and clause 2.29 of the SBC/DB. If the contractor does not complete the contract works, the employer's loss will be the additional cost of completing the works, if any. If the works are completed at no additional cost, there will be no loss. It should, however, be noted that there is authority in Scotland to the effect that where a breach of contract is established the pursuer is entitled to nominal damages, even if no loss can be demonstrated. This comes from the opinion of the Lord President in *Webster & Co.* v. *The Cramond Iron Co.* (1875) in which he stated that:

> [Where the] contract and the breach of it are established … that leads of necessity to an award of damages. It is impossible to say that a contract can be broken even in respect of time without the party being entitled to claim damages – at the lowest, nominal damages.

Secondary Option X7 of NEC3 provides for delay damages at the rate specified in the Contract Data from the Completion Date through to the earlier of Completion or the date when the Employer takes over the works. While Secondary Option X18 of NEC3 places a number of limits on liability, it excludes various matters, including, notably, delay damages.

Another common breach of contract by the contractor is the existence of defects in the works executed or that the works executed do not conform to the requirements of the parties' contract. Assuming the contract contains no specific mechanism under which the contractor is obliged to remedy defects, such as clauses 2.38 and 2.39 of the SBC, clauses 2.35 and 2.36 of the SBC/DB and clauses 43–45 of the NEC3, and, further, that the contractor will not do so voluntarily, the measure of the employer's loss can, ordinarily, be assessed in one of two ways. The first is the cost of the necessary repairs. The second is the difference in value between the building in the condition contracted for and the building in its actual condition, i.e. with the defective work, see *GUS Property Management Ltd* v. *Littlewoods Mail Order Stores Ltd* (1982).

A pursuer can, ordinarily, proceed on the basis of either measure. These are not the only available measures of loss and a court is not confined to making an award based on one of these measures, see *Ruxley Electronics and Construction Ltd* v. *Forsyth* (1995). It may be prudent, where possible, to proceed on the basis of both measures as alternatives. The proper measure of damages may be determined by checking one measure against the other, see *Prudential Assurance Co. Ltd* v. *James Grant & Co. (West) Ltd* (1982).

The law in relation to this matter was clarified in England in the case of *Ruxley Electronics and Construction Ltd* (1996). The House of Lords held that in assessing damages for breach of contract for defective building works, should the court decide that the cost of reinstatement would be out of all proportion to the benefit to be obtained by the innocent party by reinstatement, the innocent party's claim would be restricted to the difference in value between the building in the condition contracted for and the building in its actual condition. Whether or not the innocent party actually intends to reinstate will be relevant in determining if it is reasonable to insist upon reinstatement.

The position in Scotland was clarified in *McLaren Murdoch & Hamilton Ltd* v. *Abercromby Motor Group Ltd* (2002). Lord Drummond Young generally accepted the principle set out in *Ruxley* that ordinarily a party is entitled to claim the cost of making works conform to contract. However, he also set out exceptions to that rule. The first, as with *Ruxley*, is where the cost involved is manifestly disproportionate to any benefit that will be obtained from it. The second exception is where the other party leads evidence to show a significant disproportion between cost and benefit. In the latter situation the court considered that the balance between cost and benefit should not be weighed too finely.

It should be borne in mind that certain contracts have detailed mechanisms for assessing the sum due by one party to the other on termination, see, for example, clause 8 of both the SBC and the SBC/DB, as well as clause 9 of the NEC3. In part, at least, this deals with damages arising out of the termination. Termination is considered in Section 9.4.

Finally, it must always be borne in mind that only such losses as are consequent on the breach may be recovered. An example of this can be seen in the case of *British Westinghouse Electrical & Manufacturing Co. Ltd* v. *Underground Electric Railways Co. of London Ltd* (1912).

10.5 Finance charges

It has long been judicially recognized that, in the ordinary course of things, when contractors require capital to finance a contract they either borrow the capital and pay for the privilege, or use their own capital and, as a consequence, lose the interest which they would otherwise have earned. See, for example, *F G Minter Ltd* v. *Welsh Health Technical Services Organisation* (1980). Similarly, it has been judicially recognized that, in the construction industry, delay in payment to contractors might naturally result in them being short of working capital, thus causing them to incur finance charges. See, for example, *Ogilvie Builders Ltd* v. *City of Glasgow District Council* (1995). Whether or not such finance charges are recoverable by contractors has been the subject of considerable judicial discussion over the years, but it is now well settled, both in Scotland and England, that finance charges are recoverable in certain defined circumstances.

First, finance charges are recoverable as direct loss and/or expense under clauses 4.23–4.26 of the SBC and clauses 4.20–4.23 of the SBC/DB, but only if the requirements of that clause are satisfied. See *Ogilvie Builders Ltd*, following *F G Minter Ltd* and *Rees & Kirby Ltd* v. *Swansea City Council* (1985). Clauses 4.23–4.26 of the SBC are considered in Section 8.3. Under the terms of the NEC3, finance charges may or may not be recoverable as Defined Cost, depending on which Main Option is chosen.

Second, finance charges are recoverable as damages in a case based on breach of contract. The words 'direct loss and/or expense' are to be given the same meaning in a case of breach of contract as they would be given in a case for payment under contract, see *Ogilvie Builders Ltd*. Recovery by way of a claim based upon breach of contract (at least until the decision in *Ogilvie Builders Ltd*) proved more problematic in Scotland, with claims being unsuccessfully advanced in cases such as *Chanthall Investments Ltd v. F G Minter Ltd* (1975). It was stressed in that case, however, that in each case where this issue arises, it is a question of fact and the particular circumstances as to whether or not the loss in question was within the contemplation of the parties. This approach was approved by the Inner House of the Court of Session in *Margrie Holdings Ltd* v. *City of Edinburgh District Council* (1994). This approaches recovery by way of the second branch of the rule in *Hadley* v. *Baxendale* (1854), which is considered in Section 10.4.2. That part of the rule permits the recovery of such losses as may reasonably have been supposed to have been in the contemplation of both parties, at the time the contract was entered into, as the probable result of a breach of it. This falls to be contrasted with the first branch of the rule, namely, that where two parties have entered into a contract and there has been a breach of contract by one of the parties, the damages to which the innocent party is entitled should be such as may fairly and reasonably be considered as arising naturally from the breach.

In *Ogilvie Builders Ltd*, Lord Abernethy stated that he did not read any of the Scottish cases cited to him as indicating any general proposition that claims for finance charges, if recoverable at all, could only be recoverable under the second branch of the rule in *Hadley* v. *Baxendale*. He held that that a claim for finance charges under the first branch was relevant as a matter of law. In Scotland, claims advanced under the second branch of the rule have been held to be relevant as a matter of law, see *Caledonian Property Group Ltd* v. *Queensferry Property Group Ltd* (1992). What *Ogilvie Builders Ltd* recognized was the commercial reality that extra finance charges could arise 'naturally' from a breach of contract in the construction industry.

10.6 Interest

10.6.1 Common law

Late payment of sums admittedly due is commonplace in the construction industry. One way in which this can be addressed is by way of interest. The general rule in Scotland is that, unless a contract provides otherwise, interest will only be awarded from a date prior to the serving of a writ if the money has been wrongfully withheld. That has been the position in Scotland for some considerable time and was enunciated by Lord Atkin in *Kolbin & Sons* v. *Kinnear & Co.* (1931). His Lordship stated that:

> [I]t seems to be established that, by Scots Law, a pursuer may recover interest by way of damages where he is deprived of an interest-bearing security or a profit-producing chattel, but otherwise, speaking generally, he will only recover interest, apart from contract, by virtue of a principal sum having been wrongfully withheld and not paid on the day where it ought to have been paid.

The observations of Lord Atkin were accepted in *F W Green & Co.* v. *Brown & Gracie Ltd* (1960) as setting out the broad principle. Whilet a number of cases have dealt with the issue of interest, none of them has ever precisely said what is meant by 'wrongfully withheld'. Various views have been expressed as to its meaning, including failure to pay following the issuing of a certificate, 'negligent' under-certification and client interference in the certification process. The law of Scotland in relation to the entitlement to interest was commented upon by the Inner House of the Court of Session in *Elliott* v. *Combustion Engineering Ltd* (1997).

While a decision of the Inner House of the Court of Session on this topic is welcome, it must be said that the opinion of the court in *Elliott* does not, in reality, answer many of the questions that have been posed in this field over the years. The difficulty with the decision is that, perhaps understandably, 'wrongful withholding' is not directly defined. The court's conclusion was that modern authority indicated that, in general, interest would run on contractual debts from judicial demand (that is service of a writ), and that while there might be qualifications or exceptions to the general rule, the circumstances of *Elliott* did not fall within any such qualification or exception.

Unfortunately, therefore the extent of these qualifications or exceptions to the general rule is not fully set out in *Elliott*. It must, however, be recognized that this was not necessary to resolve the problem then before the court. In relation to the decision in *Elliott*, it is also pertinent to observe that it flowed from an arbitration in which the power of the arbitrator was to award interest if the claimant was entitled to it. That is, as the court observed, if the claimant had a right to it by the application to the circumstances of the relevant law. In *Elliott*, the arbitrator did not have a power to award interest from such date and at such rates as he saw fit. Such a power in relation to interest is found in clause 9.5 of the SBC and the SBC/DB. Such a clause is, however, no longer strictly necessary. Until recently in Scotland, an arbitrator in Scotland had no power at common law to award interest from a date prior to that of his award.

This perceived inadequacy has now been remedied by virtue of the Scottish Arbitration Rules, contained within Schedule One to the Arbitration (Scotland) Act 2010. Rule 50 now provides that an arbitrator's award may order interest to be paid on any amount in respect of any period up to the date of the award. This goes some way to removing the uncertainty left following the decision in *Elliott*.

10.6.2 The Late Payment of Commercial Debts (Interest) Act 1998

The Court of Session in *Elliott* stated that it was a matter of concern that in modern commercial contexts the law did not, in general, allow for interest to run on debts from a date earlier than judicial demand (that is the date of service of a writ) and that reform of the law on interest on debts was a matter for government. At or about the time of the decision in *Elliott*, this was a subject upon which the government had been consulting and that process resulted in the enactment of the Late Payment of Commercial Debts (Interest) Act 1998, which came into force on 1 November 1998.

The Late Payment of Commercial Debts (Interest) Act 1998 ('the 1998 Act') was introduced with a view to encouraging purchasers to pay on time and to compensate suppliers where late payment persisted. The right to claim interest is to compensate suppliers for not being able to make use of the money owed to them and to cover the cost of increased borrowing resulting from late payment. The 1998 Act provides suppliers with a statutory right to interest on late payments. The rate of interest currently prescribed is 8% over the base rate of the Bank of England. The 1998 Act operates by implying a term into contracts to which it applies to the extent that any qualifying debt carries interest at the prescribed rate. A 'qualifying debt' is simply one where an obligation to make payment of the contract price arises under a contract to which the 1998 Act applies, namely, a contract for the supply of goods and/or services where both the purchaser and supplier are acting in the course of a business. The interest to which the supplier is entitled is simple interest.

All businesses and United Kingdom Public Authorities have the right to claim interest at the statutory rate against all other businesses and United Kingdom Public Authorities. A 'United Kingdom Public Authority' is defined at length in the commencement order for the 1998 Act but, in general, it means any emanation of the State.

A supplier is free to decide whether or not to claim interest. The statutory right is not compulsory. The right to claim interest arises when a payment is late. A payment is late when it is not made by the 'relevant day'. The relevant day is the date agreed for payment or, in the event that no such date has been agreed, the last day of the period of 30 days beginning with the later of the day of the supply/performance or the date of notice to the purchaser of the amount of the debt – in practice, the invoice date.

Different rules exist where the contract requires advance payment. These are dealt with in s.11 of the 1998 Act. The principle is that the 1998 Act does not give a right to claim interest unless and until at least some of the goods have been delivered or part of the service performed. In essence, the section 11 provisions allow for the right to claim interest 30 days after delivery/performance.

Once the payment is late, interest runs at the prescribed rate from the day after the relevant day until the principal sum is extinguished by payment. Unless the supplier accepts a payment on other terms, any payment received goes first to extinguish or reduce the accrued interest. A claim for interest is made by the supplier informing the purchaser, once the payment is late, that they are claiming interest. Notification can be in any fashion but it would appear prudent to make such a claim in writing. A claim for interest need not be made immediately. The ordinary rules of prescription will apply. These are considered in Section 9.9.

It is, of course, common to find standard terms and conditions providing for interest to run on late payment. In recognition of that, the 1998 Act provides that, where arrangements have already been made, the statutory right to interest will not apply. To prevent purchasers abusing their right to agree arrangements with a supplier, any contractual remedy must be what is termed a 'substantial remedy'. This term is defined by s.9 of the 1998 Act. A remedy for late payment is 'substantial' if it is sufficient to compensate the supplier for the cost of late payment or to deter late payment, and it is fair and reasonable to allow the remedy to oust or vary the statutory interest that would otherwise apply.

In determining whether or not a remedy satisfies the fair and reasonable test, regard is to be had to the benefits of commercial certainty; the relative strength of bargaining power between the parties; whether the term was imposed by one party to the detriment of the other; and whether the supplier received an inducement for agreeing to the term. If the contractual remedy is not a substantial remedy, it is void.

The scope of the 1998 Act was extended by the introduction of the Late Payment of Commercial Debts (Scotland) Regulations 2002 ('the 2002 Regulations').

The 2002 Regulations introduce a right to a fixed sum by way of compensation for the costs suffered by suppliers arising from late payment. This fixed sum is based on the size of the debt. The 2002 Regulations also provide that a representative body may bring proceedings on behalf of small and medium-sized enterprises in the Court of Session where standard terms used by the purchaser include a term varying or excluding the statutory interest in relation to contracts to which the 1998 Act applies. 'Small and medium-sized enterprises' and 'representative body' are defined in the Regulations.

The scope of the 1998 Act was further extended on 29 March 2013 by the introduction of the Late Payment of Commercial Debts (Scotland) Regulations 2013 ('the 2013 Regulations').

The 2013 Regulations, which apply to contracts entered after 16 March 2013, provide for a maximum payment period of up to 30 days where the purchaser is a public authority. In other cases the payment period can be up to 60 days or longer if agreed by the parties to the contract and provided if it is not grossly unfair to the supplier. The 2013 Regulations also provide for a period of either up to 30 days, or longer if expressly agreed by the parties and if it is not grossly unfair to the supplier, for a purchaser to confirm that the goods or services they have received from the supplier conform with the contract before the payment period commences. There is also a right created for suppliers to compensation for the reasonable costs of recovering a debt incurred if that amount exceeds the fixed charge sum.

10.6.3 Interest under the SBC, the SBC/DB and the NEC3

The entitlement to interest in respect of Interim Certificates is dealt with by clause 4.12.6 of the SBC and clause 4.9.5 of the SBC/DB. If the Employer fails properly to pay the amount, or any part thereof, due to the Contractor under the conditions by the final date for its payment, the Employer is obliged to pay to the Contractor, in addition to the amount not properly paid, simple interest thereon for the period until such payment is made. Payment of such simple interest is treated as a debt due to the Contractor by the Employer.

The rate of interest payable is 5% over the base rate of the Bank of England current at the date the payment by the Employer became overdue. This rate should be contrasted with the statutory rate provided for by the 1998 Act. It is conceivable that it could be argued that it is not a substantial remedy in the context of s.9 of the 1998 Act.

Clause 4.15.7 of the SBC and clause 4.12.10 of the SBC/DB make similar provisions in respect of sums due under a Final Certificate, whether those sums are due to the Employer or the Contractor. In each case, any payment of simple interest under the clause in question shall not, in any circumstances, be construed as a waiver by either party of their right to proper payment of the principal amount due.

Clause 51.2 of the NEC3 provides that if an amount due to the Contractor is late, interest is to be paid on the late payment at the rate specified in the Contract Data provided by the Employer. Interest is calculated from the date on which the outstanding payment should have been paid until the date when payment is actually made and is to be included in the first assessment after the late payment is made. Interest is calculated daily and is compounded annually.

10.6.4 Interest on damages

In terms of the Interest on Damages (Scotland) Act 1958, as amended by the Interest on Damages (Scotland) Act 1971, where a court grants a decree for payment by any party of a sum of money as damages, the court's order may include provision for payment by that party of interest on the whole or any part of the amount of damages for the whole, or any part, of any period between the date when the right of action arose and the date of the court's order. The court also has a discretion as to the rate or rates at which such interest is to be paid. The mere fact that a right of action arose on a particular date prior to decree does not, of itself, justify an award of interest from that date, see *James Buchanan & Co. Ltd* v. *Stewart Cameron (Drymen) Ltd* (1973).

In *Macrae* v. *Reed and Mallik Ltd* (1961) the Inner House of the Court of Session stated that the discretion conferred upon the court by the 1958 Act must be exercised on a selective and discriminating basis and that the exercise of that discretion was open to review on the question as to whether the circumstances of the case warranted the course taken. The Court also held that interest from a date earlier than the date of decree could be allowed only on damages awarded for loss suffered before the date of decree and where such loss could be definitely ascertained.

10.7 Interdict

Where it can be demonstrated by a party that a legal wrong is continuing or that they are reasonably apprehensive that such a wrong will be committed, they are entitled to seek interdict against the wrong, see *Hay's Trustees* v. *Young* (1877). If the wrong has been completed, and it cannot be contended that there is a likelihood of it recurring, interdict will not be granted, see *Earl of Crawford* v. *Paton* (1911).

Both permanent and interim interdict can be granted in either the Court of Session or the Sheriff Court. In practice, few actions in which interdict is sought proceed beyond the interim interdict stage. The grant or refusal of interim interdict is often determinative of the issue between the parties.

Assuming the pursuer can satisfy the court that they have title and interest to bring the action and that they are confronted by, or threatened with, a wrong on the part of the defender, interim interdict will still only be granted if the balance of convenience favours the pursuer. To meet this test, the pursuer must demonstrate a cogent need for interim interdict, see *Deane* v. *Lothian Regional Council* (1986).

Although a detailed examination of the law of interdict is beyond the scope of this book, in the context of building contracts its availability as a remedy should not be overlooked. The remedy is available should there be a continuing, or reasonably anticipated, breach of contract.

10.8 Withholding payment

10.8.1 General

Having examined the issue of payment in Chapter 8, it is appropriate to consider the remedies that are open to a party under a building contract who is, on the face of it, obliged to make payment, but has reasons for not doing so. In Scotland, there exist two distinct and separate remedies, namely, retention and compensation. These are frequently confused. The term 'set-off' is often used in place of compensation. Here, we will consider retention and compensation in the context of payment obligations. It should, however, be noted that retention applies not only to obligations to pay but also to all other obligations incumbent upon a party under a contract. The wider application of retention is considered in Section 10.9.1. Finally, we will consider the statutory right of paying less as contained within s.111 of the 1996 Act.

10.8.2 Retention

The principle of retention is, perhaps, best illustrated by the opinion of Lord Shand in *Macbride* v. *Hamilton & Son* (1875) in which he stated:

> [I]n cases of mutual contract a party in defence is entitled to plead and maintain claims in reduction or extinction of a sum due under his obligation where such claims arise from the failure of the pursuer to fulfil his part of the contract.

For retention to operate, both claims must arise from the one contract. Retention, when considered in the context of withholding payment, should not be confused with compensation. Retention is, in effect, a form of security, whereas compensation extinguishes a debt, in whole or in part.

Retention has long since been a favoured remedy in building contract disputes, see, for example, *Johnston* v. *Robertson* (1861). In that case, the employers were entitled to plead in defence a claim for liquidated damages for non-completion against the contractors' claim for the balance of the contract price and payment for extra works.

In Scotland, there was at one time authority which suggested that the general rule in respect of retention may not apply to the case of a building contract which contained provision for payment by instalments, it being doubted whether the employer had any right to withhold payment of an instalment by virtue of a claim against the contractor, see *Field & Allan* v. *Gordon* (1872). That position has, however, been accepted to be incorrect. Unless it is shown in clear and unequivocal words that the parties had agreed in the contract that the common law right of retention was to be excluded, that right would be available in respect of breaches of contract, see *Redpath Dorman Long Ltd* v. *Cummins Engine Co. Ltd* (1981). Retention is considered further in Section 10.9.1.

10.8.3 Compensation

The essence of compensation is that sums are due at the same time by parties to each other. Where each party owes the other a sum of money, compensation can operate to extinguish, or partly extinguish, the debts. Certain prerequisites must be satisfied. First, the debts must be due at the same time. A debt that is due at a future date cannot be set off against one that is presently due, see *Paul & Thain* v. *Royal Bank* (1869). Second, each debt must be what is termed 'liquid'. A liquid debt is one that is for a readily ascertainable amount and is not disputed. A claim for damages is not a liquid debt, see *National Exchange Company of Glasgow*. v. *Drew and Another* (1855). In certain, narrow circumstances, however, there may be exceptions to this general rule, allowing a liquid debt to be postponed because the debtor has an illiquid claim against the creditor, provided the two contract debts are closely linked, see *Inveresk* v. *Tullis Russell Papermakers Limited* (2010). There must also be what is termed *concursus debiti et crediti*, which is that each party must owe money and be owed money in the same capacity. An example of this is the case of *Stuart* v. *Stuart* (1869) in which it was held that the defender, as an individual, could not plead in compensation certain alleged counterclaims competent to him as his father's executor.

In Scotland, there has been legislation governing compensation for over 400 years, see the Compensation Act 1592.

10.8.4 Paying Less under the 1996 Act

In construction contracts governed by the 1996 Act (see Section 1.2.2) a party to a construction contract may seek to pay less than the sum notified to them by the payee, provided they have given an effective notice in terms of s.111(3).

Such notice of intention to pay less than the sum notified requires to state how the sum has been calculated and, in effect, will provide the reasons as to why it is considered that a lesser sum is payable. The notice must be given not later than the prescribed period before the final date for payment. Parties are free to agree what that prescribed period is to be, but if they do not, the Scheme for Construction Contracts applies. An effective notice is not retrospectively undermined by a change of circumstances, for example, a pay less notice issued based on an entitlement to deduct liquidated damages.

Paragraph 10 of Part II of the Schedule to the Scheme provides that any notice of intention to pay less than the notified sum shall be given not later than seven days before the final date for payment under the contract. The notice of intention to pay less can form part of the notice that is required under s.110A of the 1996 Act, being the notice which specifies the amount that the payer considers to be due or to have been due at the payment due date, together with a note of the basis upon which that amount is calculated.

10.8.5 Paying Less under the SBC, the SBC/DB and the NEC3

Clause 4.13 of the SBC contains provisions consistent with the 1996 Act regarding the giving of a notice of intention to pay less in relation to a sum which is due under an Interim or Final Certificate. Not later than five days before the final date for payment of an amount due under a Certificate, the Employer may give a written notice to the Contractor which specifies the sum that is considered to be due at that date and the basis on which that sum has been calculated. It is immaterial that the amount then considered to be due may be zero. Similar provision exists under clause 4.10 of the SBC/DB.

Option Y(UK)2 of the NEC3 also contains provisions consistent with the 1996 Act, providing that a notice of intention to pay less must be given no later than seven days before the final date for payment.

See Sections 8.1.4–8.1.6 for detailed commentary on the Pay Less provisions in the SBC, the SBC/DB and the NEC3.

10.9 Suspending performance

10.9.1 General

The principle of retention, considered above in Section 10.8.2 in the context of withholding payment, has a wider application. That wider application emanates from what is known as the mutuality principle. It is perhaps best shown in the opinion of Lord Benholme in the building contract case of *Johnston* v. *Robertson* (1861). Lord Benholme stated that:

> One party to a mutual contract, in which there are mutual stipulations, cannot insist on having his claim under the contract satisfied unless he is prepared to satisfy the corresponding and contemporaneous claim of the other party to the contract.

Accordingly, where the common law right of retention is open to a party, such as a contractor, they are entitled to suspend performance when confronted by an employer who refuses to pay. It should be noted that it is possible to contract out of the common law right of retention, see *Redpath Dorman Long Ltd* v. *Cummins Engine Co. Ltd* (1981). That can only be achieved by the use of clear and unequivocal words in the parties' contract.

While each case will depend on its own individual facts and circumstances, caution should be exercised before deciding to withhold performance, as it is possible that a wrongful suspension by a Contractor may amount to a repudiatory breach, acceptance of which would entitle the Employer to rescind the contract, as discussed above at Section 10.2, see *Mayhaven Healthcare Limited* v. *David Bothma and Teresa Bothma T/A DAB Builders* (2009).

10.9.2 Suspension of performance under the 1996 Act

Section 112 of the 1996 Act as amended provides a right to suspend performance for non-payment where a sum due under a construction contract is not paid in full by the final date for payment and no effective notice of intention to pay less has been given. In those circumstances, the person to whom the sum is due is entitled to suspend performance of any or all of their obligations under the contract to the party by whom payment ought to have been made. This right is without prejudice to any other right or remedy open to the party entitled to payment. This would allow them to raise separate proceedings for payment, should they so wish. The right to suspend performance does not deprive the entitled party of any other rights competent to them. The right to suspend performance under the 1996 Act only arises in the event of non-payment.

The right may not be exercised without first giving to the party in default at least seven days' notice of intention to suspend performance. The notice must state the ground or grounds upon which it is intended to suspend performance. The contract can stipulate that a period in excess of seven days' notice of intention to suspend performance must be given. In practice, employers under a main contract (and main contractors in a sub-contract) will insist upon a greater period of notice. The right to suspend performance ceases when the party in default makes payment in full of the amount due. Subsection 112(3A) provides that the party in default is liable to pay to the party exercising the right to suspend performance under the 1996 Act a reasonable amount in respect of the costs and expenses reasonably incurred by that party as a result of the exercise of the right. Any period of suspension of performance is disregarded in computing the time taken by the party to complete the works. Not only does this apply to the party exercising the right to suspend, but also to any affected third party.

10.9.3 Suspension under the SBC, the SBC/DB and the NEC3

The Contractor's statutory right of suspension is provided for by clause 4.14 of the SBC. A written notice of intention to suspend must be given to the Employer, with a copy to the Architect. If the failure to pay continues for seven days after that notice is given, the Contractor may suspend performance of their obligations under the contract to the Employer until payment in full occurs. A suspension under clause 4.14 is not a default by the Contractor under clause 8.4.1, nor is it a failure to proceed regularly and diligently with the Works, another contractor default, under clause 8.4.2.

By virtue of clause 2.29.6 of the SBC, a delay arising from a suspension by the Contractor of the performance of their obligations under the contract pursuant to clause 4.14 is a relevant event which may entitle the Contractor to an extension of time, see Section 6.5.2. Further, clause 4.14.2 enables the Contractor exercising the right to recover a reasonable amount in respect of costs and expenses reasonably incurred as a result of exercising the right. Similar provision is also made in the SBC/DB, see clauses 2.26.5 and 4.11.

Clause Y2.4 of Option Y(UK)2 of the NEC3 provides that exercise by the Contractor of his statutory right to suspend performance is a compensation event, to be dealt with in accordance with clauses 60–65.

10.10 Delictual claims

10.10.1 General

One party can owe a duty to another in the absence of a contractual relationship. In the context of building contracts, for example, a sub-contractor owes certain duties to the employer, see *British Telecommunications plc* v. *James Thomson & Sons (Engineers) Ltd* (1999).

Liability in delict in construction projects is most likely to arise under the law of negligence or the law of nuisance, though claims may also arise in relation to breach of statutory duty.

10.10.2 Losses recoverable

Broadly speaking, in order to establish a claim in negligence the pursuer must show that:

- the defender owed the pursuer a duty of care in respect of the type of loss in question;
- this duty was breached;
- the breach of duty caused the pursuer's loss; and
- the loss is not too remote.

It should be noted that the precise rules on remoteness of damage differ between claims based on breach of contract and those based on delict, see *Koufos* v. *C Czarnikow Ltd* (1967). In delict, the losses recoverable are those reasonably foreseeable to the defender at the time of the negligent act, see *Allan* v. *Barclay* (1864). In breach of contract cases, on the other hand, the losses recoverable are those foreseeable at the time the contract is entered into, see Section 10.4.2. The reasoning behind this distinction is that in a contract there is the opportunity for one party to obtain protection against a particular type of potential loss by directing the other party's attention to it before the contract is made. In cases arising out of delict there is no such opportunity.

10.10.3 Economic loss

As a general rule, the losses claimed in delict must not be too remote. This means that damages for personal injury, death and loss of or physical damage to property (and economic loss flowing from such loss of or physical damage to property) arising from a breach of duty would normally be recoverable. However, the right to recover economic loss in the absence of physical damage is a particularly problematic area. A detailed examination of the issue is beyond the scope of this book, but the following is a brief overview.

A convenient starting point, which illustrates the type of situation in which matters of this nature arise, is the decision of the House of Lords in the Scottish case of *Junior Books Ltd* v. *The Veitchi Co. Ltd* (1982). In this case the pursuers owned a factory. They entered into a contract with builders for, among other things, the laying of flooring in the factory's production area. The builders sub-contracted this work to the defenders, who were specialist flooring contractors. The pursuers subsequently raised an action against the defenders, seeking damages for loss allegedly sustained as a result of their negligent workmanship. This loss included the cost of replacing the floor surface, allegedly defectively laid; storing goods and moving machinery during the period of replacement; paying wages to employees unable to work during this period; and fixed overheads which would produce no return during this time.

The pursuers also claimed for loss of profit sustained by the temporary closure of the business. They argued that the defenders, as specialists, knew what products were required; were alone responsible for the composition and construction of the flooring; must have known that the pursuers had relied upon their skill and experience; and must be taken to have known that if they did the work negligently, the pursuers would suffer economic loss in requiring to expend money to remedy the resulting defects. The pursuers did not argue that actual or prospective danger to persons or property arose from the state of the flooring. If they had done so there would have been a duty of care under the principles laid down by *Donoghue* v. *Stevenson* (1932).

The defenders argued that the case was irrelevant in law. They contended that the law did not make them liable in delict for the cost of replacing the floor or for economic or financial loss consequent upon that replacement. They argued that while they were under a duty of care to prevent harm being done to property or persons by their faulty work (in accordance with *Donoghue*), they had no duty of care to avoid such faults being present in the work itself. They argued that for the court to hold otherwise would extend the duty of care owed by manufacturers and others far beyond the limits to which the courts had previously extended them; and that a manufacturer's duty not to make a defective product set a standard of care which was much less easily ascertained than that for a duty not to make a dangerous product.

The Inner House of the Court of Session and the House of Lords rejected that argument. They held that there was sufficient proximity between the parties so as to give rise to the relevant duty of care relied on by the pursuers. Further, they held that there were no considerations in this particular case to negate, restrict or limit that duty of care. Pure economic loss was the sort of loss which the defenders, standing in the relationship to the pursuers which they did, ought reasonably to have anticipated as likely to occur if their workmanship was faulty.

The law in England in relation to economic loss now rests with the decision of the House of Lords in *Murphy* v. *Brentwood District Council* (1991), in which it was held that the defendants, who had negligently approved plans that contained erroneous calculations submitted by the builders constructing the plaintiff's house, owed no duty of care to the plaintiffs. The consequence of the plans being incorrect was that the plaintiff, upon selling the house, was unable to obtain the full market value. In this case it was clear, according to the court, that there was no proximity between the plaintiff and the defendants. In essence it could not be said that the defendants had assumed any responsibility to the plaintiff in respect of the plans which they approved.

Further policy considerations are evident in their lordships' decision, namely the 'floodgates' argument and the fear that had a duty been imposed the court would have introduced a transmissible warranty of quality into property transactions which was a legislative matter for Parliament. A further consequence of the decision in *Murphy* was the restriction of *Junior Books·Ltd* to its own special facts, mirroring the court's reluctance in previous cases, such as *D & F Estates Ltd* v. *Church Commissioners for England* (1989), to apply the decision in *Junior Books Ltd.*

It was held in *Murphy*, which was also followed in the subsequent case of *Department of the Environment* v. *Thomas Bates & Son* (1990), that foreseeability of harm based on *Donoghue* principles will not of itself generally be sufficient to impose a duty of care for economic loss. Rather, the courts now have to determine whether there was sufficient proximity between the parties to justify the imposition of a duty of care and also whether it is fair, just and reasonable to impose a duty, see *Caparo Industries plc* v. *Dickman* (1990). It is not, however, necessary to demonstrate that the imposition of a duty of care is 'fair, just and reasonable' where it is established that there is an assumption of responsibility by one party combined with reliance by the other, see *Henderson* v. *Merrett Syndicates Ltd* (1995). Where such a duty of care is held by the court to exist, the recoverable heads of economic loss can prove to be quite extensive, see *Conarken Group Limited* v. *Network Rail Infrastructure Limited* (2011).

It should be noted that it is possible to include contract terms to exclude liability for economic loss and that these will usually be given effect to by the courts so long as they do not fall foul of the reasonableness test contained in the Unfair Contract Terms Act 1977, see *Robinson* v.*PE Jones (Contractors) Limited* (2011) and more recently in *Elvamite Full Circle Limited v. AMEC Earth & Environmental (UK) Limited* (2013).

The tests set out in *Caparo* and *Henderson* have been followed by the Scottish courts, see, for example, *The Governor and Company of the Bank of Scotland* v. *Fuller Peiser* (2002) and *Royal Bank of Scotland plc* v. *Bannerman Johnstone MacLay* (2005).

The law in England in relation to economic loss was reviewed by the House of Lords in *White and Another* v. *Jones and Others* (1995). While the case addresses the issue of whether a solicitor, who negligently drew up a will, owed a duty of care to disappointed prospective beneficiaries, it is submitted that the principles enunciated by the court are of general application. The court expounded the view that proximity was to be assessed by considering whether it could be said that the party causing the loss assumed responsibility in whatever form to the party suffering that loss. The court also reaffirmed their commitment to allowing recovery by analogous extension only. That is, the court will look to determine whether recovery has been permitted in similar situations before allowing recovery in the case under consideration. The court in *White and Another* reiterated their opposition to any *carte blanche* extension of the law in relation to economic loss. In essence, the law will only be allowed to develop by increment rather than by quantum leap.

Notwithstanding the above, it must be borne in mind that *Junior Books Ltd*, being a House of Lords' decision in a Scottish case, is still binding upon Scottish courts. It has not been overruled by any of the subsequent House of Lords' decisions in English cases. Instead, its applicability has been stated to be confined only to the very limited circumstances which pertained to the facts of that case. However, there have been several decisions of the Scottish courts which have at least indicated that, while the House of Lords' decisions in *Murphy* and *D & F Estates Ltd* will be of very high persuasive authority to a Scottish court, it is perhaps rather premature to assume that such cases will be followed unquestioningly by Scottish courts or that, as some commentators would suggest, *Junior Books Ltd* is dead and buried, see *Parkhead Housing Association Ltd* v. *Phoenix Preservation Ltd* (1990) and *Scott Lithgow Ltd* v. *GEC Electrical Projects Ltd* (1992). To conclude this brief overview of economic loss, it is perhaps worth considering one of the major policy restrictions often cited as the principal reason for imposing restrictions on the recovery of economic loss, namely the 'floodgates' argument. The floodgates argument should not be misunderstood as being a reflection of the courts' unwillingness to countenance a multitude of claims against one party. Rather the floodgates argument is the courts' unwillingness to allow liability in an indeterminate amount for an indeterminate time to an indeterminate class. The courts' formulation of the test of proximity will, inevitably, filter out those claims where liability is indeterminate.

Chapter 11
Sub-contractors and Suppliers

11.1 Introduction

On any large construction project it is not uncommon for the majority of the works to be performed by sub-contractors. Indeed it is not unheard of for all the works to be sub-contracted by the main contractor. In turn, many sub-contractors will themselves engage sub-sub-contractors. The principles regarding the formation of a building contract, considered in Chapter 3, apply equally to sub-contracts. Similarly, many of the issues considered in Chapters 4 and 5 will apply in a sub-contract situation. This chapter will deal with the types of sub-contractor and the relationship between employer, main contractor and sub-contractor, and finally will outline some of the issues which frequently arise in practice.

11.2 SBCC Standard Sub-Contracts

The SBCC published in 2011 an updated suite of standard sub-contract conditions for use depending on the form of SBC main contract used and the scope of the sub-contractor's design responsibilities (if any). These are:

- *Standard Building Sub-Contract Conditions (SBCSub/C/Scot 2011 Edition)* – for use where the main contract is the SBC and the sub-contractor has no design responsibilities;
- *Standard Building Sub-Contract with Sub-Contractor's Design Conditions (SBC-Sub/D/C/Scot 2011 Edition)* – for use where the main contract is the SBC with Contractor's Designed Portion and the sub-contractor is to design all or part of the sub-contract works;
- *Design and Build Sub-Contract Conditions (DBSub/C/Scot 2011 Edition)* – for use where the main contract is the SBC/DB and whether or not the sub-contract works include design by the sub-contractor.

There are separate standard forms of sub-contract agreement for use with each of these sets of conditions.

The SBCC has also published a Short Form of Sub-Contract (ShortSub/Scot 2011 Edition) for use where the main contract is an SBCC contract and the

MacRoberts on Scottish Construction Contracts, Third Edition. MacRoberts.

sub-contract works are small or straightforward and the sub-contractor has no design responsibility.

11.3 NEC3 Standard Sub-Contracts

The suite of revised NEC3 contracts published in April 2013 contains the following forms of sub-contract for use where the main contract is the NEC3 Engineering and Construction Contract (ECC):

- the *NEC3 Engineering and Construction Subcontract (ECS)* – this follows the format of the ECC and so contains Core clauses, Main Option clauses for pricing options A–E, Schedule of Cost Components and Contract Data;
- the *NEC3 Engineering and Construction Short Subcontract (ECSS)* – this can also be used where the main contract is the NEC3 Engineering and Construction Short Contract (ECSC) and is intended for low risk and straightforward sub-contract works. As the name suggests, it is a simple form containing sub-contract clauses and Contract Data.

11.4 Nominated and domestic sub-contractors

Under the JCT and the SBCC standard forms of main contract published prior to 2005, sub-contractors were typically either domestic or nominated. A nominated sub-contractor would submit his quotation to the architect or quantity surveyor and in turn be nominated on behalf of the employer if successful. A domestic sub-contractor is usually invited by the main contractor to tender competitively. The SBCC published the Sub-Contract DOM/A/Scot in 1997 as the standard form for domestic sub-contracts in Scotland where the main contract was governed by JCT 98. Nominated sub-contracts have not featured in any of the editions of NEC3 or its predecessors.

In 2005, the JCT and the SBCC issued a new suite of documents, which swept away reference to 'nominated sub-contractor' and the disappearance of the term has continued with the subsequent revisions published in 2011. As a consequence, in some respects the distinction between the two forms of contract may be of historical interest only; but certainly not completely, given that the nature of continuing or potential liabilities under existing contracts may depend on whether a sub-contract is nominated or not. We will therefore consider the distinction, albeit briefly.

What are the principal differences between nominated and domestic sub-contracts? In the first place, an important difference exists between the respective payment provisions. Under clause 35.13.1.1 of JCT 98 the Architect was obliged to identify that portion of the sum due under any Interim Certificate which related to work carried out by Nominated Sub-Contractors. Thereafter, the Architect was required to inform each Nominated Sub-Contractor of the amount of any interim or final payment allocated to their work. Prior to any further certification in favour of the Contractor, the Contractor was obliged to provide the Architect with reasonable proof of payment to the Nominated Sub-Contractor. No equivalent provisions existed under JCT 98 in respect of payments to domestic sub-contractors.

The second important difference between nominated and domestic sub-contracts was the ability of the Nominated Sub-Contractor under JCT 98 to secure payment direct from the Employer. Clause 35.13.5 of JCT 98 provided that where the Contractor had failed to provide reasonable proof of payment to a Nominated Sub-Contractor of the amount included in an Interim Certificate, the Architect should issue a certificate to that effect stating the amount in respect of which the Contractor had failed to provide proof; the amount of any future payment otherwise due to the Contractor was reduced by the amount due to Nominated Sub-Contractors which the Contractor had failed to pay; and the Employer made payment of the relevant amounts direct to the Nominated Sub-Contractors concerned. The obligation on the part of the Employer to make direct payment only arose if the Employer had entered into the direct agreement, NSC/W/Scot, with the Nominated Sub-Contractor.

No direct payment could be made if, at the date when the deduction and payment to the Nominated Sub-Contractor would otherwise be made, the Contractor had become bankrupt, had become 'apparently insolvent' or had had a winding-up order made. This left open the question whether the direct payment provisions could be operated upon the appointment of an administrator or receiver to the Contractor.

Given that much, if not all, of the work performed by Nominated Sub-Contractors was of a specialist nature, what was the liability of the Contractor to the Employer in the event of default by the Nominated Sub-Contractor? Where there was delay on the part of the Nominated Sub-Contractor (or nominated supplier), which the Contractor had taken reasonable steps to reduce, clause 25.4.7 of JCT 98 provided that this was a Relevant Event for the purposes of the Contractor securing an extension of time.

Clause 35.21 of JCT 98 provided that the Contractor had no liability to the Employer in four specified situations:

- for the design of any Nominated Sub-Contract works insofar as they were designed by the Nominated Sub-Contractor;
- for the selection of the kinds of materials and goods which had been selected by the Nominated Sub-Contractor;
- for the satisfaction of any performance specification or requirement insofar as that was included in the Nominated Sub-Contract works;
- for the provision of information by the Nominated Sub-Contractor in reasonable time in order that the Architect could comply with the relevant provisions of the main contract.

11.5 Named and specialist Sub-Contractors

Mention should also be made of 'listed' and 'named' sub-contractors. Clause 3.8 of the SBC provides that, in certain circumstances, work must be carried out by one of a number of persons named in a list which is either in or annexed to the Contract Bills. The work in question will have been priced by the Contractor, and the selection of the person to carry out the work is at the sole discretion of the Contractor. This procedure has the benefit for the Employer that certain specialist work will be carried out by suitably experienced Sub-Contractors.

Provision is also made in the SBC/DB (Schedule Part 2, Part 1, paragraph 2) for the Employer's Requirements to state that 'Named Sub-Contract Work' is to be executed by a named person who is to be employed by the Contractor as a Named Sub-Contractor. If that option is chosen, the Contractor is obliged to enter into a sub-contract with the Named Sub-Contractor but shall remain responsible for their work. If there is a *bona fide* reason why a sub-contract cannot be entered into, the Employer is obliged to either remove the reason for the inability so that the Contractor can enter into a sub-contract as a change which amends the Employer's Requirements, or omit by a Change the Named Sub-Contract Work and issue instructions for the execution of the work. As a Change, the Contractor would be entitled to additional payment. Termination of the Named Sub-Contract is subject to the Employer's consent (not to be unreasonably withheld or delayed). If termination occurs, the Contractor shall complete the Named Sub-Contract Work and may be entitled to payment (unless such termination resulted from the Contractor's default, act or omission or the Employer's consent was not obtained).

The optional Named Sub-Contractor provisions mentioned above apply only to the SBC/DB. However, in May 2012, the SBCC published a Named Specialist Update for use with the SBC. If used, this enables an Employer to name individual specialists as domestic sub-contractors for identified parts of the works by appropriate entry in the Contract Particulars. Typically, this provision will be used where the work is of a specialist nature and allows the Employer the opportunity to select a sub-contractor with the relevant expertise.

The key features of the Named Specialist Update are:

- The Employer has the freedom to choose his sub-contractor for the specialist work.
- The Employer can take control of the timing and involvement of the sub-contractor to meet his requirements.
- The Employer's right may be limited to the specialist pre-named in the contract documents (or their replacements), or, for provisional sum work, may also extend to post-naming.
- The Contractor has a right of reasonable objection (within 7 days) in the case of the naming of any replacement specialist, or if post-naming applies.
- The Contractor has to employ the Named Specialist using the relevant SBCC Standard Building Sub-Contract and cannot impose his own terms and conditions.
- The Contractor carries all the risk for the Named Specialist save where:
    - the Named Specialist becomes insolvent;
    - the Contractor exercises his right of objection.

In those circumstances the Contractor may be entitled to an extension of time and loss and expense.

If the Contractor becomes entitled to terminate a Named Specialist's employment or to give notice of a specified default which could become grounds for termination, he requires to notify the Architect/Contract Administrator beforehand and, if requested, consult with the Employer.

This is not a step back to nomination of sub-contractors which featured in JCT 1998 (and earlier) suites of contracts where there was a direct contractual link between the Employer and nominated sub-contractor. There was a perception in some quarters that the wholesale removal of nomination provisions left a gap since nomination had the advantage to Employers of allowing them to specify who actually performed the relevant sub-contract works, and also gave the Contractor some limited protection in the event of a failure by that sub-contractor (which protection was not conferred by the 'listed sub-contractor' provisions). The Named Specialist Update perhaps goes some way to addressing that gap.

The NEC3 contains no corresponding provisions in respect of named or specialist sub-contractors.

11.6 Direct payments to sub-contractors

Apart from the provisions on direct payment discussed above in the context of nominated sub-contracts, in certain circumstances (usually for commercial considerations), employers agree to pay sub-contractors directly. This should not be done in the absence of agreement with the contractor, whether under the contract or otherwise, and any direct payment provision should not only confer a right on the employer to make a direct payment to a sub-contractor in certain defined circumstances, but must also allow the employer to deduct an equivalent amount from sums otherwise due to the main contractor. Otherwise, the employer's primary liability to pay the contractor is not satisfied by the direct payment. An illustration of the issues that can arise is the Extra Division decision of *Brican Fabrications Ltd* v. *Merchant City Developments Ltd* (2003). In this case, Merchant City, as employer, agreed to pay Brican direct (as sub-contractor) on account of what turned out to be the latter's well-grounded concerns as to the solvency of the main contractor. The sub-contract entered into between the main contractor and Brican provided that the main contractor assented to the direct payment arrangement between Merchant City and Brican. The main contractor thereafter went into liquidation and the parties to the action could not agree whether Merchant City had agreed to pay Brican direct or whether the agreement was that Merchant City would deduct from sums due to the main contractor that portion which the main contractor owed to Brican and pay it direct to Brican, as agent of the main contractor. The Extra Division preferred the former.

Care should also be taken in relation to direct payments where there is a likelihood of the contractor's insolvency, since such payments could in certain circumstances be challenged by the contractor's insolvency practitioner as an unfair preference under section 243 of the Insolvency Act 1986 (see *British Eagle International Airlines Ltd* v. *Compagnie National Air France* (1975)).

An alternative mechanism for ensuring security of payments to sub-contractors and which is growing in popularity is that of Project Bank Accounts. See Section 8.7.

11.7 Privity of contract

11.7.1 General

Ordinarily there is no direct contractual relationship between the employer and the sub-contractor, and the individual contracts which make up the contractual chain between sub-contractor and employer are (subject to collateral warranties and to the circumstances described in Section 11.7.2) enforceable only by the parties to such contracts. This principle is known as privity of contract.

The law would, however, in certain circumstances, permit an employer to sue a supplier direct, should the supplier have given certain assurances or warranties as to, for example, the fitness for purpose of the supplier's product, notwithstanding the fact that the contract for the sale of the product was with the contractor appointed by the employer, see *Shanklin Pier Ltd* v. *Detel Products Ltd* (1951) and *British Workman's and General Assurance Co.* v. *Wilkinson* (1900).

Another exception is where appropriate rights are assigned by a main contractor to a sub-contractor, see *Constant* v. *Kincaid & Co.* (1902). Assignation is considered in Chapter 12. In the absence of a direct contractual relationship, or an assignation of rights, neither employer nor sub-contractor can sue the other under contract. A practical example of this is that, in the absence of the former direct agreement, NSC/W/Scot, an employer and nominated sub-contractor would not be able to enforce any rights against the other, for example, direct payments (see above).

It should be noted that (if so provided for in the Contract Particulars) clause 2.26 of the SBCSub/C/Scot (with similar provisions in DBSub/C/Scot and SBC-Sub/D/C/Scot) specifically requires any sub-contractor to give a collateral warranty in favour of the Employer, purchaser, tenants and/or funder within 14 days from receipt of a notice from the Contractor. The SBCC has published the following forms of sub-contractor collateral warranty for use with its sub-contract forms:

- Sub-Contractor/Funder collateral warranty (SCWA/F/ Scot 2005 (October 2007 revision));
- Sub-Contractor/Purchaser and Tenant collateral warranty (SCWA/P&T/Scot 2005 (October 2007 revision));
- Sub-Contractor/Employer collateral warranty (SCWA/E/Scot 2005 (October 2007 revision)).

Clause 7E of the SBC and the SBC/DB imposes an obligation on the Contractor to procure for the Employer collateral warranties from a Sub-Contractor to purchasers, tenants and/or funders in the relevant forms described above, subject to amendments proposed by the relevant Sub-Contractor and approved by the Contractor and Employer (such approval not to be unreasonably withheld or delayed), if so provided for in the Contract Particulars. It is not uncommon, however, for employers to insist on bespoke forms of collateral warranties rather than those published by the SBCC, by making appropriate amendments to clause 7E and annexing such bespoke forms. In that event, consequential amendments must of course also be made to the sub-contracts.

As mentioned in Section 13.2.1, there are no equivalent provisions in the NEC3 for the provision of collateral warranties and so, if warranties are required, this will require to be addressed in both the main contract and sub-contract by means of Z clauses.

11.7.2 *Jus quaesitum tertio*

The above statement on privity of contract is, however, qualified where a *jus quaesitum tertio* has been created by the contract. The creation of such a right will give a third party (the '*tertius*') a right to sue under the contract, notwithstanding that it is not a party to it. In order to create the right, the contract must expressly, or by implication, confer a benefit on the *tertius* or a class of persons of which the *tertius* is a member. Unless an intention on the part of the contracting parties to create a *jus quaesitum tertio* in favour of a third party is expressed or can be inferred from the terms of the contract, no such right will be created. See *Scott Lithgow Ltd* v. *GEC Electrical Projects Ltd* (1992) and *Strathford East Kilbride Ltd* v. *HLM Design Ltd* (1997). In relation to third party rights under the SBC, see Section 13.5.

In England, the courts, for a period, permitted companies within the same corporate group to pursue losses in contract notwithstanding the fact that the losses in question were sustained by another connected company. This approach was accepted where the entity sustaining the loss had no direct course of action against the wrongdoer. This was to avoid a 'legal black hole' preventing the loss being recoverable. In the Scottish case of *Clark Contracts Ltd* v. *The Burrell Co. (Construction Management) Ltd* (2002) the court held that the existence of the *jus quaesitum tertio* in favour of another group company of the defenders did create a direct course of action, and as such the English authorities could not be relied upon by the defenders in support of a counterclaim for damages in respect of losses allegedly sustained by the defenders' sister company. Lord Drummond Young, in *McLaren Murdoch & Hamilton Ltd* v. *The Abercromby Motor Group Ltd* (2003), stated that the *jus quaesitum tertio* is of limited utility. His Lordship took the view that the decision of the House of Lords in *Alfred McAlpine Construction Ltd* v. *Panatown Ltd (No. 1)* (2001), albeit an English case, was wholly consistent with the principles of Scots law and that Scots law should adopt the general rule in that case as described by Lord Clyde. See also Section 12.3 for a more detailed analysis.

11.7.3 Delict

Prior to certain case law in the late 1980s, and in particular, *D & F Estates Ltd* v. *Church Commissioners for England* (1989) and *Murphy* v. *Brentwood DC* (1991), it had been understood that an employer could sue a sub-contractor direct under delict and recover economic loss, see *Junior Books Ltd* v. *The Veitchi Co. Ltd* (1982). Collateral warranties emerged as a result of *D & F Estates Ltd* and *Murphy*. (See also Section 13.2.) Market forces dictated that any perceived vacuum in the law of negligence be filled by the law of contract. Developers, owners and funders of large commercial developments need the ability to sue professional team members and/or specialist sub-contractors.

This branch of the law of negligence has nevertheless continued to develop in both Scotland and England. In *White and Another* v. *Jones and Others* (1995), the House of Lords in an English appeal decided by a majority that a solicitor owed a duty of care to beneficiaries under a will which had been negligently drawn up. In the Scottish case of *Scott Lithgow Ltd* v. *GEC Electrical Projects Limited* (1989), Lord Clyde allowed to proceed to proof a case in which an employer sued a domestic sub-contractor for recovery of economic loss stemming from allegedly defective wiring. He held that nomination was not a necessary factor before a duty of care could arise, but it was an important element where it did exist. Where it does exist it obviously serves to point towards the degree of proximity which is required for the employer to succeed. For a further discussion on the law of delict, see Section 10.10.

It must be fair, just and reasonable for a duty of care to exist. This issue was addressed by the House of Lords in the Scottish case of *British Telecommunications plc* v. *James Thomson & Sons (Engineers) Ltd* (1999). In that case the employer sued a sub-contractor in delict in respect of losses sustained as a consequence of a fire breaking out in the employer's premises for which the employer held the sub-contractor responsible. The sub-contractor had been engaged by the main contractor on the same terms and conditions of contract as those ruling between the employer and the main contractor. The insurance provisions in the main contract made it clear that damage caused in the way suggested by the employer was to be covered by an insurance policy which the employer was bound to take out. In short, the damage in question was one of the specified perils under the main contract. As such, it was contended by the sub-contractor that it would not be fair, just or reasonable to impose a duty on them to avoid such damage. Their Lordships, however, attached significant weight to the fact that the insurance arrangements in the main contract afforded any nominated sub-contractor the benefit of a waiver by the relevant insurers of any right of subrogation which they may have against the nominated sub-contractor but no such provision existed for the benefit of domestic sub-contractors. The unanimous decision of the court was that it would be fair, just and reasonable to impose a duty of care on the domestic sub-contractors to the employer. See further *European and International Investments* v. *McLaren* (2001) and *Tartan American Machinery Corp* v. *Swan & Co.* (2004).

11.8 Relationship between main and sub-contracts

It is common for main contractors to attempt to incorporate by reference the terms of the main contract into the sub-contract. This practice of wholesale incorporation is not to be encouraged and frequently leads to disputes between the parties. The degree of incorporation can vary. Many main contractors attempt to incorporate their own programme into the sub-contract, see *Scottish Power plc* v. *Kvaerner Construction (Regions) Ltd* (1998).

The effect of incorporation of main contract terms was considered in *Babcock Rosyth Defence Ltd* v. *Grootcon (UK) Ltd* (1998). In this case the sub-contractor raised an action against the main contractor. The main contract incorporated a modified form of the ICE Conditions of Contract (Fifth Edition). The issue for the court was whether or not clause 66, the arbitration clause, formed part of the sub-contract. The main contractor maintained that the ICE Fifth Edition was incorporated into the sub-contract, subject to the express qualifications made and to its adaptation for practical effectiveness in the sub-contractual relationship. To that extent the main contractor's submissions did not go as far as those made in *Parklea Ltd* v. *W & J R Watson Ltd* (1988). In the latter case the court rejected the contention that the whole provisions of the main contract were to be incorporated *mutatis mutandis* into the sub-contract. In *Babcock Rosyth Defence Ltd* the defenders acknowledged that certain of the main contract provisions would have no place in the sub-contract. The judge, Lord Hamilton, stated:

> When parties make reference to a set of conditions designed primarily for use in another contract but do not expressly adapt those conditions to meet the circumstances of their own relationship, it is often difficult to determine with confidence the contractual effect. Where, on the one hand, the circumstances demonstrate a plain common intention to incorporate terms, albeit expressed in language designed primarily for another purpose, the court will, where it is possible to do so without substantially rewriting the parties' bargain, give effect to the parties' plain common intention by incorporating terms subject to appropriate linguistic adaptation ... Where, on the other hand, the common intention is not plain or there are major difficulties about linguistic adaptation, the result will be otherwise. Even in cases where incorporation subject to linguistic adaptation is possible and appropriate, there may yet remain a question as to the extent to which conditions are so incorporated.

Lord Hamilton held that the parties had plainly intended that the ICE Fifth Edition should apply to some extent, albeit with appropriate linguistic adaptation. He was not satisfied, however, that it was sufficiently clear that the parties intended to incorporate the arbitration clause into the sub-contract. To avoid ambiguity, therefore, parties should make it clear which particular main contract provisions are to be incorporated into the sub-contract and to what extent.

Similar difficulties were encountered by the pursuers in *Watson Building Services* v. *Harrison* (2001) when they unsuccessfully contended that the adjudication provisions of the main contract had been incorporated by reference into their sub-contract with the defenders. See also the discussion in Section 3.6.

11.9 Restrictions on sub-contracting

It is unusual, and in most cases impractical, for a main contract to contain a blanket prohibition on sub-contracting. However, most standard forms of contract allow the employer to retain some level of control over sub-contracting. For the position on sub-contracting (in legal terms, 'delegation') under the common law, see Section 12.7.

Clauses 3.7–3.9 of the SBC, clauses 3.3 and 3.4 of the SBC/DB and clause 26 of the NEC3 impose certain obligations on the contractor in relation to sub-contracting.

The restrictions in the SBC and the SBC/DB are broadly similar and permit the sub-letting of any part of the Works, provided that the consent of the Architect or the Employer (as the case may be) is obtained. Such consent is not to be unreasonably delayed or withheld. Clause 3.7.1 of the SBC and clause 3.3.1 of the SBC/DB make it clear that the Contractor will remain wholly responsible for carrying out and completing the Works, notwithstanding any subletting. Clause 3.3.2 of the SBC/DB and clause 3.7.2 of the SBC each contain equivalent provisions in respect of sub-contracting, respectively the design of the Works and any Contractor's Designed Portion. Clause 3.9 of the SBC and clause 3.4 of the SBC/DB also provide that 'where considered appropriate' the Contractor shall engage the Sub-Contractor using the relevant version of the SBCC Standard Building Contract (in the case of the SBC) or the SBCC Design and Build Contract (in the case of the SBC/DB). It is not clear if this leaves the decision as a matter for the discretion for the Contractor or whether a refusal to use the relevant SBBC sub-contract could justify the Employer withholding consent to the sub-letting. At any rate, whatever form of sub-contract is entered into, clause 3.9 of the SBC and clause 3.4 of the SBC/DB specify certain mandatory terms which any such sub-contract must contain. These include provisions in respect of the passing of title of materials delivered to site (see Section 11.13); exclusion of any liability of the Employer to the Sub-Contractor; payment of interest by the Contractor to the Sub-Contractor on late payments; and delivery of Sub-Contractor warranties, where applicable (see Section 11.7).

The conditions under which the Contractor may sub-contract under the NEC 3 are contained in clause 26 of that contract. Clause 26.1 makes it clear that sub-contracting does not affect the Contractor's primary responsibility for performance. Before appointing a Sub-Contractor, the Contractor must first obtain the Project Manager's acceptance of the proposed Sub-Contractor and (except where agreed as not required or an NEC contract is proposed) the proposed sub-contract conditions. A reason for not accepting a Sub-Contractor or sub-contract conditions is that the appointment will not allow the Contractor to do all the work necessary to complete the Works or (in the case of the sub-contract conditions) do not include a statement that the parties to the sub-contract shall act in a spirit of mutual trust and co-operation. Although on the face of it, these grounds for non-acceptance appear not to be exhaustive, they need to be read along with clause 60.1(9) which provides that it is a compensation event where the Project Manager withholds an acceptance for a reason not stated in the contract. So it follows that withholding consent for a reason other than that specified in clause 26 may give rise to time and cost consequences. See also the discussion on sub-contracting under the NEC3 in Section 5.2.4.

11.10 Main contractor's discount

It is common for sub-contracts to allow the main contractor a discount on the price of the sub-contract works. Unless the parties to a sub-contract make an express provision which connects the main contractor's ability to deduct discount with prompt payment, the courts are likely to view the discount as being no more than a reduction in the sub-contract price, see *Team Services plc* v. *Kier Management and Design Ltd* (1993).

11.11 Suppliers

Contracts of supply or sale are regulated by the Sale of Goods Act 1979, as amended by the Sale and Supply of Goods and Services Act 1994. In terms of s.14 of the 1979 Act as amended by s.(1) of the 1994 Act, there is an implied term that the goods supplied under a contract of supply or sale are of 'satisfactory quality'. The quality of goods is deemed to include their state and condition and, in appropriate cases, their fitness for all the purposes for which goods of the kind in question are commonly supplied, their appearance and finish, their freedom from minor defects, their safety and their durability. Fitness for purpose can also include supplying goods that comply with Building Regulations, see *Lowe & Anor* v. *W Machell Joinery Ltd* (2011).

If the purchaser has made known the purpose of the goods, then the implied condition of fitness for purpose can only be discharged by evidencing that the purchaser did not rely on, or that it is unreasonable for him to rely on, the skill or judgement of the seller. The Court of Appeal case of *BSS Group Plc* v. *Makers (UK) Ltd* (2011) held that the onus is on the seller to show that this exception applies, and it may be difficult where the seller is a 'specialist' dealer.

11.12 Retention of title clauses

Suppliers' terms and conditions commonly include a retention of title clause. The object of such a clause is to protect the supplier against the insolvency of its customer by delaying the passing of ownership of the goods in question to the customer until payment has been made. Otherwise, the ownership of the goods will normally transfer upon delivery, by virtue of s.17 of the Sale of Goods Act 1979.

Section 19 of the Sale of Goods Act 1979 permits a seller to retain ownership of goods, notwithstanding delivery to the purchaser, in the event that the parties to the contract expressly provide that change of ownership is to be conditional. The most obvious condition will, of course, be as to payment of the price. Section 19 is a restatement of the position at common law and under the Sale of Goods Act 1893. The period between the mid-1970s and 1990 saw considerable litigation on the subject of retention of title clauses, starting with *Aluminium Industrie Vaassen BV* v. *Romalpa Aluminium Ltd* (1976) and ending with *Armour* v. *Thyssen Edelstahlwerke AG* (1990). In the latter case, the House of Lords held that 'all sums' retention of title clauses were effective in Scotland, as had been the case in England for some time. Prior to the decision by the House of Lords, the Scottish courts had restricted the applicability of retention of title clauses to the extent that they reserved title to the seller of goods in the event that the purchase price for those goods had not been paid. The courts had refused to give effect to retention of title clauses which purported to reserve title to the seller until all sums due by the purchaser to the seller, including sums due in respect of other goods, had been paid.

A retention of title clause will not protect the unpaid supplier in the event of the contract of sale being governed by s.25(1) of the Sale of Goods Act 1979, that is where a third party has purchased the relevant goods in good faith and without notice of the retention of title clause. Section 25(1) provides that:

> Where a person having bought or agreed to buy goods obtains, with the consent of the seller, possession of the goods or the documents of title to the goods, the delivery or transfer by that person, or by a mercantile agent acting for him, of the goods or documents of title, under any sale, pledge, or other disposition thereof, to any person receiving the same in good faith and without notice of any lien or other right of the original seller in respect of the goods, has the same effect as if the person making the delivery or transfer were a mercantile agent in possession of the goods or documents of title with the consent of the owner.

The application of s.25 is illustrated by the case of *Archivent Sales and Developments Ltd* v. *Strathclyde Regional Council* (1985). In that case the supplier delivered goods to site and payment was made by the employer to the main contractor. The main contractor failed, in turn, to make payment to the supplier, who sought to recover the goods from the employer on the basis of the retention of title clause in its contract with the contractor. The supplier's claim to ownership failed because the existence of the supplier's retention of title clause in the contract of sale to the main contractor had not been brought to the employer's attention.

There are a number of other practical difficulties in enforcing retention of title clauses, not least of which is that the clause cannot be founded upon in a question with the building owner where the materials have been incorporated into the structure, provided that there is no element of bad faith on the part of the building owner, see *Archivent Sales and Developments Ltd*. This arises from the principle of Scots law that all buildings and fixtures pass into the ownership of the party who has title to the ground upon which they are erected, see *Brand's Trustees* v. *Brand's Trustees* (1876). Whether an article attached to a structure (as distinct from remaining moveable property) becomes, by virtue of the principle of accession, a fixture and thus part of the structure is a matter of fact to be determined by the circumstances of the case, see *Scottish Discount Co. Ltd* v. *Blin* (1986).

On a further practical level, the seller will need to identify their goods and, if necessary, distinguish them from other similar goods. A seller of identifiable items bearing serial or batch numbers is likely to enjoy greater success in enforcing a retention of title clause than a supplier of sand or bricks. In the latter situation (in the absence of an 'all sums' retention of title clause) even if the seller can identify the bricks they supplied, that will not be enough unless they can connect particular quantities of unfixed brick with particular unpaid invoices.

Retention of title disputes commonly present themselves in insolvency situations. The supplier must either ensure that the retention of title clause triggers automatically by insolvency, obliging a customer to immediately identify any remaining stock and provide possession of it, or communicate to the administrator/liquidator as soon as possible that they intend to rely on their retention of title clause, see *Sandhu v Jet Star Retail Limited* (2011).

11.13 Supply of goods by sub-contractors

A distinction falls to be drawn between materials supplied by a supplier and those supplied under a sub-contract, see *Thomas Graham & Sons* v. *Glenrothes Development Corporation* (1967). As seen above, title to goods or materials supplied by a supplier will normally pass to the contractor upon delivery to site, unless title has been retained by the supplier under a retention of title clause. In contrast, in the case of a sub-contract, i.e. a contract for the supply of goods and services, the common law applies and ownership passes when the goods or materials are fixed to the structure, see *Stirling County Council* v. *Official Liquidator of John Frame Ltd* (1951).

The conceptual difference between goods or materials supplied under a contract of sale and those supplied under a contract for the supply of goods and services is further illustrated by the way in which the SBC deals with the purchase of off-site materials. Clause 4.17 of the SBC provides that if the Architect is of the opinion that it is expedient to do so, the Employer may enter into a separate contract for the purchase from the Contractor or any Sub-Contractor of any materials or goods prior to their delivery to site. The equivalent provision to the SBC clause 4.17 in the SBC/DB is clause 4.15.

If such a contract is entered into, the materials or goods cease to form part of the contract and the Contract Sum is adjusted accordingly. Such a provision is particularly useful in the case of major equipment which is being manufactured by a specialist supplier in its premises and which needs to be paid for prior to delivery to site. This has the result of characterizing the transaction as a contract of sale, which is subject to the Sale of Goods Act 1979, which allows ownership to pass prior to delivery. Clause 4.17 of the SBC falls to be contrasted with the corresponding clause in JCT 2011 which allows the amount stated as due in an interim certificate to include the value of 'listed items' before their delivery to site, provided that the Contractor demonstrates that after payment, such listed items will become the property of the Employer. For the reasons stated above, under Scots law it would not be possible to demonstrate that ownership had effectively transferred in such circumstances, and hence the difference in approach under the SBC.

Clause 70.1 of the NEC3 provides that whatever title the Contractor has to plant and materials (i.e. items intended to be included in the works) outside the Working Areas passes to the Employer if the Supervisor has marked it for the contract in question. For the reasons stated above, it is unlikely that this clause will be enforceable where the contract is subject to Scots law, and so if the intention under such a contract is to transfer ownership in off-site materials to the Employer, a Z clause of similar effect to the SBC clause 4.17 will be required.

In terms of clause 2.15.2 of the Standard Building Sub-Contract Conditions for use in Scotland, where the value of unfixed materials or goods which have been delivered to or placed on or adjacent to the works has been included in any Interim Certificate issued under main contract, and the amount properly due by the Employer to the

Contractor has been discharged, then such materials or goods become the property of the Employer, and the Sub-Contractor agrees that they cannot deny that such materials or goods are and have become the property of the Employer. Should the Contractor pay the Sub-Contractor for any such materials or goods prior to the Employer having first discharged his obligation to pay the Contractor for same, then clause 2.15.3 of the Standard Building Sub-Contract provides that the materials or goods become the property of the Contractor. These provisions meet the terms of clause 3.9.2 of the SBC (and clause 3.4.2 of the SBC/DB) which sets out certain mandatory provisions which must be contained in any sub-contract let under the SBC.

Clause 70.2 of the NEC3 ECS provides that whatever title the Sub-Contractor has to plant and materials passes to the Contractor if it has been brought within the Working Areas (subject to title passing back if the relevant item is removed from the Working Areas with the Contractor's permission). A back-to- back provision is contained in clause 70.2 of the NEC3 so that title to an item delivered to site by the Sub-Contractor will pass to the Contractor and then immediately to the Employer. This is of course subject to the Sub-Contractor having title in the first place, since it is only 'whatever title' he has which is passed on. It should also be noted that, in contrast to SBC and SBC/DB, title to materials delivered to site will pass regardless of payment.

11.14 Adequate mechanism for payment

The amendments to the 1996 Act made by the 2009 Act resulted in changes to the mechanism for payment intended to benefit sub-contractors. As discussed in Section 8.1, the 1996 Act prohibited 'pay when paid' clauses except in certain circumstances; however, the main contractor could circumvent this by inserting a 'pay when certified' clause into the sub-contract, resulting in payment under the sub-contract being dependent on an equivalent amount being certified under the main contract. This normally extended to the release of sub-contract retention, which was usually conditional on the issue under the main contract of the practical completion certificate or making good defects certificate.

The new section 110(1A) of the 1996 Act introduced by the 2009 Act provides that a construction contract does not provide an adequate mechanism for determining what payments become due where that contract makes payment conditional upon (a) the performance of obligations under another contract; or (b) a decision by any person as to whether obligations under another contract have been performed. In other words, parties will not be permitted to include 'pay when certified' clauses in their contracts. For a more detailed discussion on this topic, see Section 8.1.7.

The release of retention, particularly the second tranche of retention funds (or 'final payment') has traditionally been a thorny issue for sub-contractors. Under earlier editions of the SBC it was widely accepted that the issue of a certificate was a condition precedent to a contractor's right to demand payment (*Costain Building and Civil Engineering Ltd* v. *Scottish Rugby Union* (1993)). This proposition was recently reinforced in the unreported Paisley Sheriff Court decision of *Clark Contracts Ltd* v. *Brite1 Fund Trustees Ltd* (2013). In that case the contractor applied for final payment nine years after practical completion in circumstances where no final certificate had been issued by the employer's architect. The employer argued that the contractor's claim had prescribed as the five-year prescriptive period ran from the point when the final certificate ought to have been issued. The contractor claimed that the debt remained contingent until the issue of the final certificate or an award from an arbitrator or a decree of the court (*Karl Construction Ltd* v. *Palisade Properties plc* (2002)). The court agreed and rejected the argument that the claim had prescribed. The question is whether the new regime has simplified the process of retention release.

The amendments to the 1996 Act to outlaw 'pay when certified' clauses mean that a sub-contract must now contain a specific date for the release of retention, and such release can no longer be dependent on the issue of practical completion and/or making good defects certificates under the main contact. The SBCC has amended its standard sub-contracts to comply with these amendments to the 1996 Act. Under clauses 4.15 and 4.16 of the SBC Sub/C/Scot, the first half of the retention becomes due for release on practical completion of the sub-contract works, or where the main contract works are in sections, upon completion of the sub-contract works in that section. The release of the remaining retention is linked to the 'Retention Release Date' specified in the Sub-Contract Particulars. This is the date when practical completion of the sub-contract works as a whole has been achieved, and there are no outstanding defects in the sub-contract works. There is also a 'Minimum Retention Amount' contained in the Sub-Contract Particulars, so if the calculated retention is less than that amount, then no retention is to be retained and if the retention amount falls below the minimum following release of the first tranche of retention following practical completion, then the whole retention is released at that time.

SBC Sub/C/Scot also provides that (where specified in the Sub-Contract Particulars), on or before the date of commencement of the sub-contract works the Sub-Contractor shall provide to the Contractor and thereafter maintain a Retention Bond from a surety approved by the Contractor. In those circumstances deduction of retention in respect of interim payment will not apply.

Under the NEC3 ECS, retention is deductible if Secondary Option X16 is chosen. Under clause X16 the first half of the retention falls to be released on the completion or taking over of the whole of the sub-contract works, with the second half being released on issue of the Defects Certificate in respect of the sub-contract works. Since neither event is contingent on certification under the main contract, these provisions comply with the amended 1996 Act.

Chapter 12
Assignation, Delegation and Novation

12.1 Introduction

This chapter considers three separate methods of transferring rights and obligations, namely, assignation, delegation and novation.

Assignation is a method of transfer of incorporeal moveable property, such as rights arising under contract. While the employer's heritable interest in the property formed by a building project may be transferred to a third party by sale or lease, in order to provide the purchaser or tenant with an interest in the building contract, it is usual to assign that interest. In most standard forms of building and engineering contract, specific provisions are made in respect of assignation. The specific provisions of the SBC and the SBC/DB and of the NEC3 in this regard are considered in Sections 12.5 and 12.6 respectively.

A typical contract contains both rights and obligations. In England, it is trite law that the obligations under a contract, otherwise known as the burden of the contract (as opposed to the benefit), cannot be assigned without the consent of the party entitled to enforce those obligations, see *Linden Gardens Trust Ltd* v. *Lenesta Sludge Disposals Ltd and Others* (1994). The position in Scotland may be different insofar as it is a matter of some controversy in Scotland as to whether obligations, as well as rights, under a contract are capable of being assigned. It has been contended, though the matter is not free from doubt, that obligations under an executory contract (i.e. in which obligations remain to be performed) can be assigned, provided that they do not involve an element of *delectus personae* (a specific choice of natural or legal person), see *Cole* v. *Handasyde & Co.* (1909); McBryde, *The Law of Contract in Scotland* (third edition, 2007), paras 12–42 and 12–43. *Delectus personae* is considered in Section 12.4 in this volume. In *Scottish Homes* v. *Inverclyde District Council* (1997), it was suggested that there is no rule in Scots law that would prevent a contracting party substituting another in his place 'both as regards performance and the benefits of the contract'. However, see the Scottish Law Commission 'Discussion Paper on Moveable Transactions', June 2011 (paras 4.15 and 4.16) which suggests that Scots law is broadly the same as English law in this respect and that contractual obligations cannot be assigned without consent, regardless of *delectus personae* (though no authority is given for this proposition). The Discussion Paper also suggests (again without reference to authority) that doubt expressed in *Scottish Homes* that Scots law is the same as

MacRoberts on Scottish Construction Contracts, Third Edition. MacRoberts.

English law in relation to assignation of obligations is misplaced. Strictly speaking, it is the rights (and perhaps the obligations) under a contract which are assigned, rather than the contract itself, and so while certain rights and obligations may involve elements of *delectus personae*, and are thus not assignable without consent, certain others may not and so (at least arguably) may be assigned. An express provision in a building contract that prohibits assignation, or only permits assignation with consent, will override the common law. In those circumstances, *delectus personae* is of no relevance. In view of the uncertainties relating to assignation of obligations, in practice the usual method adopted for transferring obligations is novation, see Section 12.8.

12.2 Common law

In the absence of any express provision of the contract governing assignation, the common law will apply.

No particular wording is required to constitute an assignation under Scots law as long as it effects a transfer, see *Carter* v. *McIntosh* (1862). However, the general principle is that an assignation of a right must be intimated to the person against whom the right may be enforced before the assignation is effective against that person. For example, where a contractor assigns his right to receive payment of retention monies from the employer, the assignee (i.e. the party in whose favour the assignation is granted) must ensure that such assignation is intimated to the employer. Failure to intimate assignation may prevent rights being effectively transferred, see *Laurence McIntosh Ltd* v. *Balfour Beatty Group Ltd* (2006), where the resulting relationship was analysed in terms of *ad hoc* agency and meant there was no title to sue. It follows that in the event of competing assignations, these rank in priority of date of intimation, not date of execution. This reinforces the importance of intimating without delay since, for example, the insolvency or bankruptcy of the assignor after execution of the assignation but prior to intimation could result in the assignee's rights being defeated by those of the insolvency practitioner or trustee to whom the rights in question will be assigned under law.

The assignee of a right is entitled to no greater benefit in respect of that right than the assignor. This is often expressed by way of the Latin maxim *assignatus utitur jure auctoris* (literally, the assignee exercises the right of the grantor), which means that the assignee can never be in a better position than the assignor. Thus, any defences available to the debtor in respect of the claim by the assignor will also be available against a claim by the assignee, see *Scottish Widows Fund* v. *Buist* (1876). For example, it is a complete defence to demonstrate that a debt was settled before the right to sue for that debt was assigned. An example of this arose in *Smiths Gore* v. *Reilly* (2001), where it was held that an 'empty vessel' had been assigned which could not ground a meaningful claim. There may be some doubt, however, as to whether in certain circumstances, a counterclaim available against the assignor can also be pled in defence against the assignee, see *Binstock, Miller & Co* v. *E. Coia & Co Ltd* (1957); *Alex Lawrie Factors Ltd* v. *Mitchell Engineering Ltd* (2001).

12.3 *Effect of assignation upon claims*

Questions often arise in the context of construction contracts as to the effect of an assignation upon claims. This issue was considered by the House of Lords in the Scottish case of *GUS Property Management Ltd* v. *Littlewoods Mail Order Stores Ltd* (1982) and in the English cases of *Linden Gardens Trust Ltd* v. *Lenesta Sludge Disposals Ltd* (1994), *Darlington BC* v. *Wiltshier Northern Ltd* (1995) and *Alfred McAlpine Construction Ltd* v. *Panatown Ltd (No. 1)* (2001).

GUS Property Management Ltd concerned the assignee's right to claim losses arising from a delict where that right has been validly assigned. A building in Glasgow, owned by a company named Rest Property Co. Ltd (henceforth Rest), was damaged in the course of building operations at a neighbouring property. Rest transferred the property to GUS Property Management (henceforth GUS), a related company, for its full book value, ignoring the cost of repairing the damage. GUS carried out repairs to the property and Rest then assigned to GUS all claims competent to them arising out of the building operations on the neighbouring property. Relying on the assignation, GUS raised an action for damages against the neighbouring proprietors and those involved in the building operations. The defenders argued that because the property had been transferred for its full book value, the assignor had not sustained any loss at all, and thus, based on the principle *assignatus utitur jure auctoris*, the assignee could not in turn recover damages for any such loss. The House of Lords, overruling the Inner House of the Court of Session, held that such a defence was not sustainable and refused to allow the claim to fall into some kind of legal 'black hole'. Rest would have been able to sue for damages at the time of the assignation, following the general rule in *Gordon* v. *Davidson* (1864), that the owner of a property damaged as a result of a delict does not lose the right to sue on parting with that property. The fact that the transfer price had been fixed for internal accounting purposes did not affect the true loss suffered by Rest, and Rest had assigned its right to sue for that loss to GUS.

While *GUS Property Management Ltd* is a case dealing with a delictual claim, it is submitted that the principles set out in it are equally applicable to claims under contract or for breach of contract. Although the decision was based on a specific set of circumstances, it appears that the court would reject a defence to a claim by an assignee on the ground that no loss has been suffered because the assignor (i.e. the original developer) sold for full value. Such 'no loss' arguments have been resisted by both the Scottish and English courts, see *McLaren Murdoch & Hamilton Ltd* v. *The Abercromby Motor Group Ltd* (2003) and *Darlington BC* respectively. This seems to be the case whether the transaction is between related companies, as in *GUS Property Management Ltd*, or at arm's length, as in *Technotrade Ltd* v. *Larkstore Ltd* (2006).

The appeal in *Linden Gardens Trust Ltd* was heard by the House of Lords with the appeal in *St Martin's Property Corporation Ltd* v. *Sir Robert McAlpine & Sons Ltd* (1994). Both cases concerned an exception to the rule that a contracting party may only recover its own losses under the contract.

In each case the contract was subject to the JCT 63 conditions. Clause 17 of JCT 63 (as with clause 7.1 of the SBC) prohibited assignation of the contract by the Employer without the Contractor's written consent. The House of Lords held that, on a true construction of the contracts, the wording of clause 17 prohibited assignation by the Employer, without the Contractor's consent, of the benefit of the contract and the assignation of any cause of action. On the facts this meant there had been no valid assignation in either case. However, it was held that the original Employer was nonetheless entitled to recover from the Contractor the loss sustained by the purchaser. As the contract was expressly not assignable without the Contractor's consent, the House of Lords deemed that the Employer and the Contractor should be treated as having contracted on the basis that the Employer would be entitled to enforce his contractual rights against the Contractor for the benefit of third parties who would suffer from defective performance.

The exception established by *Linden Gardens Trust Ltd* was revisited in *Alfred McAlpine Construction*. Panatown employed McAlpine to build an office block and car park on a site owned by Unex Investment Properties Ltd (henceforth Unex), a member of the same group as Panatown. McAlpine granted a duty of care deed (nowadays usually called a collateral warranty) to Unex. When Panatown's claim for substantial damages reached the House of Lords, it was held by a majority that the exception did not apply because of the existence of the duty of care deed, which gave Unex a direct contractual claim against McAlpine. When the matter came back before it in *Alfred McAlpine Construction Ltd* v. *Panatown Ltd (No. 2)* (2001), the Court of Appeal emphasized that the direct route must be taken if one exists (such as in a duty of care deed).

The decision in *Alfred McAlpine Ltd* was considered in the Outer House of the Court of Session in *McLaren Murdoch & Hamilton Ltd* v. *The Abercromby Motor Group Ltd* (2003) and was adjudged to be 'wholly consistent with the principles of Scots law'. The view was taken that the exception should be conferred as a matter of general legal policy, rather than based on any considerations of the intent of the parties, and *jus quaesitum tertio* was ruled out as a general solution as this would require the contracting parties to set out to benefit a third party, and identify that third party, at the time of making the contract.

In *Scottish Widows Services Ltd* v. *Kershaw Mechanical Services Ltd and Building Design Partnership* (2011) the Inner House of the Court of Session considered the effect of an assignation of a collateral warranty granted by Scottish Widows Fund and Life Assurance Society (henceforth the Society) in favour of its subsidiary Scottish Widows Services Ltd (henceforth Services). Services was the occupier of the newly constructed Scottish Widows HQ building in Edinburgh and in addition to an assignation of the collateral warranty had also been granted by the Society an assignation of its interest under a sub-lease. Even though the sub-lease imposed no obligation on Services to carry out repairs, the Inner House held that Services was entitled to recover under the assigned collateral warranty in respect of the costs of rectifying defects to the building, in the event that liability on the part of the defenders was established. See also Section 13.2 in relation to collateral warranties.

Although there may be an element of policy underpinning some of the decisions considered above, their focus is on the extent of the assignor's rights in respect of the losses in question, and thus the scope of the rights in turn assigned, rather than a review of the principle of *assignatus utitur jure auctoris*.

12.4 *Delectus personae*

A contract which involves *delectus personae* is one where a party to the contract has entered into it in reliance upon certain qualities possessed by the other. In such circumstances, the contract cannot be performed by a third party and, consequently, the obligation to perform cannot be assigned without consent, see *Anderson* v. *Hamilton & Co.* (1875). Authorities in relation to delegation (see Section 12.7) may assist in identifying whether or not a contract contains an element of *delectus personae*.

If the contract does not place reliance upon a special skill of one of the parties, no element of *delectus personae* exists, see *Cole* v. *Handasyde & Co.* (1910). Even if the contract involves *delectus personae*, and is thus not assignable in full, certain rights arising out of that contract may be assignable. For example, an accrued right to payment of a sum of money under the contract may be assigned, notwithstanding that other rights and obligations under the same contract cannot, see *International Fibre Syndicate Ltd* v. *Dawson* (1901).

12.5 Assignation under the SBC and the SBC/DB

The assignation provisions of the SBC and the SBC/DB are to be found in clauses 7.1 and 7.2. Clause 7.1 provides that neither the Employer nor the Contractor may assign the contract or any rights thereunder without the written consent of the other. If the Contract Particulars state that clause 7.2 applies, that clause entitles the Employer to assign certain rights after practical completion. Where clause 7.2 does apply, then if the Employer alienates by sale or lease, or otherwise disposes of his interest in the Works, he may, at any time after the issue of the certificate of practical completion, assign to the party acquiring his interest in the Works his right, title and interest to bring proceedings, in his name as Employer, to enforce any of the rights of the Employer arising under or by reason of breach of the contract.

This provision recognizes the practice that certain Employers will transfer their interest in property once practical completion has been certified. Should it apply, however, the provisions of clause 7.2 are of limited use. It is only the Employer's right, title and interest to bring proceedings which may be assigned, and there may be occasions where the Employer wishes to assign prior to practical completion (e.g. an assignation in security to a funder at contract signing). The wording is a little odd and in some respects otiose, given that an assignation of rights under the building contract would, *ipso facto*, confer the right to raise proceedings to enforce such rights, and the proceedings would in common practice normally be brought in the name of the assignee (though in law the assignee may competently sue either in his own name or that of the assignor, see McBryde, *The Law of Contract in Scotland* (third edition, 2007), para 12–79). For that reason it is common to find clauses 7.1 and 7.2 amended in individual cases to reflect more conventional assignation wording. There appear to be no reported cases applying the provisions of clause 7.2 (or its predecessor, clause 19.1 of JCT 98).

12.6 Assignation under the NEC3

Perhaps somewhat surprisingly,the NEC3 is silent in respect of assignation (save in the context of termination), and in particular there is no express prohibition on assignation by either the Employer or the Contractor. In the absence of such provisions, it is likely that those rights which do not involve *delectus personae* are capable of being assigned without consent. Thus the Employer may assign its rights to enforce performance and the Contractor may assign its rights to receive payment. On the other hand, it is unlikely that, without the consent of the other, the Employer may assign its obligation to make payment (given the importance of the creditworthiness of the Employer) or that the Contractor may assign its performance obligations. Having said that, it is not uncommon for an Employer to remove all doubt by adding a Z clause expressly permitting prohibiting assignation by the Contractor.

12.7 Delegation

The delegation of building work, through the use of sub-contractors (and, where appropriate, design sub-consultants) is commonplace within the construction industry. Although delegation, in the sense of sub-contracting, raises similar issues to assignation, the two concepts should not be confused. In the case of assignation, it is the assignee, rather than the original contracting party, who is bound to perform the contract (where obligations are assigned) or who may enforce rights directly against the other party. With delegation, the original contracting party remains bound, albeit that the performance of that party's contractual obligations is carried out by another party. Delegation does not affect the underlying contractual relationship and has the effect of adding another link to the contractual chain.

As with assignation, the existence or otherwise of *delectus personae* will determine whether or not the obligation of performance can be delegated. The work content of the contract will assist in determining whether or not delegation is competent. Work which consists chiefly of manual labour has been held to contain no element of *delectus personae*, see *Asphaltic Limestone Concrete Co. Ltd and Another* v. *Corporation of the City of Glasgow* (1907). However, it has been recognized that the execution of repair work on and in residential properties is of a character which might well be the subject of *delectus personae*, see *Scottish Homes* v. *Inverclyde DC* (1997).

The question of *delectus personae* in building contracts generally was considered in *Karl Construction Ltd* v. *Palisade Properties plc* (2002) in which the court stated that: 'In general, it is clear that a building contract involving complex work will be personal to the contracting parties. That applies particularly if detailed administrative or management work is called for, or if elements of design are involved.'

The SBC and the SBC/DB contemplate delegation. Clause 3.7 of the SBC permits the sub-letting of any part of the Works, provided that the consent of the Architect is obtained. Such consent is not to be unreasonably delayed or withheld. The underlying principle of responsibility resting with the original contracting party is reflected in clause 3.7.1, which states that the Contractor will remain wholly responsible for carrying out and completing the Works, notwithstanding any subletting. Similar provisions are contained in clause 3.3.1 of the SBC/DB, save that consent is to be provided by the Employer. Clause 3.3.2 of the SBC/DB contains equivalent provisions in respect of sub-contracting the design of the Works (as does clause 3.7.2 of the SBC in relation to any Contractor's Designed Portion).

The conditions under which the Contractor may sub-contract under the NEC3 are contained in clause 26 of that contract. For a detailed summary of these conditions, see Section 11.9.

On sub-contracting generally, see Chapter 11.

12.8 Novation

Novation is, broadly, the substitution of a new party for an existing party to a contract, with the result that the new party assumes the rights and obligations under the contract. Novation requires the consent of all of the parties to the original contract. That consent need not be express and can be inferred from the conduct of the parties, see *McIntosh & Son* v. *Ainslie* (1872). Having said that, proving that novation has occurred in the absence of writing may in practice cause considerable difficulty as shown in the case of Camillin, *Denny Architects Ltd* v. *Adelaide Jones & Co Ltd* (2009), where the court held that there was insufficient evidence to prove that the contract had been novated.

Strictly speaking, novation has the effect of simultaneously extinguishing the original contract and creating a new contract. The effect of this must be carefully considered by the parties before proceeding with novation. For example, claims for breach of the original contract may be extinguished by novation, see *Hawthorns & Co. Ltd* v. *Whimster & Company* (1917).

In view of the problems surrounding the assignation of obligations as described in Section 12.1, novation is the method normally chosen where the intention is to transfer both rights and obligations under a contract. Novation is often used in design and build contracts, where the design consultants are initially engaged by the employer for the purposes of the pre-contract design but the appointments are then novated to the contractor at the same time as the design and build contract is entered into. Under such a novation agreement, the contractor assumes, in substitution for the employer, the rights against and obligations to the consultant under the consultancy appointment, normally with an express provision (subject to an agreed accounting in respect of fees) that this is from the date of commencement of the services. The consultant then becomes, in effect, the sub-consultant of the contractor. The usual intention behind this practice is to avoid delays by allowing design to proceed prior to appointment of the contractor (and thus obtain earlier planning permission, for example), to reduce costs and simplify the tender process by requiring tendering contractors to base tenders on a single design, and to ensure that the selected contractor assumes responsibility for the pre-contract design. The practice does create potential problems, which require to be addressed with care. Some of these are illustrated in *Blyth & Blyth Ltd* v. *Carillion Construction Ltd* (2002).

Blyth & Blyth provided consulting engineering services to their employer, and after some months this was formalized by a deed of appointment, which expressly provided that the agreement covered the period from when services were first provided. Under that deed, the employer instructed Blyth & Blyth to enter into a tripartite novation agreement, the effect of which was to substitute Carillion, the main contractor, as a party to the deed of appointment in place of the employer. When Blyth & Blyth raised an action for payment of fees, Carillion counterclaimed for losses arising from alleged breaches of duty, some of which related to the period before the novation agreement was executed. It was held that the novation agreement was to be construed as an agreement under which Blyth & Blyth were to provide future services to Carillion, and Carillion obtained an assignation of the employer's pre-novation rights against Blyth & Blyth. Since the building contract imposed full design responsibility on the contractor, the employer had suffered no loss arising from the defective pre-novation services. There was therefore no claim which could be assigned and so Carillion were not entitled to recover losses relating to the period before novation.

This case illustrates the importance, as far as the contractor is concerned, of ensuring that the novation agreement contains express undertakings by the consultant to the contractor in relation to pre-novation services, so that the contractor is able to rely upon such direct undertakings and recover its own losses rather than base a claim for pre-novation failures on the assigned rights of the employer. This would be of particular importance where, as is usually the case in such circumstances, the construction contract retroactively imposes on the contractor the responsibility for pre-novation design even although the contractor has had no involvement in such design. A number of standard industry forms of novation agreement have been published since *Blyth & Blyth* in order to avoid the problems thrown up by that decision, most notably by the Construction Industry Council and the British Property Federation. These forms vary in their approach but the common theme is to avoid the contractor's rights in respect of the pre-novation services being treated as an assignation from the employer and to provide that the consultant's obligations in respect of such services are directly owed to the contractor.

From the employer's perspective, it may be necessary to ensure that he retains some recourse against the consultant notwithstanding the novation, for example, if the consultant has been responsible for preparation of the Employer's Requirements under a SBC/DB contract (or the equivalent under another form of contract). This can be done by express provision within the body of the novation agreement or (more commonly in practice) by a separate collateral warranty by the consultant to the employer. However, an example of the risks to the contractor arising from such a collateral warranty is *Oakapple Homes (Glossop) Limited v DTR (2009) Limited (In Liquidation) & Ors* (2013). In that case, a collateral warranty was granted by a design consultant to the original employer following the novation of the design consultant's appointment to the contractor. The collateral warranty stated: 'The Consultant has no liability hereunder which is greater or of longer duration than it would have had if the Beneficiary had been a party to the Appointment as joint employer.' The consultant, DTR, contended, as a defence to a claim by the employer under the collateral warranty, that it was entitled under the above provision to rely on the contributory negligence of the contractor under the novated appointment. However, the Technology and Construction Court rejected that contention and held that the damages due to the employer as beneficiary under the collateral warranty could not be reduced to take account of contributory negligence by the contractor because an employer under a construction contract is not liable for the negligence of the contractor. See also Section 13.3.5.

In common with most other standard forms, the SBC, the SBC/DB or the NEC3 do not make any provision for novation of design consultants' appointments from employer to contractor, so if such novation is intended, bespoke amendments will be needed to the building contract conditions to impose an obligation on the contractor to enter into the novation agreement. It would also be prudent for the employer to include in the building contract tender documents a copy of all consultancy appointments to be novated, so that when pricing tenders, each tendering contractor is aware of the risks and responsibilities arising from the terms of such appointments. For example, are the consultant's obligations and scope of services back-to-back with those assumed by the contractor under the design and build contract? Or are there any liability caps in the consultancy appointment, either in time or amount, which result in residual risk for the contractor?

Consequential amendments may be required to the terms of the consultancy appointment with effect from the novation, if only to recognize that certain provisions applicable to a consultant/employer relationship require modification to reflect a consultant/contractor relationship.

Although novation of design consultancy appointments is the most common instance of novation in a construction contract context, it is not the only example. There may be a novation from the original employer to a new employer, such as in the case of a corporate reorganization or asset sale. Any such novation would require the consent of the contractor and in most cases (such as clause 7.2 of the SBC and the SBC/DB) a typical clause permitting assignation of the contract by the Employer would not be wide enough to cover novation. The effect of ancillary contracts would also need to be considered, such as contractor parent company guarantees and performance bonds, to ensure that the benefit under these passed to the new employer.

Novation may also be of relevance where the contractor becomes insolvent. One of the options open to the employer, and the insolvency practitioner responsible for the affairs of the contractor, is the novation of the original contract to a substitute contractor, whether on the same terms as the existing contract or on varied terms. In construction insolvency, a true novation, that is where the substitute contractor steps into the position of the original contractor and the contract continues as if the substitute contractor had been the original contractor, is most unlikely. Considerations such as liability for defects and the potential liability for liquidated and ascertained damages will militate against true novation. In most cases, therefore, the transaction will be a conditional novation, in terms of which the substitute contractor takes on only certain limited obligations, for example, the obligation to complete, and does not assume responsibility for antecedent breaches. This may have an attraction to the employer as the lesser of two evils by mitigating its losses arising from the termination, particularly if these losses are unlikely to be recoverable in full from retention or bonding arrangements, etc.

Under clause 8.3.2 of the SBC and the SBC/DB it would be possible to reinstate the Contractor's employment following termination for the specific purpose of effecting novation to a substitute contractor.

Chapter 13
Rights for Third Parties

13.1 *General*

There may be a range of parties that have a commercial or financial interest in the design and construction of a building, both during the design and construction process and following completion of the development. Although the identity of those having an involvement in a particular project will depend on the nature and use of the building being constructed, those having an interest will often include developers, funders, purchasers and tenants. Not all of those persons will be a party to, and therefore be in a position to rely upon, the building contract and other construction agreements entered into in connection with the development. For those reasons, parties with an interest in the development, and who are not party to the principal construction agreements, will need to consider how best to protect their respective interests.

The decision of the House of Lords in *Murphy* v. *Brentwood DC* (1991) reinforced the need for rights to be created in favour of such third parties. For a further discussion of *Murphy* and related cases, see Sections 10.10.3 and 11.7.3.

This chapter provides a practical overview of three ways in which rights in connection with a building contract can be created for the benefit of third parties having an interest in the building, namely, collateral warranties; third party rights schedule; and assignation.

13.2 *Collateral warranties*

13.2.1 Introduction

A practice had arisen in the early 1980s of requiring contractors, sub-contractors and construction professionals to acknowledge, by contractual means, duties of care to parties with whom they had otherwise no contractual relationship, for example, tenants, funders and subsequent owners. The perceived inadequacies and uncertainties of a remedy based upon delict led to the practice of creating, by means of a collateral warranty, a contractual nexus, which would otherwise be absent. The purpose of a collateral warranty is to impose, by contract, duties and obligations on the part of

MacRoberts on Scottish Construction Contracts, Third Edition. MacRoberts.

the contractor, sub-contractor or consultant in favour of a third party who is not the original building owner or employer but who may nevertheless suffer loss in the event of a construction or design defect.

Collateral warranties essentially deal with the apportionment of risk. Negotiation of the terms of collateral warranties has become widespread, with the need to satisfy the conflicting interests of the beneficiary and the grantor. Bodies within the construction industry, including the Scottish Building Contract Committee, have made various attempts to satisfy these conflicting interests without the need for protracted negotiation by producing standardized forms of collateral warranty.

The use of collateral warranties in the construction industry has continued and there is now included in both the SBC and the SBC/DB forms an option for the contractor to provide collateral warranties, see Contract Particulars Part 2 and clause 7. Interestingly, the NEC3 includes an option for the provision of collateral warranties only in its Professional Services Contract (under Secondary Option X13), leaving any requirement for collateral warranties in the other forms to be dealt with by bespoke amendment in an appropriate Z clause. Despite the number of collateral warranties being granted in the construction industry, there have been (at least until recently) relatively few reported cases involving collateral warranties, see, for example, *Hill Samuel Bank Ltd* v. *Frederick Brand Partnership* (1995); *Glasgow Airport Ltd* v. *Kirkman & Bradford* (2007); *Langstane Housing Association Limited v Riverside Construction (Aberdeen) Limited and others* (2009); and *Scottish Widows Services Limited v Kershaw Mechanical Services Limited and another* (2011). However, as discussed in more detail below, there have been a number of important recent cases, see *West and another v Ian Finlay & Associates* (2014); *Oakapple Homes (Glossop) Limited v DTR (2009) Limited (In Liquidation) and others* (2013); and *Parkwood Leisure Limited v Laing O'Rourke Wales and West Limited* (2013).

13.2.2 Interests in obtaining warranties

The reasons why parties involved in a construction project require collateral warranties vary depending upon the nature of their interest in the project.

Developer

In a typical commercial development, the developer of the project will intend either to realize their investment at, or shortly after, completion of the project by disposal to a third party purchaser, or to grant a leasehold interest to one or more tenants. In either case the marketability of the development will demand that collateral warranties from the contractor and the consultants are available to the purchaser and/or tenants.

Funder

While the funder will normally be protected by a heritable security over the development, they will wish to preserve a right of recourse against any party whose actions may diminish the value of that security. In addition, the funder will wish to have the option of ensuring that the development is completed (and the value of the security therefore maximized) in the event of the developer becoming insolvent prior to completion. This is achieved by exercising 'step-in rights', see Section 13.3.6.

From a funder's perspective, a collateral warranty from the quantity surveyor or other consultant administering payment is also important. Interim valuations under the building contract will normally be carried out by the quantity surveyor or other consultant on behalf of the developer and will, ordinarily, be reflected in payments to the contractor. In turn, drawdowns from the funder's loan to the developer will typically reflect the amount of such valuations and payments, hence the need for the funder to secure some degree of comfort in relation to the actions of the quantity surveyor.

Purchaser/tenant

Since the decision in *Murphy* it is clear that, except in very special circumstances, a purchaser or tenant will have no claim in delict against a contractor or consultant with whom the purchaser or tenant has no contract, hence the need for purchasers and tenants to protect themselves against the manifestation of latent defects caused by faulty design or construction. The principle of *caveat emptor* (buyer beware) applying in contracts for the sale of heritable property prohibits any recourse against a seller, and the obligations undertaken by tenants under a typical commercial full repairing lease will require them to make good all damage to the leased property howsoever caused (notwithstanding that it may be due to a latent defect) without recourse to the landlord. The case of *Royal Insurance (UK) Ltd* v. *Amec Construction Scotland Ltd and Others* (2005), an action raised pursuant to a collateral warranty, demonstrates the potential commercial significance to a tenant of the rights under a collateral warranty.

Sub-contractors

Collateral warranties from domestic sub-contractors in favour of interested third parties may be regarded as 'belt and braces'. In other words, primary liability will lie with the main contractor under their collateral warranty and additional warranties from sub-contractors will need to be enforced only in the event that recovery cannot be made from the main contractor, notwithstanding that liability is established. However, a sub-contractor's collateral warranty may also fulfil an important purpose of ensuring, by means of step-in rights to the employer, that the sub-contractor remains committed to price and to performance in the event of termination of the main contract. See also Section 11.7.1.

Design consultants

It is common practice for external professional consultants, engaged in relation to a traditional form of contract, to be required to give collateral warranties to third party beneficiaries (i.e. purchaser, funder and tenants). The same requirement for collateral warranties usually applies under a design and build contract to professional consultants engaged by or novated to a design and build contractor, in addition to the warranties to be granted to relevant beneficiaries by the contractor. This again may be regarded as a belt-and-braces approach, but with perhaps some justification in this instance, given the risk of the contractor becoming insolvent and the availability of professional indemnity insurance (in most cases) to back up claims made under such warranties.

Where a design consultant appointment is novated to the contractor, it is also important to the employer to procure a collateral warranty in its favour from the consultant to cover both pre-novation and post-novation design, particularly if pre-novation design responsibility is not fully assumed by the contractor. For other potential pitfalls arising from the novation of consultants' appointments, see Section 12.8.

13.3 Typical clauses

The following are examples of typical clauses found in collateral warranties. However, these are by no means exhaustive and the drafting of provisions intended to have the same effect may differ considerably from one warranty to another.

13.3.1 Standard of care

> The [Contractor/Sub-Contractor] warrants that in the design of the [Works/Sub-contract Works] it has exercised and will continue to exercise the reasonable skill, care and diligence to be expected of a suitably qualified and competent designer experienced in undertaking design for projects of a similar size, scope and complexity to the [Works/Sub-contract Works].

From the point of view of the grantor, it is important that the standard against which its responsibilities will be measured for the purposes of the collateral warranty is no higher than the standard of the primary or underlying contract.

13.3.2 Prohibited materials

> The [Contractor/Sub-Contractor] undertakes that it has not used and will not use materials in the Works other than in accordance with the guidelines contained within the edition of "Good Practice in Selection of Construction Materials" (British Council for Offices) current as at the date of the Building Contract.

Until fairly recently, the practice for many years had been to specify a list of materials not to be used, or specified for use, in the development. These lists were often prepared with little thought and lawyers were criticized (often with justification) for drafting collateral warranties containing lists of prohibited materials without any regard for the nature of the project or the particular properties of a prohibited material. A blanket prohibition may not be appropriate for a particular material, depending on the nature of the project and the intended use of the material. The matter was brought to a head when a manufacturer of calcium silicate bricks successfully obtained an interim interdict against a local authority preventing it from including calcium silicate bricks in the list of deleterious materials in its collateral warranties. It is now common practice to use a more general clause.

Parties should be clear as to whether the grantor is warranting that the materials have not been used during the construction of the development, or whether it is warranting only that they have used reasonable skill, care and diligence to ensure that they have not been used, or have not been specified for use as the case may be. It should also be borne in mind that references to 'good building practice' are potentially subjective and, in terms of scope, very wide.

Knowledge of the unsuitable nature of the material is usually stated to be tested at the time of either specification or use of the material in question. There is some danger in a contractor or sub-contractor accepting the time of use as the relevant benchmark, given that the material may have been capable of being used legitimately or its unsuitable properties may not have been known at the time of specification, and yet the contractor/sub-contractor will be in breach of the warranty if that state of affairs changes prior to the use of the material in the development.

13.3.3 Limitations on proceedings

> No action or proceedings for any breach of this Agreement shall be commenced against the [Contractor/Sub-Contractor] after the expiry of [12] years from the later of the date of termination of the Building Contract or the date of practical completion of the Works or the last section of the Works (where the Works are being completed in sections) as certified under the Building Contract.

A clause of this type is particularly important where the grantor does not enter into the collateral warranty until after practical completion. Failure to include such a clause could potentially leave the grantor owing contractual duties to a third party of longer duration than those owed to the client under the primary contract as, in the absence of such a provision, the prescriptive period may not then commence until the date of execution of the collateral warranty, notwithstanding that the design or construction failure giving rise to breach of the warranty occurred prior to the execution of the warranty.

It is sometimes otherwise difficult to see a rational justification for imposing a limitation period in a collateral warranty which reduces the grantor's period of exposure to claims (at least in the case of latent defects) to a period usually significantly less than the statutory prescriptive period (see Section 9.9). The period of 12 years is often chosen for no apparent reason other than its familiarity under English law. Ultimately it is a commercial matter. The grantor will argue that, as the beneficiary is being granted rights beyond those otherwise available to it at law, there is sound commercial sense for the grantor to place limitations on the period during which such rights may be exercised. Pragmatic considerations may also apply. For instance, a funder may consider it unnecessary to require a limitation period longer than the period for repayment of the loan. Clauses of this type fall to be construed strictly, see *Port Jackson Stevedoring Ltd* v. *Salmond & Spraggon (Australia) Pty Ltd ('The New York Star')* (1980).

13.3.4 Net contribution

> The liability of the Consultant for costs under this Agreement shall be limited to that proportion of the Beneficiary's losses which it would be just and equitable to require the Consultant to pay having regard to the extent of the Consultant's responsibility for the same and on the basis that [list names of other Consultants] shall be deemed to have provided contractual undertakings on terms no less onerous than this Clause [ ] to the Beneficiary in respect of the performance of their services in connection with the Development and shall be deemed to have paid to the Beneficiary such proportion which it would be just and equitable for them to pay having regard to the extent of their responsibility.

The above is commonly known as a 'net contribution' clause and is intended to alleviate what are regarded (at least in the eyes of grantors and their insurers) as the harsh consequences of joint and several liability. This arises where damage to the beneficiary is caused as a result of a breach of duty by more than one party. If each breach of duty has materially contributed to the same damage to the beneficiary, the beneficiary is entitled to recover the losses arising from such damage from any or all of the parties in breach. Thus one sub-contractor may (in the absence of a net contribution clause) be pursued for the whole of the loss, notwithstanding that other consultants or contractors may have contributed to that loss.

The clause creates the fiction that all the relevant parties are deemed to have granted collateral warranties (whether, in reality, they have or not) to the beneficiary. To the extent that they have a responsibility, they are deemed to have already paid their fair share of the recoverable loss or damage suffered by the beneficiary. The relevant grantor is then left liable only for its share of the loss on the basis that the other grantors are deemed to have paid their contribution. Care needs to be taken when agreeing the terms of a net contribution clause in a consultant collateral warranty in the context of a design and build contract, that the design and build contractor is not named as one of the parties 'deemed' to have provided similar undertakings and paid the relevant proportion according to its responsibility. Since the design and build contractor's responsibility is for the whole of the design, its 'deemed' payment will be for the whole of the claim and so the consultant who has the benefit of such a net contribution clause could unintentionally end up escaping all liability as a consequence.

The foregoing passage also appeared in the second edition of this book and was the subject of adverse criticism by Lord Tyre in the case of *The Royal Bank of Scotland Plc v. Halcrow Waterman Ltd* (2013). The net contribution clause considered in that case was not materially different from the model clause set out above. The defenders, who were structural engineers and had granted a collateral warranty in favour of tenants of the development, contended that, notwithstanding the absence of an express reference to the contractors in the clause, the contractor's contribution to the tenant's loss following the discovery of defects should still be taken account of in assessing the defenders' contribution to the overall losses. Lord Tyre rejected that argument, but in considering *obiter* the above passage he said:

> In my reading of the clause, the reference to 'the extent of their responsibility' at the end of the clause is not a reference to responsibility for design but rather a reference to the same kind of responsibility as has previously been referred to in the clause, i.e. responsibility for the claimant's loss. The contractor would not, therefore, if named as one of the parties, be deemed to have paid more than his just and equitable proportion of the total loss and the consultant with the benefit of the clause would not escape liability for his own just and equitable proportion.

But if the contractor is a design and build contractor responsible in terms of the underlying building contract for the design carried out by the consultants who are acting as his sub-contractors, is it not the case that the 'just and equitable proportion' of the total losses arising from a design defect for which the contractor is deemed liable under a net contribution clause is the whole of these losses? This seems consistent with the observation by Lord Drummond Young in *Scottish Widows Services Ltd* v *Harmon/CRM Facades Ltd (in liquidation)* (2010), and repeated by Lord Tyre, that a net contribution clause in effect restricts joint and several liability by limiting a co-obligant's liability to a fair assessment of the consequences of his own breach of contract. In the case of a design and build contractor, a breach by the consultant results, at least in most cases, in a breach by the contractor. It may well be that a court may not accept such a consequence of including a design and build contractor in a net contribution clause on the grounds that it offends commercial common sense, but we must respectfully disagree with Lord Tyre that it is not an arguable position.

A net contribution clause is intended to reflect the principles of section 3 of the Law Reform (Miscellaneous Provisions) (Scotland) Act 1940 which gives the court the power to apportion damages against joint wrongdoers, and the clause may, on the face of it, seem equitable. However, unlike the 1940 Act, the clause places on the beneficiary the onus of obtaining similar warranties from the other parties 'on terms no less onerous' and also the risk that recovery may not in fact be made from one of the relevant parties, notwithstanding that liability of the other parties is calculated on the basis that it is deemed to be so made.

There have been some concerns expressed regarding evidence that may be admitted by the courts and the potential of a decree being granted against an unrepresented party. Some professional indemnity insurers will insist upon the inclusion of a net contribution clause in collateral warranties before extending cover to liabilities arising out of them. There are also attempts to extend such clauses to appointments, which are normally met with resistance.

Reliance on a net contribution clause contained in the standard ACE conditions of appointment was challenged in *Langstane Housing Association Limited v Riverside*

Construction (Aberdeen) Limited and others (2009). The challenge was made on the basis that the net contribution clause fell foul of the Unfair Contract Terms Act 1977, being, it was claimed, unusual and controversial in that it significantly altered the common law position by placing the risk of insolvency of the contractor or one of the other consultants onto the employer. As such, so the argument ran, it was for the engineer to demonstrate that it was fair and reasonable to include the clause in the contract. The judge did not consider net contribution clauses to be either unusual or controversial. He further held that the net contribution clause was neither an exclusion nor a restriction of liability for the purposes of the Unfair Contract Terms Act, being designed instead simply to ensure that the engineers were held liable only for the consequences of their own breach and not (due to joint and several liability) for the breaches of others. In *West and another v Ian Finlay & Associates* (2013), the architect's appointment with the homeowners contained a net contribution clause providing that the architect's 'liability for loss or damage will be limited to the amount that it is reasonable for [the architect] to pay in relation to the contractual responsibilities of other consultants, contractors and specialists appointed by [the homeowners]'. The architects had design responsibilities for the main contractor's work. The court at first instance held that, notwithstanding the net contribution clause, the architect had liability for loss and damage caused by the main contractor (which had gone into liquidation). The reasoning was based on the particular circumstances of the case, where the homeowners had employed a number of contractors themselves in addition to the main contractor, and which were outside of the arrangement with the architect. The court considered that these contractors could be distinguished from the main contractor, and it was only the former who were 'other' contractors for the purposes of the net contribution clause. Although overturned on appeal, the lesson is that a properly drafted net contribution clause should specifically name the 'other' parties.

13.3.5 Equivalent rights of defence

> The [Contractor/Sub-Contractor] shall be entitled in any action or proceedings by the Beneficiary to rely on any limitation in the [Building Contract/Sub-Contract] and to raise the equivalent rights in defence of liability as it would have against the [Employer/Contractor] under the [Building Contract/Sub-Contract], had the Beneficiary been named as [Employer/Contractor] under the [Building Contract/Sub-Contract].

The intention of this clause is to ensure that liability under the collateral warranty is co-extensive with that under the primary contract. For example, the contractor or sub-contractor would be able to raise in defence of liability any rights of retention, set-off or counterclaim that it has against the employer under the building contract (see also Section 13.3.10). However, in the drafting of this type of clause (which may be worded in a number of different ways) care must be taken to ensure that it does not confer a 'no loss' right of defence on the grantor. The limits to this clause were also recently illustrated in the case of *Oakapple Homes (Glossop) Limited v DTR (2009) Limited (In Liquidation) and others* (2013). In that case, the beneficiary under a collateral warranty from the architect (DTR) claimed that fire damage was due to the negligent design of DTR. The collateral warranty contained a typical 'equivalent right of defence' clause (the 'employer' being the contractor, following novation of the appointment). DTR claimed that there had been contributory negligence on the part of the contractor and that DTR was entitled to raise this as a defence to a claim under the collateral warranty. However, the court rejected this argument on the basis that while such a clause limits the liability of the warrantor to that under the underlying contract, it did not entitle DTR to defend a claim based on contributory negligence of the contractor, since an employer is not liable for the contributory negligence of its contractor. See also Section 12.8.

13.3.6 Step-in rights

The rights contained in this clause allow the beneficiary to step into the shoes of the employer in certain pre-defined situations, typically insolvency events or where the grantor seeks to exercise its right to determine the principal contract.

Step-in provisions will normally appear in a funder's collateral warranty, though in some situations they may also be appropriate in a purchaser's collateral warranty and (in the case of sub-contractors) in an employer's collateral warranty. Step-in allows the beneficiary to step into the shoes of the employer under the primary contract by way of novation. Novation is considered in Section 12.8. It is generally activated where the beneficiary wishes to prevent the grantor from terminating the primary contract or otherwise wishes to ensure completion of the project following default by the developer or (in the case of sub-contractors' collateral warranties) default by the contractor.

The beneficiary will normally be obliged to assume all the outstanding and future obligations of the employer, for example, payment of outstanding and future fees. The beneficiary has the right, but ordinarily not the obligation, to step-in upon the termination event or default occurring. This means that there will generally be a specified time after such event within which the beneficiary, having received all of the relevant information to allow them to reach an informed view, is required to decide whether they wish to exercise their right or not. Grantors will wish to keep this period to a minimum as they are otherwise prevented from terminating during this period and so will require to continue to perform their obligations, even though they may not be getting paid.

Collateral warranties containing step-in provisions will normally be tripartite, with the employer, beneficiary and grantor being signatories. The reason for the employer being a party is that it acknowledges that the grantor will not be in breach of its obligations to the employer by reason of the grantor complying with its obligations to the beneficiary in respect of step in.

13.3.7 Assignation/obligation to enter into further warranties

A beneficiary will normally require that a collateral warranty in its favour is capable of being assigned without the beneficiary having to obtain the prior consent of the grantor. In the ordinary course of events, the grantor, to limit their exposure, will attempt to limit the number of occasions upon which the collateral warranty may be assigned. Where relevant, both grantor and beneficiary should check the provisions of the grantor's professional indemnity insurance cover when considering the number of assignations as most insurers will wish to restrict the permitted number of assignations, especially where the collateral warranty contains few restrictions on liability.

For the reasons described in Section 13.6, an assignation of a collateral warranty may be of limited value in certain circumstances and a second purchaser or tenant will often prefer a new warranty in its favour. This is notwithstanding that in *Scottish Widows Services Limited v Kershaw Mechanical Services Limited and another* (2011) the Court of Session held that the assignee was entitled to recover under an assigned collateral warranty.

13.3.8 Professional indemnity insurance

> The [Contractor/Sub-Contractor] shall maintain professional indemnity insurance in an amount of not less than [ ] for every occurrence or series of occurrences arising out of any one event with insurers of substance and repute in the UK Insurance Market until [12] years from the date of Practical Completion under the Building Contract, provided always that such insurance is available to the [Contractor/Sub-Contractor] at commercially reasonable rates. The [Contractor/Sub-Contractor] shall immediately inform the Beneficiary if such insurance ceases to be available at commercially reasonable rates in order that the [Contractor/Sub-Contractor] and the Beneficiary can discuss the means of best protecting the respective positions of the Beneficiary and the [Contractor/Sub-Contractor] in the absence of such insurance. As and when it is reasonably requested to do so by the Beneficiary, the [Contractor/Sub-Contractor] shall produce for inspection documentary evidence that its professional indemnity insurance is being maintained.

The well-advised grantor will, where it maintains professional indemnity insurance in respect of its obligations in terms of the collateral warranty, refer any collateral warranty it proposes to enter into to its professional indemnity insurers for comment. Where the grantor has responsibility for design, it will normally be required to maintain professional indemnity insurance for a number of years after practical completion. For the same reasons as mentioned in Section 13.3.3, a period of 12 years would seem to be the norm. Further qualifications to this obligation often include the requirement for insurance to be available at commercially reasonable rates and on reasonable terms and conditions.

It is worth noting that there are few, if any, insurers prepared to underwrite any form of absolute risk. Professional indemnity insurance traditionally covers the 'legal liability' of the insured, i.e. in the case of a professional, the exercise of reasonable skill and care. Any voluntary assumption of a greater duty by the insured in contract is generally covered by way of an extension to the policy. This can either be a general endorsement within certain parameters, or require 'approval' of each contract by the insurer.

13.3.9 Intellectual property licence

Most collateral warranties will contain a non-exclusive licence by the grantor in favour of the beneficiary in respect of the use of intellectual property rights in the design documentation prepared relative to the development. This may be particularly important, for example, to a purchaser who wishes to construct an extension to the property consistent with the existing design. It should be noted in this context that most grantors of collateral warranties would usually seek to carve out from the scope of the licence the right to re-produce the original design in any extension, limiting the licence to use of the drawings as a reference point for the original design only. Grantors, particularly consultants, may seek to make the licence conditional upon all fees being paid, but the well-advised beneficiary will strongly resist this.

13.3.10 No greater duties or liabilities

In *Safeway Stores Ltd* v. *Interserve Project Services Ltd* (2005, a collateral warranty clause in the following terms was considered: 'The Contractor shall owe no duty or have no liability under this deed which [is] greater or of longer duration than that which it owes to the Developer under the Building Contract.'

Not surprisingly, it was decided that this clause permitted the contractor to set off against claims made against it under the collateral warranty sums that were owed to it under the primary building contract. For that reason, while the principle of this type of clause is normally acceptable to beneficiaries, it is not uncommon for them to seek to expressly exclude from the clause (and also from clauses of the type referred to in Section 13.3.5) rights of retention, set-off and counter-claim. Grantors will of course attempt to resist such exclusion.

13.3.11 Limitations

A grantor often wishes to restrict, either in *quantum* or nature, the losses for which it may be liable under the collateral warranty. For example, it is common, particularly in the standard forms of warranties (and is often demanded by professional indemnity insurers) that the grantor's liability be restricted to the reasonable cost of repair, renewal or reinstatement of the development, to the extent attributable to the grantor's breach. Any liability, for example, for loss of use would therefore be excluded. However, to successfully restrict the grantor's liability, any such limitation must be clearly expressed, see *Glasgow Airport Ltd* v. *Kirkman & Bradford* (2007).

13.4 Effects of the 1996 Act on collateral warranties

Are collateral warranties agreements for the carrying out of construction operations in terms of s.104(1)(a) of the Housing Grants, Construction and Regeneration Act 1996?

Conflicting views have been expressed on the answer to this question. While a number of legal commentators argued when the 1996 Act came into force that it would extend to collateral warranties, the prevailing view since then has probably been that it does not. However, that view has been challenged by the recent decision of the Technology and Construction Court in *Parkwood Leisure Limited v Laing O'Rourke Wales and West Limited* (2013). In that case it was held that a collateral warranty may constitute a 'construction contract' for the purposes of the 1996 Act and that as a result the beneficiary may be entitled to bring adjudication claims under the 1996 Act. The court placed importance on the direct link between the collateral warranty and the underlying contract and the fact that the warranty under consideration contained future obligations in respect of work not yet carried out at the time of signing. However, it emphasized that not all collateral warranties will be construed as construction contracts under the 1996 Act; the test is whether the contract is for 'the carrying out of construction operations' for the purposes of s.104(1) of the 1996 Act and an important factor will be whether the grantor undertakes to carry out future works as opposed to warranting that past works have been carried out in accordance with the underlying contract. In the latter case it is unlikely that the collateral warranty will be construed as a construction contract. As an English case, it is only of persuasive authority in Scotland. Nevertheless, since the decision is unwelcome to contractors and consultants (and their PI insurers) the immediate consequence of the decision is likely to be that those parties may demand greater precision in the scope of obligations contained in collateral warranties. It may also mean more reliance on third party rights as an alternative to collateral warranties (see Section 13.5).

13.5 Third party rights schedule

13.5.1 Introduction

A third party rights schedule is a schedule forming part of a primary contract, such as a building contract, which sets out a series of rights which can, in the circumstances set out in the primary contract, be conferred on third party beneficiaries without the need for a separate contract to be entered into with the third party beneficiary. In a construction context the rights in a third party rights schedule will typically be very similar to the rights that are contained in a collateral warranty.

In England, the Contracts (Rights of Third Parties) Act 1999 (which does not apply in Scotland) provides a legal basis for the use of third party rights schedules as an alternative to collateral warranties. Although, under the common law doctrine of *jus quaesitum tertio* it has always been legally possible in Scotland (unlike in England) for the parties to a contract to confer rights in favour of third parties, provided certain criteria are fulfilled, it was not until third party rights schedules were adopted in England that they were considered for use in Scotland.

Both the SBC and the SBC/DB forms provide for third party rights pursuant to clauses 7A and 7B and Part 5 of the Schedule, as an alternative to the provision under clause 7C of collateral warranties from the Contractor to purchasers, tenants and/or funders. The parties in whose favour such rights are to be granted require to be set out in the Contract Particulars. The Contract Particulars further provide that, if in relation to an identified person it is not stated whether third party rights or collateral warranties will apply, then the former will apply. Compared to collateral warranties, the use of third party rights schedules is a relatively recent innovation and it remains to be seen whether in time these will replace collateral warranties as the medium of choice for providing rights in favour of third parties.

The NEC3 makes provision for third party rights under secondary option clause Y(UK)3. The terms capable of being enforced and the parties entitled to do so are to be set out in the Contract Data. However, it should be noted that this clause is limited to the exercise of rights under the Contracts (Rights of Third Parties) Act 1999, which does not apply to Scotland. This means that if the contract is governed by Scots law and the intention is to grant third party rights under the *jus quaesitum tertio*, then either specific amendments will require to be made to Y(UK)3 or a separate Z clause added.

13.5.2 Creation of the rights

To create third party rights in Scots law under the *jus quaesitum tertio*, the parties to the contract need to show an intention to create such rights in favour of a particular person or class of persons and to make those rights irrevocable. For a more detailed discussion, see Section 11.7.2 and the cases mentioned therein.

13.5.3 Advantages

There are a number of perceived advantages to using third party rights schedules. From the beneficiary's perspective, probably the most compelling benefit is that where third party rights are properly conferred on a beneficiary, the parties to the primary contract will (in the absence of an express right to do so) be unable to change the primary contract so as to interfere with the third party rights. In the absence of express terms in a collateral warranty prohibiting them from doing so, it would be open to the parties to the primary contract to adjust it in such a way as may affect the rights being granted under the collateral warranty.

Third party rights schedules could reduce the requirement to prepare and complete large numbers of collateral warranties in major projects. Instead the grantor would agree to one third party rights schedule which will contain many of the general collateral warranty clauses outlined above.

Third party rights can be prepared at the same time as the main contract and the beneficiary can be identified as a specific legal entity or as a class of persons (e.g. tenants). This means that third party rights can be made available for beneficiaries that have not yet been identified or that are not in existence at the time the building contract is entered into.

From the perspective of the grantor, third party rights may be more attractive than a collateral warranty as such rights would not be a 'construction contract' under the

1996 Act and so any claim under such rights could not made by way of adjudication. See Section 13.4 above.

13.6 Assignation

It is not unusual for developers to offer to assign to a purchaser or tenant their rights against the contractor and/or consultants responsible for constructing and designing the development (and express provision is made for this in clause 7.2 of the SBC and the SBC/DB). However, assignation is arguably an unsatisfactory substitute for a collateral warranty or third party rights schedule. This is because the principle *assignatus utitur jure auctoris* applies to assignation, i.e. the assignee stands in the shoes of the assignor, can acquire no better rights than the assignor, and is subject to any defence available against the assignor. Although judicial decisions of both the English and Scottish courts suggest that claims in such circumstances would not be allowed to fall into a legal 'black hole' (see Section 12.3), these are arguably based more on considerations of policy rather than legal principle.

Chapter 14
Insurance

14.1 *Insurance: General principles*

14.1.1 Introduction

Construction insurance is a specialized and complex subject, a comprehensive exploration of which is beyond the scope of this chapter. This chapter deals only with the essentials of the subject, in the context of the SBC, the SBC/DB and the NEC3 insurance provisions. However, most standard form building contracts contain insurance provisions which are broadly similar.

In the majority of building contracts the contractor undertakes to indemnify the employer for loss and liabilities arising from death of or injury to persons and loss of or damage to property, and the contractor will be obliged to maintain employer's liability and public liability insurance to cover the risk of such loss or liability occurring.

Such types of insurance fall within the category of 'liability' insurance. In other words, the insurance will cover the liability which the insured party has to a third party as a result of the insured event.

Most construction contracts will also expressly deal with the other common category of insurance, namely, property insurance. In the context of construction contracts this type of insurance will typically cover the contract works, site materials, plant and equipment. The obligation to take out and maintain such insurance may be dealt with in differing ways. Thus, while in most cases insurance will be in the joint names of employer and contractor, some contracts may provide that the obligation to take out such insurance is that of the employer, while others may impose that obligation on the contractor.

It should also be borne in mind that the characteristics of certain types of contract may demand more extensive insurance requirements. For example, insurance may need to be taken out in certain contracts to cover business interruption, fortuitous pollution, marine claims and/or professional indemnity. At the same time, not all risks are insurable (or at least not under conventional policies). An obvious example of this is a construction defect not involving a design error. While this may be insurable under a specialist latent defects policy (see Section 14.5), it will not be covered by a standard property insurance policy or a professional indemnity policy.

MacRoberts on Scottish Construction Contracts, Third Edition. MacRoberts.

14.1.2 Definition of insurance

Generally speaking, a contract of insurance is a contract whereby, for a consideration (normally involving payment of a premium), the insured obtains a benefit (usually payment of money) upon the happening of a certain event in respect of which there is uncertainty as to either whether it will happen or when it will happen. The insurance must be 'against something', see *Prudential Insurance Co.* v. *IRC* (1904).

14.1.3 Legal characteristics of insurance

The requirement that insurance must be 'against something' is generally taken to mean that the insured must have an 'insurable interest'. This means that the insured must have a pecuniary interest in the subject matter of the insurance so that upon the occurrence of the insured event the insured has, as a result, either himself suffered a loss or incurred a legal liability.

The other fundamental principle to which all insurance contracts are subject is that of *uberrimae fidei*, or utmost good faith. This principle requires each party to make a full disclosure of all material facts which may influence the other party in deciding to enter into the contract. A failure to disclose such material facts may render the policy void, which in practical terms would allow the insurer to refuse to meet a claim. This would apply even in the absence of fraudulent intent.

14.1.4 Joint names insurance

A construction contract will usually require insurance in respect of the works to be in the joint names of the contractor and the employer, and where appropriate, may also include as co-insured funders or other third parties having an insurable interest. Under a joint names policy, each co-insured has its own rights under the policy and is entitled to claim under the policy in respect of its own interest. This is to be distinguished from noting a party's interest on the policy. In the latter case, the party in question may have a right to share in the insurance proceeds (provided of course that it has an insurable interest) but it cannot make a claim under the policy. In addition, while it is usually thought that subrogation rights against a co-insured under a joint names policy are excluded (subject to the comments in Section 14.1.5), there is no such implied exclusion in relation to a party whose interest is noted on the policy.

Examples of contractual requirements to maintain joint names policies are those to be effected under Insurance Options A–C of the SBC and the SBC/DB and clause 84.2 of the NEC3 (see Sections 14.2.4 and 14.3.1). Where a joint policy is also a composite policy, then in the event of fraud or non-disclosure by one of the named insureds, the rights of the other insured parties (provided they are not also guilty of fraud or non-disclosure) will survive. If the joint policy is not specifically taken out as a composite policy (e.g. as required by the SBC and the SBC/DB), there is a risk that the insurer may be able to avoid the whole policy if there has been fraud or non-disclosure on the part of one of the co-insureds.

14.1.5 Subrogation

The principle of subrogation is common to all insurance contracts which involve the insurer indemnifying the insured in respect of a loss or a liability. Subrogation means that the insurer is entitled to exercise any remedy which may have been exercisable by the insured in respect of the insured event. In practice, it means that the insurer can pursue a claim (in the name of the insured) against a third party who may be responsible, either wholly or partly, for the insured loss. Such a right is subject to the insurer having made payment in respect of the insured's claim and to subrogation rights not having been excluded by any express contractual term. It was until recently thought to be a settled rule of law that that subrogation rights are not available against a party who is a joint insured under a joint names insurance policy, see *Petrofina (UK) Ltd* v. *Magnaload Ltd and Another* (1984). However, the position is perhaps not quite so certain following the decision in *Tyco Fire & Integrated Solutions Ltd v. Rolls-Royce Motor Cars Ltd* (2008). Certain 'obiter' remarks made by the Court of Appeal (i.e. made in passing and non-binding) are to the effect that there is no rule of law that a co-insured under a joint names policy (or its insurer by way of subrogation) cannot sue another co-insured in respect of damage covered by the policy and that the position must be determined on a true construction of the underlying contract. The practical lesson is that to avoid any doubt the contract should expressly require that the policy exclude subrogated rights against co-insureds.

14.1.6 Indemnities and insurance

It is important to recognize the distinction between indemnity and insurance. A building contract will normally contain provisions in terms of which the contractor will undertake to indemnify the employer against the occurrence of certain events. The contract will also impose obligations on either or both of the parties relative to the insurance of risks. There is a cross-over between these indemnity and insurance obligations, insofar as insurance may be required to be taken out against the risks covered by certain indemnities, but the obligations are not necessarily co-extensive. Risks covered by a particular indemnity may not necessarily be insurable (see Section 14.2.2) or an indemnified risk may be covered by insurance but the indemnifier may still have a liability for any insurance deductible or for any claim in excess of the insured amount. These are matters which it is prudent for the party giving the indemnity to check. Further, the parties may agree that a certain risk be covered by insurance and that neither party should have liability, notwithstanding fault.

Clauses 6.1 and 6.2 of the SBC and the SBC/DB broadly impose an obligation on the Contractor to indemnify the Employer against two matters: first, against death and personal injury arising out of or in the course of or caused by the carrying out of the Works, except to the extent that the death or injury is due to the act or neglect of the Employer or any of the Employer's Persons or of any Statutory Undertaker; and second, against property damage (other than damage to the Works) arising out of or in the course of or by reason of the carrying out of the Works and to the extent that the damage is due to any negligence, breach of statutory duty, omission or default of the Contractor or any of the Contractor's Persons.

It should be noted that the Contractor's liability for property damage arises to the extent due to his negligence and/or default whereas his liability for death or personal injury is not so limited. In either case, loss or damage arising from the act or neglect of the Employer, or those for whom he is responsible, will be excluded and so should normally be insured against separately by the Employer.

The NEC3 deals with indemnities rather differently by specifying in clause 80.1 those risks which are Employer's risks and providing in clause 81.1 that the risks not carried by the Employer are carried by the Contractor. Under clause 83, each party indemnifies the other against claims due to an event which is his risk, such liability being reduced to the extent that events at the other party's risk contributed to such claims. In this way, it follows that the Contractor is liable to indemnify the Employer for death and personal injury and for loss or damage to third party property arising from the Works, save to the extent they were due to the negligence or breach of statutory duty of the Employer.

It should be noted that a standard public liability policy (even where it purports to include a 'contractual liability' extension) will cover only the insured's liability to compensate a third party in delict or under contract where the contractual liability is co-extensive with liability in delict. It follows that a standard public liability policy will not cover an insured's contractual liability to compensate a third party for 'pure' economic loss for which there would otherwise be no liability in delict, see *Tesco Stores Ltd* v. *Constable and others* (2008). A carefully worded extension to the policy would be required to cover such loss. This illustrates the need for caution in agreeing to accept indemnities which go beyond the standard death or injury and property damage indemnities. For example, a wide indemnity by the contractor to the employer might require the contractor to indemnify the employer in respect of the employer's contractual liabilities to third parties, which would not be covered by the contractor's public liability policy. *Tesco* was followed in *MJ Gleeson Group plc* v. *AXA Corporate Solutions S.A.* (2013) in which it was held that a public liability policy taken out by a main contractor did not, in the absence of damage to property, cover the defective work of a sub-contractor. This was notwithstanding that the policy contained an extension of cover to include the insured's liability arising out of the defective workmanship of subcontractors, including the cost of making good defective workmanship. The court held that this extension had to be read as an extension of the public liability cover under the policy, which required damage to property, and did not provide stand-alone cover for a subcontractor's defective workmanship. This case again illustrates that, in the absence of express wording to the contrary, the courts will be slow to construe any extension to a public liability policy as providing cover beyond liability for third party claims for death, personal injury or loss of or damage to property.

14.2 Insurance under the SBC and the SBC/DB

14.2.1 Introduction

The SBC and the SBC/DB impose (or, in some cases, give an option to impose) obligations to take out and maintain insurance covering the following risks:

- personal injury and death (see Section 14.2.2);
- damage to property (other than the Works) arising from the Contractor's default (again, see Section 14.2.2);
- 'non-negligent' damage to property (other than the Works) (see Section 14.2.3);
- damage to the Works (see Section 14.2.4).

Insurance against the risk of loss of liquidated and ascertained damages (formerly an option pursuant to clause 22D of JCT 98) is no longer to be found within the SBC or the SBC/DB, largely due to the limited scope of cover which has been available and the low level of use (though see the comments on Project Insurances in Section 14.6). Excepted Risks are excluded from the obligation to insure (see Section 14.2.9).

There are certain types of insurance cover not provided for by the SBC or SBC/DB, for example, latent defects insurance (see Section 14.5). A well-advised party to a construction contract will consider whether any such (and other) risks should be covered by insurance.

14.2.2 Insurance against injuries to persons or damage to property

Under clause 6.4 of the SBC and SBC/DB, the obligation to insure against the death of, or injury to, any person or loss of, or damage to, any property arising out of or in consequence of the execution of the Works is imposed on the Contractor. This reflects the indemnity given by the Contractor under clauses 6.1 and 6.2 for such injury, damage or loss (see Section 14.1.6). The cover is usually contained in two separate policies, namely, a public liability policy and an employer's liability policy.

Insurance relating to personal injury or death of an employee of the Contractor must comply with the Employers' Liability (Compulsory Insurance) Act 1969 which specifies a statutory minimum level of cover, currently £5 million, see regulation 3(1) of the Employers' Liability (Compulsory Insurance) Regulations 1998.

Insurance cover in respect of the death of, or injury to, other persons and loss of, or damage to, property will be effected under a public liability policy. The minimum amount of public liability cover in respect of any one occurrence should be stated in the Contract Particulars. The insurance must remain in force until practical completion.

14.2.3 Clause 6.5.1 insurance

There is a further option open to the Employer under clause 6.5.1 of the SBC and the SBC/DB. This insurance need only be taken out by the Contractor if the Contract

Particulars so specify and it covers the potential liability of the Employer to third parties which would not normally be met by the Contractor's public liability insurance nor covered by their indemnity under clause 6.2.

This insurance relates to damage to property, other than the Works, and to Site Materials, caused by certain specified risks, namely collapse; subsidence; heave; vibration; weakening or removal of support; or lowering of ground water arising out of, or in the course of carrying out, the Works. This cover is most commonly required when there is neighbouring property susceptible to damage by any of these risks.

Under clause 6.5.1 fault does not need to be established, but there are a number of exceptions which reduce considerably the scope of this insurance cover, namely:

- injury or damage for which the Contractor is liable under clause 6.2 (which should be insured under clause 6.4.1);
- injury or damage attributable to errors or omissions in the designing of the works (which, where either the SBC/DB or the SBC with Contractor's Designed Portion is used, should be covered by professional indemnity insurance under clause 6.12 of the SBC/DB or the SBC);
- injury or damage which can reasonably be foreseen to be inevitable, having regard to the nature of the work to be executed or the manner of its execution;
- injury or damage which is the responsibility of the Employer to insure where Insurance Option C (insurance of existing structures) applies;
- injury or damage arising from war risks or the Excepted Risks (see Section 14.2.9);
- injury or damage directly or indirectly caused by or arising out of pollution or contamination during the period of insurance, save in respect of a sudden identifiable, unintended and unexpected incident; and
- injury or damage which results in costs or expenses being incurred by the Employer or any other sums being payable by the Employer in respect of damages for breach of contract, save to the extent which they would have attached in the absence of any contract.

14.2.4 Insurance of the Works

The SBC and the SBC/DB provide three options ('Insurance Options') for insuring the Works. A choice must be made in the Contract Particulars, while certain provisions apply regardless of which option is selected. Under both forms, the Insurance Options are set out in Schedule Part 3.

The contract calls for a Joint Names Policy for All Risks Insurance. Both terms are defined in clause 6.8 of the SBC and the SBC/DB. Where the policy is in joint names, both the Contractor and the Employer are to be named as composite insured and either of them may make a claim under the policy in its own name. The definition also makes it clear that the policy must provide that the insurer has no right of subrogation against a named insured (see Section 14.1.5).

The All Risks Insurance which either the Contractor or Employer, as the case may be, is obliged to take out should provide cover against any physical loss or damage to work executed and Site Materials, but excluding the costs necessary to repair, replace or rectify:

- property which is defective due to wear and tear, obsolescence, deterioration, rust or mildew;
- any work executed or any Site Materials lost or damaged as a result of its own defect in design, etc; and
- loss or damage caused by or arising from the consequences of war, invasion, rebellion, nationalization, disappearance or shortage (if such disappearance or shortage is only revealed when an inventory is made or is not traceable to an identifiable event), or an Excepted Risk.

14.2.5 All Risks Insurance by the Contractor

If Option A of the SBC or the SBC/DB is selected, the Contractor must take out and maintain a Joint Names Policy for All Risks Insurance for the full reinstatement value of the Works plus the percentage, if any, to cover professional fees stated in the Contract Particulars. Under the SBC and the SBC/DB, this Joint Names Policy must be maintained up to the date of issue of the certificate or statement of practical completion or the date of termination of the employment of the Contractor, whichever is the earlier. Where the contract is silent as to the duration of the obligation on the Contractor to maintain the Joint Names Policy, the obligation will cease on practical completion, notwithstanding the Contractor's continuing defects liability obligations, see *TFW Printers Ltd* v. *Interserve Project Services Ltd* (2006).

An alternative open to the Contractor, and which is widely used in the industry, is to use an existing annual policy which complies with the obligations in Schedule Part 3. However, the policy must still be a Joint Names Policy and the Contractor must provide documentary evidence that the policy is being maintained and, when so required, supply for inspection the policy itself and the premium receipts.

14.2.6 All Risks Insurance by the Employer

The second option, contained in Option B, is for the Employer to take out the Joint Names Policy for All Risks Insurance on the same terms and for the same period as described above. There is a corresponding provision to that contained in Option A entitling the Contractor to take out the Joint Names Policy if the Employer fails to do so.

14.2.7 Existing structures

The third option, Option C, applies where the contract is for the alteration of, or extension to, existing structures owned by the Employer or for which they are responsible. In this case the Employer takes out and maintains the Joint Names Policy for All

Risks Insurance for the Works themselves (as per Option B) but must also maintain a Joint Names Policy for existing structures to cover the cost of reinstatement, repair or replacement of loss or damage due to one or more of the Specified Perils. If the Employer fails to take out either of these two insurances the Contractor is entitled to do so and recover the cost of the premiums. The Specified Perils are defined as fire, lightning, explosion, storm, tempest, flood, bursting or overflowing of water tanks, apparatus or pipes, earthquake, aircraft and other aerial devices or articles dropped therefrom, riot and civil commotion, but excluding Excepted Risks.

Option C, and its interaction with the Contractor's obligations to indemnify the Employer against property damage to the extent that it is due to default of the Contractor, has produced some interesting results. The equivalent to Option C under JCT 63 stated that existing structures and the contents thereof were at the sole risk of the Employer. In the Scottish appeal of *Scottish Special Housing Association* v. *Wimpey Construction UK Ltd* (1986), the House of Lords held that the effect of this wording was that the Employer was bound to insure the property against the risk of damage by all of the Specified Perils, including fire, and that the liability for risk of such damage rested with the Employer, notwithstanding that the fire was caused by the negligence of the Contractor.

Although the SBC and SBC/DB do not expressly state that the existing structures are to be at the Employer's sole risk as regards the Specified Perils, clause 6.2 (which follows clause 20.2 of JCT 98) states that where insurance Option C applies, the Contractor's obligation thereunder to indemnify the Employer against damage to property excludes loss or damage to any insured property caused by a Specified Peril. The effect is that the Employer continues to bear the whole risk of damage to the existing structure caused by the Specified Perils notwithstanding the Contractor's negligence.

In the House of Lords decision in *British Telecommunications plc* v. *James Thomson & Sons (Engineers) Ltd* (1999), it was held that domestic sub-contractors may be under a duty of care to the Employer to prevent damage to existing structures. In that case the duty was implied as subrogation rights for nominated sub-contractors were removed, whereas they were preserved for domestic sub-contractors. In *Kruger Tissue (Industries) Ltd* v. *Frank Galliers Ltd* (1998), it was held that a negligent Contractor could be liable for consequential losses suffered by an Employer notwithstanding that the damage occurred to an existing structure covered by a Joint Names Policy. This was on the basis that consequential losses were not included in the losses where the Employer is required to insure under what is now Option C.

Whilst at one time it may not have been clear whether the Employer had a right of indemnity against a negligent Contractor where insurance has been effected under Option C, it does now appear clear that the Employer has no obligation to indemnify the Contractor against claims by third parties against the Contractor, in respect of this same damage, see *Aberdeen Harbour Board* v. *Heating Enterprises (Aberdeen) Ltd* (1990).

In *Tyco Fire & Integrated Solutions Ltd* v. *Rolls-Royce Motor Cars Ltd* (2008), the Court of Appeal held that Rolls-Royce could recover damages from a negligent contractor responsible for causing flood damage to existing facilities, even although Rolls-Royce had failed to comply with its obligation to insure the existing facilities against specified perils. On a construction of the contract, Rolls-Royce were not obliged to insure against loss from specified perils caused by negligence, the obligation to insure did not extend to an obligation to include the contractor as a named insured, and the contractor had undertaken to indemnify Rolls-Royce against damage arising out of its negligence.

14.2.8 Claims

The consequences of loss of or damage to the works under the SBC and the SBC/DB vary according to which insurance Option has been selected.

If the Contractor has taken out insurance under Option A, the Contractor must give notice of the loss or damage to the Employer (and where the SBC is used, also to the Architect) and, following any inspection by the insurers, the Contractor must restore the damaged work and repair or replace any lost or damaged Site Materials and remove and dispose of debris. The Contractor must authorize the insurers to pay any monies to the Employer, who then passes them on to the Contractor by way of interim certificates. It should be noted that under the SBC/DB payment of such insurance monies will be by instalments under Alternative B even if staged payments under Alternative A otherwise apply under the contract.

No other sums are payable to the Contractor by virtue of the loss or damage. This would appear to exclude a claim for loss and expense as a Relevant Matter under clause 4.24 of the SBC and clause 4.21 of SBC/DB, though an extension of time may be granted if the loss or damage was occasioned by any of the Specified Perils (see clause 2.29.10 of the SBC and clause 2.26.9 of the SBC/DB).

Where the All Risks Insurance has been taken out by the Employer under Option B, there is a similar procedure for the Contractor to give notice, but any restoration, replacement or repair work and (when required) removal and disposal of debris is to be treated as a Variation under the SBC and a Change under the SBC/DB and valued accordingly. In this case, any shortfall between the insurance proceeds and the cost will be made up by the Employer. Under Option A, in contrast, the Contractor will only receive such monies as the insurers pay out.

If there is material loss or damage to existing structures which have been insured by the Employer, there is an option under paragraph C.4.4 of Schedule Part 3 of the SBC and the SBC/DB for either party to terminate the employment of the Contractor if it is just and equitable so to do. Any dispute on this point may be referred to dispute resolution. If no notice of termination is served, or where the notice of termination is not upheld at dispute resolution, then the Contractor must restore, replace or repair the loss or damage, such work being treated as a Variation (or a Change, in the case of the SBC/DB).

14.2.9 Excepted Risks

The provisions of clauses 6.1, 6.2 and 6.4.1 of the SBC and the SBC/DB are each subject to clause 6.6 which excludes from the Contractor's liability personal injury to or the death of any person or any damage, loss or injury caused to the Works or Site Materials, work executed, the site or any other property by the effect of an Excepted Risk. These are defined in clause 6.8 of the SBC and the SBC/DB as (a) ionizing radiations or contamination by radioactivity from any nuclear fuel or from any nuclear waste from the combustion of nuclear fuel, radioactive, toxic, explosive or other hazardous properties of any explosive nuclear assembly or nuclear component thereof; (b) pressure waves caused by aircraft or other aerial devices travelling at sonic or supersonic speeds; and (c) any act of terrorism that is not within the Terrorism Cover required to be taken out and maintained under the contract. The last mentioned category was added by the SBCC Terrorism Cover Update 2010. See Section 14.2.10.

14.2.10 Terrorism

The common difficulty in obtaining terrorism cover wider than that generally available under the Pool Reinsurance Company scheme (Pool Re Cover) was recognized by the publication of the SBCC Terrorism Update 2010 which made amendments to the SBC and the SBC/DB. Clause 6.10 of each contract now provides that to the extent that the joint names policy for work and Site Materials excludes loss or damage caused by terrorism, the Contractor (or the Employer, where Options B or C apply) shall take out such terrorism cover as specified in the Contract Particulars. Where Option A is used, the following applies. Where the specified terrorism cover is Pool Re Cover, the cost of that cover and its renewal is deemed to be included in the Contract Sum provided that, if at any renewal of the cover there is a variation in the rate on which the premium is based, the Contract Sum shall be adjusted to reflect such variation. Where the terrorism cover specified is other than Pool Re Cover, the cost of such cover and its renewal is added to the Contract Sum; however, where the Employer is a local authority and there is an increase at renewal of terrorism cover (of any type) of the rate on which the premium is based, the Employer may instruct the Contractor not to renew the terrorism cover; and if such an instruction is given then the replacement or repair by the Contractor of works or Site Materials which are subsequently subject to loss or damage by terrorism shall be treated as a Variation (or Change under the SBC/DB).

14.3 Insurance under the NEC3

14.3.1 General

Consistent with its overall style, the NEC3 adopts a somewhat different approach to the SBC and the SBC/DB by setting out core requirements in relation to insurance and allowing the parties flexibility to modify these by means of the Contract Data. These core requirements are contained in clauses 84–87 and the Insurance Table. The Contractor is required to provide the insurances stated in the Insurance Table except any insurance which the Employer is to provide as stated in the Contract Data. Thus, under the Insurance Table the Contractor is to insure against (a) loss of or damage to the works, Plant and Materials; (b) loss of or damage to Equipment; (c) liability for loss of or damage to property (other than the works, Plant and Materials and Equipment) and for bodily injury or death (other than in respect of a Contractor's employee) caused by activity in connection with the contract; and (d) liability for death of or bodily injury to a Contractor's employee in the course of their employment in connection with the contract.

Alternatively, if the Contract Data so provides, any of the above insurances and/or any additional insurances may require to be taken out and maintained by the Employer rather than the Contractor. This is the equivalent means of requiring SBC Insurance Option B or C as described in Section 14.2.

The Contract Data may also require the Contractor to provide additional insurance as required by the Contract Data. This is often the means used to require the Contractor to maintain professional indemnity insurance (but see the comments in Section 14.4).

The insurance requirements in the NEC3 are somewhat more terse than the equivalent provisions in the SBC and the SBC/DB. For example, though clause 84.2 requires insurances to be in the joint names of the parties, there is no definition of joint names insurance, so the insurance does not require to be composite (see Section 14.1.4). Rather than define the required cover with reference to all-risks insurance, the cover required under clause 84.2 is for events which are at the Contractor's risk until the Defects Certificate or a termination certificate has been issued. Since clause 81.1 provides that the risks not carried by the Employer are carried by the Contractor (see Section 14.1.6), the obligation to provide cover for 'events which are at the Contractor's risk' might seem unduly wide. There is no provision requiring joint policies to exclude subrogation rights against co-insureds (see Section 14.1.5) although clause 85.2 does require insurance policies to include a waiver of subrogation rights against directors and other employees of every insured, except in the event of fraud.

Clause 85.4 provides that any amount not recovered from an insurer is borne by the Employer for events which are at his risk and by the Contractor for events which are at his risk. As mentioned in Section 14.1.6, events which are the Employer's risk are set out in clause 80.1, and these may be supplemented by additional Employer's risks specified in the Contract Data. Thus, for example, loss of or damage to the works, plant and materials caused by war, etc, strikes, riots and civil commotion, or radioactive contamination is an Employer risk so that if such risks are excluded from insurance cover, then the liability lies with the Employer. This is equivalent to (but not exactly the same as) the SBC notion of Excepted Risks (see definition in Section 14.2.9).

Where the insured event is an Employer risk, this is a Compensation Event under clause 60.1(14) but otherwise the Contractor will have no entitlement to any extension of time or additional cost for the occurrence of an insured event.

The NEC 3 has no corresponding provisions to those in the SBC and the SBC/DB dealing with the consequences of an insured event, other than the briefly stated obligation contained in clause 82.1 that until the Defects Certificate has been issued and unless otherwise instructed by the Project Manager, the Contractor promptly replaces loss of and repairs damage to the works, Plant and Materials.

14.3.2 Delay caused by insured event

An issue to bear in mind is the absence of relief to the Contractor under the NEC3 for delays arising from loss or damage to the works or materials caused by an insured event. As mentioned in Section 14.2.8, clause 2.29.10 of the SBC and clause 2.26.9 of the SBC/DB provide that loss or damage occasioned by any of the Specified Perils (see definition in Section 14.2.7) is a Relevant Event for the purposes of an extension of time. So, for example, in the event of damage to the works caused by fire, the cost of reinstatement would be covered by the All Risks insurance policy, while the consequential delay in completion would be subject to an extension of time. There is no corresponding provision in the NEC3 so that, unless such risk is expressly stated in the Contract Data to be an additional Employer's risk pursuant to clause 80.1, any delay arising from an insured event, such as loss or damage caused by fire, is the Contractor's risk.

14.4 Professional indemnity insurance

With the continuing demand for design and build packages, professional indemnity or design liability insurance for contractors has become an integral part of the building contract structure in terms of spreading risk.

Under a design and build contract the contractor assumes responsibility for both design and construction of the project and, in the absence of in-house designers, will sub-contract design to one or more professional firms of architects, structural engineers, services engineers or the like. Nonetheless, the employer's contract remains with the contractor alone, albeit a prudent employer would also seek collateral warranties from each member of the design team.

In the event of a defect in design, the employer's primary claim will lie against the contractor and it is for this reason that the contractor would be well advised (if not obliged under the contract) to maintain professional indemnity insurance to cover any such claim. The contractor or his insurers may have a right of recovery against the design consultants and/or their professional indemnity insurers. It should be tied to any limitation of actions period.

An obligation on the Contractor to take out and maintain professional indemnity insurance is contained within clause 6.12 of the SBC (where there is a Contractor's Designed Portion) and the SBC/DB.

It is worth noting that the extent of the Contractor's liability under clause 2.17.1 of the SBC/DB in respect of any inadequacy in design is 'the like liability to the Employer … as would an architect or as the case may be other appropriate professional designer holding himself out as competent to take on work for such design'. The comparable obligation in the SBC where there is a Contractor's Designed Portion is to be found in clause 2.19.1.

This design warranty therefore equates the duty of the contractor to that of a professional designer, namely the duty to exercise reasonable skill and care. This removes any term, which may otherwise be implied, that the design is fit for the purpose for which it is required. Most insurers are not prepared to cover such a fitness for purpose warranty given by a design professional, and similarly are not prepared to cover such a warranty from a contractor with design responsibility.

Professional indemnity policies are usually issued on a 'claims made' basis. This means that any claim will be dealt with under the policy in force during the year in which the claim is made. For that reason it is prudent for the contractor to maintain professional indemnity insurance long after completion of a project. Clause 6.12 of the SBC (and of the SBC/DB) provides that such insurance should be maintained for the period stated in the Contract Particulars. For most projects this will be a period of 10 or 12 years.

The amount of professional indemnity insurance cover should ideally be sufficient to cover any loss likely to result from a defect in design. Parties should, however, be aware that this is a continuing annual obligation. Notwithstanding the usual qualifications relating to availability of cover and the terms of renewal, parties should take care when agreeing the level of cover, bearing in mind the potentially volatile nature of the insurance market.

It is of course possible to limit the potential liability of the contractor and indeed clause 2.19.2 of the SBC and clause 2.17.2 of the SBC/DB permit the parties to insert a ceiling of liability for loss of use, loss of profit or other consequential loss arising in respect of the liability of the Contractor for inadequacies in design. The NEC3 also provides a means under Secondary Option X18 to limit the contractor's liability for Defects due to its design which are not listed on the Defects Certificate to the amount stated in the Contract Data.

Although there are no express requirements in the NEC3 core clauses or in the Insurance Table in respect of professional indemnity insurance, clause 84.1 does make provision for the Contract Data to specify any additional insurances to be provided by the Contractor. However, if professional indemnity insurance obligations are to be added in this way, drafting changes may be required to clause 84.2 which suggests that all insurances (including additional insurances stated in the Contract Data) are to be in the joint names of the parties and to provide cover until a Defects Certificate or a termination certificate has been issued, neither of which requirements is appropriate for professional indemnity insurance.

Furthermore, for the reasons stated above in relation to 'fitness for purpose', there may be doubts as to the extent to which a professional indemnity policy will cover the Contractor's design liabilities unless Secondary Option X15 is used, i.e. limitation of the Contractor's liability for his design to reasonable skill and care.

It is also important to understand the distinction between professional indemnity policies which provide cover on an 'each and every claim' basis and those where cover is in the aggregate for each period of insurance (which may or may not be subject to one or more reinstatements). It is more common for professional indemnity insurance maintained by design and build contractors to be on an aggregate basis, while such insurance is usually taken out by professional consultants on an each and every claim basis. Both types of cover will often also have lower levels of cover, usually in the aggregate, for certain types of claim, such as those related to pollution, asbestos or contaminated land.

14.5 Latent defects insurance

This is a separate type of policy which the employer itself can take out to cover physical damage to the works regardless of who is legally at fault. The policy will normally run from practical completion of the works for ten years (it is otherwise known as 'decennial insurance'), and covers major physical damage or inherent defects in the structure which threaten its stability. The defect must be inherent (i.e. existing from the outset) but not discovered until after completion of the building.

As such a policy is intended to cover major damage, regardless of who is responsible, premiums tend to be high and this may explain why latent defects insurance is not currently widely used in the UK. Further, such policies do tend to contain exclusions which have not found favour in the UK. Also, from the contractor's perspective, unless such a policy contains a waiver of subrogation (available at additional cost), it is of little benefit to them. Such insurance does, of course, have the advantage that, in the event of a design fault becoming apparent, it is not necessary for the employer to prove breach of contract and/or negligence on the part of the contractor and/or design consultant. Although consequential loss extensions are available at additional cost, it should be noted that these policies will normally only cover physical damage to the building. Therefore a claim for negligent design, for example, on account of a building having a smaller net internal area than that set out in the employer's requirements, would have to be dealt with by a claim in the usual way against the contractor or designer who produced the drawings.

14.6 Project insurance

Anecdotal evidence would appear to suggest that it is becoming more common, at least in major infrastructure projects, for employers to arrange the key insurances, i.e. contract all risks and public liability. This is often referred to as 'project insurance' or 'owner controlled insurance programme' (OCIP). Under such arrangements, the employer will arrange the key insurances and meet the premium costs. Such insurance would normally name the employer and the contractor, and possibly also sub-contractors and consultants and third parties such as lenders, as co-insured. The purported advantages for the employer include retaining control over placing and managing insurance and claims, savings in premium costs, control over deductibles, ensuring all relevant parties are joint insured, the ability to tailor the insurance to meet project requirements, and avoiding both overlaps and gaps in cover. This type of arrangement would require amendments to the SBC and the SBC/DB since it goes beyond Option B, but could fit into the NEC3 model by means of appropriate entries in the Contract Data.

Another advantage of employer-arranged insurance (and this would also apply where the SBC Option B is followed in isolation) is that where the employer takes out all risk cover, it can include, as an extension, cover for the losses arising from delay resulting from loss or damage caused by an insured event (often called either delay in start up (DSU) or advance loss of profits (ALOP) insurance). This would cover such matters as loss of profit, loss of revenue or rent, or debt servicing costs up to the insured amount and would normally be subject to a deductible (in the form of an initial period of the delay for which no recovery is made). The insurance would plug the gap (subject to the deductible period) arising from loss of recovery of liquidated damages where the contractor is entitled to an extension of time under the SBC and the SBC/DB in relation to damage caused by Specified Perils (see Section 14.2.8). The employer's recovery would be the actual loss which it demonstrates to the insurers that it has sustained as a consequence of the delay and not simply the loss of the right to claim liquidated damages. It should be noted that such cover is normally only available in conjunction with an all risks policy arranged by the employer and not where such policy is arranged by the contractor.

It is possible that the growing uptake of Building Information Modelling (see Section 1.7), and its emphasis on collaboration in the design process, may lead to an increase in the use of project insurance. The UK Government has also encouraged the consideration of integrated project insurance in public sector projects, as illustrated by its Construction Strategy published in May 2011, one of the aims of which was to 'explore with the industry the use of integrated project insurance to support new procurement models', and the promotion of a number of pilot projects using integrated project insurance.

Chapter 15
Litigation

15.1 Introduction

Disputes frequently arise under construction contracts. In this chapter, we will consider certain of the available methods by which disputes can be resolved. Two methods of dispute resolution are, however, so significant that they merit chapters in their own right. In Scotland, the majority of construction industry disputes were, traditionally, resolved by arbitration. That arose from the commonplace insertion of arbitration clauses into construction contracts, and the right of a party to such a contract to insist upon arbitration in such circumstances. Arbitration is considered in Chapter 17. Nevertheless, the Scottish courts frequently become involved in construction contract disputes, with many cases that are ultimately resolved by arbitration having started off in the courts, whether in the Court of Session or the Sheriff Court.

The 1996 Act introduced a statutory right to adjudication. Adjudication was not a creation of the 1996 Act; it had been available in Scotland in relation to certain disputes arising under the DOM/1/Scot and DOM/2/ Scot forms of sub-contract and the SBC/DB. Adjudication is considered in Chapter 16.

There are many other forms of dispute resolution, either formally involving the determination of a third party or informally by negotiation. These are dealt with in Chapter 18.

15.2 The litigation process

15.2.1 Introduction

Prior to the advent of adjudication, as we now understand it, in the absence of an arbitration agreement, or where parties have waived the arbitration agreement, the courts in Scotland traditionally resolved building contract disputes. While the number of disputes reaching the courts has dropped following the introduction of adjudication, the courts are, on the whole, becoming the preferred forum of final determination due to the perceived cumbersome nature of arbitration in Scotland.

The SBC has, as a default provision, legal proceedings as the means of dispute resolution unless arbitration is specifically contracted for. Articles 8 and 9 of the SBC contain provisions related to arbitration and court proceedings. If the parties wish to

MacRoberts on Scottish Construction Contracts, Third Edition. MacRoberts.

resolve disputes by arbitration, then the Contract Particulars require to be completed to state that article 8 and clauses 9.3–9.5 of the Conditions apply and the words 'do not apply' which appear in the Contract Particulars need to be deleted.

There is no corresponding default provision in the NEC3. Parties incorporate one of two available dispute resolution provisions within their contract, either Option W1 or Option W2, the latter to be used in the UK where the 1996 Act applies. Under both provisions, disputes are to be referred to adjudication in the first instance. If parties wish disputes to be determined finally by arbitration, then they must agree to specify that the tribunal is arbitration in the Contract Data, Part 1.

If court action is proceeded with, the action can be raised either in the Sheriff Court, which is the local court for each area, or the Court of Session, which is based in Edinburgh. Typically, small to medium-sized claims are pursued in the Sheriff Court while larger value or more complex claims are pursued in the Court of Session. In which Sheriff Court the action can be pursued depends upon the rules of jurisdiction. For example, the Sheriff Court local to the defender's place of business, or that of the pursuer's business address where the sums sued for are to be paid, are two possible grounds of jurisdiction. The Civil Jurisdiction and Judgments Act 1982 sets out the various grounds of jurisdiction. Parties may agree in the building contract which courts have jurisdiction. A detailed consideration of the law relative to jurisdiction is beyond the scope of this book.

Claims with a monetary value of up to £5,000 must be pursued in the Sheriff Court. Above this figure, the Court of Session and the Sheriff Court have concurrent jurisdiction in respect of monetary claims. As a result of a review of the Scottish civil justice system, and Lord Gill's 2009 report, the Scottish Civil Courts Review, a major programme of reform was recommended. Following on from that, the Courts Reform (Scotland) Bill is passing through its parliamentary stages and sets out the framework within which the court rules will be changed. A key proposal is raising the monetary limit below which actions must be raised in the Sheriff Court, rather than the Court of Session, from £5,000 to £150,000.

Within the Court of Session, a choice needs to be made by the pursuer as to whether to initiate proceedings in the Commercial Court or under the ordinary procedure. The court rules provide for a more flexible procedure to be adopted by the Commercial Court.

15.2.2 The procedure

The rules of procedure of the Sheriff Court are set out in the Sheriff Courts (Scotland) Act 1907, the First Schedule of which contains the Ordinary Cause Rules 1993. The rules of procedure of the Court of Session are set out in the Act of Sederunt (Rules of the Court of Session 1994) 1994 Schedule 2, which contains the Rules of the Court of Session 1994. Both of these are amended from time to time.

A detailed examination of court procedure is beyond the scope of this book, however, certain important features of Court of Session and Sheriff Court actions are considered below.

The initiating writ or summons

An action for payment in the Court of Session, whether a commercial action or under ordinary procedure, is initiated by a summons. An ordinary cause in the Sheriff Court is initiated by an initial writ. In each case, these set out details of the remedy sought by the pursuer, the relevant facts and circumstances and the legal basis of the claim.

Sheriff Court

In the Sheriff Court a defender, ordinarily, has 21 days from service upon him of an initial writ to defend the action. The 21-day period can be shortened with the permission of the Sheriff. A defender defends the action by lodging a notice of intention to defend. If the defender does not lodge a notice of intention to defend within that period, the pursuer may seek a decree.

If a notice of intention to defend is lodged, the defender must lodge defences 14 days after the 21-day period of notice has expired. If the defender also has a claim against the pursuer in certain defined circumstances, a counterclaim setting this out may be lodged and dealt with at the same time as the principal action. There then follows a fixed period in which each party expands on their pleadings (the initial writ, defences and any counterclaim and answers to it) to respond to the other side's case and focus the issues between them. This is known as the period of adjustment. After the period of adjustment, the court will hear both parties at an Options Hearing to determine further procedure.

Further procedure may take the form of a debate, which is a hearing to deal with legal issues. These could include arguments that one party's case is not specific or detailed enough to give the other party fair notice of the party's position or arguments that even if one party proved everything it offered to prove, it would still have no legal entitlement to the remedy sought. The debate may resolve the whole action or it may lead to refinement of a party's case. A proof or a proof before answer may be ordered after, or instead of, a debate. The purpose of a proof is to try the factual issues of the case by hearing evidence from witnesses. Where legal arguments have been reserved and are to be heard at the conclusion of the evidence, the hearing is known as a proof before answer.

Incidental applications for matters such as the recovery of documents or interim decree are made by way of written application to the court known as a motion.

If a commercial action is raised in the Sheriff Court, the initial steps are the same as with an ordinary action. However, the defender only has seven days after the expiry of the period of notice in which to lodge defences. Thereafter, a Case Management Conference is arranged for no sooner than 14 days and no later than 28 days after the expiry of the period of notice. The sheriff has power in commercial actions to make whatever orders he thinks fit for the progress of the case. The sheriff's purpose is to secure the expeditious resolution of the action. The sheriff will remain proactive throughout the duration of the case.

The Court of Session

The ordinary procedure of the Court of Session is broadly similar to that of the Sheriff Court. Once the 21-day period of notice has expired, the pursuer in the action requires formally to lodge the Summons with the Court of Session for calling. This is a procedural step as opposed to a hearing of the case in court. Once the action calls, which means it appears in the rolls of court, the timetable for the action starts running. The party proposing to defend an action must enter appearance within three days of the action calling. This involves attending at the court office and formally notifying them that the action is being defended. The period for lodging defences is seven days from the date of calling.

Once defences are lodged, the pursuer is required, within 14 days, to lodge an open record, a document which comprises the summons and defences. Once this is lodged, the court fixes a start date for the period of adjustment which then runs for eight weeks. The adjustment period can be extended by the court, on application by one of the parties to the action. At the end of the adjustment period, parties can opt to go to Debate, known in the Court of Session as a procedure roll hearing, a proof or a proof before answer.

The Commercial Court of the Court of Session has a somewhat more flexible procedure. There is no automatic right to adjustment of pleadings. The procedure and progress in the case are under the direct control of the commercial judge who is proactive in his handling of the case. A case will normally be dealt with by the same judge throughout. The case first calls in front of this judge for a preliminary hearing within 14 days of the expiry of the period for lodging of defences. Thereafter, there will follow a series of hearings designed around the requirements of the case. A proof, proof before answer, debate or some alternative procedure may follow. Incidental applications are made by written motion.

It is important to be aware of the Court of Session Practice Note No. 6 of 2004 regarding commercial actions, introduced in November 2004. This includes directions as to the pre-action stage. It sets out the aim of having the matters in dispute discussed and focused in pre-action correspondence between the solicitors for both parties. It is stated in the Practice Note that the commercial action procedure works best where issues have been investigated and ventilated before the action is raised. The Practice Note directs that the pursuer should fully set out, in correspondence to the defender, the nature of the claim and the factual and legal background on which it proceeds, supply all documents relied upon and disclose any expert's report commissioned, prior to raising a commercial action. The defender is expected to provide a considered and reasoned reply and to disclose any document or expert report on which they rely. It is suggested by the Practice Note that parties may wish to consider whether all or some of the dispute may be amenable to some form of alternative dispute resolution. A failure to comply with the Practice Note may result in awards of expenses against the non-compliant party.

The Practice Direction reflects a stated desire to reserve the Commercial Court for cases in which there is a real dispute between the parties which requires to be resolved by judicial decision rather than by other means, and to enable an early ventilation of the issues in dispute. While it is a shadow of the English Civil Procedure Pre-Action protocols, it reflects the same desire to make litigation in the courts the final step in the resolution of claims as opposed to the first.

15.2.3 Protective measures

The right to seek protective measures to secure the claim or the subject matter of the dispute, pending the outcome of proceedings, may be exercised at the outset of Sheriff Court or Court of Session proceedings, or at a later date by application to the court. In this section we consider the principal protective measures available in Scotland.

Arrestment and inhibition

The principal protective measures are the rights to arrest or inhibit pending the outcome of the action, known as arrestment and inhibition on the dependence.

Arrestment is a means of attaching money or other moveable property of the defender in the hands of a third party. So, for example, a pursuer in an action might attach any funds at credit in the defender's bank account by placing an arrestment in the hands of the bank. That freezes the money in that account and prevents the bank from releasing it to the defender. Warrant to arrest can be granted in either the Court of Session or the Sheriff Court.

Inhibition is similar but attaches heritable property of the defender preventing the defender from selling it. Since The Bankruptcy and Diligence etc. (Scotland) Act 2007 ('the 2007 Act'), which came into force in April 2008, warrant to inhibit can be granted in either the Court of Session or the Sheriff Court. The 2007 Act makes significant changes to the law of diligence. Following a series of cases involving Article 1 of the First Protocol to the European Convention on Human Rights, see *Karl Construction Ltd* v. *Palisade Properties plc* (2002), *Advocate General for Scotland* v. *Taylor* (2004), *Barry D Trentham Ltd* v. *Lawfield Investments Ltd* (2002), *Fab-Tek Engineering Ltd* v. *Carillion Construction Ltd* (2002), *Gillespie* v. *Toondale Ltd* (2005) and *F G Hawkes (Western) Ltd* v. *Szipt Ltd* (2007), the 2007 Act sets out as law the practice adopted by the court in these cases, particularly *Karl Construction*.

The new regime is contained in section 169 of the 2007 Act (which introduces new sections 15A–5N of the Debtors (Scotland) Act 1987). An application for warrant for diligence is made by motion accompanied by a form setting out the basis upon which diligence is sought. Often there is also some reference to the basis upon which diligence is sought in the written pleadings.

There is a new statutory test to justify the need for diligence. A party applying for a warrant must show:

1. that he has a *prima facie* case on the merits (a relevant and persuasive case on the face of the facts as set out);

2. that there is a real and substantial risk that enforcement of any decree to follow in the action would be defeated by the debtor being insolvent, on the verge of insolvency, or likely to dispose of his assets; and

3. it is reasonable in the circumstances for diligence to be granted.

The rules do envisage an opposed motion hearing to determine the application for diligence. However, warrant for diligence on the dependence can still be sought on an *ex parte* basis (i.e. without the other party being heard). If it is granted in such circumstances, a hearing at which both parties are invited to attend will be fixed to take place a short period after the warrant has been granted. This is commonly called the Section 15K hearing, which is a reference to the provision which deals with recall of diligence on the dependence. It is expressly provided that if the statutory test is not met, the diligence should be recalled.

Interim possession of property

Another protective measure available in certain circumstances in the Court of Session is provided by s.47(2) of the Court of Session Act 1988. This allows the court to make orders regarding the interim possession of property which is the subject of a court action. It is a powerful remedy in that the court may be asked to make an interim order at a very early stage of an action, without having to wait for written pleadings to be finalized or evidence to be heard. The decision is made on the basis only of the written cases and legal argument. This is because the decision is only interim and it is open to the court, at the end of the process, to reverse the interim order.

An ultimately unsuccessful attempt to use the remedy was made in *Scottish Power Generation Ltd* v. *British Energy Generation (UK) Ltd and Another* (2002). In that case, the pursuer sought to have sums of money, said to have been overpaid, placed by the defenders into a designated account to be held in trust in order to protect the funds from the claims of the first defender's creditors. The Inner House, on appeal, confirmed the basis on which the court should exercise its discretion to grant an interim order. Quoting Lord President Hope (as he then was) in *Mackenzie's Trustees* v. *Highland Regional Council* (1994), the Inner House held that the question must depend on the balance of convenience, namely the nature and degree of the harm likely to be suffered on either side by the grant or refusal of the interim order. Regard should also be had to the relative strength of the cases put forward by each party as one of the factors to be considered in determining where the balance of convenience lies. Following *Church Commissioners for England* v. *Abbey National plc* (1994), the Inner House confirmed that in order to justify an interim order, the person seeking it must establish a *prima facie* case that an obligation exists, that there is a continuing or threatened breach of that obligation and that the balance of convenience favours the making of the order sought.

The s.47(2) remedy was successfully used in *V Tech Wabag UK Ltd* v. *Morgan Est (Scotland) Ltd* (2002). The test applied by the court was whether a valid legal case had been made out and consideration of the balance of convenience. In relation to balance of convenience, the court considered the relative strengths of the parties' cases and the maintenance of the status quo. The judge was prepared to treat the status quo as being the making of payments under the construction contract as they fell due. An exception to this may be if there are serious doubts about the solvency of the party to whom payments were to be made, although that consideration was not relevant in this particular case.

In *Purac Ltd* v. *Byzak Ltd* (2005), the court considered a s.47(2) application on the basis of three tests, namely, whether the pursuers had set out a *prima facie* case; whether the balance of convenience favoured the making of the order; and the need to maintain the integrity of the parties' contractual arrangements (or what had been referred to in *V Tech Wabag UK Ltd* as maintaining the contractual status quo).

In this case, the application failed on the third test because the defenders argued there was a right of retention in any contract where there were mutual obligations between the parties. This allows one party to withhold performance until the other performs its obligations. The contractual arrangements between the parties required the right of retention to prevail over any obligation to pay. This case therefore restricted the circumstances in which it will be possible to obtain an order. See also the brief summary of the law of retention in Section 10.9.1.

Recovery of documents

Under the Administration of Justice (Scotland) Act 1972, section 1, an order may be sought from the court for the inspection, photographing, preservation, custody and detention of documents and other property where the documents or property appear to be relevant to any question which might arise in a court action which is likely to be brought. The order may also cover the production and recovery of the documents or property, taking of samples or carrying out experiments. Applications for such an order are sometimes referred to as 'dawn raids', as they tend to be used where there is a fear that documents or property will be destroyed and so are sought at short notice and, often, without the person in possession being given any prior notice.

The test for granting a section 1 application was confirmed in *Pearson* v. *Educational Institute of Scotland* (1997), namely that proceedings are likely to be brought and, in relation to those proceedings, that the person making the application has a *prima facie*, intelligible and stateable case.

The procedure for making an application under the 1972 Act is to apply by petition to the Court of Session. The petition requires to set out a list of the documents or other property which the applicant, known as the petitioner, wishes to be subject to the order, the address of the premises where these are likely to be and the facts on which the petitioner relies in believing that if the order is not granted, the listed documents would cease to become available. The petition is accompanied by an affidavit (a sworn witness statement) from the petitioner which supports the statements made in the petition. The petitioner also requires to give an undertaking to the court that: (1) they will comply with any court order for payment of compensation if recovery of the listed items causes loss; (2) they will commence a court action within a reasonable time; and (3) they will not use any information obtained for any purpose other than the court action they intend to raise.

It is possible, with the court's agreement, for an application to be heard outwith the presence of the party holding the documents or property. Most applications under this procedure are dealt with in this way. In *The British Phonographic Industry Ltd* v. *Cohen, Cohen, Kelly, Cohen & Cohen Ltd* (1983), it was held that an application can be granted outwith the presence of that party if the documents are essential to the petitioner's case and are at risk of destruction or concealment. If the respondent has lodged a caveat (see below), a section 1 order can be granted without the respondent having the opportunity of being heard.

There are separate provisions in the court rules related to recovery of evidence in court actions which have already been raised. This is known as a commission and diligence for recovery of documents and is the equivalent in Scotland of the English process of discovery, albeit the former is more restricted in scope. Prior to a proof or proof before answer being allowed, the only documents that may be recovered are those required to allow a party to make more specific what is already included in their case or to allow specific replies to the other side's case, see *Moore* v. *Greater Glasgow Health Board* (1978).

An application for recovery of documents under this procedure would be accompanied by a specification of documents which is a list of the documents, property or information which the party wishes to recover.

Suspension and interdict

Suspension is used to stop unlawful conduct taking place. It would most commonly be used in the context of suspension of a court decree where, for some reason, it has been invalidly obtained. It is often used in conjunction with the remedy of reduction where the decree would be reduced or set aside. Suspension can deal only with past actions and prevents them taking effect.

Interdict is the Scottish equivalent of an injunction in England. Its purpose is to prevent a party from carrying out an activity where their doing so would be unlawful or would infringe the rights of another party. It can also be used where there is a wish to maintain the status quo until a decision is made in a case. Interdict prevents a future action from taking place. It can only be used where action is threatened or where there is a reasonable apprehension that that action will be taken. It does not require the defender to do anything but prevents them from doing so thereby maintaining the status quo. See also Section 10.7.

Both suspension and interdict can be obtained on an interim basis at the outset of a case with the final decision on permanent suspension and interdict being taken at the conclusion of the matter.

Reinstatement of possession or specific relief

Where a party who is respondent in an action has done something which the court could have prohibited by interdict, there is provision in section 46 of the Court of Session Act 1988 to allow the court to order that respondent to take a positive action in order to reinstate the pursuer or petitioner in the action to the position he would have been in had the interdict been obtained. This can include an order to reinstate possession or the granting of specific relief.

Caveats

Caveats are a form of early warning procedure. They are documents lodged with the court to allow the party lodging them to be given notice of applications for orders being made where, in the absence of a caveat, those orders could be granted without any notice being given.

Caveats can be lodged to give notice only of certain orders, as specified in the court rules. These include interim interdict, orders for sequestration (bankruptcy) of an individual, orders for the appointment of a liquidator or an administrator to a company and other interim orders.

It is not possible, by lodging a caveat, to gain notice of arrestment or inhibitions being sought.

15.2.4 Appeals

In the Sheriff Court, parties have, in certain defined circumstances, a right of appeal to the Sheriff Principal or to the Court of Session. Leave (or permission) to appeal is necessary in certain circumstances. A further appeal from the Sheriff Principal to the Court of Session may also be competent. The appeals procedure in the Sheriff Court is regulated by the Sheriff Courts (Scotland) Act 1907 as amended by the Sheriff Courts (Scotland) Act 1971.

It should be noted that the Courts Reform (Scotland) Bill (see Section 15.2.1) provides for the establishment and rules of the Sheriff Appeal Court, to deal with all civil appeals from the Sheriff Court. The right to take an appeal directly from the Sheriff Court to the Court of Session will cease. Sheriffs Principal will become Appeal Sheriffs, and provision is made for the appointment of a President and Vice-President with administrative functions.

The appeal function of the Court of Session is exercised by the Inner House. If an appeal is taken from a Court of Session judge's decision, it is known as a Reclaiming Motion. As with the Sheriff Court, leave to appeal may be required in certain circumstances.

Finally, an appeal from the Inner House of the Court of Session to the Supreme Court may be competent. The appeals procedure in the Court of Session is regulated by the Rules of the Court of Session 1994. The method by which an appeal to the Supreme Court can be taken is regulated by the Court of Session Act 1988 and the Supreme Court Rules 2009.

Chapter 16
Adjudication

16.1 *Introduction*

While it was appreciated that Part II of the Housing Grants, Construction and Regeneration Act 1996 (henceforth 'the 1996 Act') would have a very significant impact upon dispute resolution in the construction industry, when the first edition of this book was published, there had been very little experience of the operation of the 1996 Act in practice. In the first edition, only two pages of text were devoted to adjudication. Such has been the impact of adjudication that there have since been hundreds of decisions from the courts and numerous books written on the subject.

The 1996 Act came into force on 1 May 1998. It was, arguably, the most significant piece of legislation to affect the construction industry for decades. The legislation sought to address certain long-standing problems within the industry, as set out by Sir Michael Latham in his 1994 Report 'Constructing the Team', namely serious payment problems affecting many in the construction industry, particularly smaller firms; and the problem of the costs and delays in resolving construction disputes. Adjudication and payment are, by their nature, inextricably linked. Payment is considered in Chapter 8.

As far as adjudication itself is concerned, the aim of the 1996 Act was to offer a quick means of resolving disputes. The 1996 Act introduced a right to adjudication as a means of dispute resolution for construction contracts as those contracts are defined by the 1996 Act, which include, by virtue of amendments made to the 1996 Act by the 2009 Act all construction contracts, whether wholly in writing, partly in writing or wholly oral. This is considered in Section 1.2.2.

16.2 *The scope of Part II of the 1996 Act, as amended by the 2009 Act*

To properly appreciate the scope of Part II of the 1996 Act requires an understanding of sections 104–107. These sections are discussed in Chapter 1.

Section 108, as amended, enshrines the right to refer a dispute to adjudication. It provides that:

1. A party to a construction contract has the right to refer a dispute arising under the contract for adjudication under a procedure complying with this section. For this purpose 'dispute' includes any difference.

MacRoberts on Scottish Construction Contracts, Third Edition. MacRoberts.

2. The contract shall include provision in writing so as:

   (a) to enable a party to give notice at any time of his intention to refer a dispute to adjudication;
   (b) to provide a timetable with the object of securing the appointment of the adjudicator and referral of the dispute to him within 7 days of such notice;
   (c) to require the adjudicator to reach a decision within 28 days of referral or such longer period as is agreed by the parties after the dispute has been referred;
   (d) to allow the adjudicator to extend the period of 28 days by up to 14 days, with the consent of the party by whom the dispute was referred;
   (e) to impose a duty on the adjudicator to act impartially; and
   (f) to enable the adjudicator to take the initiative in ascertaining the facts and the law.

3. The contract shall provide in writing that the decision of the adjudicator is binding until the dispute is finally determined by legal proceedings, by arbitration (if the contract provides for arbitration or the parties otherwise agree to arbitration) or by agreement. The parties may agree to accept the decision of the adjudicator as finally determining the dispute.

3A. The contract shall include provision in writing permitting the adjudicator to correct his decision so as to remove a clerical or typographical error arising by accident or omission.

4. The contract shall also provide in writing that the adjudicator is not liable for anything done or omitted in the discharge or purported discharge of his functions as adjudicator unless the act or omission is in bad faith, and that any employee or agent of the adjudicator is similarly protected from liability.

5. If the contract does not comply with the requirements of subsections (1) to (4), the adjudication provisions of the Scheme for Construction Contracts apply.

6. For Scotland, the Scheme may include provision conferring powers on courts in relation to adjudication and provision relating to the enforcement of the adjudicator's decision.

The 1996 Act gives a party to a construction contract the right to refer a dispute arising under the contract to adjudication under a procedure complying with section 108. While the 1996 Act allows parties to agree their own adjudication provisions, care is required not to fall foul of section 108(5). It is to be noted that if any bespoke adjudication provisions in the contract do not comply with the requirements of sections 108(1) – 108(4), all of the Scheme will apply. The Scheme will not just apply to the extent that the bespoke provisions do not meet the requirements of the 1996 Act. Accordingly, when considering adjudication, it is always important to determine whether any bespoke adjudication provisions apply because they comply with the requirements of the 1996 Act, or whether the Scheme applies (even if the latter only arises because the bespoke provisions do not apply, by virtue of section 108(5)).

The 2009 Act inserted section 108(3A) into the 1996 Act, which now requires construction contracts to include provision for the English common law 'slip' rule, allowing adjudicators to correct clerical or typographical errors in their decisions arising by accident or omission. A new paragraph 22A has been inserted into the Scheme to a similar effect, allowing adjudicators to correct clerical or typographical errors within their decisions. The Scheme, however, additionally requires for all corrections to be made within five days of the date upon which the adjudicator's decision is delivered to the parties. A five-day period is also stipulated by the NEC3 in its equivalent provision, see option W2.3(12). The SBC and the SBC/DB incorporate the Scheme.

It is perhaps noteworthy that, despite the concluding words of section 108, the Scottish Scheme did not confer any new powers on the courts.

16.3 *The notice of adjudication*

The 1996 Act and Scheme require a written notice to start the adjudication procedure which must be sent to every other party to the contract. It would be difficult to overemphasize the importance of this notice and the need for its terms to be considered with great care. The notice has four basic purposes:

- It informs the other parties to the contract that there is a dispute and the nature of the dispute, but it cannot create the dispute.
- It informs the adjudicator (when appointed) of the dispute.
- It provides information to allow the appointing body to select an appropriate person to act as adjudicator.
- It defines the jurisdiction of the adjudicator.

The nature and brief description of the dispute should be fairly general to avoid restricting the issues unduly.

In *Ken Griffin and Another (t/a K & D Contractors)* v. *Midas Homes Ltd* (2000), His Honour Judge Lloyd said that in considering the validity of the Notice of Adjudication, the essential questions will include whether a dispute has arisen. 'A dispute is not lightly to be inferred.' Paragraph 1(3) of the Scheme requires the dispute to be defined. In this case, the Notice of Adjudication dated 3 May took the form of a letter from the Referring Party's solicitors to the Respondents' solicitors. Within the letter, reference was made to letters to the Respondents dated 11 and 13 April. The 3 May letter said that as no payments had been forthcoming from the Respondents in respect of either of the attached letters 'a dispute now exists between our client and your client and this dispute will be referred to adjudication'. The Judge refused to enforce the adjudicator's decision. He highlighted the fact that as at 3 May there were a number of issues in dispute: valuation of the sub-contract account, whether the Respondents had properly determined the sub-contract, and damages for loss of profit in respect of the wrongful determination. The Respondents wrote to the Referring Party's solicitors and stated that they had absolutely no idea 'from the notice which of the numerous items were in fact being referred to the adjudicator'. This request for clarity was ignored.

Although it is possible to give a Notice of Adjudication by reference to other correspondence, the Scheme requires the dispute to be defined. A degree of precision is required. The issue of whether a dispute has arisen between the parties is considered further in Section 16.10.2.

The courts have construed strictly the requirement that notice can be given at any time, see, for example, *John Mowlem & Co. plc* v. *Hydra-Tight Ltd* (2000) in which His Honour Judge Toulmin decided that a clause in a standard ICE contract which provided a mechanism called a Notification of Dissatisfaction (which delayed a referral to adjudication for four weeks during which time the parties had the opportunity to meet and resolve their differences) contravened the entitlement contained in the 1996 Act to adjudicate at any time. This case has had an impact upon other ICE standard forms which provide for a Notice of Dissatisfaction procedure as a pre-condition for a dispute to arise. It is generally believed that, whatever good intention lay behind the Notice of Dissatisfaction provisions in ICE conditions, they resulted in the adjudication provisions not being 1996 Act compliant.

Care has to be taken to comply with any conditions relating to service of the Notice of Adjudication. In *Primus Build Ltd* v. *Pompey Centre Ltd & Anor* (2009) the contract provided for notices to be delivered personally or to be sent by fax. The notice was sent by post and while it was held that this did not breach the contractual notice provisions (delivered personally meant actual delivery to an appropriate individual and the method of achieving that was irrelevant), it was observed that if there had been a breach of the notice provisions this would have affected the validity of the adjudication.

16.4 Appointment of the adjudicator

While almost anyone could act as an adjudicator, as no qualifications are required, it is an extremely important appointment. Parties should try to get the right person for the job.

If there is no agreement as to who should act as adjudicator, and no specified nominating body in the contract, the adjudicator will be chosen by an Adjudicator Nominating Body. The Scheme definition of an Adjudicator Nominating Body is somewhat circular. The early drafts of the Scheme listed 16 organizations that were to be the only approved Adjudicator Nominating Bodies, but this was dropped because every future amendment to the list would require a further statutory instrument. Different disputes require different skills and the wider the choice of adjudicator, the better chance of the right person being selected.

If an application is made to an Adjudicator Nominating Body, then they must communicate the selection of an adjudicator to the Referring Party within five days of receiving a request to do so, see paragraph 5(1) of the Scheme.

It is advisable to find out the prospective adjudicator's terms and conditions (including fees) for acting, but this should not delay the start of the adjudication.

If no adjudicator is named, but an Adjudicator Nominating Body is, then application must be made to that body for the nomination of an adjudicator. Most nominating bodies require payment of a fee before they will act, as well as completion of an application form.

Once the nomination has been made, the Referring Party should contact the person nominated and request confirmation that the person is willing to act, as the person requested to act as adjudicator must indicate within two days of receiving the request whether or not he is willing to act, failing which a party can apply to the Adjudicator Nominating Body to select a different person to act as adjudicator.

The terms of paragraph 10 of the Scheme should be noted. It provides:

> Where any party to the dispute objects to the appointment of a particular person as adjudicator, that objection shall not invalidate the adjudicator's appointment nor any decision he may reach in accordance with paragraph 20.

Therefore while an objection to the appointment must be made as soon as the appointment is made, it will not invalidate the appointment.

An adjudicator nominated prior to the issue of a Notice of Adjudication has been held to have no jurisdiction to act notwithstanding that an earlier Notice of Adjudication existed (the change being of limited importance compared to the dispute as a whole) and that there was no prejudice suffered by the Respondent. See *Vision Homes Ltd* v. *Lancsville Construction Ltd* (2009).

16.5 The referral notice

'Referral notice' is arguably a misnomer, as it is a full submission with supporting documentation, rather than a 'notice'. It should include details of the contract (including the parties) the background to the dispute, the relevant facts of the dispute plus supporting documents, the contractual and legal basis of the claim, comment on what the other side's position appears to be and should set out the redress the Referring Party wishes the adjudicator to grant. This should mirror the redress set out in the Notice of Adjudication.

A distinction has been drawn between (i) the Referral Notice for the purpose of starting the Adjudication timetable and to refer the dispute and (ii) the supporting documents to which reference is made to establish whether or not the referring party's case has a sound basis, see *KNN Coburn LLP* v. *G.D. City Holdings Ltd* (2013).

16.6 Conduct of the adjudication

16.6.1 The adjudicator's powers

Paragraph 13 of the Scheme sets out the wide-ranging powers of the adjudicator. It provides:

The adjudicator may take the initiative in ascertaining the facts and the law necessary to determine the dispute, and shall decide on the procedure to be followed in the adjudication. In particular, he may –

(a) request any party to the contract to supply him with such documents as he may reasonably require including, if he so directs, any written statement from any party to the contract supporting or supplementing the referral notice and any other documents given under paragraph 7(2);
(b) decide the language or languages to be used in the adjudication and whether a translation of any document is to be provided and, if so, by whom;
(c) meet and question any of the parties to the contract and their representatives;
(d) subject to obtaining any necessary consent from a third party or parties, make such site visits and inspections as he considers appropriate, whether accompanied by the parties or not;
(e) subject to obtaining any necessary consent from a third party or parties, carry out any tests or experiments;
(f) obtain and consider such representations and submissions as he requires, and, provided he has notified the parties of his intention, appoint experts, assessors or legal advisers;
(g) give directions as to the timetable for the adjudication, any deadlines, or limits as to the length of written documents or oral representations to be complied with; and
(h) issue other directions relating to the conduct of the adjudication.

The Scheme gives the adjudicator the necessary express powers to enable him to ascertain the facts and the law. Whether and to what extent an adjudicator should do so is a matter of some controversy among adjudicators, given the difficulties to which it can lead in the relatively short period normally available to conclude the adjudication. However, where the adjudicator makes use of the power conferred on him, he must only take the initiative to the extent that it is necessary on the submissions he has received. The Adjudicator should also take care not to fall foul of the rules of natural justice and, if he does take the initiative on any matter, should give parties an opportunity to comment on any new material.

16.6.2 Time limit for decision

Paragraph 19 sets out the deadlines imposed upon the adjudicator. It provides:

1. The adjudicator shall reach his decision not later than

   (a) twenty-eight days after the date of the referral notice mentioned in paragraph 7(1);
   (b) forty-two days after the date of the referral notice if the referring party so consents; or

(c) such period exceeding twenty-eight days after the referral notice as the parties to the dispute may, after the giving of that notice, agree.

2. Where the adjudicator fails, for any reason, to reach his decision in accordance with sub-paragraph (1) –

   (a) any of the parties to the dispute may serve a fresh notice under paragraph 1 and shall request an adjudicator to act in accordance with paragraphs 2 to 7; and
   (b) if requested by the new adjudicator and insofar as it is reasonably practicable, the parties shall supply him with copies of all documents which they had made available to the previous adjudicator.

3. As soon as possible after he has reached a decision, the adjudicator shall deliver a copy of that decision to each of the parties to the contract.

A very strict time limit is imposed by the Scheme. There is no leeway and if a decision is not reached within the requisite timescale, it will be invalid.

The leading case from the Inner House is *Ritchie Brothers (PWC) Ltd* v. *David Philp (Commercials) Ltd* (2005). David Philp was the employer and Ritchie Brothers was the main contractor under a building contract. Disputes arose between the parties which were referred to an adjudicator. An undated Referral Notice was issued to an adjudicator by Ritchie Brothers on 18 September. The 28-day period therefore expired on 16 October. However, due to delays with the postal service, the adjudicator did not receive the Referral Notice until 23 September.

On 21 October David Philp's solicitors wrote to the adjudicator indicating that the adjudicator had no power to issue a decision after 16 October, and that any decision reached after that date by the adjudicator would be a nullity. By letter, also dated 21 October, the adjudicator wrote to both parties requesting an extension of the period for reaching his decision until at least 23 October. Subsequently, on 23 October, the adjudicator wrote to the parties indicating that he had reached a decision and this decision was intimated to the parties on 27 October.

In the Outer House, Lord Eassie held that the adjudicator's failure to reach a decision within the 28-day period did not bring the adjudicator's jurisdiction to an end. He expressed the view that the provisions of paragraph 19 of the Scheme were directory rather than mandatory. He was satisfied that this provision illustrated Parliament's intention that, once initiated, the process of adjudication should be completed. Neither party had served a fresh adjudication notice. Lord Eassie held that the decision should be enforced. David Philp appealed.

Ritchie Brothers' primary submission was that, notwithstanding the statutory time limit, the adjudicator's jurisdiction only came to an end when a party to the dispute served a fresh Notice of Adjudication. In the alternative, relying on the reasoning of Lord Wheatley in *St Andrews Bay Development Ltd* v. *HBG Management Ltd* (2003), they argued that a failure to reach a decision within the time limit was a technical error which was not serious enough to invalidate the adjudicator's decision.

David Philp argued that the adjudicator's jurisdiction expired at the end of the 28-day period. In circumstances where the adjudicator is unable to reach a decision within 28 days, he can avoid losing his jurisdiction by requesting and obtaining an extension within the 28 days.

The appeal was allowed. In the leading opinion, the Lord Justice Clerk held that the statutory time limit set out in the Scheme is mandatory and the adjudicator's decision was null and unenforceable. Paragraph 19 of the Scheme provides that the adjudicator 'shall reach his decision' not later than 28 days after the Referral Notice (subject to possibilities of extension agreed by the parties). Applying the natural meaning to the words in paragraph 19 of the Scheme provides a clear time limit. This provides certainty as to the extent of the adjudicator's jurisdiction. If Parliament had a contrary intention, then this could have been expressed in plain terms.

The Lord Justice Clerk also rejected Ritchie Brothers' alternative submission that a failure to reach a decision within the 28-day time limit was not a fundamental error. The Lord Justice Clerk specifically rejected the reasoning in *St Andrews Bay Development Ltd* on that particular point. That decision did not provide any hard and fast criterion by which a court could determine for how long after the time limit a failure to reach a decision can be considered to be merely a technical failure, or in what circumstances the jurisdiction of the adjudicator could be said to come to an end.

Following a number of Scottish and English decisions which did not provide clear guidance on the law in this area, this decision provided Appeal Court guidance in Scotland that the 28-day time limit is mandatory in Scheme adjudications. The decision sends a clear message to adjudicators that their decision must be reached within the 28-day period. If this is not possible, then they must request and obtain an extension prior to the expiry of the period for reaching the decision. If they fail to obtain such an extension, their decision will be invalid. This line of reasoning has been followed in the English Technology and Construction Court case of *Epping Electrical Co. Ltd* v. *Briggs & Forrester (Plumbing Services) Ltd* (2007).

While *Ritchie Brothers (PWC) Ltd* deals with the situation where a decision is not reached within the requisite time period, it does not deal expressly with the situation where the decision is reached timeously but it is not then immediately communicated to the parties. Some have argued that as the statutory requirements only refer to reaching a decision, not communicating it to the parties, some short delay in communicating it to the parties is permissible without such delay affecting the validity of the decision. Lord Wheatley in *St Andrews Bay Development Ltd*, in comments not expressly criticized by the Appeal Court, was firmly of the view that the obligation to reach a decision must include a contemporaneous duty to communicate the decision to the interested parties. Not to require such an interpretation of the obligation to reach a decision would render the whole purpose of the legislation meaningless. This is very much in line with the need for certainty stressed in *Ritchie Brothers (PWC) Ltd*. While it may still be arguable that a decision reached timeously but communicated a very short period thereafter may be valid, it is submitted that the best and safest course is to ensure that the decision is not only reached but also communicated to the parties within the relevant timescale. A delay of 74 hours between reaching the decision and communicating the decision was held to be too long and resulted in the decision being unenforceable, see *Lee* v. *Chartered Properties (Building) Ltd* (2010).

16.6.3 More than one dispute

Paragraph 8(1) of the Scheme provides that an adjudicator can deal with more than one dispute arising at the same time if all parties to the disputes agree.

It is important to distinguish between on the one hand, different disputes and on the other, different aspects of a single dispute. This has greatly exercised the minds of judges in England. For example, in *David McLean Housing Contractors Ltd* v. *Swansea Housing Association Ltd* (2003) His Honour Judge Lloyd considered when a dispute is more than one dispute. His starting point was paragraph 8(1) of the (English) Scheme which precludes the reference of more than one dispute to adjudication. In the event the Referring Party wishes to refer more than one dispute, then the consent of the other party is required. In this case, consent was not given by the Respondents.

His Honour Judge Lloyd stated that one had to consider the Notice of Adjudication in its context. When this was examined it was plain that the real dispute was what payment ought to have been made as a result of Application 19. This contained various elements which were set out in the Notice of Adjudication. The Notice was valid in referring the dispute about the payment to be made and could not be decided without considering each element. The Judge applied what he called a 'benevolent interpretation' and said that the Notice did not refer more than one dispute. It referred one dispute, namely, 'How much should I be paid or how much should I have been paid on Application 19?'

It is considered that in Scotland this approach is one which is consistent with the wording of the Scheme, since Paragraph 20 provides that 'The adjudicator shall decide the matters in dispute and may make a decision on different aspects of the dispute at different times.' This wording is not used in the English and Welsh Scheme. A Notice that refers more than one dispute is invalid. The appointment of an adjudicator in consequence of it is similarly invalid, unless the other party has nonetheless clearly and knowingly accepted the Notice or the appointment so that there exists consent for the purpose of paragraph 8 of the Scheme.

Paragraph 8(2) of the Scheme provides that the adjudicator may, with the consent of all parties to those contracts, adjudicate at the same time on related disputes under different contracts. While multi-party adjudication is possible if consent is obtained, it is relatively rare, perhaps because of the 1996 Act timetable, the reluctance of some parties to agree to it, or because of the difficulty in ensuring that contracts are clearly 'back to back'. See also Section 16.13.

16.6.4 Resignation of the adjudicator

Paragraph 9 of the Scheme provides that an adjudicator may resign at any time by giving written notice to the parties. An adjudicator must resign where 'the dispute is the same or substantially the same as one which has previously been referred to adjudication, and a decision has been taken in that adjudication'. There have been various cases regarding whether a dispute is the 'same or substantially the same' as a previous dispute and these are considered in Section 16.10.2.

16.7 The decision

16.7.1 Reasoned decision

Paragraph 22 of the Scheme states that the adjudicator must provide reasons for his decision if requested to do so by one of the parties to the dispute.

The Scheme does not specify the timescale within which a party must ask for reasons or the period within which the adjudicator needs to produce his reasons.

Practical considerations suggest that a party should ask for reasons before the decision is issued in order that the adjudicator knows that reasons are required and in order that he produces them with or shortly after the decision. The reasons for the decision can be vitally important in considering whether there are grounds for challenging an adjudicator's decision or resisting its enforcement. It is important that reasons are produced with, or very quickly after, the decision.

16.7.2 Compliance with the adjudicator's decision

Paragraph 21 of the Scheme provides:

> In the absence of any directions by the adjudicator relating to the time for performance of his decision, the parties shall be required to comply with any decision of the adjudicator immediately on delivery of the decision to the parties in accordance with paragraph 19(3).

This is a default provision should the adjudicator fail to specify a timetable for compliance with the decision.

16.7.3 Effect of the decision

Paragraph 23 of the Scheme deals with the effects of the decision. Sub-paragraph 2 provides:

> The decision of the adjudicator shall be binding on the parties, and they shall comply with it, until the dispute is finally determined by legal proceedings, by arbitration (if the contract provides for arbitration or the parties otherwise agree to arbitration) or by agreement between the parties.

The adjudicator's decision is a temporary decision which is put into effect and must be complied with by the parties until the dispute is finally resolved as indicated by arbitration, litigation or agreement.

Paragraph 23 of the Scheme also provides that the adjudicator may, if he thinks fit, order any of the parties to comply peremptorily with his decision or any part of it.

16.7.4 Registration of the decision

The Scheme provides that a party or the adjudicator can, if they so wish, register the adjudicator's decision for execution in the Books of Council and Session. On request, the other party must consent to such registration by subscribing the decision before a witness.

On the face of it, this provision is better than its English counterpart which requires the enforcing party to go to court in order to enforce the adjudicator's decision. Once a decision has been registered in this way it does not require a court decree to confirm that the aggrieved party is entitled to enforce a decision. However, for some, this paragraph does not go far enough. The non-complying party is required to sign the decision before a witness to record his consent to registration. If the other party refuses to sign the award, the enforcing party will need to go to court. In practice, therefore, this route to enforcement is seldom, if ever, used.

16.8 Adjudicator's fees and costs/expenses of the parties

Paragraph 25 of the Scheme provides:

1. The adjudicator shall be entitled to the payment of such reasonable amount as he may determine by way of fees and expenses incurred by him and the parties shall be jointly and severally liable to pay that amount to the adjudicator.

2. Without prejudice to the right of the adjudicator to effect recovery from any party in accordance with sub-paragraph (1), the adjudicator may determine the apportionment between the parties of liability for the payment of his fees and expenses and such determination shall be binding upon the parties unless any effective contractual provision in terms of section 108A(2) of the Act applies.

Paragraph 25(1) effectively means that if one party refuses to pay their share of the fees and expenses, the adjudicator can claim the whole amount from the other.

There have been a number of cases in which payment of the adjudicator's fees have been challenged.

In *Linnett* v. *Halliwells LLP* (2009), Halliwells refused to pay the adjudicator arguing that they had objected to the adjudicator's jurisdiction and maintained their jurisdictional challenge throughout the adjudication. Halliwells claimed that as they had not agreed to the appointment of the adjudicator, they did not agree to pay for his work in the adjudication. Appointment as adjudicator was a contract and there was no concluded contract between Halliwells and the adjudicator. The judge agreed that adjudication is a contractual arrangement and the ability of an adjudicator to obtain fees depends on there being a contractual right to payment under the adjudicator's agreement with one or both of the parties. Halliwells' response made it clear that they were objecting to jurisdiction and would not agree the terms. On that basis no contract was formed between Halliwells and the adjudicator on the adjudicator's terms of engagement. However, the judge looked to the general position in relation to an arbitrator for assistance. An arbitrator is entitled to reasonable remuneration from the parties for work done. The parties are jointly and severally liable for fees and expenses. If the arbitrator does not have jurisdiction, he still may have a claim for fees based on the fact that work was carried out at the request of the parties, or one of them. In the same way, if an adjudicator is appointed, the parties by participating in the adjudication and therefore requesting the adjudicator to act, enter into a contract with the adjudicator, formed by their conduct. There would be implied terms that the parties would be liable to pay the reasonable fees and expenses of the adjudicator and would be jointly and severally liable to do so.

The circumstances of payment to an adjudicator who had issued an unenforceable decision were considered by the Court of Appeal in *PC Harrington Contractors Limited* v. *Systech International Limited* (2012), in which it was held that an adjudicator's fees were not payable because his decision was unenforceable on the grounds of breach of natural justice. The court considered the terms of the Scheme for Construction Contracts which applied in that contract. The court took the view that there was nothing in the contract to indicate that the parties had agreed that they would pay for an unenforceable decision or services rendered in preparation for such a decision. It was also significant that the adjudicator's fees were not to be paid in instalments and that the Scheme for Construction Contracts already provided for the adjudicator not receiving payment when his/her appointment was revoked as a result of default or misconduct. A breach of natural justice was considered by the court to amount to default or misconduct.

In relation to the amount of the adjudicator's fees, in *Stubbs Rich Architects* v. *W H Tolley & Son Ltd* (2001) the judge applied the criteria of a reasonably competent solicitor when assessing adjudicator's fees, i.e. how long it would take a solicitor to read the files and write his decision. On appeal, his decision was reversed as he was not comparing like with like; if comparative evidence was relevant, it should not have been a solicitor, rather an expert architect/adjudicator, and the court should be reluctant to substitute its own views of what constitutes reasonable hours. With reference to section 108(4) of the 1996 Act, the judge held that the adjudicator's fees may be challenged if, and only if, the adjudicator has acted in bad faith.

In *Fenice Investments Inc* v. *Jerram Falkus Construction Limited* (2011), the level of the adjudicator's fees was challenged. The court's view was that in such circumstances it should take a robust view with what it described as a 'considerable margin of appreciation' given to the adjudicator to recognize the speed at which the work requires to be undertaken and that routine challenges to adjudicators' fees would discourage potential adjudicators from acting, The court stated:

> Accordingly, in relation to hourly rates, provided that the rate claimed is not clearly outside an overall band of reasonableness, there will be no basis to interfere, even if it could be shown that a different adjudicator, especially an adjudicator with different qualifications, may have charged less or even significantly less.

and

> As for time spent, challenges in other areas of professional fees are usually not on the basis that the hours claimed were not worked but that the particular task took too long or that unnecessary work was done. But again, leeway needs to be afforded here because on a tight schedule different adjudicators may approach their task in different way, or order their work differently. And as for allegedly unnecessary work, it is important to bear in mind paragraph 20 of the Scheme. The adjudicator is entitled to take into account ' … matters under the contract which he considers are necessarily connected with the dispute.' Given the principles set out above, the party to an adjudication which is considering a fees challenge will need to give careful consideration as to whether there is any realistic basis for disputing the fees claimed. It is to be expected that that in the usual run of cases there will not.

This case is a clear discouragement to parties seeking to challenge the amount charged by an adjudicator.

Some adjudicators attempted to protect themselves against non-payment by including in their terms of engagement a provision that they will not issue their decision until payment of their fees is made. However, the RICS Guidance Notes for Surveyors Acting as Adjudicators forbids adjudicators holding on to their decisions until their fees are paid unless they have the agreement of the parties. Further, the case law on the time limits for reaching and communicating decisions also makes this unworkable (see Section 16.6.2).

As noted above, paragraph 25(2) of the Scheme provides that an adjudicator may by direction determine the apportionment between the parties of liability for his fees and expenses. See also Secondary Option clause W2.3(8) of the NEC3 which provides that the Adjudicator may in his decision allocate his fees and expenses between the Parties.

There is, however, no express provision in the statutory framework allowing an adjudicator to include in his decision an award in relation to the costs or expenses which the parties have themselves incurred in relation to the adjudication.

There had been conflicting authority on this point, see *John Cothliff Ltd* v. *Allen Build (North West) Ltd* (1999); *Northern Developments (Cumbria) Ltd* v. *J & J Nichol* (2000); *Bridgeway Construction Ltd* v. *Tolent Construction Ltd* (2000); *Yuanda (UK) Co. Ltd* v. *WW Gear Construction Ltd* (2010); and *Profile Projects Ltd* v. *Elmwood (Glasgow) Ltd* (2011). A degree of clarity was provided in this area by the introduction of section 108A to the 1996 Act by the 2009 Act. Section 108A provides that contractual provisions relating to the apportionment of liability for adjudication costs and expenses are ineffective unless:

1. Parties include a provision in the construction contract in writing allowing the adjudicator to allocate his fees and expenses between the parties; or

2. Parties include a provision in writing after the giving of notice of intention to refer the dispute to adjudication.

Thus, as long as these requirements are met, parties may include contractual provisions whereby one party is required to pay the other's adjudication costs, including their legal expenses. This new provision provides clarity though does not go as far as some commentators would have wished to outlaw these clauses altogether.

The courts have recently been willing to consider the costs incurred in adjudication as damages recoverable as a result of a breach of contract. See *The Board of Trustees of National Museums and Galleries on Merseyside* v. *AEW Architects and Designers Limited* (2013). In this case, following an adjudication in which the employer was unsuccessful, the employer sought to recover damages against the architect including the adjudicator's fees and the Museum's legal and expert costs incurred in defending the adjudication. The Museum argued that the adjudicator's fees were caused by AEW's breach of contract. AEW argued that the Museum had fought the adjudication knowing that the costs would not be recoverable in subsequent proceedings and that the court proceedings should not be allowed as a backdoor method of recovering legal costs. The judge agreed with the Museum and held AEW liable for these costs. The judge formed the view that as adjudication was 'a fact of life' in construction contracts, it was reasonably foreseeable that an adjudication could be initiated by AEW in relation to the dispute over design responsibility. The judge also decided that there was a sufficient causative link between the defaults of AEW and the costs of the adjudication, a link which would only have been broken if the Museum had acted unreasonably or if their solicitors had acted negligently in advising the Museum that it had an arguable defence in the adjudication. Emphasis was also put on the fact that if AEW had done its job correctly there would have been no need for adjudication as the design issues would not have arisen and therefore no reason for the incurred costs.

16.9 Liability of the adjudicator

Paragraph 26 of the Scheme provides that:

> [An] adjudicator shall not be liable for anything done or omitted in the discharge or purported discharge of his functions as adjudicator unless the act or omission is in bad faith, and any employee or agent of the adjudicator shall be similarly protected from liability.

This protects the adjudicator from being sued by the parties unless he acts in bad faith.

16.10 Enforcement proceedings

16.10.1 Introduction

Generally, it is considered that the courts have taken a robust, purposive attitude to the enforcement of adjudicators' decisions, and will generally enforce any decision made by an adjudicator with jurisdiction to make that decision.

In Scotland, the method of enforcement is to raise court proceedings (an action for payment) and the practice has been to raise proceedings in the Commercial Court of the Court of Session. The court can be asked to shorten the period of notice which is otherwise 21 days. This can then be followed by a motion to the court for summary decree at the earliest opportunity if there is no defence to the action, or a part of it, disclosed in the defences. If a defence is stated, an early legal debate is normally sought unless there are issues of fact which need to be resolved. In the latter case some form of proof at which evidence is led will be required. However, even if the payment action is successful at first instance, the decision can be appealed (without leave) which can delay enforcement considerably. Early disposal of an appeal should normally be sought if one is acting for a Referring Party.

A petition for Judicial Review would be an appropriate procedure where a party wants to review a decision of an adjudicator, see, for example, *Karl Construction (Scotland) Ltd* v. *Sweeney Civil Engineering (Scotland) Ltd* (2002).

Notwithstanding the courts' general reluctance to interfere with decisions made by adjudicators with jurisdiction to make those decisions, challenges to adjudicators' decisions have been commonplace since the introduction of adjudication. Indeed, respondents have been known to make challenges systematically on the chance that they may find an argument to resist enforcement. Most commonly, such challenges are argued on the basis that the adjudicator has exceeded his jurisdiction or, alternatively, that the adjudicator's decision is in breach of the rules of natural justice. Such practices undoubtedly inform the courts' attitude in this area, as illustrated in the view expressed by the court in *Carillion Construction Ltd* v. *Devonport Royal Dockyard Ltd* (2005) '[that it] … should only be in rare circumstances that the courts will interfere with the decision of an adjudicator'. Nonetheless, against this background, the courts will uphold proper challenges to adjudicators' decisions.

16.10.2 Jurisdictional challenges

Probably the most regularly stated defence to enforcement proceedings is a challenge on jurisdictional grounds. An adjudicator's jurisdiction is derived from his appointment by the agreement of the parties. The adjudicator will only have jurisdiction to determine a dispute referred to him arising out of a construction contract as defined in the 1996 Act. If any requirement is absent, his decision is a nullity and not binding on the parties. In fact, many cases contain some sort of jurisdictional challenge. Below we examine some of the most common jurisdictional challenges raised in adjudication proceedings. The starting point is the early case of *Macob Civil Engineering Ltd* v. *Morrison Construction Ltd* (1999). Mr Justice Dyson (as he then was) said:

> If his decision is wrong, whether because he errs on facts or the law, or because in reaching his decision he made a procedural error which invalidates his decision, it is still a decision on the issue. Different considerations may well apply if he purports to decide a dispute which was not referred to him at all.

In *Macob Civil Engineering Ltd* it was alleged that the adjudicator had committed a procedural error in that he had decided that the mechanism for payment in the contract was not 1996 Act compliant. He found that the Scheme provisions on payment applied. It was also argued that as the arbitration clause included referral of disputes regarding an adjudicator's decision, enforcement of the adjudicator's award should be postponed pending the outcome of the arbitration.

Mr Justice Dyson rejected both arguments and held that despite the attack, 'an adjudicator's decision which appears on the face of it to have been properly issued will be binding and enforceable in the court whether or not the merits or the validity of the decision are challenged'. To adopt any other approach would be 'to drive a coach and horses through the Scheme'.

Matters have progressed significantly since Mr Justice Dyson wrote the words quoted above and they must now be treated with care. The courts have since made it clear that they will uphold proper jurisdictional challenges. Accordingly, a decision which is not valid because of lack of jurisdiction or breach of the rules of natural justice will not be enforced. From a practical perspective, any party wishing to object to the jurisdiction of an adjudicator should give consideration to both the timing and ground, or grounds, of their objection. Raising no objection and then participating in the adjudication can confer jurisdiction and may result in the waiver of any right to object on jurisdictional grounds. See *Brims Construction Ltd* v. *AZM Development Ltd* (2013) and *Glendalough Associated SA* v. *Harris Calnan Construction Co. Ltd* (2013). Raising specific jurisdictional objections can have the effect of waiving all other jurisdictional grounds available, but not specified, during the adjudication, see *Allied P&L Limited* v. *Paradigm* (2009). On the other hand, making a general jurisdictional objection, while it may be valid, is not without risk – much may come down to the exact wording of the objection, see *GPS Marine Contractors Ltd* v. *Ringway Infrastructure Services Ltd* (2010). It is, of course, open to parties to assert lack of jurisdiction and thereafter to take no part in adjudication proceedings, but this can result in other substantial risks.

It should also be noted that, though a successful jurisdictional challenge may result in the unenforceability of an adjudicator's decision, where the objecting party has participated in the adjudication following objection, they may still be liable for the fees and expenses of the adjudicator, notwithstanding the lack of an express contract between the adjudicator and the objecting party, see *Christopher Michael Linnett* v. *Halliwells LLP* (2009) and Section 16.8.

In the event of a challenge to jurisdiction, as a matter of practice the appropriate approach is for the adjudicator to enquire into his jurisdiction and insofar as he finds it to be the case that he has jurisdiction, he should continue with the adjudication unless and until the court orders otherwise. In the event that the adjudicator considers that he does not have jurisdiction, he should resign. See also paragraph 38(3) of Justice Ramsey's decision in *HG Construction Ltd* v. *Ashwell Homes (East Anglia) Ltd* (2007) in this regard.

Examples of commonly made jurisdictional challenges are set out below.

No construction contract between the parties

Chapter 1 considered the definition of 'construction contract' for the purposes of the 1996 Act. In *Project Consultancy Group* v. *Trustees of the Gray Trust* (1999) it was argued in defence of enforcement proceedings that the contract in question was not a 'construction contract' within the meaning of the 1996 Act as it had either (1) been entered into pre-1 May 1998 or (2) had never been entered into at all. Complicated evidence had been led before the adjudicator as to what had and had not been agreed between the parties. The adjudicator determined first that it was a construction contract and then made a monetary award.

Mr Justice Dyson distinguished this case from *Macob Civil Engineering Ltd* on the basis that a different test might require to be applied where the adjudicator purports to decide a dispute not referred to him at all, that is when he was determining his own jurisdiction. It was suggested that such an approach would enable every responding party to raise spurious jurisdictional issues, but the judge considered these fears were exaggerated. In this case it was found that there was sufficient doubt as to whether there had been offer and acceptance and therefore a contract. If there was no contract, the adjudicator could not have jurisdiction. The judge declined to award summary judgment to enforce the decision.

The question whether the adjudicator has the necessary jurisdiction is not itself a dispute arising under a construction contract. An adjudicator has no power to decide his own jurisdiction.

The abolition by the 2009 Act of the requirement for a construction contract to be in or evidenced by writing, as discussed in Section 1.2.2, is of particular importance to adjudication. This change is partly designed to prevent jurisdictional objections being made unnecessarily to impede adjudications. There is a risk, however, that considerable time and expense may now be spent ascertaining the terms of any oral or partly oral contract and what exactly is in dispute as a matter of contract.

Matter not referred to adjudicator/failure to exhaust referral

An adjudicator should exercise what jurisdiction he has and reach a decision on any matter properly referred to him, see *Ballast plc* v. *The Burrell Company (Construction Management) Ltd* (2003). An adjudicator cannot decide a matter which has not been referred to him. He must decide all the matters referred to him, see *S L Timber Systems Ltd* v. *Carillion Construction Ltd* (2002).

In *F W Cook Ltd* v. *Shimizu (UK) Ltd* (2000), four items in the final account were referred to adjudication, but no request for payment was actually set out in the Referral Notice. The adjudicator attempted to assess a monetary value on each of the four items and Cook tried to enforce the decision. The judge took a different view and said that Cook had sought to achieve a decision on the specific items with the hope that other items in the final account might be negotiated. The adjudicator had not made any decision on payment and would have exceeded his jurisdiction if he had done so. Summary judgment was refused.

An adjudicator must also not underestimate his remit. In the case of *RBG Limited* v. *SGL Carbon Fibers Limited* (2010), an action for enforcement was successfully defended on the basis that the adjudicator had failed to exhaust his jurisdiction. The Notice of Adjudication referred a dispute regarding the payment of sums set out in various invoices, plus interest. SGL argued that the adjudicator required to consider their defence, namely, that they had previously made overpayments to RBG, so that, even if the invoice amounts were due, no further payment was required to be made by them. While this was permitted by the applicable contractual payment mechanism (in this case the NEC3 Option C), the adjudicator came to a decision with regard only to the sums sought in RBG's invoices. The adjudicator declined to consider the overpayments at the earlier stage of the works, as these, he thought, would have to be dealt with in a separate adjudication. Lord Menzies considered that, even if the question of overpayments did not expressly fall within the scope of the dispute, as defined in the Notice of Adjudication, it would nonetheless fall within the scope of the adjudication. An adjudicator, it was held, must consider any relevant defence on which the respondent relies. See also *Pilon Ltd* v. *Breyer Group Ltd* (2010) and *KNN Coburn LLP* v. *GD City Holdings Ltd* (2013).

No dispute

Sindall Ltd v. *Solland Interiors Lts and Others* (2002) considered the issue of whether a dispute had arisen. Solland determined Sindall's employment in December 2000 on the grounds of failure to proceed regularly and diligently with renovation works at a property in Mayfair, a notice of default having been served. The adjudicator appointed to determine the extension of time issue decided that Sindall was entitled to an extension of time to October 2000 in respect of events to August 2000. In January 2001 Sindall sought a further award of extension of time. The contract administrator requested further information which Sindall provided with a letter stating that a formal response was required within seven days. Adjudication proceedings were commenced with Sindall seeking:

- a declaration that the determination was wrongful;
- a declaration that it was entitled to an extension of time to the date of determination.

The adjudicator found in favour of Sindall on both points and enforcement proceedings were raised. Solland argued that the adjudicator had acted without jurisdiction because, as at the date of service of the adjudication notice, there was in fact no dispute between the parties regarding the extension of time question.

His Honour Judge Lloyd decided that no dispute had arisen on the basis of Sindall's letter, enclosing further information in support of the claim for an extension of time requesting a response within seven days. The contract administrator should have been given sufficient time to make up his mind before the inference could be drawn that the absence of a substantive reply meant that there was a dispute. However, the adjudicator had jurisdiction in relation to the dispute because the dispute before him concerned the determination issue. As an integral part of this was the time within which the works should have been completed, it involved considering the extension of time claim.

The judge favoured a wide interpretation of 'dispute' to encompass all matters that are contentious between the parties at the relevant time. Specifically, the judge said that the courts should endeavour to adopt a pragmatic approach to adjudication as opposed to a legalistic approach.

In 2003, two cases appeared to indicate a shift in the attitude of the courts to the definition of dispute for the purposes of adjudication, which had until that point appeared to be different from the definition of dispute for the purposes of arbitration.

First, in the case of *Beck Peppiatt Ltd* v. *Norwest Holst Construction Ltd* (2003), the judge quoted with approval the words of His Honour Judge Lloyd in *Sindall Ltd*:

> For there to be a dispute for the purposes of exercising the statutory right to adjudication it must be clear that a point has emerged from the process of discussion or negotiation that has ended and that there is something which needs to be decided.

However, the judge decided that he did not see any conflict between this approach and the approach of the Court of Appeal in *Halki Shipping Corporation* v. *Sopex Oils Ltd* (1998) where it was said that there is a dispute once money is claimed unless and until the defendants admit that the sum is due and payable.

In *Orange EBS Ltd* v. *ABB Ltd* (2003), the judge applied the tests set out in *Beck Peppiatt Ltd.* Part of the dispute related to the final account. Orange submitted a final account on 2 December 2002, but served a notice of adjudication on 6 January 2003. Orange's contract had been terminated in July 2002, but it had taken no further steps between July and December. ABB instructed an investigator to consider the final account and suggested that they would be able to respond by 20 January and if no agreement had been reached within seven days thereafter ABB indicated that they would be willing to submit to adjudication.

ABB said that there could be no dispute because the contractual machinery under DOM/1 in relation to the time given for ABB to consider the final account had not run its course before the Notice of Adjudication was served. Orange said that the effect of repudiation was to bring the sub-contract to an end and thus the contractual mechanism for payment of sums due fell away. Applying the *Halki* test, the fact that ABB had not admitted the claim or paid meant that a dispute had arisen. Applying the *Sindall Ltd* test was more difficult. Notwithstanding the Christmas industry shutdown and the fact that ABB had made what they thought was a reasonable alternative suggestion in relation to the timetable, the judge concluded that by 6 January sufficient time had elapsed for ABB to have both evaluated the claim and to have concluded any discussions and/or negotiations with Orange.

From the point of view of a Referring Party this approach has much to commend it at a time when there had been a growing trend for potential respondents to adjudication to either try to put off the fateful day of adjudication or to set up a possible jurisdictional challenge on the basis that there was not a dispute at the time of the adjudication notice. This trend was evidenced by no positive rejection of a party's case or claim but merely asking for more information or for an allegedly reasonable time to consider it. It is always a matter of fact and degree when a dispute can be said to have arisen following the submission of a claim.

In *R G Carter Ltd* v. *Edmund Nuttall Ltd* (2002), His Honour Judge Seymour refused to enforce the decision of an adjudicator since the adjudicator had no jurisdiction. When Nuttall commenced adjudication proceedings, the notice included a claim for an extension of time based on a claim document prepared in May 2001. When the Referral Notice was served, it included a delay analysis prepared by an expert on behalf of Nuttall, which made a claim for an identical extension of time, however, the justification for the extension was different to that put forward in the May claim.

The judge was required to decide whether the dispute which had been adjudicated upon and in respect of which a decision was given, was the dispute which was the subject of the Notice of Adjudication.

The judge rejected the submission that the dispute should be identified by reference, at least principally, to what was being claimed. Nuttall suggested that it was enough that the extension of time sought was always the same and it was irrelevant that the facts and arguments relied upon in the expert report were significantly different from the facts and arguments relied upon in the previous claim. The judge considered that for there to be a dispute, there must have been an opportunity for the protagonists each to consider the position adopted by the other and to formulate arguments of a reasoned kind. He felt that some form of rejection of a party's claim was required. The judge said that 'a party can refine its arguments and abandon points not thought to be meritorious' but it cannot 'abandon wholesale facts previously advanced'. Such wholesale abandonment would prevent a party from asserting that the 'claim' or 'dispute' remained the same.

In *Fastrack Contractors Ltd* v. *Morrison Construction Ltd and Another* (2000) Morrison were the main contractors for the construction of a new leisure arena. Fastrack were the brickwork sub-contractors. The case concerned second adjudication proceedings (the first concerned interim application 12). The Notice of Adjudication provided:

> [T]he disputes to be referred are the issues as to the [Claimant's] rights to payment … under the following headings/descriptions: measured work; scaffold variations; other variations; dayworks; storm damage; prolongation costs, and loss and expense arising from delay and disruption caused to the sub-contact works by the [Defendants'] breaches of contract; a fair and reasonable extension of time for completion of the sub-contract works; loss of profit as a result of repudiation; additional overheads; such other sums as the Adjudicator deems appropriate.

The Notice of Adjudication sought payment of approximately £483,000. The Referral Notice claimed approximately £479,000.

Morrison argued that the only dispute in existence at the time the Notice of Adjudication was served was in relation to application 13 and there were significant differences between the sums claimed in that application for measured works, variations, scaffolding, preliminaries, disruption and damages.

The adjudicator decided the issues were materially the same as application 13 so there was a pre-existing dispute, and he awarded payment of approximately £120,000.

Morrison defended enforcement proceedings. His Honour Judge Thornton decided 'the dispute' was whatever claims, heads of claim, issues, contentions or causes of action were in dispute at the moment that the Referring Party first intimated an adjudication reference. All the issues in the Notice of Adjudication had been referred by Fastrack to Morrison, had been rejected by Morrison and had therefore ripened into disputes by the time the Notice of Adjudication was served.

The authorities in relation to this issue were considered in detail in *Amec Civil Engineering Ltd* v. *The Secretary of State for Transport* (2004). Lord Justice Jackson identified seven propositions regarding the constitution, or otherwise, of a dispute, namely:

1. The word 'dispute' which occurs in many arbitration clauses and also in section 108 of the [1996 Act] should be given its normal meaning. It does not have some special or unusual meaning conferred upon it by lawyers.

2. Despite the simple meaning of the word 'dispute', there has been much litigation over the years as to whether or not disputes existed in particular situations. This litigation has not generated any hard-edged legal rules as to what is or is not a dispute. However, the accumulating judicial decisions have produced helpful guidance.

3. The mere fact that one party (whom I shall call 'the claimant') notifies the other party (whom I shall call 'the respondent') of a claim does not automatically and immediately give rise to a dispute. It is clear, both as a matter of language and from judicial decisions, that a dispute does not arise unless and until it emerges that the claim is not admitted.

4. The circumstances from which it may emerge that a claim is not admitted are protean. For example, there may be an express rejection of the claim. There may be discussions between the parties from which objectively it is to be inferred that the claim is not admitted. The respondent may prevaricate, thus giving rise to the inference that he does not admit the claim. The respondent may simply remain silent for a period of time, thus giving rise to the same inference.

5. The period of time for which a respondent may remain silent before a dispute is to be inferred depends heavily upon the facts of the case and the contractual structure. Where the gist of the claim is well known and it is obviously controversial, a very short period of silence may suffice to give rise to this inference. Where the claim is notified to some agent of the respondent who has a legal duty to consider the claim independently and then give a considered response, a longer period of time may be required before it can be inferred that mere silence gives rise to a dispute.

6. If the claimant imposes upon the respondent a deadline for responding to the claim, that deadline does not have the automatic effect of curtailing what would otherwise be a reasonable time for responding. On the other hand, a stated deadline and the reasons for its imposition may be relevant factors when the court comes to consider what is a reasonable time for responding.

7. If the claim as presented by the claimant is so nebulous and ill-defined that the respondent cannot sensibly respond to it, neither silence by the respondent nor even an express non-admission is likely to give rise to a dispute for the purposes of arbitration or adjudication.

The position was developed further in *Cantillon Ltd* v. *Urvasco Ltd* (2008). Urvasco engaged Cantillon to carry out demolition, piling and other works. Disputes arose which were dealt with in adjudication. These concerned Cantillon's claims for extension of time and loss and expense. Sums were awarded to Cantillon by the adjudicator but Urvasco refused to pay, claiming, among other things, that the adjudicator had no jurisdiction in relation to Cantillon's claim for a 13-week extension of time. It was argued that Cantillon had claimed for a specific 13-week period and associated costs. Given this, Urvasco argued that the adjudicator did not have jurisdiction to allow costs for a different 13-week period.

Justice Akenhead considered the issue of how and when a dispute can arise. He considered that courts should not adopt an overly legalistic analysis of what the dispute between the parties is. It cannot be said, he noted, that this is necessarily defined or limited to the evidence or arguments submitted by either party to each other before the referral to adjudication. It was said that the responding party can put forward any defence in the adjudication, whether argued previously or not. It followed from that, therefore, that the adjudicator could rule not only on that defence but also on the ramifications of that defence in so far as it impacts upon the fundamental dispute. Where parties put forward arguments, the adjudicator could not be said to be going off on a frolic of his own if he addressed these. Accordingly, in this case, the adjudicator had considered the dispute to be related to a claim for loss and expense for 13 weeks due to a piling variation, not loss and expense for a specific 13-week period. Urvasco's defence had been that the losses claimed could not be recovered because they related to a later period. That was effectively an acceptance that there were losses and the judge found that the adjudicator could deal with that issue, it having been raised.

The *Cantillon* case was followed in *Quartzelec Limited* v. *Honeywell Control Systems Limited* (2008). Quartzelec argued that a defence raised by Honeywell during the course of the adjudication, but which had not been raised prior to the Notice of Adjudication, could not form part of the dispute which was referred to the adjudicator. His Honour Judge Davies agreed with the assessment in *Cantillon*, considering that, where the dispute referred to adjudication was one involving a claim to be paid money, it was difficult to see why a respondent should not be entitled to raise any defence open to him to defend himself against that claim, irrespective of whether the defence had been raised prior to the adjudication – subject, of course, to considerations of natural justice. It was held that, not only did the adjudicator have jurisdiction to consider such a defence, moreover, if he failed to do so, he would not properly have been performing the task he was appointed to do. He would not have been acting in accordance with natural justice, as the respondent would not have been heard on all his defences put forward.

The mere fact that correspondence issued prior to an adjudication is marked 'without prejudice' may not prevent a dispute crystallizing for the purposes of adjudication. In *RWE Npower plc* v. *Alstom Power Ltd* (2009) parties had entered into three contracts for the repair of boilers around the same time which incorporated the Scheme for Construction Contracts. Three adjudications took place with the third adjudication dealing with early claims being referred to in letters marked without prejudice. A subsequent 'open' claim was made by RWE and negotiated by the parties. Alstom attempted to argue that RWE could not prove that a dispute had crystallized because of this. The judge held that even if the earlier correspondence was privileged, when the circumstances were viewed objectively it was clear that parties were in dispute regarding the claim for extension of time.

The same or substantially the same dispute

In *Skanska Construction (UK) Ltd* v. *The ERDC Group Ltd and Another* (2003), Skanska sought to have an adjudication suspended by challenging the adjudicator's jurisdiction to hear the dispute. The adjudication was the second one brought by ERDC against Skanska, and Skanska claimed it centred on a dispute, which was 'the same or substantially the same' as the first dispute. Accordingly, Skanska said that it could not be adjudicated upon and invited the adjudicator to step down. He refused.

The first adjudication had arisen over an interim application, whereas the second concerned ERDC's final account submission. Skanska argued that both disputes concerned the quantification of the loss and expense element of ERDC's claim. ERDC argued that it was quite different to the interim valuation dispute, albeit that it did concern similar claims and sums. Since the first adjudication, significant further information and documentation had come to light and been exchanged. The second adjudication centred on different sub-contract clauses and would proceed upon a different basis.

The adjudicator and the judge agreed with ERDC's arguments. Skanska's petition was refused. The judge stated that in the second adjudication a different stage in the contract had been reached; different contractual provisions applied; considerably more information might be available by the date of issue of the final account; and different considerations and perspectives might apply.

Error in adjudicator's decision

In the case of *Bouygues (UK) Ltd* v. *Dahl-Jensen (UK) Ltd* (2001), the adjudicator dealt with cross-adjudications. Dahl-Jensen issued a Notice to Adjudicate claiming sums for additional work and the cost of delay and disruption.

Bouygues issued a Notice to Adjudicate claiming repayment of sums they claimed they had overpaid to Dahl-Jensen, liquidated damages for delayed completion and damages for costs incurred as a result of the determination. This claim was treated effectively as a counterclaim to Dahl-Jensen's claim.

The adjudicator considered both claims. In carrying out his calculations, the adjudicator took a gross sum which included 5% retention and deducted it from sums paid which were net of retention. The effect of this was to release retention before it was due to be released. The practical effect was that instead of finding the sum of £141,254 was due to Bouygues, the adjudicator awarded the sum of £207,700 to Dahl-Jensen.

Bouygues went to court to set aside the adjudicator's decision and to substitute it with an award in their favour. Neither party had made any submissions on the issue of retention. Dahl-Jensen argued that if there was a mistake, it was in the adjudicator's calculations, and not a mistake in his decision to deal with a dispute that was outside his jurisdiction.

The court decided that the adjudicator plainly made a mistake. However, he did not purport to determine that Dahl-Jensen were entitled to the release of the retention. Rather, his mistake was an arithmetical one with which the court could not interfere. Effectively, the adjudicator had answered the right question, but in the wrong way. The judge concluded:

> [T]he court should bear in mind that the speedy nature of the adjudication process means that mistakes will inevitably occur, and, in my view, it should guard against characterising a mistaken answer to an issue that lies within the scope of the reference as an excess of jurisdiction.

By comparison, in *Bloor Construction UK Ltd* v. *Bowmer Kirkland (London) Ltd* (2000), a mistake had been made by the adjudicator in his decision which was dated 9 February, and sent to the parties on 11 February. He had not taken into account in his calculations payments made to date by the main contractor. The adjudicator then realized his mistake, wrote to the parties on 11 February along with a corrected decision, still dated 9 February.

Bloor sought summary judgment. His Honour Judge Toulmin reached a practical decision that, in the absence of a specific agreement to the contrary, there will be an implied term that an adjudicator can correct, clarify or remove an error or accidental omission, provided that it is done within a reasonable time, and causes no prejudice to the other party. However, that decision appears to proceed at least in part, by way of analogy with the power given to an arbitrator in England and Wales to correct arbitration awards, see section 57 of the Arbitration Act 1996, which does not apply to Scotland.

While an arithmetical error in an adjudicator's award has been capable of being corrected under this slip rule, it can only apply in circumstances where the adjudicator accepts that there is an obvious error which he is prepared to correct. It must also be a genuine slip that failed to give effect to the adjudicator's first thoughts. An adjudicator who goes further than a mere correction by recalculating the sums due using a different method or by wholly reconsidering and redrafting substantive parts of his decision will leave parties with an unenforceable decision. See *CIB Properties Ltd* v. *Birse Construction Ltd* (2004); *YCMS Ltd (t/a Young Construction Management Services)* v. *Grabiner & Anor* (2009); *O'Donnell Developments Ltd* v. *Build Ability Ltd* (2009); and *Rok Building Ltd* v. *Celtic Composting Systems Ltd (No. 2)* (2010).

A degree of clarity has been introduced to this area by amendments to the 1996 Act. The 2009 Act added section 108(3A), which requires construction contracts now to reflect the English common law 'slip' rule, allowing adjudicators to correct clerical or typographical errors in their decisions arising by accident or omission. A new paragraph 22A has been inserted into the Scheme to a similar effect, allowing adjudicators, on their own initiative or on the request of a party, to correct clerical or typographical errors within their decisions. The Scheme additionally requires, however, that the correction is made within five days of the date upon which the adjudicator's decision is delivered to the parties. A five-day period is also stipulated by the NEC3 in its equivalent provision, see Secondary Option clause W2.12(12).

Certain issues do still arise. The contractual provision required by the 1996 Act, unlike the corresponding paragraph of the Scheme, does not contain a time limit within which an adjudicator is required to make any correction (though it was held in *YCMS* that it would be an exceptional and rare case in which a revision can be made more than a few days after the decision). Further, both the 1996 Act and the Scheme allow adjudicators to correct 'a clerical or typographical error arising by accident or omission'. However, no guidance is given as to what exactly this wording encompasses. Does it, for example, include arithmetical errors? While courts may no longer require to decide upon cases where adjudicators have made clear errors in their decisions, they may require now to determine cases in which corrections have been made by an adjudicator and the party who does not benefit as a result challenges those corrections.

Human rights

Challenges have been made based on human rights arguments. Reference is made to Section 16.12 for details on this.

16.10.3 Natural justice

The statutory requirement for adjudicators to act impartially is set out at paragraph 12 of the Scheme which provides:

> The adjudicator shall –
>
> (a) act impartially in carrying out his duties and shall do so in accordance with any relevant terms of the contract and shall reach his decision in accordance with the applicable law in relation to the contract; and
> (b) avoid incurring unnecessary expense.

While the Scheme is silent on the procedure to be adopted by the adjudicator, and there is no requirement for the Respondent to submit anything, to fulfil the requirement to act impartially, in practice, the adjudicator will ask for a response to the Referral Notice.

There is now a considerable body of case law which attaches critical importance to the need for adjudicators to act impartially and in accordance with basic rules of natural justice and fairness. However, the courts have recognized that any alleged breach of such rules has to be viewed in the context of the nature of adjudication and its tight timescales, see *Ardmore Construction Ltd* v. *Taylor Woodrow Construction Ltd* (2006) in which Lord Clarke held that:

> It is now settled law that adjudicators have to observe principles of natural justice in reaching their decisions. Nevertheless, as the case law has developed, the courts

> have taken a realistic and pragmatic approach to such questions by emphasising that the nature of the process, and in particular the strict time limits within which adjudicators are constrained to operate, require that insubstantial or technical breaches of natural justice should not be taken merely to delay or avoid payment and the taking of such points should certainly not be encouraged by the courts.

The courts have, for example, held that an adjudicator agreeing to determine a complex case with voluminous and new material in the tight timescales of the adjudication process does not necessarily amount to a breach of natural justice. It is a matter for the adjudicator in each case to decide whether he can fairly reach a decision within the timescale and the careful use of his procedural powers or seeking agreement on extensions of the timescale may be sufficient to ensure fairness. See *CIB Properties Ltd* v. *Birse Construction Ltd* (2004). This position is supported in *The Dorchester Hotel Ltd* v. *Vivid Interiors Ltd* (2009), in which it was held that, notwithstanding that 37 lever arch folders of papers and a 90-page referral had been intimated shortly before the Christmas holidays, only in the rarest of cases would the court intervene and this case was not such a case. It was, however, noted that the adjudicator's obligation to determine the matter fairly was a continuing one. See also *Bovis Lend Lease Ltd* v. *Trustees of the London Clinic* (2009).

In *Discain Project Services Ltd* v. *Opecprime Development Ltd* (2000), His Honour Judge Bowsher declined to enforce an adjudicator's award where the adjudicator had failed to consult with one party upon submissions which had been made by the other party. The judge stated that he found it 'distasteful' and could not bring himself to enforce an adjudication decision which had been arrived at in that way.

It does not follow that the courts will interpret every instance of a party to an adjudication being denied an opportunity to submit or respond as a breach of natural justice. In *GPS Marine Contractors Ltd* v. *Ringway Infrastructure Services Ltd* (2010), in addition to a jurisdictional challenge, Ringway also submitted that there had been a breach of natural justice in that the adjudicator had not taken into account an uninvited submission that they had made. It was held, however, that there was nothing 'obviously unfair' about the adjudicator's decision not to allow Ringway's further submission: '[In] the context of a rapid summary procedure leading to a temporarily binding decision, the adjudicator was entitled and needed to limit the number of rounds of submissions.'

This approach was approved in *AMEC Group Ltd* v. *Thames Water Utilities Ltd* (2010) which involved a similar set of circumstances. It was argued that the adjudicator had not properly considered a submission made just over two days before his decision was due. The court held that:

> As has been pointed out elsewhere, it is becoming very common for parties in adjudication to believe that they are in some way entitled to respond to every submission put in by the other party. In my view, unless the contract or the relevant adjudication rules expressly permit it, they do not have such an entitlement. Adjudication is not intended to resolve disputes by reference to innumerable rounds of submission or pleadings.

The power of an adjudicator to use his initiative in investigating the facts and the law must be exercised in accordance with principles of natural justice and must be read in conjunction with the adjudicator's duty to act impartially. Some adjudicators interpret their duty to act impartially as a prohibition on private meetings or conversations with one party only. It is certainly the safer course to adopt. If an adjudicator procures any information from whichever source and certainly if it might have a bearing on his decision, both parties should be informed of this so that they are given an opportunity to comment on it. Despite the pressure of time on adjudicators, with the result that it may be almost impossible to give parties the opportunity to be heard on every possible topic within the time available, it would be prudent for an adjudicator to record what information has been obtained and to pass it on to the parties in sufficient time to enable them to comment before he reaches his decision.

In *RBG Limited* v. *SGL Carbon Fibers Limited* (2011) the adjudicator had on one occasion written to the parties to advise them that he intended to use his own knowledge in relation to an aspect of the dispute and requested any responses within 24 hours, which was the morning of the day the decision was due to be issued. Although the judge ultimately decided this case on a jurisdictional argument (see Section 16.10.2), he did question what chance there would have been of the adjudicator changing his mind in light of any comments received at this late stage.

In *Woods Hardwick Ltd* v. *Chiltern Air-Conditioning Ltd* (2001), Woods Hardwick were engaged to provide architectural services at a development. Disputes arose between the parties and Woods Hardwick claimed for unpaid fees and additional work. They commenced adjudication proceedings. After two site meetings the adjudicator consulted Woods Hardwick's representatives, two of Chiltern's sub-contractors, the local authority's litigation department and RIBA's legal helpline. He did not inform Chiltern either that he had obtained information from those sources or of its content. Chiltern was not given an opportunity to comment upon the information obtained. The adjudicator issued his decision dismissing Chiltern's defence and awarding Woods Hardwick most of the sums claimed. Woods Hardwick sought to enforce the decision and submitted a detailed witness statement from the adjudicator which indicated that he had taken an adverse view of Chiltern's performance and of the representatives who appeared at the meetings. The judge dismissed the proceedings on the basis that the adjudicator acted in a manner which could be easily perceived as partial in approaching one side without informing the other, in seeking additional information from third parties and in then making adverse findings against the party left in ignorance. This was compounded by the adjudicator's voluntary provision of a particular type of witness statement. The adjudicator's witness statement effectively confirmed in the judge's mind his lack of impartiality.

The case of *Balfour Beatty Construction Ltd* v. *The Mayor and Burgess of the London Borough of Lambeth* (2002) concerned a late completion dispute. The adjudicator adopted a 'collapsed as-built' method of analysis which had not been advanced by the parties but did not present his analysis to the parties for their comment. His Honour Judge Lloyd held that the adjudicator should have invited comments on whether the as-built programme he had drawn up was a suitable basis from which to derive a retrospective critical path. He should have informed parties of the methodology that he intended to adopt, or sought observations from them as to the manner in which it or any other methodology might reasonably and properly be used in the circumstances to establish or test Balfour's case. The adjudicator should not have used his powers to make good fundamental deficiencies in the material presented by one party without first giving the other party a proper opportunity of dealing with that intention and the results. Lambeth was entitled to have the dispute decided on the material provided by Balfour Beatty, either originally or in answer to the adjudicator's requests, not on a basis devised by the adjudicator. The judge said that if an adjudicator uses his powers to find out more about the facts or to form the opinion that a different principle should be applied, he should tell the parties what he has found, and the potential implications of those findings. Constructing, or reconstructing, a party's case for it without confronting the other party is such a potentially serious breach of the requirements of both impartiality and fairness that the decision was invalid, so the enforcement application was dismissed. This case illustrates that while an adjudicator can take an inquisitorial approach to ascertaining the facts, the adjudicator must act with care and even-handedly. See also *ABB Ltd* v. *BAM Nuttall Limited* (2013).

While it may not be necessary for an adjudicator to obtain further comment from the parties in a case where he applies his own knowledge and experience to assess the factual and legal submissions advanced by the parties (see *Petition of Mr and Mrs Jack Paton for Judicial Review* (2011)), it would, however, be appropriate to make known to the parties and call for their comments in a case where the adjudicator uses his own knowledge and experience to advance or apply propositions of fact or law not canvassed by the parties.

In *Costain Ltd* v. *Strathclyde Builders Ltd* (2004), Costain raised enforcement proceedings against Strathclyde Builders, seeking payment of sums found due by an adjudicator, which were various amounts deducted as liquidated and ascertained damages. Summary decree was sought by Costain.

The adjudicator had sought and was granted an extension to the deadline for reaching his decision as he wished to discuss one point in particular with his legal adviser. Neither the terms nor the result of his discussions with the legal adviser were made known to either party.

Strathclyde Builders argued that the advice given was material to which the adjudicator would probably have attributed significance in reaching his decision. Consequently, his failure to disclose the substance of the advice, and invite comments, prior to reaching his decision was a breach of the principles of natural justice.

The judge refused to grant summary judgment. He decided that it was not clear whether the one particular matter which the adjudicator indicated he intended to discuss with his legal adviser was adequately covered by the parties' submissions. It was immaterial that no actual prejudice was demonstrated, the mere possibility of prejudice was sufficient. The judge stated that if confidence in the system of adjudication was to be maintained it was important that adjudicators' decisions should be free from any suspicion of unfairness. It was not an answer to say that an adjudicator's decision could be re-opened at the end of the contract by arbitration or litigation. Basic standards of fairness should be applied to adjudicators and vigorously enforced.

In *Highlands and Islands Airports Ltd* v. *Shetland Islands Council* (2012), the adjudicator sought an informal view from Senior Counsel in a brief telephone conversation on the interpretation to be given to a clause within the parties' contract. No formal legal opinion or advice was sought and no fee was charged. The adjudicator chose not to disclose this to the parties prior to his decision being issued. In fact the existence of this informal advice only came to light by chance after the adjudication. During the enforcement proceedings Lord Menzies held that, notwithstanding the accepted facts that the adjudicator had already formed his own view on the issue prior to seeking confirmation from Counsel and that this was an informal request for advice and that no fee had been paid for it, this was legal advice and, given that the issue upon which advice was sought was central to the matter before the adjudicator, his failure to disclose the request and subsequent advice to the parties with a request for their comments amounted to a breach of natural justice.

It is quite common for adjudicators to seek legal or other specialist advice. The position is now clear. If matters are raised, parties must be given the opportunity to comment. Even if new matters are not raised, parties should still be informed of the content of the discussions. Failing which, a breach of natural justice may be held to be established and the courts will not enforce the adjudicator's decision.

The adjudicator's decision in *Jacques (t/a C&E Jacques Partnership)* v. *Ensign Contractors Ltd* (2009) to ignore a previous null and void decision was held not to be a breach of natural justice. It was held that doing so could not be said to be irrational or perverse and the adjudicator had not failed in any material sense to apply the rules of natural justice. His decision was enforced.

Considering a previous adjudicator's decision has been held not to amount to a breach of natural justice, see *Arcadi's UK Limited* v. *May and Baker Limited t/a Sanofi* (2013).

In the Scottish case of *Atholl Developments (Slackbuie) Ltd Re Application for Judicial Review* (2010), the Court of Session has given helpful guidance on the practical application to adjudication of the principles of natural justice. Lord Glennie summarized the main principles to be applied:

(i) Decisions of adjudicators are to be enforced unless there is good reason to refuse enforcement. It is no defence to enforcement, and no ground for reduction, to say that the adjudicator had erred in fact or in law or, unless it resulted in manifest unfairness, in procedure.
(ii) Where there has been a breach of natural justice, the court will interfere, but it will only do so in the plainest of cases.
(iii) The nature of the process means that the court will not be overly critical of the reasoning put forward by the adjudicator for his decision. It must be intelligible, and it must show that he had considered the issues before him and reached his decision on those issues for reasons which are explained in his decision (see also *Whyte & MacKay* v. *Blyth and Blyth Consulting Engineering Limited* (2013)). But the reasons need not be explained in great detail, nor need he refer to each document or each submission put before him (see also *PIHL UK Limited* v. *Ramboll UK Limited* (2012)).
(iv) There is a presumption of regularity or of propriety (see *RBG Limited* v. *SGL Carbon Fibers Limited* (2010) above). In other words, it will be assumed, unless the contrary is shown, that the adjudicator has looked at all the relevant

materials and given to them such consideration as he considers practicable (having regard to the pressure of time) or appropriate.

(v) The adjudicator is not to be criticised if the scrutiny given by him to a document which comes in at a late stage of the adjudication is less thorough than might have been the case had the document been part of the original submissions to him, see *AMEC Group Ltd* v. *Thames Water Utilities Ltd* (2010).

16.10.4 Adjudicator's bias

Enforcement of an adjudicator's decision can also be resisted on the grounds of bias. In *Volker* v. *Holystone Contracts Ltd* (2010), Holystone argued that the adjudicator's decision was unenforceable because, among other things, during the adjudication proceedings, Volker had wrongly made the adjudicator aware of a without prejudice offer of settlement. This, Holystone argued, resulted in the adjudicator being biased against them. Justice Coulson rejected this argument. The without prejudice offer had been made at a meeting to which both parties had referred in the adjudication. The court applied the test in the Court of Appeal case of *Re Medicaments and Related Classes of Goods (No. 2)* (2001), namely, whether those circumstances would lead a fair-minded and informed observer to conclude that there was a real possibility, or a real danger, of bias.

The court found that there was no doubt that a fair-minded and informed observer would not reach such a conclusion in the current case and any suggestion to the contrary was considered entirely unrealistic. The adjudicator made clear in his decision that he was indifferent to the fact that an offer had been made, and that the bulk of his decision had already been completed by the time he was told about it.

In response to alleged unconscious bias, it was held that in any construction dispute it would in fact be expected that negotiations had occurred. This was particularly so where questions of liability had been dealt with and the only remaining question was *quantum* (as in the *Volker* case). Reference was also made to old arbitration cases, in which the arbitrator would be made aware of the existence of a sealed envelope containing a without prejudice offer. There was no suggestion there that the arbitrator was biased due to his knowledge of an offer having been made.

In an earlier case, *Specialist Ceiling Services Northern Limited* v. *ZVI Construction* (2004), the adjudicator had been made aware of the existence of a without prejudice offer to settle and the fact that a breakdown accompanied this offer (though he did not see the breakdown). In that case, unlike in *Volker*, liability was still in dispute. The adjudicator declined to resign, as he did not consider that his impartiality was affected by this knowledge. He expected without prejudice negotiations to be taking place and noted that, in his experience, offers were often made for commercial reasons to avoid the need for adjudication and were not necessarily due to any admission that sums were in fact considered due. In this case, it was found that there was no indication of bias or unfairness in the adjudicator's decision and the decision was therefore enforced.

A different issue arose in *Fileturn Ltd* v. *Royal Garden Hotel Ltd* (2010). An application for enforcement of an adjudicator's decision was resisted on the basis that the adjudicator had been a director of the firm of claim consultants who represented Fileturn in the adjudication. The court held, however, that the number of occasions on which the adjudicator had acted when his previous firm was involved was only a small proportion of his practice. He therefore did not depend on that business. The adjudicator had had no interest in his previous firm since 2004. The court saw no difficulty with representatives of parties being well known to the decision-maker, whether that be a judge or an adjudicator. In specialist courts, like the Technology and Construction Court in England and Wales, this is a frequent occurrence. Adjudicators are professional people with their own codes of conduct and that should not present a problem. Taking this into account, it was considered 'inherently unlikely' that a fair-minded and informed observer would conclude that the adjudicator's involvement with the firm six years previously would give rise to bias in this instance. In fact, the court considered any assertion to the contrary to be 'fanciful speculation'.

In *Makers UK Limited* v. *The Mayor and Burgesses of the London Borough of Camden* (2008), the allegation was of apparent bias, due to a telephone call having taken place with the adjudicator to check his availability to act. An application was then made to the RIBA, suggesting the adjudicator be appointed if available. The court found no apparent bias in this case, but did suggest that, as practical guidance, parties and adjudicators should limit unilateral contact before, during and after an adjudication, due to the potential of any such contact being misconstrued. If there is to be any such contact, it is better to be in writing rather than verbal.

In *AMEC Capital Projects* v. *Whitefriars City Estates* (2004), the allegation was again one of apparent bias, this time based on the adjudicator having previously made a decision on the same issue. This argument was not upheld and it was said that there needed to be something of substance to lead to the conclusion that there is bias. To succeed in relation to bias will require clear factual evidence supporting the allegation, as well as a clear apprehension, on reasonable grounds, of potential bias.

An adjudicator having been appointed in an adjudication some three years previously involving the referring party and the fact that he conducted a mediation with the referring party days before his appointment in an adjudication was also held not to be sufficient evidence of bias where he had no personal knowledge of either party, was not selected by either party and had no connection to the subject matter of the dispute, see *Andrew Wallace Ltd* v. *Jeff Noon* (2009).

16.10.5 Severability of adjudicators' awards

Partial enforcement of an adjudicator's decision has been held to be competent; see *Homer Burgess Limited* v. *Chirex Limited* (2000) where Lord MacFadyen stated that:

> The alternative to my granting reduction was for me to hear submissions identifying that part of the adjudicator's decision that was within his jurisdiction, and enforce it to that extent only, by granting decree for payment in the pursuers' favour restricted to the sum reflecting the *intra vires* part of the decision. In my view either of the two suggested courses would be competent.

It has also been recognized in Scotland that the policy of encouraging the speedy provisional resolution of construction disputes might support this approach in particular cases where, for example, the adjudicator has fallen into error in relation to *quantum* alone and that the fact that parties had contracted for a decision by an adjudicator did not prevent the court severing parts of his decision if he has determined separate disputes in the one decision. See *Carillion Utility Services Limited* v. *SP Power Services Limited* (2011).

Although there is English authority which suggests that it is not possible to sever part of an adjudicator's decisions in cases where there was only one dispute referred (see *Cantillon Ltd* v. *Urvasco Ltd* (2008) and *Quartzelec Ltd* v. *Honeywell Control Systems Ltd* (2008)), it appears that judicial opinion on this matter is not settled and may be amenable to change, see *Pilon Ltd* v. *Breyer Group PLC* (2010) and *AMEC Group Ltd* v. *Thames Water Utilities Ltd* (2010).

16.10.6 Set-off against other adjudicator's decisions

The vast majority of the cases dealing with set-off against an adjudicator's decision concern the situation where an adjudicator has awarded one side a sum of money and the other party seeks to set off a separate claim for damages or delay. The general rule in such cases is that the attempted set-off is viewed as an attempt to frustrate the 1996 Act and will, in the absence of a valid withholding (now pay less) notice, ordinarily be struck at by the courts. See *Allied London & Scottish Properties plc* v. *Riverbrae Construction Ltd* (1999); *VHE Construction plc* v. *RBSTB Trust Co Ltd* (2001); and *Ferson Contractors Ltd* v. *Levolux AT Ltd* (2003); and more recently *Windglass Windows Ltd* v. *Capital Skyline Construction Ltd & Anor* (2009).

That is the case even if the claim is in the process of being prosecuted in an adjudication or even if a subsequent adjudication award has been issued but not yet become due. See *Interserve Industrial Services Ltd* v. *Cleveland Bridge UK Ltd* (2006) in which the court stated that:

> Where the parties to a construction contract engage in successive adjudications, each focused upon the parties' current rights and remedies … the correct approach is as follows. At the end of each adjudication, absent special circumstances, the losing party must comply with the adjudicator's decision. He cannot withhold payment on the ground of his anticipated recovery in a future adjudication based upon different issues. I reach this conclusion both from the express terms of the Act, and also from the line of authority referred to earlier in this judgment.

This position was approved in *YCMS Ltd (t/a Young Construction Management Services)* v. *Grabiner* & *Anor* (2009) where the court stated that:

> [The] Courts have from 1998 onwards taken the view that Adjudicators' Decisions are to be enforced summarily and expeditiously unless there is a valid jurisdictional or natural justice ground which renders enforcement inappropriate. There is,

> perhaps unfortunately, nothing in the HGCRA which legislates for setting off one adjudicator's decision against another. It is in those circumstances that the dictum of Jackson J in the Interserve case is so apposite … I see no good reason to depart from the approach adumbrated by Jackson J in the Interserve case. I do not consider that the fact that a Third Decision has been reached which on its face allows to the Defendants a net recovery is a special circumstance which justifies departing from the general rule that valid adjudicators' decisions should be enforced promptly.

The court in YCMS did, however, identify that things might be different if there were effectively simultaneous adjudications and decisions. That position was considered further in *HS Works Ltd* v. *Enterprise Managed Services Ltd* (2009), where Justice Akenhead identified the steps which required to be considered before it would be permissible to set off one decision against another:

> (a) First, it is necessary to determine at the time when the Court is considering the issue whether both decisions are valid; if not or if it cannot be determined whether each is valid, it is unnecessary to consider the next steps. (b) If both are valid, it is then necessary to consider if both are capable of being enforced or given effect to; if one or other is not so capable, the question of set off does not arise. (c) If it is clear that both are so capable, the Court should enforce or give effect to them both, provided that separate proceedings have been brought by each party to enforce each decision. The Court has no reason to favour one side or the other if each has a valid and enforceable decision in its favour. (d) How each decision is enforced is a matter for the Court. It may be wholly inappropriate to permit a set off of a second financial decision as such in circumstances where the first decision was predicated upon a basis that there could be no set off.

On the facts of that case Justice Akenhead determined that both decisions were enforceable, that the parties and the court were required to give effect to both and that the net effect of these decisions should be reflected in his order. The same result was held to apply in *JPA Design and Build Ltd* v. *Sentosa (UK) Ltd* (2009).

Attempts to suggest that set-off should be excluded where one of the decisions was an arbitration award and that such decisions 'trump' an interim decision of an adjudicator have been rejected. It was held in *Workspace Management Ltd* v. *YJL London Ltd* (2009) that both decisions were binding and as such neither had a greater status than the other.

Two exceptions to the general rule that set-off cannot be relied upon by the 'losing' party to prevent or reduce enforcement of an adjudicator's award have been identified. The first is where the express terms of the contract permit it; the second is where the nature of the adjudicator's decision permits it. See most recently in *Beck Interiors Ltd* v. *Classic Decorative Finishing Ltd* (2012).

In relation to the former exception, it has been held that though matters will always turn on the precise terms of the set-off provisions, it would require very clear terms to permit set-off against a sum otherwise due. As a result, it is thought that in practice it would be very rare for such provisions to defeat an adjudicator's award, see *Squibb Group Ltd* v. *Vertase FLI Ltd* (2012).

In relation to the latter exception, there have been instances where set-off has been permitted where the adjudicator's decision simply determines the value of a sum due in line with the contract machinery or where the decision instead addresses the operation of the contract machinery.

In *Shimizu Europe Ltd* v. *LBJ Fabrications Ltd* (2002), the adjudicator ordered that an invoice in a particular sum should be sent by the claiming party to the responding party, in order that the contractual machinery for the payment of the invoice could then be operated. The judge held that the adjudicator's decision recognized that the sum was not yet due in accordance with the contract mechanism, and that the most that could be said was that the adjudicator had decided what was due 'and, as it were, plugged the amount into the contractual mechanism'. As a result, because that contractual mechanism allowed the service of a withholding notice before the sum became finally due, the judge found that the withholding notice in that case provided a defence to LBJ's claim for summary judgment.

In *Conor Engineering Ltd* v. *Les Constructions Industrielles de la Mediterranée* (2004), it was held that the adjudicator's decision amounted to a declaration as to when final payment was due under the contract, as opposed to being an order for payment. A withholding notice served under the contractual mechanism could therefore amount to a defence to enforcement.

In *R&C Electrical Engineers Ltd* v. *Shaylor Construction Ltd* (2012), it was held that the adjudicator's decision that a sum was payable expressly in accordance with a particular clause of the sub-contract permitted the unsuccessful party to rely on that clause and other related payment provisions, so as to raise a withholding notice and a counterclaim which had not been considered on its merits by the adjudicator.

See also *Thameside Construction Company Limited* v. *Mr and Mrs Stevens* (2013) where the court provided a useful summary of the relevant legal principles concerning set-off against adjudicator's decisions.

16.10.7 Adjudication awards in insolvency

In the case of *Integrated Building Services Engineering Consultants Limited (trading as Operon)* v. *PIHL UK Limited* (2010), the Court of Session considered whether the principle of 'balancing accounts' in insolvency was a defence to the enforcement of an adjudicator's decision. In December 2007, PIHL entered into a design and build sub-contract with Operon for mechanical and electrical works at two secondary schools in Aberdeen. Operon commenced adjudication proceedings against PIHL for payment of outstanding sums claimed by them. The adjudicator held that payment was due to Operon, but PIHL refused to pay. On 20 January 2010, Operon initiated an enforcement action against PIHL in the Court of Session. On 29 January, however, Operon went into administration. PIHL's defence in the enforcement proceedings was that they were entitled to refuse to make payment on the principle of balancing accounts in insolvency. In other words, as Operon were insolvent, the court should not order PIHL to pay until PIHL's counterclaims against Operon were determined. The court held that it was open to PIHL to argue the principle of balancing of accounts in insolvency as a defence to enforcement and the court therefore refused to enforce the adjudicator's decisions.

In *Richard Heis & Others as Joint Administrators of Connaught Partnerships v Perth & Kinross Council* (2013), Connaught had gone into administration in 2010 owing about £160 million to unsecured creditors. It was anticipated only about £60,000 would be available to distribute to creditors. In 2011, Connaught was awarded £835,000 in an adjudication against Perth & Kinross Council. Given Connaught's financial position, payment by the Council would amount to a final resolution of the case since there would be no prospect of them being repaid in subsequent proceedings.

Lord Malcolm considered that the adjudication regime was not intended to transfer the risk of insolvency to either party. Its purpose was to decide which party should hold money pending final resolution, on the basis both parties were solvent. He referred to the English solution that, unless the claimant's finances were in much the same state as at the date of the contract or had been made precarious because of the defendant's refusal to comply with the award, and if the claimant was insolvent, a stay of execution would usually be granted.

The provisional nature of the adjudicator's decision was important as it was a process designed to facilitate cash flow. The courts are entitled to use the equitable principle of balancing of accounts to ensure fairness. Given the circumstances of this case, enforcement was not awarded as it would have amounted to a final resolution. The court also rejected submissions that a defence of balancing accounts on insolvency requires to be raised before the adjudicator to allow it to be relied upon in enforcement proceedings.

In *J & A Construction (Scotland) Limited* v. *Windex Limited* (2013), Windex defended an action for enforcement of an adjudicator's award on the basis that the latest accounts of J & A Construction showed an excess of liabilities over assets. This, Windex argued, would amount to grounds for winding up of the company as it evidenced that they were insolvent or verging on insolvency. They said the equitable principle of balancing of accounts in bankruptcy should be applied to prevent enforcement or, at the least, that there should be an enquiry into J & A's financial position.

Lord Malcolm made reference to Lord MacFadyen's ruling in *S L Timber Systems Ltd* v. *Carillion Construction Ltd* (2002) where a defence to enforcement based on allegations of insolvency failed and to Lord Hodge's opinion in *Integrated Building Services* v. *Pihl* in which it was said that the principle of balancing of accounts would be 'very difficult or almost impossible' to operate when an insolvency was not demonstrated by a formal legal act. These previous judgments indicated the court should be wary about refusing enforcement in the absence of 'clear or uncontested evidence of insolvency'.

The starting point was said to be to assess whether a company was commercially able to pay its debts given that a company could be creditworthy and entitled to trade even if, on paper, its liabilities exceeded its assets. All relevant circumstances were to be considered. A balance sheet deficiency should not necessarily prevent enforcement of an adjudicator's decision. Unless Windex could demonstrate there was undisputed insolvency, then the policy of the adjudication regime pointed to immediate enforcement of the award.

16.11 Proceedings following adjudication

It is important to know what matters have been referred to and decided upon by an adjudicator where there are time bar provisions in relation to proceeding with litigation or arbitration for a final decision, see *Castle Inns (Stirling) Ltd* v. *Clark Contracts Ltd* (2005).

In cases where proceedings are raised to finally determine matters, it will not ordinarily be competent to seek recovery in those proceedings of the fees and outlays of the adjudicator where it is contended that the adjudicator was in error, see *Castle Inns (Stirling) Ltd* v. *Clark Contracts Ltd* (2005). However, it should be noted that that case does not deal with the situation where the decision of the adjudicator is reduced or declared void and unenforceable because of lack of jurisdiction or breach of the rules of natural justice.

In *City Inn Ltd* v. *Shepherd Construction Ltd* (2002), the onus of proof following the adjudication was considered in the context of a dispute concerning the interpretation of a clause which required the contractor to notify the architect, on receipt of any instructions, of the effect on the contract sum, the completion date and any anticipated loss and expense. It was held that an adjudicator's decision in relation to an extension of time did not affect the onus of proof in subsequent litigation or arbitration. This remained with the party claiming the entitlement to the full extension of time to establish its entitlement as a whole, not just that over and above that awarded by the adjudicator. The judge stated:

> As has been observed in a number of cases, the function of adjudication, as contemplated in the 1996 Act, is to provide a speedy means of reaching a binding interim determination of disputes arising under construction contracts. It goes no further than that ... It is, in my view, no part of the function of an adjudicator's decision to reverse the onus of proof in an arbitration or litigation to which the parties require to resort to obtain a final determination of the dispute between them.

16.12 Human rights

Another consideration is the implication of the Human Rights Act 1998. Does this open up an avenue for challenges to adjudicators' decisions? In *Elanay Contracts Ltd* v. *The Vestry* (2001), the judge considered the effects of the European Convention on Human Rights. Article 6 of the Convention states:

> In the determination of his civil rights and obligations ... everyone is entitled to a fair and public hearing within a reasonable time by an independent and impartial tribunal established by law. The judgment shall be pronounced publicly.

The Vestry argued that enforcement should not be ordered as the adjudication procedure was unfair. This was partly due to the fact that for much of the time the defendant's key witness was attending hospital, visiting his dying mother. The shortness of the proceedings, which were over in 35 days, added to the sense of unfairness.

The judge took a different view. He considered that Article 6 did not apply to statutory adjudication. The reason given was that while proceedings before an adjudicator determine questions of civil rights, they are not in any sense a final determination.

In the case of *Austin Hall Building Ltd* v. *Buckland Securities Ltd* (2001), it was again decided that neither the Human Rights 1998 Act nor the Convention applied. The challenge was that the adjudication system set up by the statute was itself inherently unfair because it did not give sufficient time for either or both parties to present their case; and there was no public hearing and pronouncement of the decision.

A challenge on the basis of human rights arguments was more recently taken in the case of *Whyte & MacKay* v. *Blyth and Blyth Consulting Engineering Limited* (2013). Whyte & Mackay had employed Blyth & Blyth consulting engineers to design the structure for a new bottling plant. W&M took a claim related to alleged defects to adjudication and obtained an award of almost £3m. Blyth & Blyth refused to pay on three grounds, two relating to human rights:

Article 1 First Protocol (A1P1) – Article 1 provides that a party has a right to peaceful enjoyment of their possessions. It was argued that enforcement would not promote any of the legitimate aims and purposes of the 1996 Act in that there was no issue of ensuring cash flow during the progress of the works and no need for an interim or provisional award. Unusually, it would be possible to litigate the whole case to a conclusion many years before Whyte & Mackay would sustain any loss for remedial works. Even if a legitimate aim could be established, it was said that an unfair burden would be placed on Blyth & Blyth if there was enforcement. Further, Blyth & Blyth would have no security for repayment after a final determination.

Article 6(1) – It was argued that proceedings before an adjudicator did not comply with Article 6 on the basis that the hearing and judgment were not public, adjudication subordinates the correct outcome for a quick result which can result in injustice and enforcement would require immediate payment of a substantial sum to Blyth & Blyth's prejudice in the event of Whyte & Mackay becoming insolvent.

The case ultimately succeeded on the basis of a breach of natural justice. However, in relation to the A1P1 argument, it was said that in this case there was no general or public interest served by Whyte & Mackay taking the case to adjudication, in particular because the losses would not be incurred for many years into the future. There was no need to have a quick answer on the dispute. The alleged defects were not affecting use of the building. Further if Blyth & Blyth were ultimately successful in defeating the claim, there was no guarantee they would be able to recoup the money. None of the public interest justifications for adjudication in the 1996 Act applied in this case. Even if W&M had seen an advantage in dealing with the issue in adjudication, the private interests of one party cannot justify interference in the other's A1P1 rights.

While competent, it was considered adjudication was 'unnecessary and inappropriate' in this case and enforcement of the award would place an 'unfair and excessive burden' on Blyth & Blyth.

As in *Elanay Contracts*, the Article 6(1) argument was rejected on the basis that it was inapplicable, relating to determination of rights where adjudication does not amount to a final determination.

The case was decided on an unusual set of facts so is likely to be distinguishable in future cases but it may open the door to some extent to future challenges on the basis of human rights arguments.

16.13 Adjudication in PPP contracts

As a result of the complex interaction of the contracts involved in a PFI/PPP scheme and the pass down of risk from the public sector procurer through the Project Company to the Building Contractor and Facilities Management Contractor, the dispute resolution provisions within PFI/PPP project documents are necessarily more complex than in a conventional construction contract.

The further complication is that PFI contracts are excluded from the statutory adjudication regime under the 1996 Act by means of the Construction Contracts Exclusion Orders, but that only applies to the Project Agreement and not ancillary contracts which fall within the definition of 'construction contract' for the purposes of the Act, such as the Building Contract and Facilities Management Contract. It will be important, therefore, for those acting for the Project Company to seek to agree a dispute resolution procedure which is compliant with the 1996 Act so that it can be passed down into the principal sub-contracts with minimal amendment.

In most PFI/PPP contract documents, the relevant procedures for appointment of the adjudicator and rules governing the conduct of the adjudication are set out along with procedures designed to promote consistency of decision-making at Project Agreement, Building Contract and Facilities Management Contract level.

There are three alternative contractual mechanisms which are commonly used to achieve this consistency, namely (1) conjoining of disputes under related contracts (i.e. the Project Agreement and the sub-contracts) which have already been referred to adjudication and which deal with the same issue; (2) joining of a third party under a related contract to a dispute which has already been referred to adjudication; and (3) use of the same adjudicator for disputes under related contracts on the same issue.

16.13.1 Conjoining

If this procedure is used, the contract will provide that if any dispute referred to adjudication is, in the opinion of the Project Company, the same as or connected with a dispute arising out of a related contract, the Project Company may request the adjudicator to conjoin the related dispute with the original dispute.

A disadvantage of this process is that, being a three-party procedure, it is inevitably more complex, expensive and cumbersome than a bilateral procedure. The procedure may be useful for disputes on interpretation of similar contractual provisions in the related contracts where a consistent decision is required, as opposed to extension of time, compensation or variation claims. In the latter types of claim, the Project

Company will usually ensure that the sub-contracts contain 'pay when certified' clauses, which are permitted in first-tier PFI subcontracts (see Section 8.1.5) under the Construction Contracts (Scotland) Exclusion Order 2011, and which provide the Project Company with a measure of protection in avoiding mismatching financial assessments in respect of claims by it under the Project Agreement and against it under the sub-contracts.

16.13.2 Third party procedure

Broadly, this procedure is intended to operate in circumstances where the Project Company claims that it would have, in respect of the subject matter of the dispute, a right of relief, contribution or indemnity against a third party (i.e. the public sector authority or a sub-contractor) under any related contract.

This procedure brings with it similar considerations to the conjoining procedure above in terms of timescale, cost and complexity.

16.13.3 Same adjudicator

This mechanism provides for the same adjudicator to be appointed to deal with disputes under the sub-contracts and the Project Agreement. The adjudications remain separate, unlike the other two procedures mentioned above, but if parties manage to make the timescales run close together, it is possible for such claims in practice to run concurrently. This would allow consistency of decision-making where that is necessary, and without the complexities and strict timetables involved in the other two forms of procedure. There could, however, be practical difficulties where there was a substantial gap in the timings of the separate adjudications. In *AMEC Capital Projects v. Whitefriars City Estates* (2004), the court rejected an argument that this procedure could lead to bias on the part of the adjudicator (see Section 16.10.5).

Chapter 17
Arbitration

17.1 Introduction

It seems likely that arbitration was known in Scotland before the establishment of the public courts. Its introduction has been described as contemporaneous with the foundation of Scots law, see *MacCallum* v. *Lawrie* (1810).

Recourse to arbitration continued to be frequent, as in most developed countries, even after regular courts of law had been established, particularly in mercantile matters and in a wide range of cases where the questions at issue were best suited to determination by a person with requisite skill or experience, or where it was hoped to avoid the delay, expense and publicity of procedure in the courts of law.

In the past, arbitration has been the traditional means of obtaining a final decision on disputes in the construction industry. This is because arbitration procedure can be well suited to the types of dispute that arise under construction contracts. This has been reflected in standard form contracts, such as those produced by JCT, the SBCC and ICE, which have required disputes to be resolved by arbitration. However, the absence of modern and comprehensive arbitration legislation for Scotland was, until recently, considered to be one of the reasons why fewer and fewer disputes were being referred to arbitration in Scotland. Arbitration law in Scotland was derived primarily from case law rather than being codified in statute. Accordingly, the position was not readily clear or accessible with gaps and difficulties in establishing what the law actually was. This was in contrast with many other jurisdictions where clear statutory frameworks were in place. Although Scotland did have some sets of rules regarding arbitration, sponsored by interested organizations, these were without statutory backing and thus did not have adequate force. Furthermore, while the Scottish courts have supported arbitration, their role in the past has not always been clear. In the absence of a clear statutory framework, arbitration was considered to be more time-consuming and expensive than it needed to be. As a result, parties had grown accustomed to referring disputes to the Commercial Court of the Court of Session for final determination. The Arbitration (Scotland) Act 2010 ('the 2010 Act') was introduced to address these concerns. A full discussion of the provisions of the 2010 Act is beyond the scope of this Chapter but a brief outline is given in Section 17.2.

MacRoberts on Scottish Construction Contracts, Third Edition. MacRoberts.

17.1.1 Definition of arbitration

Arbitration has been defined as the method of procedure by which parties who are in dispute with each other agree to submit their dispute to the decision of one or more persons, traditionally in Scotland described as 'arbiters', rather than resort to the courts of law. Elsewhere the generally used term is 'arbitrator' and this is the term now used in the 2010 Act.

17.1.2 Elements of ordinary arbitration

The essential elements of an ordinary arbitration are: two or more parties; a dispute or question; resolved by this method under an agreement entered into voluntarily to refer to a third party, the arbitrator; the arbitrator's jurisdiction is limited by the terms of the reference and he is subject to the supervision of the ordinary courts; and the arbitrator must decide the dispute or question submitted to him by means of one or more decrees arbitral (awards). It will thus be apparent that, in the ordinary case, arbitration arises out of contract:

> The law of Scotland has, from the earliest time, permitted private parties to exclude the merits of any dispute between them from the consideration of the court by simply naming their arbiter ... It deprives the Court of jurisdiction to enquire into and decide the merits of the case, while it leaves the Court free to entertain the suit, and pronounce a decree in conformity with the award of the arbiter. Should the arbitration from any cause prove abortive the full jurisdiction of the Court will revive, to the effect of enabling it to hear and determine the action upon its merits. See *Hamlyn & Co.* v. *Talisker Distillery Co.* (1894).

17.1.3 Contractual nature of arbitration

The decision whether or not to choose arbitration over court proceedings is entirely one for the parties and the procedure is, generally, a matter which may be dealt with contractually between the parties. They may agree at the outset of their relationship that should any disputes arise they will refer them to arbitration; they may agree when a dispute arises that it should be resolved by arbitration; or, they may agree after the commencement of a court action that they would rather proceed by way of arbitration. However, once the parties have reached agreement to refer certain disputes to arbitration then either party may enforce this agreement and a Scottish court normally has no discretion but to sist (suspend) the court action for arbitration, see *Sanderson* v. *Armour* (1922).

17.1.4 Arbitrable issues

Every matter may be made the subject of arbitration with regard to which the parties have a dispute and over which they possess a sufficient power of disposal. This general statement includes matters arising out of many different types of contract including contracts relating to land and structures erected on land (heritable property) including building contracts.

The right to arbitrate may be lost if the court process is used in a way which evidences an intention not to arbitrate. Such actions may found a plea of waiver, see *Inverclyde Mearns Housing Society* v. *Lawrence Construction Co. Ltd* (1989).

Only questions which the parties could, if they wished, determine for themselves by a legally binding contract may be submitted to ordinary arbitration. Certain questions may not be referred to arbitration, for example, matters in which the public have an interest such as the status of parties (paternity, legitimacy, marriage, divorce and domicile). There cannot be an arbitration on whether someone has committed a crime. Further, certain transactions are illegal and are not enforceable and therefore cannot be referred to arbitration.

It has been held that an arbitrator may consider the question of fraud in any case where the incidental determination of this point is necessary in order to settle the real question at issue, see *Earl of Kintore* v. *Union Bank of Scotland* (1863). The question of whether or not there is an arbitrable dispute which must be referred to arbitration arises with great frequency out of many different contracts and has produced a large number of cases.

17.1.5 Scope of arbitration

The scope of the dispute referred to arbitration will depend upon what the parties have actually agreed. Thus an arbitrator will have jurisdiction and power only in relation to those specific matters or questions which the parties have agreed to submit to him. It can be seen therefore that a court may still have jurisdiction in a dispute where certain questions have been referred to arbitration but the determination of those questions does not resolve the dispute. In such circumstances, the court may still pronounce a decree in the action provided its judgment is in line with the awards of any arbitrator, see *Hamlyn & Co.* v. *Talisker Distillery Co.* (1894). Such situations arise frequently in civil actions and it is common for courts to sist actions until questions referred to arbitration are resolved.

17.1.6 Absence of codified statutory framework for control of arbitration

Unlike that of other countries, including England and Wales, the law of arbitration in Scotland was, until recently, based almost entirely on the common law developed over the centuries since the 25th Act of the Articles of Regulations 1695 and only marginally affected by statute. The most important statutory provisions were:

- Section 2 of the Arbitration (Scotland) Act 1894 – which gave the Court of Session and Sheriff Court the power to appoint an arbitrator should the parties fail to agree on a particular person.
- Section 3 of the Administration of Justice (Scotland) Act 1972 – which gave an arbitrator the power to state a case, on the application of either party, on a point

of law, to the Court of Session in order to obtain its opinion. While the arbitrator had a discretion to decide whether or not to state a case, if he refused to do so he had to provide a statement of his reasons and the Court of Session could still direct that he state a case for its opinion. The power to state a case could be excluded by the parties in which event the only remedy open to a party dissatisfied with the outcome was to seek judicial review or reduction of the award but these remedies were only available in very limited and unusual circumstances.

- Section 66 of the Law Reform (Miscellaneous Provisions) (Scotland) Act 1990 – effectively adopted the UNCITRAL Model Law on International Commercial Arbitration for international arbitrations held in Scotland.

The lack of a codifying statute setting out a modern and comprehensive Scots law on arbitration was the subject of much controversy. Many of the recognized deficiencies in the Scots law of arbitration, including the absence of a common law power to award damages or to award interest before the date of decree arbitral, had to be addressed by practical means. For example, in many cases arbitrators were given appropriate powers by the parties: (a) in the arbitration clause itself, (b) in the Deed of Appointment or Submission or (c) by adopting a set of Arbitration Rules such as the Scottish Arbitration Code which was promulgated by the Scottish Council for International Arbitration, SBCC and the Chartered Institute of Arbitrators (Scottish Branch).

17.2 The Arbitration (Scotland) Act 2010

As indicated above, the problems caused by the lack of a modern statutory framework for arbitration were addressed by the 2010 Act. A full examination of the 2010 Act is beyond the scope of this chapter. However, some of the key provisions of the 2010 Act are discussed below.

The 2010 Act seeks to be a self-standing piece of legislation to be used by those who wish to adopt arbitration as a dispute resolution mechanism for final and binding decisions. It endeavours to provide answers to the vast majority of questions likely to arise throughout the arbitration process. The 2010 Act provided a model law for Scotland which is generally in line with or surpasses international standards and aims to capture the best of international practice. The 2010 Act is designed to be applicable to both domestic and international arbitrations.

In relation to arbitrations which have already commenced or arbitration agreements which have been entered into and in connection with which the parties have agreed the 2010 Act should not apply, the common law position, or any rules which the parties have agreed should apply to the arbitration, will apply. Readers are referred to the Second Edition of this book which sets out the pre 2010 Act position in more detail.

It is hoped that the 2010 Act will encourage the use of arbitration in Scotland as well as attracting international arbitrations to Scotland. It is also hoped that industries and professions will set up their own low cost arbitration schemes. The Chartered Institute of Arbitrators (Scottish Branch) has produced a set of Short Form Arbitration Rules based upon the 2010 Act provisions (with explanatory notes) which are designed principally for use where the amount of any claim or counterclaim in dispute does not exceed £25,000.

17.2.1 Commencement

The substantive provisions of the 2010 Act came into force in June 2010. Commencement and transitional provisions were quite controversial areas at Bill stage because there were no proper transitional provisions and the 2010 Act would have applied to all arbitrations commenced after the 2010 Act came into force. Clearly, certain parties would have contracted to settle their disputes by arbitration but on the pre-2010 Act regime and some thought it unfair to tie them to a new statutory regime of which they had not been aware when they agreed to settle their disputes by arbitration.

Accordingly, section 36 of the 2010 Act now contains certain transitional provisions which seek to address that concern, namely:

1. The 2010 Act does not apply to arbitration agreements entered into before the coming into force of the 2010 Act if the parties agree that the 2010 Act is not to apply to their arbitration.

2. Ministers may by order specify any date falling at least 5 years after commencement as the day on which the position in 1. above will cease to have effect.

3. Before making the order referred to in 2. above, Ministers must consult with those persons having an interest in the law of arbitration.

In section 1 the 2010 Act is introduced by certain 'founding principles'. Those principles are three in number:

1. that the object of arbitration is to resolve disputes fairly, impartially and without unnecessary delay or expense,

2. that parties should be free to agree how to resolve disputes subject only to such safeguards as are necessary in the public interest, and

3. that the court should not intervene in an arbitration except as provided by the 2010 Act.

17.2.2 Section 3 – Seat of Arbitration

Section 3 provides that an arbitration is 'seated in Scotland': (a) if Scotland is designated as the juridical seat of the arbitration by the parties/by any third party given power to so designate/where the parties failed to designate or so authorize a third party, by the Tribunal or (b) in the absence of any such designation, the Court determines that Scotland is to be the juridical seat of the arbitration. Section 3(2) makes the important point that the fact that an arbitration is seated in Scotland does not affect the substantive law to be used to decide the dispute.

17.2.3 Section 4 – Arbitration Agreement

Section 4 defines an 'arbitration agreement' as an agreement to submit a present or future dispute to arbitration including any agreement which provides for arbitration in accordance with the arbitration provisions in a separate document.

17.2.4 Section 5 - Separability

Section 5 of the 2010 Act provides that an arbitration agreement which forms (or was intended to form) part only of an agreement is to be treated as a distinct agreement. An arbitration agreement is not void, voidable or otherwise unenforceable only because the agreement of which it forms part is void, voidable or otherwise unenforceable.

A dispute about the validity of an agreement which includes an arbitration agreement may be arbitrated in accordance with that arbitration agreement.

17.2.5 Section 6 - Law Governing Arbitration Agreement

Where (a) the parties to an arbitration agreement agree that an arbitration under that agreement is to be seated in Scotland, but (b) the arbitration agreement does not specify the law which is to govern it, then, unless the parties otherwise agree, the arbitration agreement is to be governed by Scots law.

17.2.6 Sections 7 to 9: The mandatory and default rules

The 2010 Act sets out certain mandatory and default rules which are to apply to any arbitration seated in Scotland. The rules are known as the Scottish Arbitration Rules. The 'mandatory rules' cannot be modified or disapplied by an arbitration agreement, by any other agreement between the parties or by any other means.

The non-mandatory rules are called 'default rules'. A default rule applies in relation to an arbitration seated in Scotland only insofar as the parties have not agreed to modify or disapply that rule (or any part of it) in relation to that arbitration. The parties may so agree (a) in the arbitration agreement or (b) by any other means at any time before or after the arbitration begins. Parties are to be treated as having agreed to modify or disapply a default rule:

(a) if or to the extent that the rule is inconsistent with or disapplied by (i) the arbitration agreement; (ii) any arbitration rules or other document, for example, the UNCITRAL Model Law, the UNCITRAL Arbitration Rules or other institutional rules which the parties agree are to govern the arbitration or anything done with the agreement of the parties; or
(b) if they choose a law other than Scots law as the applicable law in respect of the rules' subject matter.

The numbering of the Mandatory Rules and the Default Rules in the 2010 Act is set out in Tables 17.1 and 17.2 together with the general nature of the subject matter covered by each rule. Reference should be made to the rules themselves for full details of the provisions. A brief discussion of the terms of some of the rules is to be found in Section 17.2.12.

Table 17.1 The mandatory rules

Mandatory Rule Number	Subject Matter of the Mandatory Rule
Rule 3	Arbitrator to be an individual
Rule 4	Eligibility to act as an Arbitrator
Rule 7	Failure of procedure for appointment of Arbitrator
Rule 8	Duty to disclose any conflict of interest
Rule 12	Removal of the Arbitrator by the Court
Rule 13	Dismissal of Arbitrator/Tribunal by the Court
Rule 14	Removal and dismissal of Arbitrator by the Court
Rule 15	Resignation of the Arbitrator
Rule 16	Liability etc. of the Arbitrator when tenure ends
Rule 19	Power of Arbitrator/Tribunal to rule on own jurisdiction
Rule 20	Objection to the Arbitrator/Tribunal's jurisdiction
Rule 21	Appeal against Arbitrator/Tribunal's ruling on jurisdictional objection
Rule 23	Jurisdictional referral: procedure, etc.
Rule 24	General duties of Arbitrator/Tribunal
Rule 25	General duties of the parties
Rule 42	Point of law referral to Court: procedure, etc.
Rule 44	Time limit variation by Court: procedure, etc.
Rule 45	Securing attendance of witnesses and disclosure of evidence
Rule 48	Power of Arbitrator to award payment and damages
Rule 50	Power of Arbitrator to award interest
Rule 54	Power of Arbitrator to make part awards
Rule 56	Power of Arbitrator to withhold award if fees or expenses not paid
Rule 60	Arbitrator's fees and expenses
Rule 63	Ban on pre-dispute agreements about liability for arbitration expenses
Rule 67	Challenging an Award: substantive jurisdiction
Rule 68	Challenging an Award: serious irregularity
Rule 70	Challenging an Award : legal error appeals - procedure etc.
Rule 71	Challenging an Award: supplementary
Rule 72	Reconsideration of Award by the Arbitrator/Tribunal
Rule 73	Immunity of the Arbitrator/Tribunal
Rule 74	Immunity of Arbitrator appointing institution etc.
Rule 75	Immunity of experts, witnesses and legal representatives
Rule 76	Loss of party's right to object
Rule 77	Independence of Arbitrator
Rule 79	Death of Arbitrator
Rule 82	Rules applicable to umpires

Table 17.2 The default rules

Default Rule Number	Subject Matter of the Default Rule
Rule 1	Commencement of Arbitration
Rule 2	Appointment of Arbitrator Arbitral Tribunal
Rule 5	Number of Arbitrators
Rule 6	Method of appointment
Rule 9	Tenure of office as Arbitrator
Rule 10	Challenge to appointment of Arbitrator
Rule 11	Removal of Arbitrator by the parties
Rule 17	Reconstitution of the Arbitral Tribunal
Rule 18	Arbitrators nominated in Arbitration Agreements
Rule 22	Referral to Court of point of jurisdiction
Rule 26	Confidentiality
Rule 27	Arbitrator/Tribunal deliberations
Rule 28	Procedure and evidence
Rule 29	Place of the Arbitration
Rule 30	Arbitrator/Tribunal decisions
Rule 31	Arbitrator/Tribunal directions
Rule 32	Power of Arbitrator/Tribunal to appoint clerk, agents or employees and others etc.
Rule 33	Party representatives
Rule 34	Experts
Rule 35	Powers of Arbitrator relating to property
Rule 36	Oaths or affirmations
Rule 37	Failure of party to submit claim or defence timeously
Rule 38	Failure of party to attend Hearing or provide evidence
Rule 39	Failure of party to comply with Arbitrator/Tribunal direction or Arbitration Agreement
Rule 40	Consolidation of proceedings
Rule 41	Referral of point of law to Court during arbitration
Rule 43	Variation of time limits set by the parties
Rule 46	The Court's other powers in relation to Arbitration
Rule 47	Law etc. applicable to the substance of the dispute
Rule 49	Other remedies available to the Arbitrator/Tribunal
Rule 51	Form of award by Arbitrator
Rule 52	Award treated as made in Scotland
Rule 53	Power of Arbitrator to make provisional awards
Rule 55	Draft awards
Rule 57	Arbitration to end upon award or early settlement
Rule 58	Correcting an award
Rule 59	Arbitration expenses
Rule 61	Recoverable Arbitration expenses
Rule 62	Liability for recoverable Arbitration expenses
Rule 64	Security for expenses
Rule 65	Limitation of recoverable Arbitration expenses

Table 17.2 (*continued*)

Default Rule Number	Subject Matter of the Default Rule
Rule 66	Awards of recoverable Arbitration expenses
Rule 69	Challenging an award in Court; legal error appeal
Rule 78	Consideration where Arbitrator judged not to be impartial and independent
Rule 80	Death of party
Rule 81	Unfair treatment by Arbitrator
Rule 83	Formal communications
Rule 84	Periods of time

17.2.7 Section 10 – Suspension of Legal Proceedings

Section 10 deals with the legal position in connection with the sisting (suspending) of legal proceedings for arbitration where the parties have agreed that their disputes are to be referred to arbitration.

The 2010 Act provides that the Court *must*, on an application by a party to the proceedings concerning any matter under dispute, sist those proceedings insofar as they concern that matter if:

(a) an arbitration agreement provides that a dispute on the matter is to be resolved by arbitration (immediately or after the exhaustion of other dispute resolution procedures);
(b) the applicant is a party to the arbitration agreement (or is claiming through or under such a party);
(c) notice of the application has been given to the other parties to the legal proceedings;
(d) the applicant has not taken any step in the legal proceedings to answer any substantive claim against the applicant or otherwise acted since bringing the legal proceedings in the manner indicating a desire to have the dispute resolved by legal proceedings rather than by arbitration; and
(e) nothing has caused the Court to be satisfied that the arbitration agreement concerned is void, inoperative or incapable of being performed.

Any provision in an arbitration agreement which prevents the bringing of legal proceedings is void in relation to any proceedings which the Court refuses to sist. This does not apply to statutory arbitrations.

Section 10 applies regardless of whether the arbitration concerned is seated in Scotland.

17.2.8 Section 11 – Arbitral Award to be final and binding on the Parties

A tribunal's award is final and binding on the parties and any person claiming through or under them (but does not of itself bind any third party). In particular it is made clear that an award ordering the rectification or reduction of a deed or other document is of no effect insofar as it would adversely affect the interests of any third party acting in good faith.

Section 11 does not affect the right of any person to challenge the award (a) under part 8 of the Scottish Arbitration Rules or (b) by any available arbitral process of appeal or review.

Further, section 11 does not apply in relation to a provisional award (see Rule 53), such an award being not final and being binding only (a) to the extent specified in the award or (b) until it is superseded by a subsequent award.

17.2.9 Section 12: Enforcement of Arbitral Awards

This section provides that the Court may, on an application by any party, order that a tribunal's award may be enforced as if it were an extract registered decree bearing a warrant for execution granted by the Court. This will give the arbitration award the equivalent status to a court decree in relation to enforcement.

No such order may be made if the Court is satisfied that the award is the subject of (a) an appeal under part 8 of the Scottish Arbitration Rules; (b) an arbitral process of appeal or review, or (c) a process of correction under Rule 58 of the Scottish Arbitration Rules.

No such order may be made if the Court is satisfied that the tribunal which made the award did not have jurisdiction to do so. A party may not object on the ground that the tribunal did not have jurisdiction if the party has lost the right to raise that objection by virtue of the Scottish Arbitration Rules (see Rule 76).

This section applies regardless of whether the arbitration concerned is seated in Scotland.

17.2.10 Section 13: Court Intervention in Arbitrations

Legal proceedings are competent in respect of a tribunal's award or any other act or omission by a tribunal when conducting an arbitration, only as provided in the Scottish Arbitration Rules (insofar as they apply to that arbitration) or in any other provisions of the 2010 Act.

In particular, a tribunal's award is not subject to review or appeal in any legal proceedings except as provided for in Part 8 of the Scottish Arbitration Rules.

It is not competent for a party to raise the question of a tribunal's jurisdiction with the Court except (a) where objecting to an order being made under Section 12, or (b) as provided for in the Scottish Arbitration Rules (see Rules 21, 22 and 67).

17.2.11 Section 35: Commencement and Section 36 Transitional Provisions

Commencement and transitional provisions have been discussed in Section 17.2.1.

17.2.12 The Scottish Arbitration Rules

Tables 17.1 and 17.2 set out the mandatory and default rules. A full analysis of the Scottish Arbitration Rules is beyond the scope of this chapter. However, a summary of the terms of some of the rules and how they address the perceived difficulties created by the Scots law of arbitration before the 2010 Act is contained below.

Part 1 of the Rules deals with matters such as the commencement of the arbitration, appointment of the arbitrator, the duty of the arbitrator to disclose conflicts of interest and the liability, removal and resignation of the arbitrator. It is from Part 2 onwards that, arguably, the most significant provisions are to be found.

Rule 19: Power of tribunal to rule on its own jurisdiction: MANDATORY

This rule provides that the tribunal may rule on:

(a) whether there is a valid arbitration agreement;
(b) whether the tribunal is properly constituted; and
(c) what matters have been submitted to arbitration in accordance with the arbitration agreement.

Rule 20 lays down the procedure for objecting to the tribunal because of lack of jurisdiction and for appeal of the tribunal's ruling on jurisdictional objection to the Outer House of the Court of Session. The Outer House's decision on the appeal is final.

Rule 24: General duties of arbitrator: MANDATORY

This rule provides that the tribunal must:

(a) be impartial and independent;
(b) treat parties fairly; and
(c) conduct the arbitration without unnecessary delay and without incurring unnecessary expense.

Treating the parties 'fairly' is stated to include 'giving each party a reasonable opportunity to put its case and to deal with the other party's case'. The use of the word 'unnecessary' in the context of delay and expense may give rise to some dispute about how that concept should be applied in practice.

Rule 25: General duty of parties: MANDATORY

The parties ' … must ensure that the arbitration is conducted (a) without unnecessary delay and (b) without incurring unnecessary expense'. These provisions too may give rise to dispute about what is 'unnecessary'.

Rule 28: Procedure and evidence: DEFAULT

It is for the tribunal to determine: (a) the procedure to be followed in the arbitration; and (b) the admissibility, relevance, materially and weight of evidence.

The ability to determine procedure is not new; it merely reflects the common law position. However, the powers in relation to evidence could certainly be interpreted as giving an arbitrator power beyond that available to and different from that of a Court. For example, it is unclear what approach an arbitrator will adopt in connection with hearsay or uncorroborated evidence. These and other issues are raised by the very far-reaching power given to the arbitrator. While the wide power might be argued to assist in the speedy resolution of the dispute, it may be difficult to advise what the parties' rights and obligations are in certain circumstances. Due regard must be given to other relevant provisions, such as those in Rule 45 where the Court may not order a person to give any evidence, or to disclose anything which the person would be entitled to refuse to give or disclose in civil proceedings.

Rule 32: Power to appoint a clerk, agents or employees etc: DEFAULT

The tribunal may appoint a clerk (and such other agents, employees or other person as it thinks fit) to assist in conducting the arbitration. The consent of parties is required for any appointment in respect of which 'significant' expenses are likely to arise.

This provision reflects the position which had developed in practice regarding the appointment of a clerk, invariably a lawyer, in certain arbitrations. The obtaining of expert assistance is covered in Rule 34. It is not perhaps immediately clear how the employment of ' … other agents, employees or other persons as it thinks fit' will tie in with the obligation 'to conduct the arbitration without incurring unnecessary expense'. However, it may be justified if it can reasonably be said that it will help to minimize expense overall rather than increase it. It is likely that except in extreme cases parties will be reluctant to withhold consent for fear of alienating the arbitrator. The word 'significant' may give rise to dispute.

Rule 34: Experts: DEFAULT

The tribunal may obtain an expert opinion on any matter arising in the arbitration. The parties must be given a reasonable opportunity (a) to make representations about any written expert report; and (b) to hear any oral evidence and to ask questions of the expert giving it. It is noteworthy that the parties are not given any express right to make representations about the materials or questions to be put to the expert in obtaining his opinion. That might be a good practice if it can be done without 'unnecessary delay or expense'.

It is perhaps worth noting that there is no restriction on the nature of the expert opinion which could be sought. There is no consent of the parties required. However, the arbitrator will still be obliged to avoid unnecessary delay and expense (Rule 24).

Rule 37: Failure to submit claim or defence timeously: DEFAULT

Claim. Where:

(a) a party unnecessarily delays in submitting or in otherwise pursuing a claim;
(b) the tribunal considers that there is no good reason for the delay;
(c) delay gives rise to a substantial risk that the tribunal will not be able to resolve the issues fairly or has caused or is likely to cause, serious prejudice to the other party the tribunal *must* end the arbitration insofar as it relates to the subject matter of the claim.

This power is regulated by the wording of the rule but certain interesting questions arise. If a claim is raised within the prescriptive period, would it nonetheless be appropriate to end the arbitration in certain circumstances? What precisely is meant by 'serious prejudice'?

Defence: Where a party delays in submitting a defence and the tribunal considers that there is no good reason for the delay the tribunal may proceed with the arbitration. Delay is not, in itself, to be treated as an admission of anything.

Rule 38: Failure to attend hearing or provide evidence: DEFAULT

Where:

(a) a party fails to attend a hearing upon reasonable notice or to produce a document; and
(b) the tribunal considers there is no good reason for the failure,

the tribunal may proceed with the arbitration and make its award on the basis of the evidence (if any) before it.

At first blush this might be considered a strange provision because the words 'if any' indicate that the arbitrator may make an award where there is no evidence of any kind before him. The context does not appear to suggest a legal debate where it is assumed that the facts can be established. Accordingly, despite the apparent width of the power given, it is submitted that arbitrators may be slow to make an award without at least some evidence being advanced which would support it.

Rule 39: Failure to comply with tribunal direction or arbitration agreement: DEFAULT

If a party fails to comply with an order made under this rule, the tribunal may (a) direct that the party is not entitled to rely on any allegation or material which was the subject matter of the order; (b) draw adverse inferences from the non-compliance; (c) proceed with the arbitration and make an award; or (d) make such provisional award (including an award of expenses) as it considers appropriate in consequence of the non-compliance.

Rule 40: Consolidation of proceedings: DEFAULT

Parties may agree to consolidate the arbitration with another arbitration or to hold concurrent hearings. The tribunal may not do so on its own initiative. Although not expressly stated, it appears that where the parties are different in the arbitrations, the consent of all parties would be required. The need for agreement will considerably limit the utility of this Rule.

Rule 41: Power of Court to decide point of law: DEFAULT

The Outer House of the Court of Session may, on an application by any party, determine any point of Scots law arising in the arbitration. This a default rule but if it applies in a particular arbitration the procedure set down in Rule 42 is mandatory.

Rule 42: Point of law referral: MANDATORY

An application is valid only if:

(a) the parties agreed that it be made; or
(b) the tribunal has consented to it being made and the Court is satisfied that:

   (i) determining the question is likely to produce substantial savings in expenses;
   (ii) the application was made without delay; and
   (iii) there is good reason why the question should be determined by the Court.

The tribunal may continue with the arbitration pending the determination of the application. The Outer House's determination of the question is final as is any decision by the Outer House as to whether an application is valid.

This procedure is a complete innovation of the position before the 2010 Act. Traditionally, arbitration had been seen as a process under which the arbitrator determined both disputes of fact and law between the parties (if necessary by obtaining expert assistance on any technical or legal issues outwith the arbitrator's normal area of expertise). This procedure takes place during the currency of the arbitration and it appears that such an application would be valid where the parties agree that it should be made but the arbitrator does not so agree. The Court must be satisfied on certain matters which may be the subject of dispute if one of the parties wishes to make the application with the consent of the arbitrator but the other party has no such wish.

It is difficult to see how there would be a likelihood of substantial savings in expenses if, for example, the parties have excluded recourse to the Court on a point of law after the arbitrator has made a decision. In certain circumstances it is likely to be much more expensive to ask the Court to determine the question rather than have the arbitrator determine it. Further, what does 'without delay' mean? The word delay is not qualified by a word such as 'undue' or 'unreasonable'. In some cases, if the application is not made as soon as the arbitration commences, there will inevitably be some delay in making the application. It is not clear what will be regarded as a 'good reason' when the question is, should the point be determined by the Court rather than the arbitrator? The drafting means that it is likely that these issues will be canvassed before the Courts in due course.

Rule 43: Variation of time limited set by the parties: DEFAULT

The Court may on application of the tribunal or any party vary any time limit relating to the arbitration which is imposed by the arbitration ag reement or by any other agreement between the parties. If Rule 43 applies, Rule 44, which sets out the procedure and the test to be applied by the Court, is mandatory. Such a variation may be made only if the Court is satisfied that there is no arbitral process for varying the time limit available and that someone would suffer a substantial injustice if no variation was made. The Court's decision on whether to make a variation is final. This was a controversial provision because it means that the Court can re-write what the parties have agreed.

Rule 45: Court's power to order attendances of witnesses and disclosures of evidence: MANDATORY

This rule sets out the mechanism by which a party may ensure attendance of witnesses and disclosure of evidence. The arbitrator has no power to compel attendance or disclosure but the Court does. This is a clear example of how the Court assists and supports the arbitration process.

Rule 46: Court's other powers in relation to arbitration: DEFAULT

These include power:

(a) to make an order securing any amount in dispute in the arbitration; and
(b) to grant warrant for arrestment or inhibition.

The Court may only take such action with the consent of the tribunal or where the Court is satisfied that the case is one of emergency.

Rule 48: Power to award payment and damages: MANDATORY

The tribunal is given the power to order payment of a sum of money (including a sum in respect of damages). The words in parenthesis are particularly important given the common law position that an arbitrator did not have power to assess and award damages. It should be noted that this is a mandatory rule.

Rule 49: Other remedies available to tribunal: DEFAULT

This includes the power to order rectification or reduction of any deed or other document to the extent permitted by the law governing the deed or document. This is a very important departure from the pre-2010 Act position where only the Court had such power.

Rule 50: Interest: MANDATORY

The tribunal's award *may* order that interest be paid on the whole or any part of a sum the award orders to be paid: (a) in respect of the period up to the date of the award; and (b) any period up to the date of payment.

The tribunal may specify the rate and the period for which interest is payable. Interest is to be calculated in the manner agreed by the parties (e.g. in the contract or arbitration agreement) or failing such agreement, in such manner as the tribunal determines.

Rule 53: Provisional award: DEFAULT

The tribunal may make a provisional award granting any relief on a provisional basis which it has power to grant permanently. The term 'provisional' appears to be akin to what was previously known as an 'interim award'. There was doubt at common law about whether an arbiter had power to make interim awards. However, it should be noted that the rule gives no guidance as to the circumstances in which it would be appropriate to grant a provisional award. The Courts have developed rules setting out the circumstances in which provisional (or interim) awards will be made by them which seek to balance the interests of parties and avoid difficulties at a later stage. It is not at all clear if arbitrators will adopt a similar position and accordingly this is perhaps one default rule to which particular regard should be had in deciding whether it should be modified or disapplied.

Rule 54: Part awards: MANDATORY

The tribunal may make more than one award at different times on different aspects of the matters to be determined. There was some doubt at common law as to whether an arbiter had power to make part awards. Recently, the Inner House of the Court of Session appeared to indicate that such a power did exist at common law, see *Apollo Engineering Ltd* v. *James Scott Ltd.* (2009).

Rule 58: Correcting an award: DEFAULT

The tribunal may correct an award so as to:

(a) correct a clerical, typographical or other error in the award arising by virtue of accident or omission; or
(b) clarify or remove an ambiguity in the award.

Such a correction may be made at the tribunal's own initiative (within 28 days of the award concerned being made) or upon an application by any party. Such an application is valid only if made within 28 days of the award concerned or by such later date as the Outer House or Sheriff on application may allow. No tests or criteria are set out which the Court is to apply in determining whether to allow correction at 'such later date'. It is submitted that the wording of limb (a) is problematic because, arguably, it is wide enough to cover errors beyond those of a mere clerical or typographical nature and the extent of that may leave room for dispute.

Rules 59–62 inclusive deal with the expenses (costs) of the arbitration, including the Arbitrator's fees and expenses and what expenses are recoverable by one party from the other.

Rule 63: Ban on pre-dispute agreement about liability for expenses: MANDATORY

Any agreement allocating the parties' liability between themselves for any or all of the arbitration expenses has no effect if entered into before the dispute being arbitrated has arisen. This mandatory rule would, for example, outlaw any attempt in the underlying contract to make the claimant liable for the respondent's expenses in the arbitration irrespective of the claimant's success in the arbitration. The provision was apparently inserted in light of such attempts in adjudication clauses in construction contracts.

Rule 64: Security for expenses: DEFAULT

The tribunal may order a party making a claim to provide security for recoverable arbitration expenses. This is an important new power because it was not available at common law. The rule sets down certain circumstances in which such an order may not be made, but otherwise it contains no test or criteria to be satisfied before the arbitrator should make such an order.

Rule 67: Challenging an award: substantive jurisdiction: MANDATORY

A party may appeal to the Outer House of the Court of Session against a tribunal's award on the ground that the tribunal did not have jurisdiction to make the award. Appeal to the Inner House is possible with leave of the Outer House. The Inner House's decision on such an appeal is final.

Rule 68: Challenging an award: serious irregularity: MANDATORY

A party may appeal to the Outer House against a tribunal's award on the ground of serious irregularity as defined. For a recent indication of how similar provisions have been applied in England, see *Atkins Ltd* v. *Secretary of State for Transport* (2013). Appeal to the Inner House is possible with leave of the Outer House. Here again, the Inner House's decision on such an appeal is said to be 'final'.

Rule 69: Challenging an award: legal error: DEFAULT

A party may appeal to the Outer House against the tribunal's award on the ground that the tribunal erred on a point of Scots law.

Rule 70: Legal errors appeal: procedure: MANDATORY

A legal error appeal may be made only with the agreement of the parties or with leave of the Outer House. Leave is only to be granted if the Outer House is satisfied that: (a) deciding the point will substantially affect a party's rights; (b) the arbitration tribunal was asked to decide the point; and (c) the decision was 'obviously wrong' or, where the court considers the point of law to be one of general importance, the arbitrator's decision is open to serious doubt.

The first successful application for leave to appeal on grounds of legal error was issued in October 2011 (*Arbitration Application No. 3 of 2011*) in which Lord Glennie gives general guidance for parties on the procedural aspects of such applications and how they will be approached by the court. The aim is to ensure that the procedure is speedy, simple and flexible, reflecting the founding principles set out in Section 1 of the 2010 Act. The Outer House's determination on an application for leave is final.

It was noteworthy that in that case, though leave to appeal was granted, Lord Glennie eventually held that there had been no error in law by the arbitrator. An example of the Court upholding a legal error challenge is to be found in the decision of Lord Malcolm in *Manchester Associated Mills Limited* v. *Mitchells & Butler Retail Ltd* (2013).

The Outer House may decide a legal error appeal by:

(a) confirming the award;
(b) ordering the tribunal to reconsider the award; or
(c) if it considers reconsideration inappropriate, setting aside the award.

An appeal may be made to the Inner House against the Outer House's decision on a legal error appeal with leave of the Outer House. Here again, the Inner House's decision on such an appeal is final.

It should be noted that Rule 71, a mandatory rule, contains certain important supplementary provisions regarding the challenging of awards.

A new chapter has been added to the Court of Session Rules (Chapter 100) dealing with applications and appeals under the Scottish Arbitration Rules.

Rule 76: Loss of right to object: MANDATORY

This rule provides that a party who participates in an arbitration without making a timeous objection on certain grounds may not raise that objection later. The grounds are:

- Arbitrator ineligible to act.
- Arbitrator not impartial and independent.
- Arbitrator has not treated parties fairly.
- Tribunal does not have jurisdiction.
- Arbitration has not been conducted in accordance with the arbitration agreement/Scottish Arbitration Rules/any other agreement of the parties.

Chapter 18
Other Forms of Dispute Resolution

18.1 Mediation

18.1.1 Introduction

Only a very small percentage of construction cases proceed to a decision in arbitration or litigation. The vast majority of cases settle in the run-up to the full hearing or proof. Settlement is often the result of negotiations at client or lawyer level. A more formal process of negotiating a case to settlement has developed, known as Mediation, a form of Alternative Dispute Resolution (ADR). It can be initiated as an alternative to litigation or arbitration or it may be conducted in parallel with ongoing proceedings.

The aim of mediation is to achieve a negotiated settlement in an economic and effective manner. As the process is consensual, a reference to mediation is non-binding unless the parties choose it to be binding. Parties may withdraw at any time until they have formalized the terms of any settlement reached.

The mediator is a third party 'neutral' who conducts or facilitates the process. His role is to mediate between the two or more parties to the dispute to facilitate settlement. The procedure in mediation is very flexible and a mediator will tailor it to that which is most appropriate for each individual dispute. The process is essentially one of each party assessing risk by fully understanding the other party's or parties' positions. Each party needs to undertake a process of examining their own case in the light of evidence put forward by the others, and often their experts, to assess what is the most likely outcome if the case cannot be settled and ends up being decided by a Judge or Arbitrator. The mediator assists in this process of reality testing and risk assessing of positions. For each party, it is important to consider who should attend the mediation in terms of an ability to contribute to the risk assessment process either through knowledge of the project, expertise in any technical issues, but also, essentially, someone of sufficient seniority to take a commercial view, having considered all of the arguments presented. A number of organizations have trained mediators available for appointment, such as the Centre for Effective Dispute Resolution (CEDR).

One of the advantages of mediation, apart from speed and being relatively inexpensive, is that the settlement can cover any number of matters. It is not constrained by the issues in dispute, in the way that litigation and arbitration are, or by availability of legal remedies. It may be that what a party requires is an explanation or an apology,

MacRoberts on Scottish Construction Contracts, Third Edition. MacRoberts.

to continue a business relationship, to regulate a relationship going forward or other remedy not available in the courts but in the power of the other party to provide. In construction disputes, considerations such as a party carrying out remedial work, ongoing liability for latent defects which are not the subject of the present dispute, payment for work carried out and payment due to defective work often come into play in addition to issues of allocation of responsibility as between two, and often more, parties. Creative solutions can be found and it is entirely within the control of parties what package can be put together which then results in a settlement.

The timing of a successful mediation is all important. If a reference is sought too early, the issues may not be sufficiently well prepared or focused to form a background for discussion; too late and the parties may have become entrenched in their positions. A will to constructively engage in the process and attempt to resolve the matter is necessary before progress can be made.

Good preparation for the mediation itself is essential if it is to be given the best chance to succeed. In construction disputes, this is likely to involve experts considering their own views and where these may be vulnerable to attack by the other party's or parties' experts. Often, consideration needs to be given in detail to the sums claimed requiring QS input. For example, in a defective building claim, realistic assessments of the extent of the remedial work required, the cost of this, and any consequential effects such as loss of profit or disruption to business need to be quantified. Where there are multiple parties, for example, an employer, contractor and designers, it is necessary to consider positions not just with the pursuing party but also among the various defenders. For example, if defects in a building may have been caused by both design and workmanship defects, there will be a discussion required with the employer as to the remedy it expects from the contractor and designers. However, there will also be a need for a separate discussion between the contractor and designers (and most likely their insurers) in terms of their respective contributions to any settlement. This requires good preparation in assessing likely levels of culpability for each aspect of the defects and thereafter agreeing positions on who pays what. This can involve considering the extent to which the employer itself has contributed by lack of maintenance or by having placed restraints on the design or budget at the outset. Where a resolution is anticipated to involve one party carrying out work, thought requires to be given to matters such as whether this is to be under the existing contract or as a stand-alone, when it can start, how long it will take, what liability will remain with that party going forward, quality of work, scope of work and the mechanism for any payment required, for example, as a contribution either from the employer or from the other parties to the dispute. Again, it is helpful if this can be considered in advance to allow time spent in the mediation itself to be as productive as possible.

It is also worth having available an assessment of the best and worst likely outcomes, both in respect of the principal sum and of the details of the cost of the dispute to date and likely cost going forward. Consideration needs to be given to the impact of an adverse award of costs in the event of not being ultimately successful in court or arbitration. For certain levels of dispute, the cost analysis exercise can assist in providing good guidance and parameters for settlement figures, purely taking account of commercial considerations in this respect.

Mediation can be initiated at any time by either of the parties inviting the other to accede to the process, or by having one of the mediation bodies open up dialogue. Parties may be represented by lawyers through the process, but that is not essential. Normally, the costs of the whole process are shared equally by the parties.

18.1.2 Mediation and the SBC

Article 9.1 was a significant new provision in the 2005 edition of the SBC and the SBC/DB. It provided that the parties may by agreement seek to resolve any dispute or difference arising under the contract through mediation. This would be implied in any event since parties can agree to mediate any dispute at any time. However, adding reference to it is useful in directing parties to consider this as a potential way of resolving disputes. That provision was strengthened in the SBC and the SBC/DB 2011 Edition with the clause now stating that each party is to give serious consideration to any request by the other to refer the matter to mediation. It is still not mandatory but is intended to be a strong encouragement to parties to consider mediating. This supplements the SBC Schedule Part 8 paragraph 6 and the SBC/DB Schedule Part 2 paragraph 12 (Supplemental Provisions) which provide for an escalating dispute resolution provision.

This approach is reflective of the increasing move towards early identification of disputes and their avoidance and resolution at an early stage and without the need for third party involvement. It is in line with the increased emphasis on collaborative working and the provisions on this to be found within the SBC and the SBC/DB.

18.1.3 Mediation and the NEC3

The NEC3 contract contains no reference to mediation. It is of course still open to parties to agree to refer disputes to mediation should they wish to do so.

18.1.4 The approach of the courts to compelling mediation

Mediation is developing in Scotland. The Scottish courts have become more favourably disposed towards mediation, though they have not yet embraced mediation in the way courts in other jurisdictions have. In the Court of Session, the Commercial Court Practice Note No. 6 of 2004 encourages parties to consider ADR before an action is raised. In the Sheriff Court, there are a number of mediation initiatives. These include a provision in the Sheriff Court commercial action rules that a sheriff is under a duty to secure the expeditious resolution of the action by means of a range of orders including the use of ADR. In Small Claim and Summary Actions (up to £5000) the Sheriff is under a duty to seek to negotiate and secure settlement of the claim at the first hearing of the case.

The 2009 Report of the Scottish Civil Courts Review chaired by Lord Gill recognized that mediation and other forms of ADR have a valuable role to play in the civil justice system. It recommended that the courts ensure that litigants and potential litigants are fully informed about the dispute resolution options available to them and should encourage parties, in appropriate cases, to consider ADR.

However, it stopped short of compelling parties to enter into an ADR process, the view being taken that people should have a fundamental right to have their disputes dealt with in the courts. Mediation and other forms of ADR should therefore supplement rather than be an alternative to the court system.

The courts are, however, becoming increasingly aware of the benefits of mediation. In the case of *Candleberry Ltd.* v. *West End Home Owners Association and Others* (2006), Lord Nimmo Smith observed:

> [W]e hope that we have said enough to reinforce our observations in court, that this is a dispute which ought to be resolved. It cannot be in the interests of the neighbourhood that it be prolonged, and we would encourage a resolution by compromise, perhaps with the assistance of a mediator.

This is a clear indication of the promotion of mediation by the Scottish courts, which will continue to develop. There have been a number of endorsements of mediation in the English courts which have taken a much more hands-on role. In *Dunnett* v. *Railtrack plc* (2002), Lord Justice Brooke said: 'Skilled mediators are now able to achieve results satisfactory to both parties in many cases which are quite beyond the power of lawyers and courts to achieve.'

In *Cowl and Others* v. *Plymouth City Council* (2002), Lord Woolf CJ said:

> Insufficient attention is paid to the paramount importance of avoiding litigation whenever this is possible … both sides must by now be acutely conscious of the contribution alternative dispute resolution can make to resolving disputes in a manner which both meets the needs of the parties and the public and saves time, expense and stress.

In *Reed Executive plc* v. *Reed Business Information Ltd* (2004), Lord Justice Jacob said: 'a good and tough mediator can bring about a sense of commercial reality to both sides which their own lawyers, however good, may not be able to achieve.'

In *Halsey* v. *The Milton Keynes General NHS Trust* (2004), Lord Justice Dyson said: 'But it is also right to point out that mediation often succeeds where previous attempts to settle have failed.'

In *Multiplex Constructions (UK) Limited* v. *Cleveland Bridge UK Limited* (2006), the Judge made comments in relation to mediation:

> With the assistance of the court's decision on the ten preliminary issues, it may now be possible for both parties to arrive at an overall settlement of their disputes, either through negotiation or with the help of a Mediator, who is unconnected with the court. I commend this course to the parties, if only as a means of saving costs and management time. If, however, the parties would prefer the court to resolve all remaining issues, then so be it. This court encourages sensible commercial settlements, but nevertheless stands ready to determine every issue which the parties wish to litigate.

In a later dispute between the same parties, *Multiplex Constructions (UK) Limited* v. *Cleveland Bridge UK Limited, Cleveland Bridge Dorman Long Engineering Limited* (2008), the same Judge made further comments under a 'lessons to be learned' section in his lengthy judgment. He said:

> [F]ollowing that judgment, the parties attended a Mediation, however, instead of reaching a sensible resolution at that Mediation, the parties spent the next two years litigating about two matters ... A resolution broadly along the lines of this judgment could have been arrived at by the parties at fractional cost, if both parties had instructed their advisers to go through the accounts together in a constructive spirit taking as their starting point the court's decision on issues 1 to 10. The lesson for the future which may be drawn from this litigation is that parties would be well advised to use the dispute resolution service offered by the Technology and Construction Court in a more conventional and commercial manner than has been adopted in this case. Once this court has decided questions of principle, the parties can save themselves and their shareholders many millions of pounds by instructing their advisers to agree reasonable figures for quantum, if necessary with the assistance of a Mediator unconnected with the court.

More recently, in the English Court of Appeal case of *Ali Ghaith* v. *Indesit Company UK Limited* (2012), the court was critical of the parties for not trying mediation, stating:

> It is a great pity that Indesit did not pursue the option of mediation ... [Counsel] informed us that it was not pursued because the costs had already exceeded the likely amount in issue. This is an inadequate response to the court's encouragement of mediation since a full day in this court will inevitably result in a substantial increase in costs.

The court went on:

> No-one should underestimate the new dynamic that an experienced mediator brings to the round table. He has a canny knack of transforming the intractable into the possible. That is the art of good mediation and that is why mediation should not be spurned when it is offered.

The Government has also endorsed mediation. In March 2001, the Chancellor of the Exchequer announced an 'ADR Pledge' in which all government departments and agencies made commitments, including that ADR would be considered and used in all suitable cases, where accepted by the other party. A number of government initiatives have followed since the pledge was made.

Under the English Civil Procedure Rules, there is an overriding objective requiring the court to actively manage cases, which includes encouraging the use of ADR procedures to prompt an earlier settlement.

The courts have tended to stop short of compelling parties to mediate as it is considered this would amount to an unacceptable obstruction on their right of access to the court, which would be contrary to the European Convention on Human Rights. Further, it is recognized that ADR procedures work best when parties voluntarily take part in them and that compelling people to do so would not be effective. The Pre-Action Protocol for Construction and Engineering Disputes encourages parties to consider whether some form of ADR procedure would be more suitable than litigation and, if so, agree which form to adopt. However, it expressly recognizes that no party can or should be forced to mediate or enter into any form of alternative dispute resolution.

Despite this, however, in the case of *Shirayama Shokusan Co. Ltd* v. *Danovo Ltd (No. 1)* (2004), the judge granted an order for mediation which had been applied for by the defendant but opposed by the claimant. The court considered that, under the Civil Procedure Rules, it has jurisdiction to direct ADR between the parties even if one party is unwilling. It was also considered that in the circumstances of this case, an attempt at mediation was worthwhile. The parties were likely to need to work together in future years and a number of the disputes between them appeared to be in relation to small points where it was considered mediation could be beneficial. They also had a shared commercial interest.

Where parties have contracted to mediate, the courts have been willing to enforce the contractual commitment. In *Cable & Wireless plc* v. *IBM United Kingdom Ltd* (2002), the parties' contract contained a mediation clause. It provided that if disputes arose, they should be resolved through negotiation and that, if that was not successful, an attempt should be made in good faith to resolve the dispute through ADR as recommended to the parties by CEDR.

Cable & Wireless objected to the Court action being put on hold while a mediation took place but the judge found against them for two reasons. First, the contract obliged the parties to participate in an ADR procedure and this was an obligation sufficiently certain to allow a court to ascertain whether it had been complied with. Second, it would be contrary to public policy to decline to enforce references in contracts to ADR. It was said that strong cause would have to be shown before a court could justify declining to enforce such an agreement. It would not be sufficient that an issue of construction of a long-term contract (as in this case) was involved. The judge, Mr Justice Colman, stated that parties entering into an ADR agreement should recognize that:

> Mediation as a tool for dispute resolution is not designed to achieve solutions which reflect the precise legal rights and obligations of the parties, but rather solutions which are mutually commercially acceptable at the time of the mediation.

As further evidence of the court's active encouragement of mediation, there have been a number of cases in England in which the courts have considered whether a party ought to be penalized in costs for failure to mediate. The general principle, as established by the Court of Appeal in *Cowl and Others* and in *Dunnett* is that a party who refuses to go to mediation without good and sufficient reasons may be penalized for that refusal, particularly, in respect of costs. In *Cowl and Others*, Lord Woolf said: 'Today sufficient should be known about ADR to make the failure to adopt it, in particular when public money is involved, indefensible.'

Mr Justice Lightman in *Hurst* v. *Leeming* (2003) endorsed this fully stating:

> Mediation is not in law compulsory … But alternative dispute resolution is at the heart of today's civil justice system, and any unjustified failure to give proper attention to the opportunities afforded by mediation, and in particular in any case where mediation affords a realistic prospect of resolution of a dispute, there must be anticipated as a real possibility that adverse consequences may be attracted.

Hurst was a professional negligence action. The action was found to have no merit and was dismissed by the court. The normal situation in such circumstances would be that the defendant Leeming would be entitled to his costs. The plaintiff Hurst argued that Leeming should not be entitled to costs because both before and after the action commenced, Hurst had proposed mediation but Leeming had refused. The court held that the fact that substantial costs had already been incurred would not be good reason to refuse to mediate – this was simply a factor to take into account in the mediation process. Neither was it a good reason to refuse to mediate that the action was one of professional negligence or that one party believes it has a watertight case. Where details had been provided refuting the other party's case, this would be a relevant consideration but would not, in itself, be sufficient to justify refusing to mediate.

The judge considered the critical factor to be whether, viewed objectively, a mediation has any real prospect of success. If it does not, a party may refuse to mediate. However, the word of warning was that if that ground was relied on and the court disagreed, the cost penalty could follow. The court stated that:

> The mediation process itself can and often does bring about a more sensible and more conciliatory attitude on the part of the parties than might otherwise be expected to prevail before the mediation, and may produce a recognition of the strengths and weaknesses by each party of his own case and of that of his opponent, and a willingness to accept the give and take essential to a successful mediation. What appears to be incapable of mediation before the mediation process begins often proves capable of satisfactory resolution later.

On the particular facts of this case, it was found that the refusal to mediate was justifiable. However, that was said to be exceptional.

In *Halsey*, the court gave further guidance as to how the question of whether a defendant had acted unreasonably in refusing ADR should be answered. The starting point was to say that regard should be had to all the circumstances of the particular case. Among the relevant matters to take into account are:

- the nature of the dispute;
- the merits of the case;
- the extent to which other settlement methods have been attempted;
- whether the costs of the ADR would be disproportionately high;
- whether any delay in setting up and attending the mediation would have been prejudicial; and
- whether the ADR had a reasonable prospect of success.

None of these, on its own, would be decisive and this is not an exhaustive list of factors. The factors set out in *Halsey* are now accepted as the standard test in considering whether a party has acted unreasonably in refusing to mediate. The case of *P4 Ltd* v. *Unite Integrated Solutions plc* (2006), for example, considered the question with reference to those factors.

The case of *Mr NF Burchell* v. *Mr and Mrs Bullard and Others* (2005) involved an appeal by a builder against a costs order arising from a litigation concerning work done to the property of Mr and Mrs Bullard. Once more the defendant's refusal to mediate was analysed. The offer had been made long before the action started and before substantial costs had been incurred. Again, the Judge made reference to the *Halsey* test and stated that:

> [A] small building dispute is *par excellence* the kind of dispute which, as the recorder found, lends itself to ADR. Secondly, the merits of the case favoured Mediation. The defendants behaved unreasonably in believing, if they did, that their case was so watertight that they need not engage in attempts to settle ... there was clearly room for give and take. The stated reason for refusing Mediation that the matter was too complex for Mediation is plain nonsense. Thirdly, the costs of ADR would have been a drop in the ocean compared with the fortune that has been spent on this litigation. Finally, the way in which the claimant modestly presented his claim and readily admitted many of the defects, allied with the finding that he was transparently honest and more than ready to admit where he was wrong and to shoulder responsibility for it augured well for Mediation. The claimant has satisfied me that Mediation would have had a reasonable prospect of success, the defendants cannot rely on their own obstinacy to assert that Mediation had no reasonable prospect of success.

The issue of costs as a result of failure to mediate was also considered in *Rolf* v. *De Guerin* (2011), which was described by the Judge as 'a sad case about lost opportunities for Mediation'. The case concerned a dispute between a home-owner and her builder. The relationship broke down during the build and the builder walked off site. The home-owner had the job completed by someone else and sued the builder for alleged defects and the cost of completion. She was, however,only awarded £2500 of her claim which, at its highest, had been valued at around £92,000 and was unsuccessful on several key issues. On an appeal only on the issue of costs, the Judge decided neither party should be awarded their costs. The key factor for the Judge was to consider how the parties had conducted the case. The home-owner had made it clear from an early stage that she was willing to settle. She offered a reasonable sum and proposed mediation or even a meeting to discuss settlement. The builder ignored these proposals until just before the case was heard. It appeared that one of the reasons he had been unwilling to mediate was that he wanted his day in court. The Judge held that this was not an adequate response to 'a proper judicial concern that parties should respond reasonably to offers to mediate or settle and that their conduct in this respect can be taken into account in awarding costs'.

Since *Rolf v De Guerin*, however, there have been other decisions in which a contrary approach was taken by the English courts.

In *Swain Mason and Others* v. *Mills & Reeve* (2012), the claimants had proposed mediation and the Judge had also encouraged parties to consider this. The defendant consistently declined on the basis they considered the claim to be without merit. The Judge at first instance took the view that the refusal to mediate was unreasonable and took this into account in a manner adverse to the defendant in his decision on costs. However, the Court of Appeal took a different approach. The defendant had been successful in its defence as far as allegations of breach of duty were concerned. The position it took was ultimately found to be justified on this important matter. The first instance Judge had considered that there was a real possibility that, had there been a mediation, both parties would have gained an understanding of the weaknesses of their cases. The Court of Appeal was not clear what weaknesses would have been revealed and noted that the Judge had not gone on to say that, even if they were, this could have led to a settlement. The first instance Judge had taken into consideration that an advantage of mediation would be to avoid the risk of 'collateral reputational damage' to the defendant. The Court of Appeal disagreed. It thought that some professional defendants may, quite reasonably, wish to vindicate themselves in respect of claims.

The Court of Appeal was also concerned at the first instance Judge's suggestion that a mediated settlement was 'not unrealistic'. In contrast, it characterized the parties' positions as 'a hundred miles apart'. In the circumstances, it was difficult to see how a mediation could have had reasonable prospects of success. Also, as nothing had changed in the case to necessitate a re-evaluation of liability, the initial reasonable refusal to mediate did not become unreasonable by reason of it being maintained throughout.

Attention was drawn to the Court of Appeal's statement in *Halsey* that mediation is not a panacea. It can have disadvantages as well as advantages and is not appropriate for every case.

In *Halsey*, the court wanted to be clear that parties were not to be compelled to mediate and it was a relevant factor that a party reasonably believed it had a strong case. That reasonable belief could be sufficient justification for a refusal to mediate.

The Court of Appeal in *Swain Mason* considered the fundamental question to be whether it had been shown by the unsuccessful party (in this case, the claimants) that the successful party had acted unreasonably in refusing to mediate. In this case it was thought they had not and therefore that the first instance Judge had been wrong to take it into account in his decision on costs.

In *ADS Aerospace Limited* v. *EMS Global Tracking Limited* (2012), the court took a similar approach. The defendants had claimed $16 million. Their claim was dismissed. The claimant sought a reduction in the defendants' costs entitlement to reflect the defendants' unwillingness to mediate. The Judge, again with reference back to the principles in *Halsey*, found that the onus was on the claimant to show the defendant had acted unreasonably, but that they had not established this. The relevant factors were:

- The defendant had made four attempts to initiate a discussion.
- The claimant felt strongly that it was entitled to substantial compensation and did not appear interested in a nuisance payment.

- The claimants' suggestion of mediation came less than 20 working days before the trial.
- The defendant was not unreasonable in believing it had a strong case.

In *PGF II SA* v. *OMFS Company, Bank of Scotland plc* (2012) and (2013), the court had to consider the question of costs. In that case, the claimant's solicitors had invited the defendant to participate in mediation. The defendant had not responded. The Court at first instance took this silence to be a refusal and decided it was unreasonable, applying the *Halsey* test. In relation to the *Halsey* test of whether the mediation had a reasonable prospect of success, the burden of establishing that was on the claimant but, the Judge said: 'It is not an unduly onerous burden given that it does not need to show that the mediation would have been successful, merely that it had a reasonable prospect of success.' The Judge went on to make a general comment about Mediation, stating that

> The essence of all successful Mediations is a willingness to compromise and/or the realisation that certain points are not as strong as the party believed ... In my view there was a reasonable prospect that these parties, given the essentially commercial nature of the dispute and being well advised, would have been prepared to compromise and/or would have accepted that various points raised were not as strong or certain as the open position they adopted.

In these circumstances the court at first instance considered it was unreasonable for the defendant to refuse to mediate and this was taken into account in refusing to order costs in favour of the defendant for the period after the mediation was proposed. On appeal, the Court went further and took the opportunity to endorse the advice in the *ADR Handbook* published in 2013 which sets out the steps to be considered to avoid a costs sanction following a request to mediate. These include:

- Not ignoring an offer to engage in ADR.
- Responding in writing with full reasons why ADR is not appropriate.
- Raising any shortage of information or evidence which is believed to be an obstacle to successful ADR.
- Not closing off ADR in case it is worth pursuing at a later date.

This was described as requiring 'constructive engagement in ADR rather than flat rejection, or silence'. Silence was said as a general rule to be 'of itself unreasonable'. On appeal, it was decided that 'the defendant's silence in face of two requests to mediate was itself unreasonable conduct of litigation sufficient to warrant a costs sanction, without the need for the detailed point by point analysis of Halsey guidelines.'

18.1.5 The approach of the courts to documents

The English courts have been involved in other important ancillary issues regarding mediation, including that of the use of documents prepared for the purposes of mediation in subsequent legal proceedings.

In *Aird* v. *Prime Meridian Ltd* (2007), the parties were involved in a court action which was stayed (put on hold) to allow a mediation to take place. Prior to that, the court had ordered the parties' expert architects to meet on a without prejudice basis and put together a joint statement setting out areas of agreement and disagreement between them with a view to this being used in the mediation. They did so. The parties, in the mediation agreement, agreed to 'keep confidential all information, whether oral or written or otherwise produced for or at the mediation'. The joint statement was then used in the mediation. The mediation which followed did not result in a settlement. The court action recommenced and the defendants wished to use the joint statement. The claimants objected on the basis that the statement was privileged, having been prepared for the purpose of, and to be used in, the mediation. The Court of Appeal decided that the joint statement was not privileged. It was ordered by the court and required to assist the court to exercise its case and trial management functions.

The fact that the court had made the order to assist a contemplated mediation did not alter the status and interpretation of the court's order. It was a question of fact whether the document produced was for the mediation alone or was produced to comply with the court order. As the experts in this case had removed the words 'without prejudice' from the final, signed version of the statement, the court took the view that the document was a joint statement prepared to comply with the court order. It was not privileged and could be referred to in the court proceedings.

This particular case turned on a specific rule within the English Civil Procedure Rules. However, it does highlight the need to be clear as to the basis on which experts are instructed and documents prepared and to take care as to the content of documents. If a document is prepared solely for a mediation and not for future use, it should state this on the face of the document to avoid any dubiety later.

18.1.6 The approach of the courts to confidentiality

The courts have also been required to consider issues of confidentiality in relation to mediations. In *Farm Assist Limited (in liquidation)* v. *The Secretary of State for the Environment, Food and Rural Affairs (Number 2)* (2009), a mediation took place in June 2003 resulting in a settlement agreement being signed. However, Farm Assist later sought to have this set aside on the basis that it claimed to have entered into it under 'economic duress'. In the court action, the Secretary of State for the Department of the Environment, Food and Rural Affairs (DEFRA) called the mediator to give evidence.

The mediator wrote to DEFRA to advise that given the time that had elapsed since the mediation, coupled with the fact that she had not taken any notes and had no recollection of the one-day mediation, any evidence that she would give would not be of any merit to either party. The mediator also relied on the terms of the parties' Mediation Agreement in relation to the issues of privilege and confidentiality and a clause which stated that she could not be called as a witness in any future litigation. The court commented on each aspect of this as follows:

- *'Without prejudice' privilege*: It was acknowledged that the mediation was covered by a without prejudice privilege. However, the Judge held that this was a privilege that existed between the parties and was not a privilege of the mediator. The privilege could be waived by those parties and that had happened in this case.

- *Confidentiality*: Unlike privilege, confidentiality affects not just the parties, but also the mediator and governs all of the dialogue between the respective parties. This could not be waived without the consent of all parties. The mediator could reserve the right to enforce the Mediation Agreement's confidentiality provision. However, where in the interests of justice it was necessary for evidence to be given on matters that would otherwise be held as confidential, the Court could order such evidence to be given.
- *The exclusion clause*: The parties had agreed that the mediator could not be called as a witness to give evidence 'in any litigation ... in relation to the Dispute'. The Dispute was defined as relating to work carried out by Farm Assist on behalf of DEFRA during the foot-and-mouth disease outbreak in 2001. The dispute in the current litigation was whether or not Farm Assist had entered into a settlement agreement under economic duress. The judge held that these were not one and the same.
- *The mediator's argument that she had little or no recollection of the mediation*: The Judge was not convinced by this. He held that memories can be jogged when a witness is shown evidence and anything that she could recall of what was said or done at the mediation would be important to either party's position.

The judge refused the mediator's application.

18.1.7 Cross-border mediations

The Cross-Border Mediation (Scotland) Regulations 2011 came into force on 6 April 2011 and apply to cross-border disputes, which are disputes where at least one party is domiciled or habitually resident in a different Member State of the European Union than that of another party. The Regulations therefore would not apply to disputes between parties residing in different parts of the UK, but they would apply if one party was in any part of the UK and another party was in another EU country.

The Regulations provide (Article 3) that a mediator of a relevant cross-border dispute is not to be compelled in any civil proceedings or arbitration to give evidence, or produce anything, regarding any information arising out of or in connection with that mediation. The exceptions to this are where all parties to the mediation agree otherwise.

Importantly, the Regulations also make changes to the prescription and limitation periods under the Prescription and Limitation (Scotland) Act 1973 (for an explanation of the Act, see Section 9.9). Article 5 provides that the prescriptive period calculated in relation to any relevant cross-border dispute is extended where the last day of the period would otherwise fall (1) in the eight weeks after the date that a mediation in relation to the dispute ends; (2) on the date that a mediation in relation to the dispute ends; or (3) after the date when all of the parties to the dispute agree to participate in a mediation in relation to the dispute but before the date that mediation ends. In these circumstances the prescriptive period is extended so that it expires on the date falling eight weeks after the date on which the mediation ends. The end of the mediation is defined as being the date when: (1) all of the parties reach an agreement resolving the dispute; or (2) the parties agree to end the mediation; or (3) a party withdraws from the mediation; or (4) 14 days after the mediator's tenure ends (by death, resignation or otherwise) if a replacement mediator has not been appointed.

18.2 Settlement agreements

18.2.1 General

In the event that a settlement is achieved in the course of a mediation, it is normal practice to record the terms of settlement in a formal written agreement which is then signed by parties before bringing the mediation to an end.

It is important to ensure that the settlement agreement deals with all matters in dispute between the parties and sets out fully the terms of settlement agreed. It should deal with matters such as payment of VAT, tax, interest, legal costs and the disposal of any proceedings underway in court, arbitration or any other forum. It should cover the mechanisms to implement the settlement, including who is to do what, by when and what is to happen if a party fails or delays in taking action required or a dispute develops as to the terms of settlement. It would often include provisions covering confidentiality of the terms of settlement and sometimes of the existence or content of the dispute.

18.2.2 Interpretation of settlement agreements

Of course, settlement agreements are entered into in most cases outwith the context of mediation and there have been a number of cases in the courts related to the interpretation of settlement agreements. The basic principle is that these are dealt with in the same way as any contract and interpreted in accordance with the normal rules on contract interpretation. However, particular care is needed to ensure clear drafting as there will be a 'reluctance to infer that a party intended to give up something which neither he, nor the other party, knew or could know that he had', according to Lord Bingham in *BCCI* v. *Ali* (2002). Lord Clyde in the same case said:

> Generally if they intend their agreement to cover the unknown or unforeseeable, they will make it clear that their intention is to extend the agreement to cover such cases. If an agreement seeks to curtail the possible liabilities of one party, he, if not both of them, will generally be concerned to secure that the writing clearly covers that curtailment.

In other words, it is perfectly possible in a settlement agreement to compromise future claims, for example, to reach a full and final settlement which covers not only defects currently known about but also future, latent defects. However, to do so, the wording of the settlement agreement must make it clear that this is the intention of the parties.

The issue of interpretation of a settlement agreement arose in *Point West London Limited* v. *Mivan Limited* (2012). In that case, Mivan had worked on apartments for Point West. Practical completion was achieved in July 2002. There were issues with the curtain walling and the heating and cooling systems. An agreement on the final account was reached in 2005. The agreed sum was not paid by Point West. Instead, in 2007 they entered into further discussions as Point West wanted to 'do a deal to enable Mivan to walk away'.

An exchange of letters followed. These referred to an agreement reached 'in respect of all Works carried out, and any corresponding outstanding matters'; 'achieving full and final settlement in respect of the above Works together with any and all outstanding matters'; and 'this final agreement concluded Mivan's responsibilities and obligations in respect of their Works'.

Subsequently, a tenant of one of the apartments pursued Point West in relation to the defects in the curtain walling and heating and cooling systems. It turned out that there were serious defects in these, necessitating complete replacement. This was in contrast to the position in 2007 when they were considered to be relatively minor. Point West argued that the 2007 agreement did not include a settlement of any liability to pay damages for defects, including in particular latent or unknown defects. The serious defects in the curtain walling and heating and cooling systems were therefore, they argued, not covered in the agreement.

The Judge decided that the agreement relieved Mivan from defects in the curtain walling and heating and cooling systems. This conclusion was reached based on a detailed consideration of the factual circumstances. The words used in the settlement were also scrutinized and those quoted above were thought to clearly envisage a full and final settlement of all of Mivan's responsibilities and obligations in respect of defects.

18.2.3 Interaction of settlement agreements with 1996 Act

A further consideration in construction disputes is the potential interaction of settlement agreements with the 1996 Act.

In *JB Leadbitter & Co Limited* v. *Hygrove Holdings Limited* (2012), the parties entered into a supplemental agreement. That agreement provided that it varied and supplemented the JCT building contracts entered into between the parties – Coastal Housing Group and Hygrove under a head contract and Hygrove and Leadbitter as employer and contractor.

The supplemental agreement provided for payments from Coastal under the head contract to be paid into an escrow account and sums due from Hygrove to Leadbitter under their contract to be paid out of the escrow 'provided sufficient funds exist in the escrow … '.

Certificates were issued to Leadbitter. There were no withholding notices. Hygrove failed to pay Leadbitter. They argued that Coastal had failed to pay sufficient money into the escrow account. They argued on the basis of the wording of the supplemental agreement that this meant there was no obligation on them to pay. They were effectively arguing that the supplemental agreement set up a pay when paid mechanism.

Leadbitter argued that this was contrary to the 1996 Act which outlawed pay when paid clauses. The court decided that as the supplemental agreement amended the JCT contracts, it should be treated as a construction contract and therefore was required to comply with the 1996 Act. This payment mechanism was not 1996 Act compliant and the money was therefore due to be paid.

This particular case did not concern a settlement agreement but the same principle would apply if, for example, a settlement agreement required one party to do work under the existing contract or if the agreement was framed as an amendment or supplement to the original building contract. It is therefore worth considering whether the terms of the 1996 Act are likely to impact on settlement agreements to ensure what is intended to be included within the settlement is in fact delivered.

18.2.4 Disputes under settlement agreements

The usual expectation when parties enter into a settlement agreement is that it will resolve disputes between them rather than trigger them. However, there have been a number of cases where this has not happened. In *Interserve Industrial Services Limited* v. *ZRE Katowice* (2012), Interserve and ZRE had entered into sub-contracts for scaffolding and insulation at a power station in Pembrokeshire. The sub-contract contained dispute resolution provisions which provided that final settlement of disputes was to be by arbitration.

Disputes arose regarding Interserve's entitlement to interim payments. These were resolved and documented in a Settlement Agreement. It set out provisions for payment of the 'Outstanding Sum' on achievement of milestones. It confirmed in respect of any other amounts becoming due that these would be dealt with in accordance with the sub-contract. The Agreement was said to be governed by the laws of England and Wales and the courts of England and Wales were stated to have exclusive jurisdiction in respect of any dispute arising under the Agreement. Following the Agreement, work continued on the remainder of the sub-contract. However, disputes arose about non-payment of sums said to have fallen due under the Agreement.

Interserve raised a court action, claiming entitlement to do so under the exclusive jurisdiction clause in the Agreement. ZRE applied to have the court action put on hold and the dispute referred to arbitration in accordance with the dispute resolution provisions of the sub-contract.

In reaching a decision, the Judge noted that the Settlement Agreement was not made in full and final settlement of all claims under the sub-contract and that the parties' obligations under the sub-contract remained to be performed. He considered there was an implied term in the Agreement that disputes under it would be subject to the same dispute resolution procedure as the sub contract. In doing so, he adopted the approach in *L Brown & Sons Limited* v. *Crosby* (2005). There, the parties entered into side agreements varying the terms of the contract. These did not contain separate dispute resolution provisions but, applying the 'officious bystander' test, it was considered that an officious bystander would assume that the underlying contract provisions would apply.

He had to go further though, since in the *Interserve* case there was a dispute clause giving the English courts exclusive jurisdiction. To deal with this, he made a distinction between a dispute resolution clause (such as that in the sub-contract) and an exclusive jurisdiction clause (such as that in the Agreement). He considered the sub-contract dispute resolution clause was a self-contained regime for the resolution of disputes. The exclusive jurisdiction clause supplemented that by making clear what laws should apply to any arbitration but did not trump it by substituting a new forum for disputes.

This is a surprising result. The Settlement Agreement provided not just for which law was to apply but also for which forum had jurisdiction over the dispute. Here, the courts were given exclusive jurisdiction. Normally, terms would not be implied where they are in conflict with express terms and here there was a clear express term in the Settlement Agreement which dealt with the method of resolving disputes under that Agreement.

The practical advice, in the light of this, would be to ensure that intentions are fully and clearly set out in any settlement agreement. In the unfortunate circumstances of an agreement still giving rise to disputes, this should at least ensure that the dispute is limited to the substantive issues as opposed to a preliminary (and possible expensive) skirmish over where the battle is to take place.

18.2.5 Multi-party settlements

The other issue arising in relation to settlements is in the context of a settlement between two parties in a contractual chain, say, an employer and contractor, where one of the parties may have a related claim against a third party, possibly a sub-contractor.

The ideal situation in such circumstances would be for any settlement agreed to involve all three parties. That would avoid any risk associated with the contractor being stuck in the middle with one deal being agreed up the contractual chain with the employer but ending up with a different deal down the line with the sub-contractor.

Where that is not possible, there is some guidance from the courts on what the party in the middle requires to do and what it can do to protect itself.

In terms of proving liability, a settlement between two parties in the contractual chain would not relieve a party from requiring to establish the liability of the third party in any claim, see *Fletcher & Stewart Limited* v. *Peter Jay & Partners* (1976). The settlement achieved will not help or hinder that process. As Geoffrey Laing LJ put it:

> The nature and amount of any settlement negotiated previously to that between the defendant and the plaintiff had nothing to do with the liability as between the defendant and the third party. It might have been relevant on the amount of damages to be paid by the third party to the defendant once liability on the part of the third party had been established.

In relation to quantum of the claim, the amount of any settlement previously made may be relevant evidence of the amount recoverable, see *Biggin & Co* v. *Permanite*

Limited (1951). That case concerned the question of whether a reasonable sum paid in settlement of a claim can be regarded as the proper measure of damages in a subsequent action when liability was not being disputed. It was said the settlement sum would constitute the upper limit of what could be recovered in the third party case. If reasonable, it should be taken as the measure of damages. On the question of what evidence would be necessary to establish reasonableness, this would include proof that the settlement was made on the basis of legal advice. There would also need to be evidence on what would be likely to be proved if the first case had proceeded to allow the court to consider reasonableness of the sum paid.

The basis for that approach is that the third party is taken to have foreseen that a consequence of its breach of contract would be that the party pursuing it would be liable to the other party, that that liability might give rise to litigation, and that any such litigation might be compromised resulting in loss to the pursuer. See *Bovis Lend Lease Limited* v. *R D Fire Protection Limited* (2003).

It is necessary as part of mitigation of loss for the pursuer to show both that it was reasonable to settle the claim at all and that the settlement amount was reasonable.

Where there are a number of heads of claim and if it is possible, it would be prudent to identify heads of claim and the individual treatment of each of these in the settlement as well as the amount of money allocated to each issue, see *P&O Developments Limited* v. *The Guy's and St Thomas' National Health Service Trust* (1999). In this case, the starting point was said to be a requirement to show that the third party was in breach of a duty owed. The breaches by the third party were likely to be different from those by other parties. That meant the consequences of any breaches could not be said to be identical and that questions of causation would arise in assessing damages. In this case a global settlement of £83 million had been reached by Guy's with their management contractor which Guy's then sought to recover from their project manager and M&E services engineer. The court found that even if that overall settlement sum was found to be reasonable, it did not necessarily follow that the sums then allocated to individual works contractor claims were also reasonable. This meant it did not follow that, if the reasonableness of the global sum was established, the sums then allocated to works contractors would represent Guy's loss. It was said to be necessary for Guy's to prove that the sums in fact allocated were reasonable sums to allow.

It can be difficult to achieve an allocation of a settlement between various heads of claim in the context of a negotiated settlement where often a global figure is arrived at. In that case, it should be recognized that this may present a problem when it comes to proving liability and thereafter allocation of responsibility against the third party.

However, this will not necessarily be fatal and the courts will be reluctant to allow a claim to fail on this basis. In *Bovis* v. *R D Fire Protection* above, it was said:

> However, it does not follow from these difficulties that it is impossible to allocate either an overall value to the settlement or a value of that part of the settlement that is attributable to defects in the fire protection works. Even rudimentary evidence of how the settlement was arrived at would enable it to be determined whether anything was included for fire protection works claims and as to whether the overall value of the individual component claims being settled should be pro rated or assessed in some other way. If such an apportionment or assessment is not possible

> given the nature of the settlement negotiations, that difficulty could be explained and proved by evidence … Thus, without both factual and expert evidence to support the assertion, the court cannot and will not proceed on the basis that neither a global valuation of the settlement nor an appropriate allocation of the settlement to the fire protection works claims is possible.
>
> Moreover, the court can undertake its own assessment of the value and appropriate apportionment to be placed on a settlement with only very limited material to work with.

The general principles applicable were summarized in *P&O Nedlloyd* v. *M&M Militzer & MNCH International Holding AG* (2003) as including:

1. The law encourages reasonable settlements, particularly where strict proof would be expensive.

2. In relation to the evidence required to establish reasonableness, it is relevant to prove that settlement was made on the basis of legal advice. In this case, the absence of legal advice was a factor taken into account by the Judge in finding the settlement was unreasonable in the circumstances.

3. The claimant requires to establish that the amount for which he settled was reasonable. If a claimant overlooked a point which he should have taken, the amount of the settlement will not be taken as the correct measure of damages in a subsequent action.

4. In a case where an indemnity against loss suffered due to claims is relied on, consideration will be given to whether the loss was due to a reasonable settlement of a claim which had some prospect or significant chance of success.

In terms of practical steps to take, the following are suggested:

- Legal advice should be taken.
- Records should be kept of communications during the negotiation of the settlement and they should avoid being 'inconsistent, inconclusive and contradictory' (see *P&O Nedlloyd* above, paragraph 133, in which this was the description given to communications between key people within the claimant's organization when investigating the claim).
- It should be kept in mind that those conducting the negotiation may be called upon to give evidence and so people of suitable seniority and an ability to do so should be involved. They should keep whatever records they need to remind themselves at the stage of giving evidence of what took place. Contemporaneous documents will be important evidence and are likely to be regarded as more reliable than just the recollections of a witness.
- In terms of evidence as to prospects of success in the claim, in addition to legal advice, expert reports and input from relevant technical experts are likely to be important.

- Keeping the third party informed of arguments being made, asking for their views as to any other points which can be made and inviting their participation in any negotiation can be helpful. Even if they refuse to participate, the request that they do so may pre-empt any suggestion by them that there were arguments which should have been, but were not, made.
- The potential need to give evidence on the make-up of the settlement sum as between various heads of claim should be borne in mind and records kept of any likely split between the heads of claim, even if the ultimate settlement is made on a global basis.

If it is intended to use the fact that legal advice was taken and relied on, then in showing that it was reasonable to rely on that, the party doing so will be taken to be waiving legal privilege over documents related to that advice, see *Lloyds* v. *Kitson* (1994).

18.3 Expert determination

Expert determination is a method of dispute resolution which is available either where parties include a provision to this effect in their contract or if they subsequently agree to use it. It is included in some of the standard forms of contract. An example is the Institution of Chemical Engineers' Model Forms of Contract where there is provision for disputes in relation to certain matters to be referred to an expert, the identity of whom is to be agreed between parties to the contract (failing which, appointed by a specified appointing body). The Institution of Chemical Engineers has published 'Rules for Expert Determination' which detail how the determination is to be conducted. The expert's remit is, in terms of those contracts, to decide all disputes referred to him and the parties agree to be bound by and to comply with decisions made. Provisions for expert determination are also often found in bespoke forms of contract.

The procedure (unless this is set down in the contract or in procedural rules such as in the Institution of Chemical Engineers' Rules referred to above) is flexible. The expert's remit requires to be set out by the party referring the dispute.

The identity of the expert is a matter for the parties. They can agree to a named individual (or a list of individuals) in the contract or else agree to apply to an appropriate professional body for the appointment of an expert. Normally the expert would be someone skilled in a discipline relevant to the subject matter of the contract or the dispute between the parties.

There is a distinction between judicial decisions and the decisions of experts, see *Bernhard Schulte GmbH & Co. KG* v. *Nile Holdings Ltd* (2004). Judicial decisions are made on the basis of submissions and evidence presented. Expert decisions are made (unless the contract provides otherwise) on the basis of the expert's own investigations, opinions and conclusions, regardless of what is submitted by the parties. An expert determination (as opposed to a judicial determination) is not limited by the submissions made or the evidence put forward by the parties unless the contract and the terms of reference state that this is the case. It is this distinction which is being made when phrases such as 'acting as an expert and not as an arbitrator' are used.

Perhaps surprisingly, it has been found that there is no requirement for the rules of natural justice or due process to be followed in an expert determination for the decision in that process to be valid and binding. However, a decision made due to actual bias (not just a perception of bias) on the part of the expert would be capable of being set aside.

The courts have established ground rules as to how an expert determination should be conducted, and have had to decide whether they should put court actions on hold to allow the matters in dispute to be referred to expert determination. This can be due to questions being raised as to the extent to which the court has any right, for example, to determine whether or not an expert has jurisdiction to hear a dispute.

The courts have decided they are entitled to consider questions as to whether the expert does in fact have jurisdiction to decide a dispute under an expert determination clause. This was the issue in *Barclays Bank plc* v. *Nylon Capital LLP* (2011). In that case, the agreement contained an expert determination clause stating that in the event of a dispute regarding profits, a party could refer the matter to an accountant for a determination. The reference could be made 30 days after an allocation of profits had been made. Barclays argued the expert had no jurisdiction because Nylon had not yet allocated profit.

The court distinguished the approach to an expert determination clause from that to an arbitration clause. Arbitration was usually an alternative to court for resolution of disputes. With expert determination, these clauses tended to refer only certain types of dispute to an expert, reserving other types to the court. In this case, whatever decision the expert made on jurisdiction could be challenged. That was different from his determination of a matter within his jurisdiction which was not challengeable.

In *Wilky Property Holdings plc* v. *London and Surrey Investments Limited* (2011), the court stayed (put on hold) a claim concerning the interpretation of a consultancy agreement where the agreement contained an expert determination clause to allow the matter to be decided through expert determination.

In *Toepfer* v. *Continental Grain Co.* (1974), it was said:

> When parties enter into a contract on terms that the certificate of some independent person is to be binding as between them, it is important that the court should not lightly relieve one of them from being bound by a certificate which was honestly obtained and not vitiated by fraud or fundamental mistake on the part of the certifier.

In *Jones and Others* v. *Sherwood Computer Services plc* (1992), it was said the principal ground for challenge would be that an expert had materially departed from his instructions, so that the determination is not a determination made in accordance with the terms of the contract. Material is said to be anything other than trivial or *de minimis*, meaning that it is so minor as not to make any possible difference to either party. In such cases, the determination would not be binding on the parties.

In *Nikko Hotels (UK) Ltd* v. *MEPC plc* (1991), it was said that unless the terms of the contract provided otherwise, an expert determination cannot be challenged on the ground the expert made a mistake, as long as the expert answered the question which was put to him and had not otherwise departed from his instructions.

In *Veba Oil Supply & Trading GmbH* v. *Petrograde Inc* (2002), the principle was stated (following Lord Denning in *Campbell* v. *Edwards* (1976)) that:

> If an expert makes a mistake whilst carrying out his instructions, the parties are nevertheless bound by it for the very good reason that they have agreed to be bound by it. Where, however, the expert departs from his instructions, the position is very different: in those circumstances the parties have not agreed to be bound.

It was said that once a material departure from instructions is established, the court is not concerned with the effect of that on the result. The departure itself is sufficient to render the decision non-binding. The case set out the test for establishing whether an expert has materially departed from his instructions. It was said that any departure would be material unless it could truly be characterized as trivial or *de minimis*. In considering what parties would have regarded as being material, the court will take into account the subject matter and express terms of the contract and all the relevant circumstances.

Decisions of experts can be subject to scrutiny by the courts in certain circumstances, see *Bernhard Schulte GmbH & Co. KG*. In that case, there were claims that the reference to the expert had been made outwith the timescales allowed for in the contract, that the expert acted outwith his mandate, that the expert conducted the reference unfairly and that the expert, without regard for the agreed procedure, made findings without giving one of the parties adequate opportunity to make submissions to him and did not properly take into account the submissions they did make.

The court recognized that there was a distinction between the expert making a mistake in carrying out his functions, on one hand (which was said to be part of the risk run by parties in agreeing to be bound by the expert's decision) and failures to carry out his instructions, on the other (which would mean the expert's determination had not been made under the contract and would therefore not be binding due to the failure to adhere to the contract requirements and failure to carry out the functions required of him). It was also noted that where the contract provides for the decision of an expert to be final and binding, it binds the parties as long as there is no fraud, collusion, bias or material departure by the expert from his instructions.

The courts again indicated a willingness to be involved in the regulation of expert determination in *Halifax Life Ltd* v. *The Equitable Life Assurance Society* (2007). In this case, the contract between the parties provided for certain disputes to be dealt with by a binding expert determination. A dispute arose and the parties agreed the expert's terms of reference. These included provisions that his decision would be binding, save for manifest error, and that he would provide a reasoned decision. Halifax were dissatisfied with the expert's decision and sought a declaration from the court that the determination was not binding. This was on the basis that the expert had 'materially departed from the agreed terms of reference by failing to provide any adequate reasons for his decision'. It was also said the determination contained a manifest error. In relation to the question of manifest error, the court observed that:

> If a decision is issued in a dispute where it is binding 'save for manifest error' a party wishing to challenge the decision may face insuperable difficulties if the expert is not obliged to give reasons and fails to set out the reasons for his decision.

In this case, the court considered that the expert's reasons were not sufficient to explain the conclusions reached and the case was sent back to the expert.

Manifest error has been defined in *Conoco (UK) Limited* v. *Phillips Petroleum (UK) Limited* (unreported, 19 August 1996), which was cited with approval in *Veba Oil Supply & Trading GmbH* v. *Petrotrade Inc* (2002), to mean oversights and blunders so obvious and obviously capable of affecting the determination as to admit of no difference of opinion. In relation to the provision of reasons for a decision, Mr Justice Cresswell drew on a number of cases related to other forms of tribunal. In *English* v. *Emery Reimbold & Strick* (2002), it was said: 'We would put this matter at its simplest by saying that justice will not be done if it is not apparent to the parties why one has won and the other has lost.'

In *South Buckinghamshire DC* v. *Porter (No. 2)* (2004), Lord Brown of Eaton-under-Heywood stated:

> The reasons for a decision must be intelligible and they must be adequate. They must enable the reader to understand why the matter was decided as it was and what conclusions were reached on the 'principal important controversial issues', disclosing how any issue of law or fact was resolved. Reasons can be briefly stated, the degree of particularity required depending entirely on the nature of the issues falling for decision.

The issues of error and provision of reasons were linked. Where the parties provided for a decision to be subject to review in cases of manifest error, it was essential the decision contained reasons sufficient to explain why, on each head of claim, one party won and the other lost. In the absence of this, it would not be possible to tell whether or not an error had been made.

The expert held private meetings with each party. That in itself was not criticized but it was said that if an expert proceeded in this manner, it was essential that he set out in his decision what information and evidence he had taken into account from the private meetings, and how that information influenced him in reaching his decision. The requirement to give reasons was said to mean the reasons should explain what the expert's conclusions were on the heads of claim and give adequate reasons in the circumstances. The court considered it had power to direct the expert to state further reasons to allow them to properly consider the overriding issue of whether his decision was binding on the parties.

In *Homepace Limited* v. *SITA South East Limited* (2008), the expert volunteered additional reasons and the Court of Appeal decided these showed the decision contained an error and should not be enforced.

In *Walton Homes Ltd* v. *Staffordshire County Council* (2013), the court had to consider whether an expert's interpretation of a clause in a contract for sale of land was manifestly erroneous. The court found it impossible to say the reasoning was manifestly erroneous. It was said that 'Manifest is a word which gives a very limited window of opportunity to challenge.' By reference to previous authorities it was said that it would require something like an arithmetical error or a reference to a non-existent building or something similar. The court drew support for its conclusion that there was nothing manifestly wrong with the expert's decision from the fact that the competing arguments put forward by the parties were strong on both sides.

In cases of fraud or collusion by the expert, the determination would not be binding, irrespective of whether this affected the result, see *Veba Oil Supply & Trading GmbH* and *Campbell.*

The expert does not have immunity if he has acted negligently and it would be open to the party who sustained a loss due to such negligence to pursue a claim against the expert for loss suffered. In this respect the position of the expert is different from that of arbitrators and adjudicators. See *Arenson* v. *Casson, Beckman Rutley & Co* (1977) and *Palacath* v. *Flanagan* (1985). It is of course open for the parties to agree to exclude any such liability on the part of the expert by means of an express term in the appointment, and it is common for an expert to insist on such an exclusion as a condition of him accepting an appointment.

The advantages of expert determination are that it is private, it can be cost-effective and speedy and the decision is made by a person skilled in the relevant area. The disadvantages are the lack of control which may be exercised by the courts in a situation where an expert is making a final and binding decision, given the unavailability of any appeal and the very restricted grounds for challenge available.

18.4 Early neutral evaluation

Early neutral evaluation is a process whereby parties may present their cases to an independent third party who will then give their preliminary views of the likely ultimate outcome of the case or of particular aspects of it. The evaluation of the case is non-binding but may assist parties in any negotiations or avoid unnecessary other forms of procedure. Parties agree the matters to be submitted to the process and the material to be presented to the evaluator. It is possible for this to be dealt with on the basis of only written submissions and documents or to include a hearing with oral submissions. This would be a matter either for agreement by the parties or for the evaluator to direct.

This is a process which is offered by the English courts and there is provision in their court rules for a judge to act as the evaluator. The rules provide that in what are considered to be appropriate cases, and with the agreement of all parties, the court will provide a without prejudice, non-binding evaluation of a dispute or particular issue. The judge who conducts the evaluation is then not involved any further in the case, and if the case proceeds following the evaluation it is dealt with by a different judge.

The advantages of the procedure are that at an early stage, and before significant costs are incurred, parties can be given an indication of where matters might ultimately end up, thereby facilitating negotiations and resolution. Should the view be negative for one party, it allows them to make an early decision on whether or not to proceed further and, potentially, to obtain the further evidence which is required to avoid the predicted outcome. The procedure is confidential. The disadvantage, if one party receives a negative appraisal of their prospects, is that that party has effectively 'shown its hand' and others will be aware that there exists a poor view on prospects, which could make a negotiated settlement on good terms more difficult to achieve.

18.5 Senior management review

Some contracts provide for a board or panel to be set up, containing individuals at a number of levels within the management structure of the parties to the contract. The idea of this is that if those at, say, quantity surveyor level are unable to resolve a dispute, it is then referred to senior manager or director level then, ultimately, to the managing directors/chief executives of the parties. This can take the heat out of a situation where it may be personalities rather than real issues that are getting in the way of resolution. In some contracts, these boards or panels are the first port of call when any dispute arises. The dispute is referred to the board so that the board can attempt to reach a resolution, generally within a short and specified period of time. This would be dealt with while the project was on-going and would allow the works to continue. Normally if resolution is not achieved, the dispute is referred to a more formal form of dispute resolution procedure.

The SBC Schedule Part 8 paragraph 6 and SBC/DB Schedule Part 2 paragraph 12 (Supplemental Provisions) provide for this escalating form of dispute resolution. With a view to avoidance or early resolution of disputes or differences, each party is to promptly notify the other of any matter that appears likely to give rise to a dispute or difference. The senior executives nominated in the Contract Particulars are then to meet as soon as practicable for direct, good faith negotiations to resolve the matter.

In some cases, these clauses can be misused by a party on the receiving end of a claim simply as a way to delay the claim being dealt with if this suits their purpose. At their extreme, they can themselves give rise to litigation in order to obtain rulings on whether or not the clauses are binding or represent conditions precedent to an ability to have a dispute determined in court or arbitration.

In *Cable & Wireless plc* v. *IBM United Kingdom Limited* (2002), the court considered the multi-stage dispute resolution provisions of the contract.

Stage 1 was that parties were to 'attempt in good faith to resolve any dispute or claim' through negotiations between senior executives in accordance with the procedure set out in the contract. This involved discussions at project review meetings, discussion at management level and then discussion at a more senior level of management. The procedure included time limits for each stage.

Where the management level discussions failed, parties were to 'attempt in good faith to resolve the dispute or claim through an ADR procedure as recommended to the parties by the Centre for Dispute Resolution'.

The clauses provided that neither party could initiate any legal action until the stage 1 process had been completed. It also provided that if stage 1 failed, parties 'shall seek to resolve disputes' by the ADR procedure specified.

The discussions at various levels of management failed to resolve matters and IBM wished the court action to be stayed pending ADR. Cable & Wireless declined to refer the claim to ADR, arguing that the clause was unenforceable because it lacked certainty and was no more than an agreement to negotiate. The court analysed the clause and considered there was no doubt it was the intention of parties that litigation was to be a last resort if negotiation or ADR failed. In considering whether the reference to ADR could be binding, the court considered that the parties had gone further than simply agreeing to attempt in good faith to negotiate a settlement. They had identified a particular procedure, namely, ADR as recommended by CEDR, recognized as an experienced dispute resolution provider which had published a model Mediation Procedure and Agreement, detailing procedures to be followed in any mediation.

The court contrasted this with a simple undertaking to negotiate a contract or settlement agreement. This would be insufficiently certain to allow the court to apply objective criteria to decide whether or not the parties were in compliance or breach of such a provision. If the clause had simply required the parties to 'attempt in good faith to resolve the dispute or claim', that would not have been enforceable. The Cable & Wireless/IBM clause, on the other hand, set out in more detail how the parties were to go about their attempts to reach agreement. This was thought to include steps which were sufficiently certain to allow a court to readily ascertain whether or not there had been compliance. An important consideration would be whether the obligation to mediate was expressed in unqualified and mandatory terms. Where it was, it was thought a sufficiently certain and definable minimum duty of participation should not be hard to find.

The court considered there to be extremely strong case management grounds for allowing the reference to ADR to proceed and delayed hearing the Cable & Wireless claim pending all outstanding disputes going to ADR.

In *Tang Chung Wah* v. *Grant Thornton International Limited* (2012), the claimants were seeking a court order to the effect that an arbitrator's award was of no effect because the tribunal did not have jurisdiction. The basis for this argument was that the underlying contract contained an escalating dispute clause which, they argued, operated as a condition precedent to any arbitration taking place. The requisite steps had not been taken and therefore the arbitration was premature and the award should not be enforced.

These arguments had been raised during the arbitration but the arbitration tribunal had decided that there was no contractually enforceable condition precedent to prevent it having jurisdiction. The mechanism set out in the contract included, as stage 1, any dispute being referred to the Chief Executive. The Chief Executive was to attempt to resolve the dispute in an amicable fashion and had up to one month after receipt of a request to attempt to do so. Stage 2 was a reference to a panel of three members of the Board to be selected by the Board. The panel had up to one month to attempt to resolve the dispute. Until the earlier of the date that the panel determined it could not resolve the dispute or one month after the dispute was referred to it, no party could commence any arbitration. Stage 3 was that any dispute was to be referred to and finally resolved by arbitration.

In this case, a decision had been taken to expel a member from the Grant Thornton Member Firm Agreement. This gave rise to a dispute which was referred to the Chief Executive. The Chief Executive's response was that as he had been involved in the decision to expel he did not consider he could act as an objective conciliator and therefore recused himself from the role. This triggered Stage 2 of the procedure. The Chairman of the Board requested members of the Board to advise if any of them felt able to act on a reconciliation panel. As they had previously supported the decision to expel, they considered it futile to form such a panel. No members put themselves forward and the three-person panel referred to within the dispute clause was not constituted.

The court considered whether the dispute clause constituted an enforceable obligation and whether it was a condition precedent to arbitration. On the basis of numerous authorities, the court listed the relevant guidelines:

- Agreements to agree and agreements to negotiate in good faith, without more, are unenforceable.
- Good faith is too open-ended a concept to provide a sufficient definition of what such an agreement must, as a minimum, involve and when it could objectively be determined to be properly concluded.
- However, where a provision is only one part of an otherwise enforceable contract, the court will do its utmost to find a construction which gives it effect. It may imply criteria or supply machinery sufficient to enable the court to determine: (1) what process is to be followed and when; and (2) how, without the necessity for further agreement, the process is to be treated as successful, exhausted or properly terminated.
- The court will consider each case on its own terms. The test is whether the obligations the clause imposes are sufficiently clear and certain to be given legal effect.
- In the context of a positive obligation to attempt to resolve a dispute amicably before referring a matter to arbitration, the test would be whether the provision prescribes, without the need for further agreement, a sufficiently certain and unequivocal commitment to commence a process. That would need to include the steps each party is required to take to put that process in place, sufficiently clearly defined to enable the court to determine objectively what is the minimum required of the parties to the dispute in terms of participation and when the process will be exhausted without breach.
- Where there is a negative stipulation preventing a reference to arbitration until a given event, the event needs to be sufficiently defined and it needs to be possible for a court to ascertain whether or not it had happened.

In this case, the clause was considered 'too equivocal in terms of the process required and too nebulous in terms of the content of the parties' respective obligations to be given legal effect'. The clause contained no guidance on the quality or nature of efforts to be made to resolve a dispute. This left the court unable to determine whether or not there had been compliance. In relation to the provision preventing a reference to arbitration until the panel determined it could not resolve or one month after the reference to the panel, the court considered this had to be interpreted taking into account that the purpose of the provision was to provide an end date after which any restriction on the right to arbitrate would lapse. Here, the panel had not been established as no Board member considered they could participate. However, more than a month had passed before the reference to arbitration. That was considered sufficient. The claimants had argued that as no panel had been constituted, no arbitration could be commenced. The court was not attracted to this and considered it unrealistic to consider parties could have intended the Board could indefinitely postpone the right to arbitration simply by not convening the requisite panel.

In addition to establishing the principles to be applied to such clauses, these cases highlight the potential pitfalls which accompany them. The clauses can be cumbersome and if there are too many steps, each with their own timetable, it is possible to delay the sometimes inevitable commencement of a court action or arbitration for quite some time. This can be a particular problem if there is any question of time bar looming or indeed if cash flow is tied up pending resolution. Also, the more steps there are, the more possibility for arguments as to whether each has been complied with.

It would be worth considering in each case what parties are hoping to achieve by such clauses. No clause is required to allow parties to negotiate or mediate. A clause may be considered useful to focus parties' minds on these options but they are likely to be options that would result from good management of disputes within businesses as a matter of course and the potential downside is worth consideration. If such clauses are to be included, they require to be carefully drafted to allow them to be used constructively to short-circuit disputes.

18.6 Dispute boards

18.6.1 Background

Dispute boards in the formal sense first came into being in the United States in the 1960s and 1970s. The procedure was first used on the Boundary Dam project in Washington in the 1960s in the form of a Joint Consulting Board. In 1975, there was the first Dispute Review Board on the Eisenhower Tunnel Project in Colorado. It is a procedure which, since then, has extended internationally. In 1981, the World Bank suggested a board be appointed for the El Cajon Dam and Hydro project in Honduras and, in 1990, it produced a modified FIDIC contract incorporating a Dispute Review Board procedure. In 1995, the World Bank made their use mandatory for certain projects over US$50 million, which increased its use further. The procedure was used on the 'Big Dig' tunnel project in Boston – a US$14 billion and 14-year-long contract. In the UK, it was used on the Channel Tunnel and the Docklands Light Railway projects and the Olympic Delivery Authority appointed a Dispute Resolution Board to oversee all contracts for the 2012 London Olympics.

According to the Dispute Resolution Board Foundation (DRBF), from a database of over 1200 projects since 1975, 60% of projects with a Dispute Review Board had no disputes, 98% of disputes referred to a Dispute Review Board resulted in no subsequent litigation or arbitration and worldwide use of Dispute Review Boards is growing by in excess of 15% per year. Costs of the Dispute Review Board are reported to be in the region of 0.05%–0.25% of final contract costs depending on the level of difficulties in the project and therefore the level of involvement of the Dispute Review Board.

The purpose of dispute boards is either to prevent disputes occurring or to achieve resolution of disputes quickly while work proceeds and to prevent escalation of disputes to the extent that relationships break down completely.

18.6.2 Standard form provisions, rules and procedures

Provisions regarding dispute boards are to be found in certain standard form contracts. For example, in 1995, the FIDIC Orange Book contract (and later the Red, Yellow, Gold and Silver Books) introduced the concept of a Dispute Adjudication Board (DAB) to which claims would be submitted for consideration and which would issue decisions to parties to the contract which would be binding unless a contrary decision was subsequently made in arbitration proceedings. The Red and Gold Books provide for a permanent DAB while the Silver and Yellow Books provide for the Board to be appointed on an ad hoc basis. The FIDIC procedure provides for a binding decision to be made within 84 days of referral of the dispute. Unless a notice of dissatisfaction is then issued within 28 days, the decision becomes final and binding. The International Chamber of Commerce produced a set of Dispute Board Rules in 2004 and the Institution of Civil Engineers (ICE) produced a Dispute Resolution Board Procedure in 2005.

18.6.3 Membership and role

The members of the board are appointed at the beginning of the project and remain in place until completion. There are often three members, often a mixture of construction professionals such as engineers and lawyers. In some cases, each party would nominate one member with the parties approving each other's nominees, then those two members appoint a third member as chairman. In some cases there is one person appointed – referred to in FIDIC contracts as a Dispute Review Expert (DRE). They are provided with the contract documents and other relevant information. These boards can have a roving role whereby they visit the site and the parties on a regular basis, and either identify possible future areas of dispute or deal with any matters which have arisen. Alternatively, they can be brought in as and when disputes arise to make a recommendation or decision. These boards tend to be used particularly on larger and long-running contracts.

Members of the boards are chosen for their skills and experience in both dispute resolution procedures and also the technical issues involved in the particular contract. They should also be independent of the parties. If the board are involved in visiting sites on a regular basis, then not only do they become aware of disputes at an early stage but also, if disputes do arise, they are able to gain an understanding of the issues very quickly given their knowledge of the project.

Dispute Boards are used in a number of different ways. The function of a Dispute Review Board (DRB) is to make a recommendation to the parties with which the parties may comply on a voluntary basis, though they may not be required to do so. In some cases the recommendation would be binding if neither party expressed dissatisfaction within a specified period. A Dispute Adjudication Board (DAB) issues decisions which may be binding on the parties, on an interim basis, if the contract so provides. A Combined Dispute Board (CDB) is a combination of both DRBs and

DABs and can either issue a recommendation or make a decision, depending on what the parties request in the particular situation.

There tend to be time limits built into the procedure within which the board is required to make its decision or recommendation. The aim is for this to be done quickly. The decision or recommendation is often binding only on an interim basis and the dispute can ultimately be taken to court or arbitration, should the parties so desire. However, like adjudication, it is often a quick decision that is required, even if it is rough and ready, in order to allow parties to move on and there may then be little appetite for later arbitration or court proceedings.

18.6.4 Enforcement of decisions

An important issue in relation to decisions of dispute boards is how to enforce them. The FIDIC Red Book provision is that the DAB decision 'shall be binding on both Parties, who shall promptly give effect to it unless and until it shall be revised in an amicable settlement or an arbitral award'. The final form of dispute resolution in FIDIC is international arbitration.

Where there is no notice of dissatisfaction issued within the time limits for this, FIDIC goes on to provide that the decision shall then become final and binding. In relation to enforcement of a DAB decision which has become final and binding (i.e. where no notice of dissatisfaction is issued), FIDIC provides that the failure to comply with the decision may be referred to arbitration. Where there is a notice of dissatisfaction, the arbitral tribunal has power, among other things, to open up, review and revise any decision of the DAB. This would involve a re-run of the dispute.

There is, therefore, on the face of it, a gap in that no specific provision is made for enforcement of the DAB decision in a situation where there is a notice of dissatisfaction. This gap was discussed in the Singapore Court of Appeal case of *CRW Joint Operation* v. *PT Perusahaan Gas Nagara (Persero) TBK* (2011), which dealt with enforcement of an order for payment in a DAB decision. The DAB had issued a decision requiring PGN to pay CRW US$17 million. PGN issued a notice of dissatisfaction. CRW referred the case to arbitration. One of the orders sought was for a final award enforcing the DAB's decision on the basis PGN was in breach of the contract provision requiring that parties should promptly give effect to DAB decisions. They sought immediate payment of the amount in the DAB decision. The arbitral tribunal concluded the DAB decision was binding and made a final award finding the sum awarded by the DAB was due and that PTN should make an immediate payment. They declined to open up, review or revise the DAB decision. PTN sought an order from the Singapore High Court to set aside that decision. Both the High Court and the Court of Appeal refused to enforce the award. The Tribunal could have dealt with this by way of an interim or part award pending final resolution of the parties' dispute but it had enforced by a final award without a hearing on the merits. That is where they went wrong.

The FIDIC has sought to close this loophole in the Gold Book, which provides that if a party fails to comply with a DAB decision, whether binding or final and binding (in other words, whether or not there has been a notice of dissatisfaction), the other party may refer that failure to arbitration. It is likely that this same provision will make its way into the other FIDIC forms in due course. In the meantime, the FIDIC Contracts Committee has issued a Guidance Note dated 1 April 2013 in relation to the Red, Yellow and Silver Books. This provides suggested amendments to the relevant clauses to close the loophole.

Other issues which would affect enforceability could include the DAB exceeding its jurisdiction to consider the dispute referred to it, failure to act in accord with the rules of natural justice, and failure to make a decision within the contractual time limits.

The importance of time limits is apparent from ICC case 10619, 2001 and 2002. It dealt with enforcement of an Engineer's decision. The case concerned the construction of a road in an African state. The Engineer under the FIDIC Red Book Fourth Edition (1987) had made four decisions. The employer refused to implement them. The contractor commenced an arbitration relating to a number of matters. They also sought an interim decision that the employer should give effect to the Engineer's decisions and an award for immediate payment on the basis of those decisions.

The arbitral tribunal refused to enforce two of the decisions because the contract required a decision within 84 days and the decisions were late. The other two had been made timeously. The contract required the decisions to have an immediate binding effect on parties. The tribunal therefore ordered payment in relation to these.

Chapter 19
Tax

19.1 *Value Added Tax (VAT)*

19.1.1 Introduction

This chapter assumes a basic knowledge of how VAT works. A detailed explanation of the VAT rules which apply to the provision of building and construction work is outside the scope of this chapter.

19.1.2 Legislative framework

Value Added Tax (VAT) is regulated by the Value Added Tax Act 1994 (VATA 1994), the Value Added Tax Regulations 1995 (SI 1995/2518) ('the VAT Regulations') and a number of other statutory instruments.

HM Revenue & Customs ('HMRC') publishes a number of notices which provide guidance on the application of VAT legislation. Further guidance on the application of VAT to building and construction work can be found in Notice 708, *Buildings and Construction.*

19.1.3 Rates of VAT

Most supplies relating to the construction of new buildings and works on existing buildings are standard rated for VAT purposes. Certain categories of works benefit from the application of the zero-rate or reduced rate (5%) of VAT. These are considered in Sections 19.1.5 and 19.1.6.

19.1.4 The VAT-inclusive rule

If a contract is silent on VAT, then any sums mentioned in it are deemed to be VAT-inclusive, and the contractor is unable to charge VAT in addition to the contract sum. In such circumstances, the contractor will be obliged to account for VAT to

MacRoberts on Scottish Construction Contracts, Third Edition. MacRoberts.

HMRC out of the contract sum received, with an inevitable consequence on the contractor's return. The contractor should ensure therefore that the contract provides for VAT to be payable in addition to the contract sum.

VAT is also chargeable in relation to any non-cash consideration. It is essential that the contract specifies that VAT is payable in relation to both cash and non-cash consideration.

19.1.5 Zero-rated supplies

The rules governing which supplies of construction services can be zero-rated are contained in Schedule 8 VATA 1994. The main categories of supplies which are zero-rated are as follows:

- supplies in the course of the construction of a dwelling;
- supplies in the course of the construction of a building to be used solely for a relevant residential or relevant charitable purpose;
- supplies in the course of any civil engineering works necessary for the development of a permanent park for residential caravans;
- supplies to a relevant housing association in the course of conversion of a non-residential building or a non-residential part of a building into a building or part of a building to be used for dwellings or a number of dwellings, or a relevant residential purpose;
- approved alterations to protected buildings (this category of zero-rating is being phased out and will end completely on 1 October 2015).

Further guidance on what is meant by each of the above is contained in HMRC Notice 708, *Buildings and Construction.*

Zero-rating does not apply to the services of architects, surveyors or other professionals, but see Section 19.1.8 in relation to design and build contracts.

19.1.6 Reduced-rate supplies

The rules governing the supplies of construction services which can be reduced-rated are contained in Schedule 7A VATA 1994. The main categories of supply to which the reduced rate of 5% applies are as follows:

- certain renovations and alterations to empty residential premises;
- certain conversions of non-residential buildings into qualifying dwellings;
- certain renovations of and alterations to buildings which have been empty for three or more years.

Further guidance on what is meant by each of the above is contained in HMRC Notice 708, *Buildings and Construction.*

Reduced rating does not apply to the services of architects, surveyors or other professionals, but see Section 19.1.8 in relation to design and build contracts.

19.1.7 Conditions for zero and reduced rating

As explained in Sections 19.1.5 and 19.1.6 above, a number of different types of supplies of construction services can be zero-rated or reduced-rated. Prior to applying the zero or reduced rate of VAT to a supply, it is necessary to determine whether all of the relevant conditions for the zero or reduced rate of VAT to apply to that supply have been met. The conditions which must be met in relation to each type of supply set out in Sections 19.1.5 and 19.1.6 are slightly different, but there are a number of conditions which are common to each type of supply:

- the supply should relate to the construction, renovation or alteration of a qualifying building;
- the person making the supply should be the person carrying out the construction, renovation or alteration; and
- where necessary, a valid certificate is held.

Further guidance on the application of each of the above conditions, and the specific conditions relating to each type of supply can be found in HMRC Notice 708, *Buildings and Construction*.

The conditions which must be met in order for a supply to be zero-rated or reduced-rated are normally dependent on the identity of the employer and to what use the employer ultimately intends to put the building. In relation to supplies which require a certificate to be provided to the contractor in order for zero- or reduced-rate VAT to apply to a supply, such certificate must usually be provided to the contractor before the contractor makes any supply.

If it is anticipated that any element of building or construction works should be a zero- or reduced-rated supply for VAT purposes, it would be prudent to amend the SBC, the SBC/DB or the NEC3 to ensure that the employer provides the contractor, at the appropriate time, with all information, undertakings and certificates necessary to enable the contractor to zero-rate or reduce-rate their supply.

19.1.8 Services of architects, surveyors and other professionals: design and build contracts

Zero-rating and reduced rating do not apply to the services of architects, surveyors or other professionals. Where, however, these costs are incurred under a single design and build contract which does not separately identify the design element, there is a single composite supply of the design and build project which can be zero-rated or reduced-rated, as appropriate.

19.1.9 VAT liability of goods provided under construction contracts

The basic rule is that the VAT status of building materials provided under a construction contract follows the VAT status of the construction services provided under that contract; however, this is not always the case. For example any 'white goods' or carpets provided under a contract to construct a dwelling will be standard rated for VAT purposes, not zero-rated. It is important, therefore that consideration is given to the VAT status of the various supplies to be made under the building contract and also to whether any revisals to the SBC, the SBC/DB and the NEC3 (or other form of contract) are appropriate.

19.1.10 VAT liability of supplies by sub-contractors

The VAT status of supplies made by a sub-contractor to a contractor under a building contract is not entirely straightforward.

Where the sub-contractor provides services in relation to the construction of a dwelling, then the VAT status of the sub-contractor's supply generally follows the VAT status of the main contract, i.e. it can benefit from the application of zero and reduced rate VAT. If, however, the supply to be made by the contractor under the main contract is one in respect of which a certificate is required (see Section 19.1.7), then the sub-contractor's supply will be standard rated.

Any contract between a contractor and the sub-contractor should allow for VAT to be charged on the services to be provided by the sub-contractor.

Further guidance on the VAT status of supplies made by sub-contractors can be found in HMRC Notice 708, *Buildings and Construction*.

19.1.11 VAT invoices

A VAT invoice is required in order to recover input VAT which has been incurred. The contract should therefore provide for a VAT invoice (or an authenticated receipt) to be supplied, or for the self-billing invoice procedure to be used. See Sections 19.1.16 and 19.1.17 for further information on authenticated receipts and self-billing arrangements.

19.1.12 Time of supply: when VAT must be accounted for

The tax point of a supply dictates when the party making the supply has to account to HMRC for VAT on that supply, and when the party paying for the supply can recover the related input tax.

The tax point rules which apply to building works are partly determined by reference to the payment terms contained in the contract governing those building works; whether the contract provides for a single payment or stage payments.

19.1.13 Single payment contracts

If a building contract provides for a single payment, then the normal tax point rules contained in section 6 VATA 1994 will apply to that payment. Section 6 provides that a basic tax point occurs:

- in relation to a supply of goods, when those goods are made available; and
- in relation to a supply of services, when the supply is completed.

The basic tax point rule can be displaced, and an actual tax point created, if the person making the supply issues an invoice before the basic tax point date, or payment is received. An actual tax point is also created if an invoice is issued within 14 days of the date of the supply being made.

The problem with single payment building contracts from a VAT perspective is that most building contracts relate to the provision of a mixture of goods and services, for example, the provision of wood (goods) and the provision of carpentry services (services). In that example the tax point in relation to the wood would arise at the point when the wood is delivered, or made available, i.e. when it is left on site, but in relation to the carpentry services the basic tax point would not occur until the works have been completed, which will usually be at the point of practical completion.

If a VAT invoice is issued more than 14 days after the occurrence of the basic tax point, then the time at which VAT must be accounted for is determined by reference to the basic tax point. It is possible to obtain HMRC's consent to invoices being issued more than 14 days after the occurrence of the basic tax point and HMRC will normally accept this if the invoice is issued within a month.

The fact that the basic tax point rules mean that a contractor can end up having to account for VAT prior to payment being received for the goods/services provided, is one of the reasons that larger construction contracts normally provide for staged payments and the use of either authenticated receipts or a self-billing procedure.

19.1.14 Retention under a single payment contract

Where a single payment contract provides for an amount to be retained from payment under a contract 'pending full and satisfactory performance of the contract', then Regulation 89 of the VAT Regulations provides that the time of supply in relation to the retention payment is delayed until payment is received or a VAT invoice is issued, whichever is earlier. Regulation 89 of the VAT Regulations does not apply to staged payment contracts.

The VAT liability for retention payments follows the status of the main contract (i.e. if supplies under the main contract are zero-rated, any retention payments will also be zero-rated).

19.1.15 Stage payment contracts

Where a building contract provides for stage payments to be made, then Regulation 93 of the VAT Regulations provides that the basic and actual tax point rules contained in section 6 VATA 1994 (see Section 19.1.13) do not apply, and instead a tax point occurs on the earlier of receipt of payment or the issue of a tax invoice.

In general, where a contract contains stage payment terms, it is desirable for an authenticated receipts payment mechanism to be implemented so that the contractor only becomes liable to account for VAT on the amount of the payment actually received, rather than the amount of payment requested.

19.1.16 Authenticated receipts

The authenticated receipts procedure may only be used in conjunction with a stage payment contract. It must not be used when a contractor makes a supply under a single payment contract. The procedure may also only be used if both parties agree to it.

Under the authenticated receipts procedure, the contractor submits a request for payment to the employer. When the employer agrees the amount of the payment due, it prepares a receipt for the services received and provides that receipt, together with the payment for those services, to the contractor. The contractor then authenticates the receipt and returns it to the employer. The contractor must not issue a VAT invoice.

The tax point in relation to an authenticated receipt arises at the time of payment by the employer, not upon the issue of the authenticated receipt.

The benefit to the contractor of the authenticated receipts process is that the contractor does not need to account for VAT until payment has been received and practical issues surrounding non-payment of VAT invoices and bad debt relief claims which could otherwise arise are avoided.

From the employer's point of view, the use of authenticated receipts allows input VAT to be reclaimed in the VAT period in which payment is made. There is no need for the employer to wait for the authenticated receipt before reclaiming the input VAT (though the employer must obtain the authenticated receipt from the contractor and keep a copy of it with its VAT records in order to support the input VAT claim).

The fact that the employer may have to pursue a contractor for an authenticated receipt may make the use of the authenticated receipts procedure unattractive to some employers.

If the parties wish to use the authenticated receipts procedure, then the building contract should be amended accordingly where appropriate.

19.1.17 Self-billing arrangements

The self-billing procedure can be applied to both single and staged payment contracts, but the procedure is of most use where there are regular supplies and payments between parties. It is often used in relation to payments from a contractor to a sub-contractor.

In a self-billing situation the customer prepares the supplier's invoice and then forwards that invoice to the supplier with payment. Further information regarding the operation of self-billing arrangements can be found in HMRC Notice 70/62, *Self-Billing*.

Both parties must agree to the use of a self-billing arrangement and an appropriate self-billing agreement must be entered into. A self-billing arrangement must run for a period of 12 months; it can also be tied to the term of a contract. HMRC Notice 700/62. *Self-Billing* sets out the information which must be contained in a self-billing agreement and provides a suggested form of such agreement.

If the self-billing agreement which is entered into between the parties is not valid, then the self-billing invoices issued by the customer are not valid and consequently cannot be used to support an input tax claim.

The use of a self-billing arrangement does not displace the normal VAT tax point rules, but HMRC accept that, for input tax purposes, where a self-billed invoice is issued with payment to the supplier, then a notional tax point is created, being the day following the date of issue of the invoice, and that notional tax point can be used as the relevant date for the reclaim of input VAT.

If a self-billing arrangement is to be used, then it would be sensible for appropriate revisals to be made to the building contract.

19.1.18 Disputes

If a dispute arises between the parties to a building contract which ultimately results in a financial settlement being reached, then the VAT status of that settlement must be determined; the fact that the settlement has been reached as a consequence of a dispute does not make the settlement payment automatically outside the scope of VAT.

Payments made to settle a dispute out of court, once proceedings have been commenced, are treated as follows:

- if the payment is compensatory, and does not relate directly to supplies of goods or services, then the payment is outside the scope of VAT and no VAT is chargeable;
- if the payment is consideration for a specific supply of goods or services, the payment is subject to VAT.

This treatment is generally accepted as applying also in cases where proceedings have not yet commenced, and to cases involving arbitration and adjudication.

It can frequently be difficult to determine whether a payment is compensatory or whether it relates to supplies of goods or services. It should not be assumed that any payment resulting from a dispute will be VAT-free.

The VAT inclusive rule (see Section 19.1.4) should be borne in mind in relation to the settlement of disputes and in drawing up an agreement to settle a court action, arbitration or adjudication proceeding, and agreements should provide for the payment of VAT in addition to the settlement sum and the issue of an appropriate VAT invoice to the payer.

19.1.19 Liquidated damages

Payments under liquidated damages clauses are not treated as payments for a supply for VAT purposes and as such are outside the scope of VAT. This applies whether the amount payable is specified as a fixed sum or whether it is arrived at by way of a formula.

If a payment for a supply under a contract is set off against a liquidated damages payment, VAT will still be chargeable in relation to the supply.

19.1.20 The SBC, the SBC/DB and the NEC3 provisions

Clause 4.4.1 of the SBC/DB and clause 4.6.1 of the SBC provide that the Contract Sum/Tender Price is exclusive of VAT and in relation to any payment to the Contractor the Employer is also obliged to pay the amount of any VAT properly chargeable in respect of it.

These clauses are adequate if the construction services which are to be provided are straightforward, standard-rated supplies and a standard payment mechanism is to be used. If, however, the services which are to be provided are zero- or reduced-rated, then the clauses should be revised to ensure that all information, undertakings and certificates required by the Contractor to be able to zero-rate or reduce-rate the supply are provided to the Contractor (see Section 19.1.7).

Each of the SBC/DB and the SBC provide for interim payments to be made. Consideration should be given to amending the contract, where appropriate, to provide for the use of either authenticated receipts (see Section 19.1.16) or self-billing arrangements (see Section 19.1.17).

Clause 4.4.2 of the SBC/DB and clause 4.6.2 of the SBC provide that if, after the Base Date, the supply of goods and services to the Employer becomes exempt from VAT, there is to be paid to the Contractor an amount equal to the amount of input tax on the supply to the Contractor of goods and services which contribute to the Works but which as a consequence of that exemption the Contractor cannot recover. This is intended to ensure that the Contractor is not affected by a future inability to recover input VAT in relation to the services provided under the contract. It is considered unlikely that this clause would ever come into effect without a change in legislation which resulted in a change in the VAT status of the construction services provided under the contract.

Clause 50.2 of the NEC3 provides that 'any tax which the Employer requires to pay is included in the amount due'. This is potentially slightly ambiguous.

The Guidance notes for the NEC3 also suggest that the Contractor and the Employer should make arrangements for the provision of invoices, etc. In addition, it might be prudent to insert Z clauses dealing with the provision of any information, undertaking and certificates required and also self-billing/authenticated receipts provisions, if appropriate.

19.2 The Construction Industry Scheme

19.2.1 Introduction

The rules governing the construction industry scheme (CIS) are contained within Finance Act 2004 ss.57 – 77 and the Income Tax (Construction Industry Scheme) Regulations 2005 (henceforth the CIS Regulations).

The CIS applies to payments by contractors to self-employed sub-contractors in relation to construction operations. The terms 'contractor' and 'sub-contractor' have particular meanings in the CIS, but these do not necessarily correspond with the way these terms are used in the construction industry.

For CIS purposes, a contractor is merely a person who pays another person (a sub-contractor) for construction services, and a sub-contractor is a person who receives payment. It follows that for CIS purposes, the employer under a building contract would be the contractor, and the main contractor would be a sub-contractor. A main contractor could also be a contractor for the purposes of the CIS if it engages sub-contractors.

Contractors for the purposes of the CIS include not just construction companies and building firms but can also include 'deemed contractors', i.e. businesses whose main trade is not construction-related but whose expenditure on construction operations exceeds certain prescribed limits (currently £1 million on average over a three-year period). This brings within the scheme bodies such as government departments, local authorities and many businesses normally known in the industry as 'clients'.

The CIS does not apply to employment contracts. If a contract is an employment contract, the contractor is obliged to apply PAYE and deduct income tax and National Insurance Contributions (NIC) from the payments made, and account to HMRC for the employer's NIC.

19.2.2 How the CIS operates

Under the CIS, before making a payment to which the CIS applies, the contractor must verify the payment status of the sub-contractor with HMRC, and:

- pay the sub-contractor gross, if the sub-contractor is registered for gross payment;
- pay the sub-contractor under deduction of tax at the lower rate of 20% if the sub-contractor is registered for payment under deduction; or
- pay the sub-contractor under deduction of tax at the higher rate of 30% if the sub-contractor is not registered or cannot be 'matched' by HMRC's systems.

The amount to which the CIS deduction must be applied is the gross amount of the sub-contractor's invoice less:

- VAT;
- the amount of Construction Industry Training Board levy they have paid;
- amounts paid for materials, consumables stores, fuel used (except for travelling), plant hire and manufacturing or prefabrication materials.

19.2.3 Verification

The verification process involves the contractor checking with HMRC the payment status of the sub-contractor, i.e. whether tax should be deducted and at what rate.

It should be noted that the verification process only verifies payment status and not employment status. It is the responsibility of the contractor to determine whether the sub-contractor is employed or self-employed and whether the CIS is applicable.

19.2.4 Higher and lower rates of deduction

There are two rates of deduction. Deduction at the standard rate is 20% and this applies to sub-contractors who are registered with HMRC but are not registered for gross payment. Deduction at the higher rate applies if the sub-contractor is not registered with HMRC. This is at the rate of 30%, which creates an incentive for sub-contractors to register with HMRC.

HMRC advise that the higher rate of deduction applies if the subcontractor is not registered for CIS purposes or cannot be 'matched' on HMRC's system, and will continue until the sub-contractor has contacted HMRC in order to register or resolve the matching problem. In other words, sub-contractors may suffer deduction at the higher rate because their records cannot be traced on HMRC's system, rather than because they have not registered. The cash flow implications for sub-contractors suffering the higher rate deduction may be severe.

Contracts should therefore contain provisions requiring CIS contractors to carry out the verification procedure expeditiously, and to advise sub-contractors of the outcome of the verification procedures, so that sub-contractors can deal with any problems arising.

Contracts should also require CIS sub-contractors to notify contractors as soon as possible of any changes to their registration status and require CIS contractors to notify sub-contractors if they are advised by HMRC of a cancellation of registration for gross payment or for payment under deduction at the lower rate, so that the sub-contractor can take steps to remedy the position if appropriate.

19.2.5 Retentions

The CIS applies on a payment basis. Where part of the contract price is retained by way of a retention, the contractor must consider the CIS position at the date on which the payment is actually made, and cannot pay gross if the sub-contractor is not registered for gross payment at the time the payment is made, even if they were registered at the time the work was carried out.

19.2.6 The SBC, the SBC/DB and the NEC3 provisions

Clause 4.7 of the SBC and clause 4.5 of the SBC/DB provide that, where it is stated in the Contract Particulars that the Employer is a 'contractor' for the purposes of the CIS or if at any time up to the payment of the Final Certificate (or final payment in the case of the SBC/DB) the Employer becomes a 'contractor', the obligations of the Employer to make any payment under the contract are subject to the provisions of the CIS. This makes it clear that the CIS shall be applied to payments made under the contract where the scheme is relevant, but it does not impose any obligation on either party to provide any of the information necessary to ensure that the scheme is operated correctly. It is for this reason that additional wording is normally inserted into the contract requiring the Contractor to provide the Employer with sufficient information to complete the verification process and imposing an obligation on both parties to inform the other party of any change in their CIS status.

The NEC3 makes no mention of the provisions of CIS, presumably on the basis that the parties are in any event obliged to comply as a matter of law. However, see comments in Section 19.2.4 on suggested additional provisions.

Chapter 20
Health and Safety

20.1 *Introduction*

The issue of health and safety is a significant one in the construction industry. It is an industry that is inherently dangerous by virtue of the nature of the site environment and the operations involved. It is also an industry that has a poor record in relation to accidents. While employers (in the construction sense) have duties in respect of health and safety, the most significant responsibilities in this field will fall upon contractors and sub-contractors, in their role as employer in the employer/employee sense.

A detailed examination of the law of health and safety is beyond the scope of this book. We will, however, consider in some detail the primary statute and the most significant subordinate legislation from the point of view of the construction industry.

20.2 *Common law*

At common law, employers have an obligation to provide competent staff, adequate material, a proper system of work, effective supervision and a safe place of work, see *Wilsons and Clyde Coal Co. Ltd* v. *English* (1938). They also have a duty to instruct and to take steps to ensure that instructions are carried out, see *McWilliams* v. *Sir William Arrol & Co. Ltd and Another* (1962).

The law relating to health and safety has relevance both in a civil and criminal context. Notwithstanding the fact that much, indeed the vast majority, of the law of health and safety at work in Scotland is to be found in statutory materials, the common law still has a relevance.

That relevance is particularly notable in civil cases. It is highlighted by section 47 of the Health and Safety at Work etc. Act 1974 (henceforth 'the 1974 Act'). In terms of s47(1)(a), the 1974 Act does not confer a right of action in civil proceedings in respect of failures to comply with any duty imposed by sections 2–7 of it, which sections we consider in detail below. However, it falls to be contrasted with s.47(2) whereby a breach of duty imposed by health and safety regulations shall, so far as it causes damage, be actionable, except insofar as the regulations provide otherwise.

Accordingly, in many civil cases, the common law is still of relevance, the basic duty of the employer being to take reasonable care that the employee is not exposed to unnecessary risk, see *Longworth* v. *Coppas International (UK) Ltd* (1985).

MacRoberts on Scottish Construction Contracts, Third Edition. MacRoberts.

It has been suggested that the duty breaks down into three basic parts, namely, (1) that the employer is required to provide and maintain suitable materials (i.e. plant, machinery and equipment); (2) to keep his premises safe and devise and operate a safe system of working; and (3) to exercise care in the selection of competent fellow employees.

20.3 Health and Safety at Work etc. Act 1974

20.3.1 General

The current starting point in relation to health and safety legislation is the Health and Safety at Work etc. Act 1974. This creates duties which are incumbent on employers (in the employer/employee sense as opposed to the construction sense), employees, persons in control of premises and designers and manufacturers of articles and substances. The sections imposing these duties are considered below, along with a number of the other significant provisions of the 1974 Act.

It has been observed that the general duties under the 1974 Act are deliberately similar to the duties of care giving rise to civil liability at common law. The general duties are to be found in sections 2–7 of the 1974 Act.

20.3.2 Section 2

The principal duties incumbent upon employers, insofar as their own employees are concerned, are contained within section 2 of the 1974 Act, namely:

- to ensure, so far as is reasonably practicable, the health and safety and welfare of their employees (see s.2(1));
- to ensure the provision and maintenance of plant and systems of work that are, so far as is reasonably practicable, safe and without risks to health (see s.2(2)(a));
- to provide safe systems for the use, storage and transport of articles and substances, so far as is reasonably practicable (see s.2(2)(b));
- to provide such information, instruction, training and supervision as is necessary to ensure the health and safety at work of employees, so far as is reasonably practicable (see s.2(2)(c));
- to maintain a safe place of work and provide safe access to and egress from that place of work, so far as is reasonably practicable (see s.2(2)(d));
- to provide and maintain a working environment that is, so far as is reasonably practicable, safe, without risks to health, and adequate as regards facilities and arrangements for employees' welfare at work (see s.2(2)(e)); and
- to prepare and, as often as may be appropriate, revise a written statement of their general policy with respect to the health and safety at work of their employees and the organization and the arrangements for implementing it (see s.2(3)).

The general duty laid down by s.2(1), and the more specific duties laid down in s.2(2)(a)–(e), set out in statutory form the common law obligations owed by employers to their employees, see *West Bromwich Building Society* v. *Townsend* (1983).

20.3.3 Section 3

Section 3 of the 1974 Act sets out the general duties of employers and the self-employed to persons other than their employees.

By virtue of sub-sections 1 and 2, virtually identical duties are imposed on employers and self-employed persons, whereby each is required to conduct their undertaking in such a way as to ensure, so far as is reasonably practicable, that persons not in their employment who may be affected by their work are not exposed to risks to their health or safety. It should be noted that s.3(2) also imposes a duty on self-employed persons to ensure, again so far as is reasonably practicable, that they, themselves, are not exposed to risks.

Like section 2, this section imposes absolute criminal liability, subject only to the defence of reasonable practicability, which defence relates only to measures necessary to avert the risk.

Particularly crucial in terms of this section is the phrase 'conduct their undertaking'. It was believed that an employer did not conduct his undertaking if he employed an independent contractor to do the relevant work, provided the employer neither exercised any control over the work nor was under any duty to do so. That was the substance of the decision of the English High Court in *RMC Roadstone Products Ltd* v. *Jester* (1994).

While the decision in *RMC Roadstone Products Ltd* has not been expressly overruled, it was, however, doubted in the case that sets out the present state of the law in England, namely, *R* v. *Associated Octel Ltd* (1996), a decision of the House of Lords. While not binding in Scotland, there is nothing to suggest that the courts in Scotland would view matters differently.

In *Associated Octel Ltd*, the appellant company, which operated a chemical plant, engaged a firm of specialist contractors to carry out annual maintenance and repair work. As part of the scheduled work, the contractors had to repair the lining of a tank within the appellant's chlorine plant, which involved grinding down the damaged areas of the tank, cleaning the dust from the surfaces with acetone and applying fibreglass matting to rebuild those areas. While the specialist contractor's employee was inside the tank the bulb of the light he was using broke and the electric current caused the acetone vapour to ignite. There was a flash fire and explosion, which badly burned the employee.

Associated Octel were charged with, and convicted of, an offence under the 1974 Act of failing to discharge the duty imposed on it by section 3.

Associated Octel appealed, unsuccessfully, against the conviction. The Court of Appeal held that the word 'undertaking' in s3(1) of the 1974 Act meant 'enterprise' or 'business' and, in the particular circumstances of the case, the cleaning, repair and maintenance of plant, machinery and buildings necessary for carrying on the employer's business were part of the conduct of their undertaking for the purposes of s.3(1), whether it was done by the employer's own employees or by independent contractors.

Accordingly, if there was a risk of injury to the health and safety of persons not employed by the employer, whether to the contractor's men or members of the public, and if there was actual injury as the result of the conduct of that operation, there was *prima facie* liability, subject to the defence of reasonable practicability.

It was further held that the question of control might well be relevant to the issue of whether it was reasonably practicable for the employer to give instructions on how the work was to be done and what safety measures were to be taken and, in each case, the question was one of fact and degree.

There was a subsequent (again unsuccessful) appeal to the House of Lords, who held that if an employer engaged an independent contractor to do work which formed part of the employer's undertaking, the employer was required by s.3(1) to stipulate for whatever conditions were reasonably practicable to avoid risk to the contractor's employees. Whether or not the employer was in a position to exercise control over work carried out by an independent contractor was not the decisive question under section 3, which is whether the activity in question could be described as part of the employer's undertaking, which will be a question of fact in each case.

20.3.4 Section 4

Section 4 imposes a duty on persons in control of, or concerned with, premises to ensure, again so far as is reasonably practicable, the safety of persons on those premises. The duty does not extend to employees. It covers 'non-employees' who use non-domestic premises made available to them as a place of work or as a place where they may use plant or substances provided for their use there. A similar duty is found in s.2(d) of the 1974 Act in respect of employees.

In terms of s.4(2), it is the duty of each person who has, to any extent, control of premises to which this section applies to take such measures as is reasonable for a person in his position to take to ensure, so far as is reasonably practicable, that the premises are safe and without risks to health.

Section 2 imposes an absolute duty, subject only to the limited qualification 'so far as is reasonably practicable', see *Mailer* v. *Austin Rover Group* (1989). This does not require the duty holder to take precautions against unknown and unexpected events.

The phrase 'person who has, to any extent, control of premises' has a wide meaning. Instructive in this regard is the case of *T Kilroe & Sons Ltd* v. *Gower* (1983). In that case, the appellant had obtained a contract to demolish large factory premises. In order to fulfil this contract, it had entered into an agreement with experienced demolition contractors to provide the bulk of the labour and execute much of the work on a profit-sharing arrangement. The factory inspectors discovered that asbestos de-lagging from pipes in the boiler house of the factory was taking place with the doors of the premises unsealed and opened and with no proper provision having been made for showering and decontamination of the workers involved. The appellants were convicted of failing to discharge their duty under s.4(2) of the 1974 Act.

The appellants appealed on the basis that: (1) the boiler house was a separate entity from the rest of the site and that since the work therein was being exclusively carried out by the demolition contractor and its employees, the appellant did not have any, or any sufficient, control of the premises to bring it within the section; and (b) in any event, it was not in breach of its duties since it had taken such steps as was reasonable for it to take by employing experienced demolition contractors to undertake the works.

The appeal was dismissed. It was held that there was no justification for treating the boiler house as a separate entity, even though it was a separate building. The contract was for the demolition of the whole site and in the absence of any evidence that the boiler house had been treated in some way as separate from the rest of the site, the premises referred to the whole site. It was also held that where there was such an obvious risk to health, even if the appellant believed that the demolition contractors were experts in asbestos stripping (which, on the evidence, the court did not accept), as soon as they observed the lack of expertise they should have acted, if necessary, by removing the demolition contractor from the work and assuming overall control themselves.

The difficulty with any authority relative to the 1974 Act is that, inevitably, each case turns, to a significant degree, on its own facts and circumstances. In each case, what is 'reasonably practicable' will differ. This term is considered in section 20.3.13.

20.3.5 Section 7

Section 7 of the 1974 Act imposes a duty on employees to take reasonable care for their own health and safety at work, as well as for other persons who might be affected by their acts or omissions.

20.3.6 Section 15

Section 15 of the 1974 Act is the provision under which health and safety regulations can be made. The relevance of subordinate legislation in the field of health and safety cannot be overstated. A plethora of regulations exist, sometimes covering esoteric and obscure industries and operations. An examination of these is beyond the scope of this book.

20.3.7 Sections 19 and 20

The day-to-day enforcement of health and safety legislation falls to enforcing authorities. Essentially, these are either the Health & Safety Executive (HSE) or, in certain limited cases, local authorities. By virtue of section 19, enforcing authorities are entitled to appoint inspectors who are to be such persons having suitable qualifications as the authority thinks necessary for carrying into effect the relevant statutory provisions within its field of responsibility.

The powers of inspectors are wide-ranging and are to be found in section 20 of the 1974 Act. The specific powers are to be found in s.20(2).

20.3.8 Section 21

If an inspector is of the opinion that a person is either contravening one or more of the relevant statutory provisions or has contravened one or more of those provisions in circumstances that make it likely that the contravention will continue or be repeated, he may serve upon that person what is known as an improvement notice.

The notice is required to state the opinion of the inspector in relation to the contravention, specify the provision or provisions involved, give particulars of the reasons why he is of that opinion and require the recipient of the notice to remedy the contravention or, as the case may be, the matter which is occasioning it, within such period as may be specified in the notice. In the case of an improvement notice that period can end no earlier than the period within which an appeal against the notice can be brought.

20.3.9 Section 22

This provision deals with the giving of prohibition notices. It applies to any activities which are being, or are likely to be, carried on by or under the control of any person, being activities to or in relation to which any of the relevant statutory provisions apply or will, if the activities are so carried on, apply. The section applies where an inspector is of the opinion that as carried on, or as likely to be carried on, an activity involves or, as the case may be, will involve, a risk of serious personal injury.

A prohibition notice requires to state that the inspector is of that opinion; specify the matters which in his opinion give rise to or, as the case may be, will give rise to the said risk; specify any relevant statutory provisions that are being, or will be, contravened; and direct that the activities to which the notice relates shall not be carried on, unless the matter specified in the notice and any associated contraventions have been remedied.

A direction contained in a prohibition notice shall take effect either at the end of the period specified in the notice, or, if the notice so declares, immediately.

20.3.10 Section 24

This section deals with appeals against improvement or prohibition notices.

The right of appeal is to an employment tribunal. On appeal, the tribunal may either cancel or affirm the notice and, if it affirms the notice, it may do so either in its original form or with such modifications as the tribunal thinks appropriate in the circumstances.

In the case of an improvement notice, the bringing of the appeal has the effect of suspending the operation of the notice until the appeal is disposed of. In the case of a prohibition notice, the bringing of the appeal has that effect only if, on the application of the appellant, the tribunal so directs. Essentially, without such a direction from the tribunal, an appeal against the prohibition notice does not suspend the operation of the notice.

20.3.11 Section 33

Section 33 of the 1974 Act sets out offences. The provision is both extensive and complex.

Seventeen separate offences are set out in s.33(1). These include failing to discharge a duty imposed by sections 2–7 of the 1974 Act; contravening any health and safety regulation; contravening any requirement imposed by an inspector; obstructing inspectors in the exercise or performance of their powers or duties; contravening any requirement or prohibition imposed by an improvement notice or a prohibition notice; and falsely pretending to be an inspector.

Penalties for offences under the 1974 Act are now prescribed by Schedule 3A to the Act. In terms of monetary penalty, on conviction on summary complaint (i.e. a judge sitting without a jury), for the majority of offences, the maximum fine is presently £20,000. On conviction on indictment (i.e. by a jury), the available fine is unlimited.

Imprisonment is an option open to the court in certain circumstances. On summary conviction, the maximum available term is one of twelve months. On conviction on indictment, the maximum available term is two years.

20.3.12 Section 37

The provisions of Section 37 are worthy of note. Where an offence under any of the relevant statutory provisions committed by a body corporate is proved to have been committed with the consent or connivance of, or to have been attributable to any neglect on the part of, any director, manager, secretary or other similar officer of the body corporate or a person who was purporting to act in any such capacity, that person, as well as the body corporate, shall be guilty of that offence and is liable to be proceeded against and punished accordingly. An example of this is *Armour* v. *Skeen* (1977).

20.3.13 Practicable and reasonably practicable

The concepts of 'practicability' and 'reasonable practicability' arise regularly in both the 1974 Act and also in many health and safety regulations made under the 1974 Act. It is, in fact, the latter concept that is much more commonplace.

Neither term is defined in the 1974 Act, or elsewhere. The *Oxford English Dictionary* definition of 'practicable', i.e. that which is capable of being carried out in action or that which is feasible, was applied by the Court of Appeal in *Lee* v. *Nursery Furnishings Ltd* (1945). Prefixing 'practicable' with 'reasonably' creates a qualification whereby the extent of the risk requires to be balanced against the measures necessary to avert that risk, a form of cost benefit analysis, see *Sharp* v. *Coltness Iron Co. Ltd* (1937) and *Edwards* v. *NCB* (1949) subsequently approved by the House of Lords in *Marshall* v. *Gotham Co. Ltd* (1954).

What is reasonably practicable depends upon whether the time, trouble and expense of the precautions suggested are disproportionate to the risk involved. What is reasonable for a large undertaking may be unreasonable for a small undertaking. It should be noted that 'reasonably practicable' has been held to have a narrower meaning than 'physically possible', see the decision of the Court of Appeal in *Marshall* v. *Gotham Co. Ltd* (1953).

20.4 The Construction (Design and Management) Regulations 2007

20.4.1 Introduction

The Construction (Design and Management) Regulations 2007 (henceforth 'the 2007 Regulations') came into force on 6 April 2007.

The 2007 Regulations revoke and replace the Construction (Design and Management) Regulations 1994 ('the 1994 Regulations') and revoke and re-enact with modifications the Construction (Health, Safety and Welfare) Regulations 1996.

The 2007 Regulations give effect to the requirements of Directive 92/57/EEC on the implementation of minimum safety and health requirements at temporary or mobile construction sites, save for those requirements which were implemented by the Work at Height Regulations 2005.

The 2007 Regulations are divided into five parts. Part 1 deals with the formalities of citation and commencement, interpretation and application; Part 2 sets out the general management duties which apply to construction projects; Part 3 sets out additional duties imposed where the project is notifiable; Part 4 deals with the duties relating to health and safety on construction sites; and Part 5 deals with the general matters of civil liability, enforcement, transitional provisions and revocations and amendments.

20.4.2 Regulation 4

Regulation 4 provides that no person on whom the 2007 Regulations place the duty shall appoint or engage a CDM co-ordinator, designer, principal contractor or contractor unless he has taken reasonable steps to ensure that the person to be appointed or engaged is competent.

In a similar vein, no person can accept an appointment or engagement unless they are competent. The Approved Code of Practice for the 2007 Regulations states that, to be competent, an organization or individual must have sufficient knowledge of the specific task to be undertaken and the risks which the work will entail; have sufficient experience and ability to carry out their duties in relation to the project; and recognize their limitations and take appropriate action in order to prevent harm to those carrying out construction work, or those affected by the work.

Satisfying oneself as to competence means making reasonable enquiries to check that the organization or individual is competent to do the relevant work and can allocate adequate resources to it.

For notifiable projects (see Section 20.4.10), a key duty of the CDM co-ordinator is to advise clients about competence of designers and contractors, including the principal contractor.

20.4.3 Co-operation, co-ordination and the general principles of prevention

Regulation 5 requires that every person concerned in a project, who has a duty placed upon them by the 2007 Regulations, shall seek the co-operation of any other person concerned in any project involving construction work at the same or an adjoining site and co-operate with any person concerned in any project involving construction work at the same or an adjoining site. Every person concerned in the project who is working under the control of another person shall report to that person anything which he is aware is likely to endanger the health or safety of himself or others.

By virtue of regulation 6, all persons concerned in a project on whom a duty is placed by the 2007 Regulations, shall co-ordinate their activities with one another in a manner which ensures, so far as is reasonably practicable, the health and safety of persons carrying out the construction work and affected by the construction work.

Regulation 7(1) requires every person who has a duty placed upon them by the 2007 Regulations in relation to the design, planning and preparation of a project to take account of the general principles of prevention in the performance of those duties during all stages of the project. In similar, although slightly different terms, regulation 7(2) requires every person who has a duty placed upon them by the Regulations in relation to the construction phase of a project to ensure, so far as is reasonably practicable, that the general principles of prevention are applied in the carrying out of the construction work.

The general principles of prevention are to be found in Schedule 1 to the Management of Health and Safety at Work Regulations 1999. They are as follows:

- avoiding risks;
- evaluating the risks which cannot be avoided;
- combating the risks at source;
- adapting the work to the individual, especially as regards the design of workplaces, the choice of work equipment and the choice of working and production methods, with a view, in particular, to alleviating monotonous work and work at a predetermined work-rate and to reducing their effect on health;
- adapting to technical progress;
- replacing the dangerous by the non-dangerous or the less dangerous;

- developing a coherent overall prevention policy which covers technology, organization of work, working conditions, social relationships and the influence of factors relating to the working environment;
- giving collective protective measures priority over individual protective measures; and
- giving appropriate instructions to employees.

20.4.4 Election by clients

Regulation 8 provides that where there is more than one client in relation to a project, if one or more of such clients elect in writing to be treated for the purposes of the 2007 Regulations as the only client or clients, no other client who has agreed in writing to such election shall be subject thereafter to any duty owed by a client under the Regulations, save for certain duties that relate to information in their possession (see regulations 5(1)(b), 10(1), 15 and 17(1)).

20.4.5 Regulation 9

Regulation 9 sets out the client's duty in relation to arrangements for managing projects. A client is an organization or individual for whom a construction project is carried out. Clients only have duties when the project is associated with a business or other undertaking (whether for profit or not). This can include, for example, local authorities, school governors, insurance companies and project originators on private finance initiative projects. Domestic clients are a special case and do not have duties under the 2007 Regulations.

Every client is required to take reasonable steps to ensure that the arrangements made for managing the project (including the allocation of sufficient time and other resources) by persons with a duty under the 2007 Regulations (including the client himself) are suitable to ensure that:

- the construction work can be carried out so far as is reasonably practicable without risk to the health and safety of any person;
- the requirements of Schedule 2 to the 2007 Regulations (which sets out the welfare facilities that contractors are obliged to provide) are complied with in respect of any person carrying out the construction work;
- any structure designed for use as a workplace has been designed taking account of the provisions of the Workplace (Health, Safety and Welfare) Regulations 1992, which relate to the design of, and materials used in, structures.

The client is obliged to take steps to ensure that the arrangements made for managing the project are maintained and reviewed throughout the project.

20.4.6 Regulation 10

Regulation 10 sets out the client's duty in relation to the provision of information. Every client is obliged to ensure that every person designing the structure and every contractor who has been, or may be, appointed by the client is promptly provided with pre-construction information.

Under regulation 10(2), the pre-construction information consists of all the information in the client's possession (or which is reasonably obtainable) which is relevant to the person to whom the client provides it, including:

- any information about or affecting the site of the construction work;
- any information concerning the proposed use of the structures as a workplace;
- the minimum amount of time before the construction phase which will be allowed to the contractors appointed by the client for planning and preparation for construction works; and
- any information in any existing health and safety file.

Generally speaking, the purpose of providing such information is to ensure, so far as is reasonably practicable, the health and safety of persons engaged in the construction works; those liable to be affected by the way in which it is carried out; and those who will use the structure as a workplace. It is also to assist the persons to whom the information is provided (i.e. the designers and the contractors) to perform their duties under the 2007 Regulations and to determine the resources required to manage the project.

20.4.7 Duties of designers

The duties of designers are to be found in regulation 11. Designers are those who have a trade or business which involves them in preparing designs for construction work, including variations. This includes preparing drawings, design details, specifications, bills of quantities and the specification (or prohibition) of articles and substances, as well as all the related analysis, calculation and preparatory work. This also includes arranging for employees or other people under their control to prepare designs relating to a structure or part of a structure.

Designers will include architects, civil and structural engineers, building surveyors, landscape architects, other consultants, manufacturers and design practices (of whatever discipline) contributing to, or having an overall responsibility for, any part of the design, for example, drainage engineers designing the drainage for a new development.

The definition does, however, go much further if one considers paragraph 116 of the Approved Code of Practice. That suggests that designers could include quantity surveyors who insist on specific materials; clients who stipulate a particular layout for a new building; building services designers or others designing plant which forms part of the permanent structure (including lifts, heating, ventilation and electrical systems); those purchasing materials where the choice has been left open; temporary works engineers; interior designers, including shop fitters who also develop the design, heritage organizations who specify how work is to be done in detail; and those determining how buildings and structures are altered.

The first point to make is that, by virtue of regulation 4(1)(b), a designer shall not accept an appointment as such unless he is competent. Once the designer is satisfied that he is able to do the job, he still cannot commence work in relation to a project unless any client for the project is aware of his duties under the 2007 Regulations.

Once the designer starts designing (or modifying a design) he has to avoid foreseeable risk to the health and safety of persons in five defined categories, namely, those:

1. carrying out construction work;
2. liable to be affected by construction work;
3. cleaning any window or any transparent or translucent wall, ceiling or roof in or on a structure;
4. maintaining the permanent fixtures and fittings of a structure; or
5. using a structure designed as a workplace.

In preparing the design, the designer shall eliminate hazards which may give rise to risks; and reduce risks from any remaining hazards, and in doing so shall give collective measures priority over individual measures. In preparing the design, the designer is obliged to take all reasonable steps to provide with it sufficient information about aspects of the design of the structure or its construction or maintenance as will adequately assist clients, other designers and contractors to comply with their duties under the 2007 Regulations.

20.4.8 Designs prepared or modified outside Great Britain

As the 2007 Regulations apply to Great Britain, the possibility of the design being prepared or modified somewhere else is dealt with by regulation 12. In those circumstances, the party who commissions the design, if they are established within Great Britain, or if they are not so established, any client for the project, is required to ensure that regulation 11 (duties of designers) is complied with.

20.4.9 Duties of contractors

Contractors are those who actually do the construction work and are, thus, those most at risk of injury and ill-health. Anyone who directly employs or engages construction workers or controls or manages construction work is a contractor for the purposes of the 2007 Regulations. This includes companies that use their own workforce to do construction work on their own premises.

As with designers, no contractor shall carry out construction work in relation to a project unless any client for the project is aware of his duties under the 2007 Regulations, and the contractor should not accept an appointment unless they regard themselves as competent to carry it through. Contractors are obliged to plan, manage and monitor construction work carried out by them or under their control in a way which ensures that, so far as is reasonably practicable, it is carried out without risks to health and safety.

Contractors are required to ensure that any contractor they engage is informed of the minimum amount of time which they will be allowed for planning and preparation before they begin work.

Regulation 13(4) sets out requirements in respect of the provision of information and training for every worker carrying out construction work. This includes:

- suitable site induction, where not provided by any principal contractor;
- information on the risks to the worker's health and safety, whether brought out by a risk assessment under regulation 3 of the Management of Health and Safety at Work Regulations 1999 or arising out of the conduct by another contractor of his undertaking;
- the measures which have been identified by the contractor in consequence of the risk assessment as measures he needs to take to comply with the requirements and prohibitions imposed upon him by or under the relevant statutory provisions;
- any site rules;
- the procedures to be followed in the event of serious and imminent danger to workers; and
- the identity of the persons nominated to implement those procedures.

A specific duty is set out in regulation 13(6) regarding the prevention of access to site by unauthorized persons. Work cannot start unless reasonable steps have been taken in that regard. Lastly, by virtue of regulation 13(7) provision is made in relation to welfare facilities. These are set out in Schedule 2 to the 2007 Regulations.

20.4.10 Part 3 Notifiable projects

Part 3 sets out the additional duties which arise where a project is notifiable. What this actually means is considered below. In such circumstances additional duties are incumbent upon clients, designers and contractors. Perhaps more importantly, however, it introduces us to the CDM co-ordinator and the principal contractor, whose roles are considered in detail below.

20.4.11 What is a notifiable project?

By virtue of regulation 2(3), a project is notifiable if the construction phase is likely to involve more than 30 days; or 500 person days (e.g. 50 people working for over 10 days) of construction work. All days on which construction work takes place count towards the period of construction work. Holidays and weekends do not count if no construction work takes place on these days.

20.4.12 Client's duties on notifiable projects

Where a project is notifiable, the client is obliged to appoint a CDM co-ordinator as soon as is practicable after the initial design work or other preparation for construction work has begun. After appointing a CDM co-ordinator, the client is obliged to appoint a principal contractor as soon as is practicable after the client knows enough about the project to be able to select a suitable person for such appointment.

The appointments of the CDM co-ordinator and principal contractor are to be changed or renewed as necessary to ensure that there is, at all times until the end of the construction phase, a CDM co-ordinator and principal contractor. For so long as either or both of these appointments are not made, the client is deemed to have been appointed as CDM co-ordinator or principal contractor, or both. By virtue of regulation 15, where the project is notifiable, the client is obliged to promptly provide the CDM co-ordinator with the pre-construction information envisaged by regulation 10(2).

Under regulation 16, the client is required to ensure that the construction phase does not start unless the principal contractor has prepared a construction phase plan (see Section 20.4.16) and he is satisfied that the requirements in respect of the provision of welfare facilities will be complied with during the construction phase.

Lastly, regulation 17 provides that the client is required to ensure that the CDM co-ordinator is provided with all the health and safety information in the client's possession (or which is reasonably obtainable) relating to the project which is likely to be needed for inclusion in the health and safety file.

Where a single health and safety file relates to more than one project, site or structure, or where it includes other related information, the client is required to ensure that the information relating to each site or structure can be easily identified.

After the construction phase, the client must take reasonable steps to ensure that the information in the health and safety file is kept available for inspection by any person who may need it to comply with the relevant statutory provisions; and is revised as often as may be appropriate to incorporate any relevant new information.

In the context of developers, regulation 17(4) is significant. If a client disposes of his entire interest in the structure, delivering the health and safety file to the person who acquires his interest in it and ensuring that he is aware of the nature and purpose of the file, is sufficient compliance with regulation 17(3) (i.e. the obligation set out above to keep the health and safety file available for inspection, etc.).

20.4.13 Additional duties of designers and contractors

Insofar as designers are concerned, by virtue of regulation 18, where a project is notifiable, no designer can commence work (other than initial design work) in relation to the project unless a CDM co-ordinator has been appointed for the project.

The designer is required to take all reasonable steps to provide with his design sufficient information about aspects of the design of the structure or its construction or maintenance as will adequately assist the CDM co-ordinator to comply with his duties under the 2007 Regulations, including his duties in relation to the health and safety file.

Insofar as contractors are concerned, by virtue of regulation 19, where a project is notifiable, no contractor shall carry out construction work in relation to the project unless:

- he has been provided with the names of the CDM co-ordinator and principal contractor;
- he has been given access to such part of the construction phase plan as is relevant to the work to be performed by him, containing sufficient detail in relation to such work; and
- notice of the project has been given to HSE.

Additional obligations are imposed by regulation 19(2). Every contractor shall promptly provide the principal contractor with any information which might affect the health and safety of any person carrying out construction work, or of any person who may be affected by it, which might justify a review of the construction phase plan; or which has been identified for inclusion in the health and safety file.

Every contractor is obliged to promptly identify any contractor whom he appoints or engages (e.g. a sub-contractor) in connection with the project to the principal contractor. Every contractor shall comply with any directions of the principal contractor and any site rules. Lastly, every contractor shall promptly provide the principal contractor with the information in relation to any death, injury, condition or dangerous occurrence which the contractor is required to notify or report under the Reporting of Injuries, Diseases and Dangerous Occurrences Regulations 1995.

Three further duties are imposed in relation to the construction phase plan. Every contractor shall take all reasonable steps to ensure that the construction work is carried out in accordance with the construction phase plan; shall take appropriate action to ensure health and safety where it is not possible to comply with the construction phase plan in any particular case; and shall notify the principal contractor of any significant finding which requires the construction phase plan to be altered or added to.

20.4.14 The CDM co-ordinator

The role of CDM co-ordinator was created by the 2007 Regulations. The role of planning supervisor, created by the 1994 Regulations, has gone. The CDM co-ordinator's role is to provide the client with a key project adviser in respect of construction health and safety risk management matters. They should assist and advise the client on the appointment of competent contractors and the adequacy of management arrangements; ensure proper co-ordination of the health and safety aspects of the design process; facilitate good communication and co-operation between project team members; and prepare the health and safety file.

CDM co-ordinators must give suitable and sufficient advice and assistance to clients in order to help them to comply with their duties under the 2007 Regulations, in particular, the duty to appoint competent designers and contractors and the duty to ensure that adequate arrangements are in place for managing the project.

They are also obliged to ensure that suitable arrangements are made and implemented for the co-ordination of health and safety measures during planning and preparation of the construction phase, including facilitating co-operation and co-ordination between persons concerned in the project and the application of the general principles of prevention (see Section 20.4.3).

They are required to liaise with the principal contractor regarding the contents of the health and safety file, the information which the principal contractor needs to prepare the construction phase plan and any design development which may affect planning and management of the construction work.

The CDM co-ordinator is required to take all reasonable steps to identify and collect the pre-construction information; to promptly provide, in a convenient form, to every person designing the structure and every contractor who has been or may be appointed by the client such pre-construction information in his possession as is relevant to each; to take all reasonable steps to ensure that the designers comply with their duties under regulations 11 and 18(2); to take all reasonable steps to ensure co-operation between the designers and the principal contractor during the construction phase in relation to any design or change to a design; to prepare, where none exists, and otherwise review and update, the health and safety file; and, at the end of the construction phase, to pass the health and safety file to the client.

The duty to notify the project to HSE falls upon the CDM co-ordinator by virtue of regulation 21(1). The particulars to be notified to HSE are set out in Schedule 1 to the 2007 Regulations.

20.4.15 Duties of the principal contractor

The duties incumbent upon the principal contractor are set out in regulations 22, 23 and 24.

The key duty of a principal contractor is to properly plan, manage and coordinate work during the construction phase in order that risks are properly controlled. There can only be one principal contractor for a project at any one time. However, sometimes two or more projects take place on a site at the same time. This can occur if different clients commission adjacent work, or if a client procures two truly independent, unrelated packages of work which do not rely upon one another for their viability or completion. Where overlapping projects are running on a single construction site, it is usually best to appoint one principal contractor for all of them. If this is not done, the principal contractors must co-operate and their plans must take account of the interfaces, for example, in traffic management.

The principal contractor for the project shall plan, manage and monitor the construction phase in a way which ensures, so far as is reasonably practicable, that it is carried out without risks to health or safety. This includes facilitating co-operation and co-ordination between persons concerned in the project and the application of the general principles of prevention (see Section 20.4.3).

Further duties imposed upon the principal contractor are:

- to liaise with the CDM co-ordinator during the construction phase in relation to any design or change to a design;
- to ensure that the requisite welfare facilities are provided throughout the construction phase;
- where necessary for health and safety, to draw up rules which are appropriate to the construction site and the activities on it;
- to give reasonable directions to any contractors so far as is necessary to enable the principal contractor to comply with his duties under the 2007 Regulations;
- to ensure that every contractor is informed of the minimum amount of time which will be allowed to him for planning and preparation before he begins construction work;
- where necessary, to consult a contractor before finalizing such part of the construction phase plan as is relevant to the work to be performed by him;
- to ensure that every contractor is given, before he begins construction work and in sufficient time to enable him to properly prepare for that work, access to such part of the construction phase plan as is relevant to the work to be performed by him;
- to ensure that every contractor is given, before he begins work and in sufficient time, such further information as he needs to comply punctually with his obligation in respect of welfare facilities and to carry out the work to be performed by him, so far as is reasonably practicable, without risk to the health and safety of any person;
- to identify to each contractor the information relating to that contractor's activity which is likely to be required by the CDM co-ordinator for inclusion in the health and safety file and to ensure that such information is promptly provided to the CDM co-ordinator;
- to ensure that the particulars required to be in the notice to the HSE are displayed in a readable condition in a position where they can be read by any worker engaged in the construction work;
- to take reasonable steps to prevent access by unauthorized persons to the construction site; and
- to take reasonable steps to ensure that every worker carrying out construction work is provided with a suitable site induction and with suitable information and training for the particular work to be carried out by him.

20.4.16 Construction phase plan

The construction phase plan should set out the way in which the construction phase will be managed, identifying the main health and safety issues arising. It should be specific and should not simply be an accumulation of generic risk assessments or method statements. It requires to be tailored to the project to which it relates. The matters that ought to be contained within the construction phase plan are set out in detail in Appendix 3 to the Approved Code of Practice to the 2007 Regulations.

The principal contractor's duties in relation to the construction phase plan are contained in regulation 23. Before the start of the construction phase, the principal contractor is obliged to prepare a construction phase plan which is sufficient to ensure that the construction phase is planned, managed and monitored in a way which enables the construction work to be started, so far as is reasonably practicable, without risk to health or safety, and which pays adequate regard to the information provided by the designer and the pre-construction information provided through the CDM co-ordinator.

From time to time, and as often as may be appropriate throughout the project, the principal contractor is obliged to update, review, revise and refine the construction phase plan so that it continues to be sufficient to ensure that the construction phase is planned, managed and monitored in a way which enables the construction work to be carried out without risk to health and safety, again insofar as is reasonably practicable.

Lastly, the principal contractor is obliged to ensure that the construction phase plan is implemented in a way which will ensure, so far as reasonably practicable, the health and safety of all persons carrying out the construction work and all persons who may be affected by that work.

The principal contractor is obliged to take reasonable steps to ensure that the construction phase plan identifies the risks to health and safety arising from the construction work, including the risks specific to the particular type of construction work concerned, and includes suitable and sufficient measures to address such risks, including site rules.

20.4.17 Health and safety file

To be completed and handed over to the client at the end of the project, the health and safety file should contain relevant information in relation to the structure which will be of assistance in relation to any future project carried out to it. As will have been noted above, clients, designers, principal contractors, other contractors and CDM co-ordinators all have duties in relation to the health and safety file.

20.4.18 Co-operation and consultation with workers

Regulation 24 imposes three specific duties relative to co-operation and consultation with workers.

First, the principal contractor is required to make and maintain arrangements which will enable him and the workers engaged in the construction work to co-operate effectively in promoting and developing measures to ensure the health, safety and welfare of the workers and then checking the effectiveness of such measures.

Second, the principal contractor is obliged to consult those workers (or their representatives) in good time on matters connected with the project which may affect health, safety or welfare, so far as they or their representatives are not so consulted on those matters by any employer of theirs.

Third, the principal contractor is obliged to ensure that such workers or their representatives can inspect and take copies of any information which the principal contractor has which relate to the planning and management of the project, or which otherwise may affect their health, safety or welfare. In this regard, qualifications do exist in relation to certain matters, see regulation 24(c)(i) – (v).

20.4.19 Part 4 Health and safety on construction sites

Part 4 of the Regulations sets out duties relating to health and safety on construction sites. It deals with a wide range of matters, namely, safe places of work; good order and site security; stability of structures; demolition and dismantling; use of explosives; excavations; cofferdams and caissons; reports of inspections; energy distribution installations; prevention of drowning; traffic routes; vehicles; prevention of risk from fire; emergency procedures; emergency routes and exits; fire detection and fire fighting; fresh air; temperature and weather protection; and lighting. A detailed consideration of Part 4 is outwith the scope of this book.

20.4.20 Civil liability

The position in relation to civil liability is regulated by Part 5 of the 2007 Regulations and, in particular, by regulation 45.

The position has changed, somewhat, from that which subsisted under the 1994 Regulations. In those, only regulation 10 (start of construction phase) and regulation 16(1) (requirements on principal contractor) could confer a right of action.

Of the 2007 Regulations, the obligations imposed by regulations 9(1)(b); 13(6) and (7); 16; 21(1)(c) and (l); 25(1), (2) and (4); 26 to 44; and Schedule 2 can confer a right of action in civil proceedings, insofar as the relevant duties apply for the protection of a person who is not an employee of the person on whom the duty is placed.

20.5 The SBC and the SBC/DB provisions

Under clause 2.1 of both the SBC and the SBC/DB, the Contractor is under an obligation to carry out and complete the Works in compliance with the Contract Documents, the Construction Phase Plan and the Statutory Requirements in relation to the Works. Implicitly, this will include all health and safety legislation insofar as it is relevant to the work.

Each party to the contract is required, under clause 3.23 of the SBC and clause 3.16 of the SBC/DB, to comply with the 2007 Regulations. This includes clause 3.23.2.1 of the SBC and clause 3.16.3.1 of the SBC/DB, under which the Contractor is obliged, while they remain Principal Contractor, to ensure that the Construction Phase Plan is received by the Employer before construction work is commenced and that any subsequent amendment to it by the Contractor is notified to the Employer, the CDM Co-ordinator and in the case of the SBC (where they are not the CDM Co-ordinator) the Architect/Contract Administrator.

While they are Principal Contractor, the Contractor must ensure that welfare facilities complying with Schedule 2 of the 2007 Regulations are provided from the commencement of construction work until the end of the construction phase. Where the Contractor is not the principal contractor, they are obliged to promptly inform the Principal Contractor of the identity of each sub-contractor that they appoint and of each sub-sub-contractor appointment notified to them.

The Contractor is also required, under clause 3.23.4 of the SBC and clause 3.16.5, to provide on written request by the CDM Co-ordinator, and to ensure that any sub-contractor provides, to the CDM Co-ordinator such information as the CDM Co-ordinator reasonably requires for the preparation of the health and safety file. This requirement must be complied with before practical completion can be certified, see clause 2.30 of the SBC and clause 2.27 of the SBC/DB.

Under clause 3.24 of the SBC and clause 3.17 of the SBC/DB, if the Employer appoints a successor to the Contractor as Principal Contractor, the Contractor is required, at no cost to the Employer, to comply with all reasonable requirements of the new Principal Contractor to the extent necessary for compliance with the 2007 Regulations. No extension of time is given in respect of such compliance.

The SBC/DB also recognizes that the Contractor may undertake the role of CDM Co-ordinator, in which event clause 3.16.2 provides that the Contractor shall comply with all the duties of the CDM Co-ordinator and shall prepare and deliver the health and safety file to the Employer.

20.6 The NEC3 provisions

There are a small number of clauses within the NEC3 which deal expressly with the issue of health and safety. Of the Contractor's main responsibilities, the requirement to act in accordance with the health and safety requirements stated in the Works Information is set out in clause 27.4. Notably, the programming requirements of the NEC3 make specific reference to health and safety requirements, namely, clause 31.2, which provides that any programme submitted for acceptance must show provisions for health and safety requirements. The Employer is permitted to terminate if there is a substantial breach of health and safety regulation in the circumstances outlined by clause 91.3. Precisely what constitutes a substantial breach will be a matter of facts and circumstances in every case.

Chapter 21

Regulatory Matters

Competition in Construction

21.1 Competition law in the UK: Introduction

Competition law has become a very important issue for businesses operating in the UK construction industry. Until April 2014, competition law in the UK was enforced by the Office of Fair Trading (OFT) and it regarded the UK construction industry as a priority area for enforcement. Even though the OFT's enforcement functions transferred to a new Competition and Markets Authority (CMA) in April 2014, it is expected that the CMA will continue to pay close attention to the UK construction industry contractors and members of the supply chain must therefore ensure that they know and conduct their activities in accordance with the competition rules. In this chapter, we give a practical overview of anti-competitive agreements and the types of behaviour which breach the Competition Act 1998. We also consider the penalties for breaching competition law (which includes criminal sanctions for individuals) and the OFT's leniency regime.

21.2 Competition law: Overview

The main competition legislation in the UK is the Competition Act 1998 (henceforth 'the 1998 Act'). The 1998 Act sets out two prohibitions, namely:

- the prohibition of anti-competitive agreements (the Chapter I prohibition); and
- the prohibition of abuse of a dominant position (the Chapter II prohibition).

These prohibitions are modelled on their EU counterparts, Articles 101 and 102 of the EC Treaty. While EU competition law applies in parallel to the domestic competition regime, this chapter will focus on the UK prohibitions under the 1998 Act. The Chapter I and Chapter II prohibitions are discussed in more detail below.

The 1998 Act is supplemented by the cartel provisions contained in sections 188–202 of the Enterprise Act 2002 (henceforth 'the 2002 Act'). Section 188 introduced a cartel offence into UK law, making participation in cartel arrangements a

MacRoberts on Scottish Construction Contracts, Third Edition. MacRoberts.

criminal offence. This changed the UK competition landscape considerably, as it meant that the individuals involved in cartel behaviour could now be targeted by the authorities. So with the threat of personal criminal sanctions, competition law was no longer just an issue for the businesses engaging in cartel conduct.

21.3 Penalties

21.3.1 Fines

The main sanction on a business which breaches the 1998 Act is a fine of up to 10% of its turnover (s.36(8) of the 1998 Act). The highest fine imposed to date for a breach of the Chapter I prohibition is £121 million. This fine was imposed on British Airways (BA) for its role in a price fixing arrangement with Virgin Atlantic in relation to fuel surcharges for long-haul passenger services to and from the UK between August 2004 and January 2006. BA agreed this fine by way of an early resolution agreement with the OFT, though the fine was subsequently reduced to £58.5 million. Virgin Atlantic received full immunity. However, the fines which were imposed in the Roofing cases and the English Construction case provide a better illustration of the fines which have actually been imposed on members of the construction industry for anti-competitive behaviour. In the Roofing cases, the fines (before leniency) ranged from £1,963 to £328,264. In the English Construction case (which involved businesses of all sizes) the fines before leniency and any reductions on appeal ranged from £713 to £17 million. The highest fine was imposed on Kier Group Plc. However, this was subsequently reduced on appeal to £1.7 million. See Section 21.5 for an analysis of these cases.

The key point to note is that fines are linked to the turnover of the business in question and so have to be considered in context of that business. While a five- or six-figure fine may not appear too significant in the abstract, such a fine could threaten the very existence of a local or regional contractor.

21.3.2 Third party damages

Where a third party has suffered loss due to a cartel, that third party has the right to sue cartel members for damages. There have, however, been very few private damages actions in Scotland to date. While the reasons for this are outwith the scope of this chapter, they essentially derive from the practical issues which one faces in the litigation context (including cost, uncertainty of outcome and quantification of loss). However, such issues are not insurmountable and so the risk of third party claims for damages cannot be discounted entirely. Furthermore, there is clearly a desire at both government and EU level to make private actions more accessible and therefore a more feasible remedy for businesses and consumers. Following a number of consultations over recent years, provisions were included in the draft Consumer Rights Bill (published in June 2013) which, if passed, will expand the jurisdiction of the Competition Appeals Tribunal and introduce a new collective actions regime.

21.3.3 Other consequences

There are other consequences which are worthy of note. For instance, if an agreement contains a provision which is anti-competitive, that provision is void. If the provision in question cannot be severed from the agreement, then, under s.2(4) of the 1998 Act, the whole agreement is void. There is also the issue of adverse publicity to consider. A business's reputation can be severely damaged by coverage in the mainstream and/or trade press relating to its involvement (or possible involvement) in anti-competitive activities. Also, as competition investigations require a significant amount of management input, this management resource has to be diverted away from running the business in order to deal with the investigation. So the business may continue to suffer in operational terms for the duration of the investigation.

21.3.4 Personal sanctions

While the above sanctions and consequences impact directly on the business in question, the individuals involved in cartel activities may also face personal sanctions. These sanctions include being disqualified from acting as a director for up to 15 years (see ss.9A and 9B, Company Directors Disqualification Act 1986, and OFT guidance 'Directors disqualification orders in competition cases' (OFT510)). However, the most notable personal sanction is that an individual who is found guilty of the cartel offence could receive a prison sentence of up to five years and/or an unlimited fine (see s.190 of the 2002 Act).

21.4 The Chapter I prohibition

The Chapter I prohibition (which is set out in s.2 of the 1998 Act) targets anti-competitive agreements, in other words, agreements between competitors (see OFT guidance 'Agreements and concerted practices' (OFT401)). In practical terms, there are two key questions to consider: (1) is there an 'agreement'?; and (2) is the subject matter or the behaviour to which the agreement relates prohibited?

21.4.1 Is there an agreement?

The scope of what constitutes an agreement for the purposes of the Chapter I prohibition is very wide. There is no need for the agreement to be in writing, to be formal or to be legally binding. A 'gentleman's agreement' or 'a nod and a wink' will be sufficient. The Chapter I prohibition also applies to a concerted practice – this is where, though there is no agreement as such, there is some degree of understanding between competitors as to the co-ordination of future behaviour in the market (see paragraphs 2.11–2.13 of the OFT guidance 'Agreements and concerted practices' (OFT401)). Decisions of associations of undertakings, such as trade associations, will also be caught by Chapter 1 in the sense that they represent the agreement of the members.

The important issue is not the manner in which any such agreement was reached or concerted practice engaged – rather it is the fact that the parties have removed the element of uncertainty which should normally exist between competitors and have replaced that uncertainty with certainty as to future behaviour. The business knows what its competitor is going to do and can adjust its behaviour accordingly. So the simple message is that very little is required to constitute an agreement or concerted practice for the purposes of the Chapter I prohibition.

It should also be noted that the Chapter I prohibition targets agreements which have as their object or effect the prevention, restriction or distortion of competition. So, the CMA can target agreements which are intended to be anti-competitive, but can also target agreements which have an anti-competitive effect, even though that was not the intention of the parties.

21.4.2 Is the subject matter of the agreement prohibited?

When considering the subject matter of the agreement, there are types of behaviour which are automatically illegal. These include price fixing, market sharing, bid rigging and collective boycotts (see Construction recruitment forum (CE/7510-06)) - each is very relevant to the construction industry and is regarded as a serious breach of the competition rules. There are, however, types of behaviour which may be permitted in some circumstances, but prohibited in others. These cases, which in the context of the construction industry include information exchanges, trade associations and joint ventures, all depend on the particular circumstances.

Price fixing

In terms of price fixing, the message is quite simple – it is illegal for a business to co-operate with its competitors to fix the prices at which goods and/or services will be supplied. Price fixing, however, is not just restricted to fixing the exact price of such goods and/or services. It extends to agreeing minimum prices, price ranges, discounts, allowances, rebates and other such components of the price. In short, a business should not be discussing prices or any aspect related to pricing with its competitors.

Market sharing

As for market sharing, it is illegal for a business to agree with its competitors the customers that they will serve or the geographic areas in which they will operate (see the Roofing case concerning the North-East of England (CA98/02/2005) discussed in Section 21.5.1). In effect, market sharing seeks to insulate the parties in question from competition. The basic rule is that a business should target customers or areas based on its own unilateral commercial decisions and not on the basis of agreements or understandings with competitors.

Collusive tendering

Collusive tendering is where the bidders collude to undermine the competitive tender process. As the tender process is so widely used in the construction industry, collusive tendering (or bid rigging) is a particular risk. This is borne out by the Roofing cases and the English Construction case, both of which are summarized in Section 21.5. The collusion may range from the organized and systematic sharing out of contracts between the 'bidders' (in the form of a bid rotation or quota scheme) to the practice of 'simple' cover pricing. Cover pricing is a form of bid rigging. 'Simple' cover pricing is where party A has been invited to bid but, for whatever reason, does not wish to win the work. Party A contacts one of the other bidders and asks for a cover price – in other words a price that resembles a genuine bid but which will ensure that party A does not win.

The message from the Competition Appeals Tribunal, however, is that collusive tendering breaches the Chapter I prohibition, see *Apex Asphalt and Paving Co Limited* v. *Office of Fair Trading* (2005) and *Kier Group Plc & Others* v. *Office of Fair Trading* (2011). The nature of the collusion will simply determine the severity of the breach.

In short, co-ordination between competitors in relation to a tender is illegal unless that co-ordination has been disclosed to the client in advance. Any decision whether to bid and what to bid should be a commercial decision which is made by the business unilaterally and on its own account.

Information exchanges

The issue of information exchanges is a difficult area. In some circumstances, information exchanges may actually benefit consumers; in other circumstances however, such exchanges may fall foul of the competition rules. It all depends on the circumstances and, in particular, the nature of the information which is being exchanged. It should be noted that an information exchange does not have to be formal. A casual conversation at an industry event for example could well be sufficient to give rise to an information exchange for the purposes of the competition rules.

The basic guideline is that sharing price or other commercially sensitive information will breach the Chapter I prohibition. Such information will, if disclosed to a competitor, give an insight into the disclosing party's competitive strategy and remove the uncertainty about its future behaviour in the market. There will be particular concerns where the information is confidential, where it relates to current or future activities and/or where it is detailed. By contrast, there will generally be less of a risk where information is historic, aggregated and/or purely statistical as such information is less likely to influence current or future conduct in the market.

Where the parties to the information exchange remain active on the market, there is a presumption that they have taken account of the information which has been exchanged and this presumption applies even if the parties only met on one occasion, see *T-Mobile Netherlands BV and Others* v. *Raad van bestuur van de Nederlandse Mededingingsautoriteit* (2009). Accordingly, a one-off exchange of information could be sufficient to fall foul of the Chapter I prohibition.

Trade associations

Trade associations and other industry bodies raise concerns under the Chapter I prohibition because, by their very nature, they involve a meeting of members of the same trade, most (if not all) of whom are competitors.

The main risk is that members may use trade association meetings (whether as part of the formal agenda or otherwise) as a forum in which to discuss matters or arrange actions which are anti-competitive. There is also the risk that such matters or actions are raised inadvertently in the meeting. A further concern is that the terms of membership of a trade association may be used as a means of excluding or restricting competition. Trade association membership rules must be objective, justified and open to all.

Trade association membership and attendance at meetings should therefore be treated with some caution. An individual attending a trade association meeting or industry event should know what the boundaries are in terms of competition law and, if conversation strays over those boundaries, that individual should at the very least object, remove himself from the discussions and report the issue to his employer. The steps which are taken thereafter will very much depend on that business's approach to competition compliance.

Joint ventures

In some circumstances, a business may wish to partner with a competitor for a particular project. As this would involve collaboration with a competitor, it has to be arranged and implemented with considerable care. The key issue is transparency and making sure that the client is informed and is kept informed of the collaboration. The parties should also ensure that they have a legitimate commercial justification for proceeding by way of a collaboration, and have appropriate controls in place to ensure that competition outwith the scope of the project specific collaboration is not compromised.

21.5 *The Roofing and English Construction cases*

21.5.1 The Roofing cases

The Roofing cases comprise five cases in which the OFT found that roofing contractors in particular parts of the UK had breached the Chapter I prohibition, namely, cases CA98/1/2004, CA98/01/2005, CA98/02/2005, CA98/04/2005, and CA98/01/2006. Two of the cases related exclusively to Scotland, two of the cases related to areas in England (namely the West Midlands and the North-East), and the remaining case related to Scotland and England.

The first investigation concerned the West Midlands (CA98/1/2004). Ruberoid Plc applied for leniency in Autumn 2001 on behalf of its subsidiary Briggs Cladding and Roofing Limited. As part of its leniency submission, it provided the OFT with information about collusive tendering in relation to flat roofing contracts in the West Midlands. Dawn raids (unannounced on-site inspections under the 1998 Act) were carried out in September 2002 and the infringement decision issued in March 2004. The OFT found that the parties involved had been involved in collusive tendering with the object of fixing tender prices. Putting it briefly, when a purchaser invited tender bids for a flat roofing contract, the parties would agree between themselves the prices at which each party would tender for that particular contract. As far as the customer was concerned, there was genuine competition. In actual fact, however, the bids had been rigged principally through the fixing and submission of cover prices.

The information provided by Briggs as part of its leniency submission also appeared to suggest that collusive tendering was taking place in Scotland in relation to mastic asphalt coverings for flat roofs. Accordingly, this formed the basis of a separate OFT investigation (CA98/01/2005). Dawn raids were carried out in November 2002 and an infringement decision was issued in April 2005. Again, the infringement concerned collusive tendering by way of cover pricing – in effect, the 'lead' contractor would contact the other parties and provide them with details of its proposed bid. The other contractors would then formulate their own cover bids.

The other Roofing cases, which all essentially derived from the West Midlands case, also involved the practice of cover pricing (although the North-East of England case (CA98/02/2005) also involved market sharing by the allocation of certain customer contracts). Indeed, in the West Midlands case, a number of the parties claimed that cover pricing was endemic in the construction industry and the roofing industry in general and this was cited by the OFT in a number of the roofing decisions which followed West Midlands. The parties in these cases cited various reasons for cover pricing, the main reason being that a contractor who did not submit a bid for a particular project risked being blacklisted. Other reasons included a lack of capacity to carry out the project if successful and a lack of desire to carry out that particular work. The Competition Appeals Tribunal has, however, held that these reasons are not a defence and that 'the subjective intentions of a party to a concerted practice are immaterial where the obvious consequence of the conduct is to prevent, restrict or distort competition' (*Apex Asphalt and Paving Co Limited* v. *Office of Fair Trading* (2005)).

21.5.2 The English Construction case

The English Construction case (CA98/02/2009) is certainly the largest competition enquiry undertaken in the UK to date. It commenced in 2004 when an audit manager acting for an NHS Trust contacted the OFT alleging collusion in the tender process for works at Queens Medical Centre in Nottingham. When the infringement decision was issued some five years later, 103 construction firms in England were found to have infringed the Chapter I prohibition, with total fines of £129.2 million imposed on those firms (a number of which were reduced on appeal – see below).

The main focus of the investigation and resulting decision was bid rigging, in particular cover pricing, in relation to general building projects in England. The OFT found that bid rigging had taken place on 199 tenders over the period from 2000 to 2006.

While this number of tenders may appear small relative to the size of the investigation, this was driven by the measures implemented by the OFT to seek to focus the investigation. After all, by the autumn of 2006 the OFT apparently had evidence of cover pricing involving over 1000 firms and 4000 suspect tenders. One such measure was to focus on those firms which appeared to have been involved in cover pricing in at least five tenders. The OFT also identified a small number of cases where 'loser's fees' were paid to the 'unsuccessful bidder'.

So the English Construction case made it clear that cover pricing was a form of bid rigging and, as such, was prohibited by the 1998 Act. While this was certainly consistent with the decisions in the Roofing cases, a number of the implicated parties submitted appeals to the Competition Appeals Tribunal in respect of liability and the penalties imposed.

A number of fines were reduced on appeal, see *Kier Group Plc & Others* v. *Office of Fair Trading* (2011); *Durkan Holdings Limited & Others* v. *Office of Fair Trading* (2011); and *GF Tomlinson Group Limited & Others* v. *Office of Fair Trading* (2011). In the *Kier* appeal, the Competition Appeals Tribunal found that 'simple' cover pricing breaches the Chapter I prohibition. However, it considered that simple cover pricing was distinct from and less serious than bid rigging as ordinarily understood, in that it did not actually determine the price payable by the purchaser. Accordingly, the penalties to be imposed for simple cover pricing in the cases under appeal had to take account of this and the other mitigating factors which existed.

There is a danger that, based on the Competition Appeal Tribunal's comments, some may regard cover pricing as simply a technical or trivial infringement. Such an interpretation would be unwise. As the Competition Appeal Tribunal itself stated (at *Kier*, paragraph 99):

> Cover pricing is certainly not an innocuous activity … It is an unlawful practice which at the very least may deceive the customer about the source and extent of the competition which exists for the work in question and which is capable of having anti-competitive effects on the particular tendering exercise and on future exercises.

The Competition Appeal Tribunal's comments in the *Kier* appeal relate to the level of the penalty and the manner of its calculation, taking account of the historic uncertainties surrounding cover pricing which existed in the industry. Such uncertainties no longer exist. It is clear beyond doubt now that even 'simple' cover pricing is not a legitimate practice and it is very unlikely that a contractor that engages in cover pricing now would be able to claim mitigation to the same extent as that afforded by the Competition Appeals Tribunal in the English Construction case appeals.

21.6 The cartel offence

S.188(1) of the 2002 Act makes it a criminal offence for an individual to dishonestly agree with one or more other persons that undertakings (i.e. businesses) will engage in a prohibited cartel activity. S.188(2) provides that the prohibited cartel activities are: (1) price fixing; (2) market sharing; (3) bid rigging; and (4) limitation of production and supply.

The cartel offence operates alongside the Chapter I prohibition. As noted above, the 1998 Act penalizes the business which is engaged in the anti-competitive conduct, whereas the cartel offence targets the individuals involved. A person found guilty of the offence is liable to imprisonment for a term of up to five years or an unlimited fine or both.

Despite being in force since 2003, there have only been four cartel offence convictions in the UK to date. Three of these convictions related to the involvement of three UK nationals in an international marine hose cartel. However, the individuals in question were not tried in the UK. Rather, as part of their plea agreements in the United States on cartel charges, the individuals agreed that upon their return to the UK they would plead guilty to the UK cartel offence. Two of the individuals were originally sentenced to three years in prison, with the third individual receiving a sentence of two and a half years. These sentences were, however, reduced on appeal to two and half years, two years and twenty months respectively (see *R* v. *Whittle & Others* (2008)). The fourth (and most recent) conviction concerned cartel conduct in the UK in the supply of galvanised steel tanks and was again secured by way of a guilty plea.

In May 2010, the OFT's first cartel prosecution collapsed when a substantial volume of previously undisclosed evidence came to light in the course of the trial (see *R* v. *George, Crawley, Burns and Burnett* (2010)). The collapse of the OFT's flagship prosecution raised significant doubts about the cartel offence. Some parties questioned the OFT's role as both enforcer and prosecutor. Others questioned whether the dishonesty element of the offence (which was a requirement of the offence at that time) actually made it too difficult to prosecute in practice.

Following consultation, the government announced in March 2012 that it would introduce legislation to amend section 188 of the 2002 Act to remove the dishonesty element from the cartel offence. This change became effective on 1 April 2014, such that there is now no requirement to prove in a cartel offence prosecutions that the individual acted dishonestly. A number of statutory exceptions and defences to the cartel offence have also been introduced and these are set out in sections 188A and 188B of the 2002 Act (see CMA guidance 'Cartel Offence Prosecution Evidence' (CMA9)).

The removal of the dishonesty element should be a source of considerable concern for those individuals who are engaged in cartel activities. The clear message from government is that cartel activity is and will remain a criminal offence in the UK. Therefore, any individual engaging in cartel activity is running the risk of prosecution and, if convicted, up to five years in prison.

21.7 Leniency

The OFT operated a leniency regime for a number of years and continue to be a key weapon in the CMA's armoury. While a detailed explanation of the leniency regime is outwith the scope of this chapter, its basis is quite simple – if a business confesses to and co-operates with the CMA, then the business may be entitled to a reduction in any fine which may be imposed upon it (see OFT Guidance 'Applications for leniency and no-action in cartel cases' (OFT 1495)). This acts as an incentive for cartel members to 'break ranks' and disclose the cartel to the authorities.

The level of the reduction depends on when the business approaches the CMA to confess. If the business is first to confess and there is no CMA investigation already underway, then the business is automatically entitled to 100% immunity from fines provided that it accepts the CMA's leniency conditions. Except in Scotland, Type A immunity for the first to confess will also include automatic immunity from prosecution for the individuals involved. In the British Airways/Virgin Atlantic case (see Section 21.3.1), the whistle was blown by Virgin Atlantic. It benefitted from full immunity from fines, while its executives received immunity from criminal prosecution.

Where the cartel offence may have been committed within the jurisdiction of the Scottish courts, the CMA cannot grant the individuals involved automatic immunity from prosecution. However, in terms of the Memorandum of Understanding between the OFT and the Crown Office (Memorandum of Understanding between the Office of Fair Trading and the National Casework Division, Crown Office, Scotland), the Lord Advocate will give serious weight to an CMA recommendation that conditional criminal immunity be granted.

Where the business is first to confess, but there is already an investigation underway, then 100% reductions in fines may still be available. Any such reduction is, however, at the CMA's discretion. For the second and subsequent confessors, reductions of up to 50% are available on a sliding scale. So the key issue with leniency is timing. There are significant benefits from being first to confess – however, even if you are not first, substantial reductions in penalties can still be obtained.

21.8 The Chapter II prohibition

The Chapter II prohibition prohibits abuse of a dominant position (see 1998 Act s.18 and OFT Guidance 'Abuse of a dominant position' (OFT402)). While the Chapter I prohibition focuses on a business's dealings with its competitors, abuse of a dominant position concerns a business's dealings with its customers and suppliers (i.e. vertical relationships).

A business is dominant if it can act independently and without consideration of its competitors. A market share of 50% or above will give a strong inference of dominance (*AKZO Chemie BV* v. *Commission* (1986)). The key test, however, is whether or not there is effective competition in the market (*Hoffman-La Roche* v. *Commission* (1979)).

It is important to note that being dominant in a market does not, of itself, breach the competition rules. However, if a business is dominant in a market, it has a special responsibility not to abuse or take unfair advantage of that position (*Michelin* v. *Commission* (1985)). So, for instance, a dominant business may not charge excessive prices just because it can get away with it. Such a business may not price below cost to drive out competitors or apply different prices to different customers for the same transaction. S.18(2) of the 1998 Act provides further examples of abusive behaviour, namely:

1. directly or indirectly imposing unfair purchase/selling prices or other unfair trading conditions;

2. limiting production, markets or technical developments to the prejudice of consumers;

3. applying dissimilar conditions to equivalent transactions with other trading parties, thereby placing them at a competitive disadvantage;

4. making the conclusion of contracts subject to the acceptance by the other parties of supplementary obligations which, by their nature or according to commercial usage, have no connection with the subject of the contracts.

21.9 Summary

For the vast majority of contractors and suppliers operating in the UK construction industry, the Chapter I prohibition will represent the greatest competition risk. As illustrated above, the construction industry has already been the subject of CMA enforcement activity. The features of the industry (with bidding as part of a tender process being particularly prevalent) and the fact that construction remains on the CMA's radar mean that competition compliance is essential at all levels of the contracting and supply chains. Quite simply, businesses should be making commercial decisions unilaterally and on their own account, without reference to or involvement from competitors. If they do not do so, both the businesses and the individuals involved may find themselves subject to significant sanctions.

Chapter 22

Regulatory Matters

The Bribery Act 2010

22.1 Compliance: The Bribery Act 2010

22.1.1 Introduction

In July 2011, the Bribery Act 2010 (henceforth the 'Bribery Act') came into force, showing a new determination of UK authorities in tackling bribery and corruption, both domestically and internationally. While the Bribery Act has extensive implications for all UK businesses, and international companies with a UK connection, in general, anti-bribery legislation is increasing in scope and application worldwide. The construction industry is one of the industry sectors likely to receive the focus of investigating authorities as a result of its operations through routine use of agents, contractors, sub-contractors, subsidiary arrangements, joint venture arrangements and its common cross-border and international nature.

22.1.2 OECD guidance

The Organisation for Economic Co-operation and Development (OECD), of which the UK is a member, has been at the forefront of ensuring anti-bribery measures are dealt with at an international level. The OECD Anti-Bribery Convention 1996 established legally binding standards which criminalized bribery of foreign public officials in international business transactions and provides for a host of related measures that make this effective. The Bribery Act is intended to satisfy the UK's obligations under the Convention.

The OECD has long raised awareness of the effects of bribery and corruption being allowed to operate unchecked. In the construction sectors bribery and corruption discourage investment and distort competitive conditions.

There are currently 41 state party signatories to the OECD Anti-Bribery Convention consisting of 34 OECD members and Argentina, Brazil, Bulgaria, Columbia, Russia, South Africa and Latvia. The OCED has stated that 'propriety, integrity and

MacRoberts on Scottish Construction Contracts, Third Edition. MacRoberts.

transparency in both the public and private domains are key concepts in the fight against bribery, bribe solicitation and extortion'.

Key stakeholders in response to the OECD Convention have acknowledged that bribery and corruption are significant problems for the construction sector, but have also contended that 'the issues stem at least as much from the "demand side" of bribery and corruption as from the "supply side"' (comments from the International Federation of Consulting Engineers (FIDIC) per 'Response to the Consultation Paper on the Review of the OECD Anti-bribery Instruments'). The International Federation of Consulting Engineers has levelled criticism at the OECD approach, stating:

> Some cases of bribery arise from simple greed. Others occur because some public servants are placed in positions of authority but are not paid a living wage by their government in the expectation that they will make their personal income through bribes. This is deep systemic corruption and the current OECD Convention will not rectify this problem.

There are currently 40 state party signatories to the OECD Anti-Bribery Convention consisting of 34 OECD members and Argentina, Brazil, Bulgaria, Columbia, Russia and South Africa. The OECD reports that together, the states parties account for just under two-thirds of world exports in 2012 and the parties to the Convention also account for over 70% of global outward flows of foreign direct investment.

22.2 The Bribery Act 2010: The offences

22.2.1 Offence of bribing another person: The briber

The Bribery Act sets out two situations in which a person offering a bribe is guilty of an offence.

Case 1 is contained in s.1(2) of the Bribery Act: (a) a person offers, promises or gives a financial or other advantage to another person, and (b) a person intends the advantage to either induce the improper performance of a relevant function or activity, or is intended to reward a person for the improper performance of such a function or activity.

S.1(4) states that in this situation it does not matter whether the person to whom the advantage is offered, promised or given is the same person as the person who is to perform, or has performed, the function or activity concerned. It is also does not matter whether the advantage offered, promised or given by the bribing party is offered directly or through a third party (s.1(5)).

Case 2 in s.1(3) deals with the scenario where: (a) a person offers, promises or gives a financial or other advantage to another person; and (b) that person knows or believes that the acceptance of the advantage would itself constitute the improper performance of a relevant function or activity.

In this scenario it is also irrelevant whether the advantage offered, promised or given by the bribing party is offered directly or through a third party (s.1(5)).

22.2.2 Offences relating to being bribed: Recipient

A person as recipient is guilty of an offence if any of the following cases applies:

- *Case 3*: the recipient requests, agrees to receive or accepts a financial or other advantage intending that, in consequence, a relevant function or activity should be performed improperly by the recipient or another person (s.2(2)).
- *Case 4*: the recipient requests, agrees to receive or accepts a financial or other advantage, and the request, agreement or acceptance itself constitutes the improper performance by the recipient of a relevant function or activity (s.2(3)). The recipient does not have to know or believe the performance of the function/activity is improper (s.2(7)).
- *Case 5*: the recipient requests, agrees to receive or accepts a financial or other advantage as a reward for the improper performance (by the recipient or another person) of a relevant function or activity (s.2(4)). Again, for this case the recipient does not have to believe or indeed know that the performance as envisaged is improper.
- *Case 6*: in anticipation of or in consequence of the recipient requesting, agreeing to receive or accepting a financial or other advantage, a relevant function or activity is performed improperly by (a) the recipient or (b) by another person at the recipient's request or with the recipient's assent or acquiescence (s.2(5)). For this scenario, where a person other than the recipient is performing the function or activity, it also does not matter whether that person knows or believes that the performance of the function or activity is improper. As for cases 4 and 5, it does not matter whether the recipient knows or believes that the performance of the function or activity is improper.

In all of the 'recipient' offences it is irrelevant whether recipient requests, agrees to receive or accepts the advantage directly or through a third party or whether the advantage is for the benefit of recipient or another person (s.2(6)).

22.2.3 Definition of function or activity

For a function or activity to fall under the Bribery Act it must satisfy at least one of the following conditions: (1) be a function of a public nature; (2) be any activity connected with a business; (3) be any activity performed in the course of a person's employment; or (4) be any activity performed by or on behalf of a body of persons. This is a very wide definition and means that the scope of the Bribery Act extends to both the public and private sectors.

The function or activity must also be intended to be performed in at least one of three ways: (1) performed in good faith; (2) performed impartially; or (3) the person carrying out the function or activity is in a position of trust by virtue of performing it.

22.2.4 Geographical scope

The Bribery Act seeks to extend the reach of the anti-bribery legislation beyond the borders of the United Kingdom. For all of the offences under the Bribery Act a relevant function or activity need not have a connection with the United Kingdom, and it may also be performed in a country or territory outside the United Kingdom.

22.2.5 Improper performance

A function or activity is deemed under s.4(1) of the Bribery Act to be improperly performed where: (1) it is performed in breach of a relevant expectation; and (2) there is a failure to perform the function or activity and that failure is itself a breach of a relevant expectation .

For the purposes of s.4 (1), a relevant expectation is one where a function is meant to be carried out in good faith or impartially, it should have been carried out in that manner.

Where a function or activity is carried out by a person who is in a position of trust by virtue of performing the activity, a relevant expectation is any expectation as to the manner in which, or the reasons for which, the function or activity will be performed that arises from the position of trust mentioned in that condition.

22.2.6 Reasonable person test

The Bribery Act defines the words and phrases 'function', 'activity' and 'improper performance'. Whether the elements of the defined offences have occurred will be judged by the standard of what a 'reasonable person' in the UK would expect (s.5(1)). Therefore any behaviour occurring outside of the UK will be judged by the standards of a reasonable person from the UK, not a person of the place where the bribing offence occurred. Furthermore, the Bribery Act provides that any local custom or practice is to be disregarded unless it is permitted or required by the written law applicable to the country or territory concerned. This places a high standard on international business actings, as a person from the UK will be required to exercise a standard of behaviour that may preclude other behaviours that might otherwise be accepted in foreign countries.

22.2.7 Bribery of foreign public officials

S.6 of the Bribery Act addresses instances of bribery of foreign officials, finally bringing the UK into compliance with the OECD Combating Bribery of Foreign Officials Convention, that it had ratified in 1998.

A person who bribes a foreign official is guilty of an offence if it is the person's intention to influence the public official in their capacity as a foreign public official. The person must intend to obtain or retain business or a business advantage. The Bribery Act provides that the financial or other advantage can benefit the foreign official or someone else at the official's request or permission and can be made directly between the parties or through a third party.

As before, if the foreign official is permitted by the written law of the territory in which the offence occurs to accept a 'bribe', then no offence is committed for the purposes of the Bribery Act.

A foreign public official is defined in s.6(4), as an individual holding a legislative, administrative or judicial post outside of the UK, or anyone carrying out a public function for a foreign country or for a public agency or enterprise outside the UK, and also includes an official or agent of a public international organization. Therefore, offences under s.6 can be committed anywhere in the world outside of the UK.

Under s.6 there is no requirement for the foreign public official to act improperly in the exercise of their function; it only requires that the person bribing the foreign official intends that they receive a business advantage or retain business. Some commentators have suggested that s.6 'was enacted to overcome perceived difficulties over proof of the exact scope and nature of the functions of foreign public officials that might prove an obstacle to conviction if intent to induce improper performance of that function under s.1 was required to be proved' (*The Bribery Act 2010: The Next Chapter* (2011) 6 JIBFL 340).

In this section the Bribery Act goes further than provided for under the OECD Convention. The OECD Convention provides that legislation should prohibit bribery of foreign officials where the bribe is given 'in order that the official act or refrain from acting in relation to the performance of official duties'. The UK legislation, in only requiring that the person offering the bribe intends to receive a business advantage, provides for a lower evidential burden, while clearly also setting the Bribery Act within the bounds of the OECD Convention's aims.

22.2.8 Failure of commercial organizations to prevent bribery: The corporate offence

Under s.7 of the Bribery Act a 'relevant commercial organisation' is guilty of an offence under that section if a person associated with the organization bribes another person intending to obtain or retain business and/or an advantage in the conduct of business for the organization. This is a strict liability offence.

A 'relevant commercial organisation' is defined in s.7(5) as a body or partnership incorporated or formed in the UK irrespective of where it carries on a business, or an incorporated body or partnership which carries on a business or part of a business in the UK irrespective of the place of incorporation or formation.

22.2.9 Associated person

Under s.8, a person associated with a commercial organization is a person who performs services for or on behalf of the organization. The capacity in which a person performs the services does not matter and can include employees, agents, contractors, subsidiaries and other persons providing services. The list is not exhaustive and the relationship will be considered on the facts, not merely by the label given to the relationship. This definition was 'intended to give section 7 broad scope so as to embrace the whole range of persons connected to an organisation who might be capable of committing bribery on the organisation's behalf' (Bribery Act 2010 Guidance, p. 16 at para 37). Therefore, where any employee or third party providing services bribes for the benefit of a commercial organization, that organization is liable under s.7 unless it can satisfy the conditions for a defence in s.7(2).

22.2.10 Specific Guidance: Associated persons in a construction setting

The Bribery Act 2010 Guidance published by the Ministry of Justice in February 2012 (henceforth 'the Guidance') – see Section 22.3.2 – provides specific advice that is very relevant to contractors within the building industry and their interactions with sub-contractors, joint ventures and other relationshsips.

Contractors

In a contractual situation where a contractor engages sub-contractors or there are a series of contractual relationships, an organization is only likely to exercise real control over its direct contractual counterparty. In most cases, persons who contract with that counterparty will be performing services for the counterparty and not for other persons in the contractual chain.

The Guidance therefore suggests:

> The principal way in which commercial organisations may decide to approach bribery risks which arise as a result of a supply chain is by employing the types of anti-bribery procedures ... in the relationship with their contractual counterparty, and by requesting that counterparty to adopt a similar approach with the next party in the chain.

This is a waterfall approach, with the responsibility for compliance cascading down the contractual chain.

Joint ventures

Joint ventures are now a very common feature in the construction sector, appearing in many guises but generally formed for a special purpose. The Guidance highlights that where the joint venture is a separate legal entity, a bribe paid by or for the joint venture could lead to liabilities for the parties of the joint venture, where the parties receive services from the joint venture and the bribe is intended to benefit that party. The Guidance does state that 'the existence of a joint venture entity will not of itself mean that it is "associated" with any of its members'. Furthermore, where there is a bribe on behalf of the joint venture, there is not automatic liability for the members of the joint venture by the mere fact of their involvement in the joint venture.

Each case will be decided on its merits; however, given the risk involved, a zero-tolerance policy agreed upon by all parties is the safest way to proceed.

In general, the mere fact that an organization benefits from a bribe paid does not amount to proof of intention under s.1 and s.2 of the Bribery Act. The Guidance states that 'without proof of the required intention, liability will not accrue through simple corporate ownership or investment or through the payment of dividends or provision of loans by a subsidiary to its parent'. Therefore, where an employee of a subsidiary is involved in bribing for the intended benefit of the subsidiary, this will not automatically involve the bribery of the parent or sister companies of the subsidiary. This is so even though the parent company or sister companies may benefit indirectly from the bribe.

22.3 The Bribery Act 2010: The defence

22.3.1 The defence

It is a defence in respect of conduct which would otherwise be an offence under s.7(1) for a commercial organization to prove that it had in place adequate procedures designed to prevent persons associated with the organization from undertaking such conduct.

22.3.2 Corporate defence: Adequate procedures

S.9 of the Bribery Act places on the Secretary of State the burden of publishing guidance on procedures that relevant commercial organizations can put in place to prevent persons associated with them from bribing as mentioned in section 7(1).

Publication of the Guidance was delayed by the UK Government in 2011, amid concerns raised by businesses in relation to the s.7 corporate offence which penalizes the failure of commercial organizations to prevent bribery by persons associated with them. In March, 2011, Kenneth Clarke, the Secretary of State for Justice, said:

> The guidance ... by improving clarity about its intentions ... should arm organisations of all sizes against the fears that millions of pounds must be spent on procedures, that in my opinion, no honest business will require ... [T]he ultimate aim of this legislation is to make life difficult for the minority of organisations responsible for corruption, not to burden the vast majority of decent and law-abiding businesses.

More recently, amid rumours of possible review of the Bribery Act or the Guidance, Damian Green, Minister of State for Policing and Criminal Justice, confirmed at a Commons debate on 2 September 2013 that his department has no current plans to review the Bribery Act. He further stated that 'the Ministry of Justice in association with the Department of Business Innovation and Skills is working to ensure small and medium-sized enterprises fully understand how the Act and the guidance relate to their business'. Therefore, it is possible that new guidance aimed specifically at SMEs may be published in the near future.

The Guidance, which remains unchanged since its publication in February 2012, is of general application and described as not being 'a one-size-fits all document'; that said, it is clear that the UK Government, by relying heavily on common themes running through the Guidance of 'proportionality' and 'a risk-based approach', has attempted, as they said they would, to reduce the regulatory burdens of commercial organizations. Specifically, the Guidance has sought to clarify that smaller commercial organizations are not likely to need the same robust and extensive policies and procedures as much larger organizations. The Guidance sets out six principles which a commercial organization should put in place to prevent bribery being committed on its behalf by persons associated with it.

Principle 1: Proportionate procedures
Principle 2: Top-level commitment
Principle 3: Risk assessment
Principle 4: Due diligence
Principle 5: Communication
Principle 6: Monitoring and review.

22.3.3 Proportionate procedures

All steps taken within an organization need to be proportionate to the risks that it faces, the geographical location, business sector, complexity of activities, parties involved, and so on. The procedures should be 'clear, practical, accessible, effectively implemented and enforced'.

If a commercial organization which follows such procedures is then charged with the offence of failing to prevent bribery, it would be able to show evidence of the 'adequate procedures' required as a defence.

22.3.4 Top-level commitment

The Guidance highlights that effective top-level management of bribery and corruption prevention will be a unique response to the needs and requirements of the specific organisation, taking into account the businesses' size, management structure and circumstances. The Guidance speaks of 'leadership on key measures', 'endorsement of all bribery prevention' documentation and other appropriate steps such as 'engagement with relevant sectoral organisations'.

22.3.5 Risk assessment

The risks that each organization faces depends on various factors, including the location of the carrying on of the business, the nature of the business operations, and the business model, including interactions with associated persons. Some commercial organizations may operate in a business sector that routinely offers hospitality as a marketing tool and so are at risk of greater scrutiny. Businesses involved in export and import will necessarily have greater exposure to and interaction with foreign officials, leading to increased opportunities for breaches to occur. Each business model needs to be assessed on its own merits, but each risk must be addressed to ensure compliance with the legislation. It will be important for the organization to determine what checks need to be made to assess the integrity of its business partners. The assessment requires to be thorough and reviewed on an ongoing basis as the business develops and new challenges are faced.

22.3.6 Due diligence

Due diligence will already be employed in organizations committed to good corporate governance, and the anti-bribery and corruption due diligence should fit firmly into that framework. In line with the other principles, the response has to be proportionate and risk-based. Due diligence is key where the organization uses associated persons in its business model, and due diligence should be carried out on third parties.

22.3.7 Communication (including training)

This principle is focused on ensuring that policies and procedures are embedded in an organization, and are widely understood. The intended benefits are enhanced awareness of the risks of bribery as well as consolidation of the commitment to ensure compliance. The Guidance points out that making 'information available assists in more effective monitoring, evaluation and review' of bribery processes and procedures. Training goes hand in hand with communication and serves as a means of establishing an anti-bribery culture. Again the approach is focused on actual risks and proportionality, advising that training should outline specific risks to specific posts, sectors, functions, departments and locations. Effective training should be continuous, monitored and evaluated to ensure compliance with the Bribery Act.

22.3.8 Monitoring and review

The monitoring and review of compliance procedures is necessary to ensure that they are updated and reflect the risks and challenges the business faces on an ongoing basis. Good monitoring should ensure internal control of the anti-bribery processes and it is only where an effective monitoring function keeps track of bribery risk indicators, sector-specific challenges, and other information arising from the processes or training and performance of procedures and controls, that an organization will be able to review processes to ensure that they are being regularly updated and reviewed.

The Guidance suggests that it might be appropriate to consider external verification, including independently verified anti-bribery standards maintained by industrial sector associations or other organizations. This is not, however, a blanket certification that the procedures are 'adequate' as a defence to any s.7 offence.

22.3.9 'Adequate procedures' – adequate guidance?

Applying the six principles will be crucial in underpinning any anti-corruption stance which an organization takes, and carrying out a risk assessment will be essential in determining whether a business has vulnerabilities. The Guidance together with the Bribery Act 2010: Joint Prosecution Guidance of the Director of the Serious Fraud Office and the Director of Public Prosecutions (henceforth 'the Joint Prosecution Guidance') go some way to assist. The construction industry is particularly vulnerable, and only by identifying where those risk points are will it be able to put in place appropriate policies and procedures and follow up with the necessary training and ongoing monitoring to address those vulnerabilities and mitigate the risks identified.

The question of adequacy of bribery prevention procedures will depend in the final analysis on the facts of each case, including matters such as the level of control over the activities of the associated person and the degree of risk that requires mitigation. The scope of the definition of associated person in s.8 of the Bribery Act needs to be appreciated within this context.

22.3.10 Bribery Act 2010 case studies

The Guidance sets out case studies, but explicitly states they are not part of the Guidance for the purposes of s.9 of the Act (see Section 22.3.2). The case studies address facilitation payments; joint ventures; hospitality and promotional expenditure; assessing risks; due diligence of agents; and community benefits and charitable donations. These illustrative case studies provide some assistance in understanding the scope and limits of the Guidance.

22.4 *Facilitation payments*

It is made clear in the Guidance that small facilitation payments 'could trigger either the s6 offence or, where there is an intention to induce improper conduct, including where the acceptance of such payments is itself improper, the s1 offence and therefore potential liability under s7'.

The Guidance does no more than recognize the difficulties in adhering to the law and refers specifically to the Joint Prosecution Guidance (an updated version of which was published in October 2012) which sets out the Serious Fraud Office's (but not necessarily the Scottish Crown Office's) approach to prosecutorial decision-making under the Bribery Act and examines the circumstances where a prosecution is required in the public interest and which factors would be taken into account (what the Guidance calls 'prosecutorial discretion'). This is helpful to the construction industry insofar as it is an aid to carrying out a risk assessment to determine the level of vulnerability to prosecution, and insofar as it highlights key risk areas, providing businesses with an opportunity to minimize identified risks.

The Serious Fraud Office (SFO) have highlighted that certain factors including large or repeated payments, planned-for payments which indicate the offence was premeditated, and payments made in breach of a clear company policy are more likely to result in prosecution. Where the offence consists of a single small payment, or comes to light by way of an organization's own diligence and remedial action is taken, then prosecution is unlikely.

In an open letter from the director of the SFO, dated 6 December 2012, regarding facilitation payments, the SFO stated that facilitation payments are illegal under the Bribery Act 'regardless of their size or frequency' and that 'the Serious Fraud Office stands ready to take effective action against the use of facilitation payments, regardless of where they are requested'. Moreover, the SFO published its revised policies on facilitation payments, business expenditure and corporate self-reporting in October 2012. With regards to facilitation payments, it is stated that the SFO is governed by the Code for Crown Prosecutors' Full Code Test and the Joint Prosecution Guidance, with the Joint Guidance on Corporate Prosecutions being applicable, where relevant, in deciding whether to prosecute.

The OECD is supportive of the approach taken by the UK Government in relation to the Bribery Act (see 2009 Recommendation of the Organisation for Economic Co-operation and Development) and in reference to facilitation payments have said that

> Exemptions in this context create artificial distinctions that are difficult to enforce, undermine corporate anti-bribery procedures, confuse anti-bribery communication with employees and other associated persons, perpetuate an existing 'culture' of bribery and have the potential to be abused.

22.5 *Consequences of offences under the Bribery Act*

22.5.1 Penalties

An individual guilty of the offences of giving or receiving a bribe or bribing a foreign public official is liable for imprisonment for a term not exceeding 10 years, or to a fine, or to both.

Companies or other entities guilty of an offence under sections 1, 2 or 6 are liable on conviction on indictment to a fine. Companies guilty of the crime of a commercial organization failing to prevent bribery are punishable by an unlimited fine.

In addition, a convicted individual or organization may be subject to a confiscation order under the Proceeds of Crime Act 2002, while a company director who is convicted may be disqualified under the Company Directors Disqualification Act 1986. Those commercial organizations operating in public sector procurement may also find their organization debarred from tendering where they have failed to prevent bribery, though the debarment is discretionary rather than, as first thought, mandatory under the EU Procurement Directive (Directive 2004/18). A new EU

Procurement Directive was enacted on 17 April 2014, with Member States allowed up to two years to implement the directive into national law. The new Directive will change the debarment risks for corporate entities; firstly the exclusion period will be reduced to 5 years for an entity facing mandatory debarment (unless it is otherwise stipulated by a convicting court) and secondly, where an entity provides evidence of satisfactory "self-cleaning", a public authority will have to end the exclusion period. However, until the Directive is implemented the Public Contracts Regulations 2006 remain in force.

The World Bank has increased its regulatory role in recent years and the potential for debarment and sanctions by the World Bank have become more prevalent. This is a response to wide-ranging corruption and fraud arising in projects funded by the World Bank. The World Bank estimates that as much as $40 billion of aid has been stolen from the world's most deprived countries since 2008. In 2013 the World Bank blacklisted 250 entities, which included major international firms, smaller firms and individual consultants. In particular, the World Bank has investigated over 600 cases of fraud and corruption in relation to development projects. In April 2013, the large Canadian construction firm, SNC-Lavalin was debarred for a 10 year period after corruption allegations in relation to a $3 billion bridge building project it carried out in Bangladesh. It is evident that the World Bank is committed to stepping up its no-tolerance approach to corruption and misconduct by firms of any size.

In April 2010 a Cross Debarment Agreement was signed by 5 of the world's Multilateral Development Banks (MDBs): African Development Bank Group, the Asian Development Bank, the European Bank for Reconstruction and Development, the Inter-American Development Bank Group and the World Bank Group. This agreement states that if one MDB declares a firm or individual ineligible for procurement contracts due to allegations of fraud and corruption, then all the other MDBs must do the same. Thus, cross debarment has now made it difficult for corrupt entities to do business with any MDB and the impact of this will pose significant financial and reputational risks for any firm or individual facing a single sanction from a MDB.

On 31 January 2014, the Sentencing Council for England and Wales published its definitive sentencing guidelines for financial crimes which will have effect from 1 October 2014. The guidelines provide clearer, more uniform sentencing. The guidelines provide for higher fines for larger corporate entities that will have a real financial impact on the entity to encourage future compliance with the Bribery Act. It remains to be seen if equivalent guidelines will be published in Scotland. The Criminal Justice and Licensing (Scotland) Act 2010 creates a Scottish Sentencing Council which produces sentencing guidelines for Scotland, but this part of the Act is yet to come into effect.

22.5.2 Deferred Prosecution Agreements (DPAs)

In England, Wales and Northern Ireland, Deferred Prosecution Agreements (DPAs) have been introduced through the Crime and Courts Act 2013 and have been offered by the SFO and Crown Prosecution Service since 24 February 2014. The revised Code of Practice for Prosecutors was also published in February 2014. DPAs are voluntary agreements used in cases concerning fraud, bribery and other economic crime, made between prosecutors and corporate organizations. Essentially, where a corporate body is suspected of committing a criminal offence under the Bribery Act, the prosecutor may choose to offer a DPA instead of prosecution. If the DPA is fulfilled according to its terms, then the charges will be dropped at the end of the specified period of the DPA, but should compliance with the DPA not be achieved, then this may result in prosecution. DPAs are far from an easy option, however, with the agreement itself lasting for a set period which could potentially be a number of years. Furthermore, in addition to a financial penalty, required to be 'broadly comparable to the fine that a court would have imposed ... on conviction for the alleged offence following a guilty plea' (Crime and Courts Act 2013, Schedule 17, paragraph 5(4)), conditions within a DPA may include disgorgement of profits or benefits within a specified time, reparation to victims, the creation of a thorough compliance programme and/or continued ongoing monitoring.

While the offering of DPAs is to be aligned with the Bribery Act, which is of course a UK Act, in Scotland there is no intention to make provision for DPAs. Shortly before the Crime and Courts Act 2013 received Royal Assent, a Scottish Government spokesperson said: 'While we are aware of the UK Government's plans, we do not have any immediate plans to legislate for deferred prosecution agreements for Scotland. We will continue to monitor developments in England and Wales.'

22.5.3 Self-reporting initiative

Given the wide scope of the Bribery Act, an initiative was announced in 2011 by the Scottish Crown Office, through the Serious and Organised Crime Division (SOCD), whereby businesses would self-report to the relevant authority acts of corruption or bribery in return for more lenient treatment. This initiative was made in conjunction with the Serious Fraud Office, the enforcement body for England, Wales and Northern Ireland. From July 2011 to (the extended deadline of) 30 June 2013 the Crown were willing to accept self-reports on behalf of businesses who disclosed conduct that would amount to a breach of anti-bribery legislation. The first instance of a company making use of the self-reporting regime in Scotland saw an Aberdeen-based oil and gas firm, Abbot Group, broker a settlement. The firm approached Scottish authorities in November 2012 after discovering that an overseas subsidiary had made bribes to win a contract, resulting in a benefit to the firm of the sum of £5.6 million. Accordingly a civil penalty of £5.6 million was agreed, with the firm being able to use the self-reporting initiative because they could satisfy the prosecutors that a thorough investigation had been undertaken and that sufficient processes had been put in place to ensure there would not be a re-occurrence of such an event.

On the 1 July 2013 the Crown Office reissued its guidance on the self-reporting programme which has recently been further extended to 30 June 2015.

Any decision to self-report should be taken with the approval of the board of the relevant organization, in conjunction with legal advisers, and after a thorough investigation has been carried out. Additionally, the business which is reporting must agree to disclose all details of the relevant conduct. It is a serious matter for any business to self-report and the Head of SOCD has warned '[t]his initiative is not a soft option' (*L. Miller; J.L.S.S. 2011, 56(7), 52*). Submitting a report to the SOCD is putting evidence of bribery and corruption within an organization into the hands of the Crown Prosecution. While the guidance states that the initial report and information given to the SOCD will be confidential, this may be used by the Crown in any subsequent criminal investigation and prosecution, or in any civil recovery investigation. Furthermore, if the SOCD decide that a case is to be referred to the Civil Recovery Unit (CRU), the solicitor for the business will be notified of this and it will be publicly acknowledged that the case is under consideration in accordance with the self-reporting initiative; though no further public comment will be made. Once a settlement has been reached, this will be made public, as was the case with Abbot. Therefore self-reporting could be potentially damaging to an organization's reputation. Additionally, since any recommendation by SOCD as to whether the case merits criminal prosecution or referral to the civil courts must be approved by Crown Counsel, the only benefit to the organization of a self-report may be that, according to SOCD guidance, the self-reporting organization 'will be able to rely on their self-reporting and co-operation with the Crown and law enforcement as significant mitigating factors to be taken into account by the Court'. But it still remains unclear how far this guidance can be relied on.

In England and Wales, corporations remain able to self-report to the SFO. The SFO published its revised policies on facilitation payments, business expenditure and corporate self-reporting in October 2012, which stated that in relation to self-reporting the SFO will prosecute if conviction is in the public interest and realistically probable. Although the SFO used to be known to prefer civil remedies, the new policies present a shift away from this position. While self-reporting will be considered when the decision as to whether to prosecute is made, each case will be decided individually and self-reporting does not guarantee that no prosecution will follow. However, David Green, QC, the Director of the SFO, stated in October 2013:

> If a company made a genuine self-report to us (that is, told us something we did not already know and did so in an open-handed unspun way), in circumstances where they were willing to cooperate in a full investigation and to take steps to prevent recurrence, then in those circumstances it is difficult to see that the public interest would require a prosecution of the corporate.

22.5.4 The court's approach to self-reporting settlements

In *R* v. *Innospec Ltd* (2011), where a plea bargain arrangement had been made with the SFO in relation to Innospec's corrupt activities, Lord Justice Thomas stated that: 'it would be inconsistent with basic principles of justice for the criminality of corporations to be glossed over by a civil as opposed to a criminal sanction". He said that those who commit such serious crimes as corruption must not be treated in any different way from other criminals.

What is significant in the light of the self-reporting initiative, is that the court forcefully stated it had concluded 'the Director of the SFO has no power to enter into the [plea] arrangements made and no such arrangements should be made again'.

Although this judgment was based on the law of England and Wales, it would be foolish to disregard it and the weight it will carry in the Scottish courts. Businesses should be fully aware that undertaking a self-report in relation to corruption or bribery does not immediately preclude any criminal prosecution, but where such proceedings are undertaken, any self-report and co-operation will be considered as significant mitigating factors to be taken into account. If a business has been involved in bribery, the self-reporting of such activities can create goodwill and will allow the business to manage the time frame of the self-report as well as the mechanics of the internal investigation.

22.6 *Prosecution under the Bribery Act*

The SFO's first prosecution under the Bribery Act began in September 2013, with the trial scheduled to commence on 22 September 2014. Charges were brought against Sustainable ArgoEnergy Plc, a bio-fuel investment company which entered administration in March 2012, and four individuals connected to the company (three former employees and an independent financial adviser). The individuals are alleged to have committed fraud in relation to the sale and marketing of bio-fuel investment products and three of the individuals were charged with the offences of giving and receiving a financial advantage under the Bribery Act. The company itself was not charged with the strict liability corporate offence under the Bribery Act, however, to which there is only the adequate procedures defence available. When deciding whether to prosecute a company, the SFO considers the Code for Crown Prosecutors and accordingly considered in this case if there was sufficient evidence to provide a realistic prospect of conviction against the company under this charge, and whether a prosecution was necessary in the public interest. (In Scotland, the prosecutor is the Crown Office and Procurator Fiscal Service (COPFS) and they consider the COPFS Prosecution Code.) It is possible that since Sustainable ArgoEnergy Plc had entered administration, the SFO considered prosecution not to be in the public interest. We are therefore still awaiting much needed clarity by way of guidance from court decisions under the Bribery Act. However, with the SFO stating in September 2013 that they have eight cases currently under investigation and David Green, head of the SFO, further stating (in the keynote speech at the 31st Cambridge International Symposium on Economic Crime, on 2 September 2013) that 'If the public interest requires more corporate prosecutions, then such a change is high on my wish list', it seems probable we will receive this clarity soon.

22.7 *Construction industry risk profile*

22.7.1 Introduction

The unique nature and operation of the construction sector mean that it is more exposed to corruption risks than other sectors. Indeed, 49% of respondents to the second survey into corruption in the industry from the Chartered Institute of Building (CIOB) stated that they believe corruption is either fairly or extremely common within the UK construction industry (The Chartered Institute of Building, *A Report Exploring Corruption in the UK Construction Industry*, September 2013), while Transparency International's Bribe Payers Index report from 2011 found construction to be the most corrupt sector worldwide.

22.7.2 Country risk

More than two thirds of the world's nations score below 50% on Transparency International's Corruption Perception 2013 Index within the public sector, meaning those countries have very high levels of perceived corruption. Often foreign governments do not have or promote anti-bribery aims. The construction sector operates internationally (frequently engaging with the public sector), often in cross-border transactions increasing the sector's risk profile. Where there are cross-border elements, there are often more complex regulatory, tax and legal implications which mean there is a greater risk of corrupt practice. Moreover, within the UK itself, a recent survey found that 55% feel the UK Government is not doing enough to prevent corruption, while 50% feel the UK construction industry's efforts are insufficient (see the CIOB Report referred to in Section 22.7.1).

22.7.3 Transaction risk

Construction projects are often on a joint venture basis (see Section 22.2.10.2), and/or involve a long supply chain of subcontractors. The presence of these associated parties increases the risk for companies operating in the sector by imposing challenges to ensuring consistency of standards. Many construction projects require interactions with government officials through planning permissions, building consents, regulation and public procurement, and so increasing the potential opportunities to fall foul of the Bribery Act. The pre-qualification and tendering stages are perhaps most at risk of corruption, with price fixing, the leaking of tender assessment procedures to preferred parties and of course the payment of bribes to win contracts being notable examples of corrupt practices which may occur.

22.7.4 Business opportunity risk

The Guidance highlights that business opportunity risks may arise in high value projects, with projects involving many contractors or intermediaries, or with projects which are not apparently undertaken at market prices. The construction sector is extremely competitive and many projects are high value, thus increasing the possibility of increased corrupt behaviour in return for high rewards. Furthermore, this competition has increased in recent years, with the construction industry being a major victim of the economic climate, meaning that there could be a temptation within the industry to adopt corrupt practices in order to obtain work or as a cost-saving measure.

22.7.5 Business partnership risk

Due to the sector's international and fragmented nature, certain relationships may involve higher risk, including the use of agents or other consultants who deal with foreign public officials; consortia or joint venture partners; and relationships with politically exposed persons where the proposed business relationship involves, or is linked to, a prominent public official. The recent findings of the Mahon Tribunal in Ireland in relation to corrupt practices concerning the acquisition of planning permission show that such practices continue to occur in countries that are otherwise compliant across different sectors.

22.7.6 The sector response to the Bribery Act

An overwhelming majority of those surveyed for the recent CIOB report displayed commitment towards tackling corruption; 77% of respondents for the CIOB 2013 Report stated that they believe it is very important to tackle the issue of corruption, while 18.5% believe it is fairly important.

22.7.7 UK Contractors Group

The UK Contractors Group currently represents over 30 leading contractors operating in the UK on construction specific issues. Its members are estimated to account for £33 billion of construction turnover, around a third of the UK's construction total output. They have published an anti-bribery code of conduct with which all its members are required to comply, specifically setting out a zero tolerance approach to bribery and corruption.

22.8 The SBC, the SBC/DB, the NEC3 and the Bribery Act

Clause 8.6 of both the SBC and the SBC/BD provide that the Employer shall be entitled by notice to the Contractor to terminate the Contractor's employment under the contract in question or any other contract with the Employer if, in relation to the contract in question or any other such contract, the Contractor or any person employed by him or acting on his behalf shall have committed an offence under the Bribery Act 2010, or, where the Employer is a Local Authority, shall have given any fee or reward the receipt of which is an offence under Section 66 of the Local Government (Scotland) Act 1973 or any re-enactment thereof.

Termination on this ground is equivalent to termination for default on the part of the Contractor, see Section 9.4.4. It should be noted that for an offence under the Bribery Act to constitute a ground for termination under clause 8.6:

- the offence may relate to a different contract with the same Employer;
- the offence may have been committed by an employee or agent, or even sub-contractor, of the Contractor. This does not appear to require knowledge by the Contractor (which is not required for an offence under s.7 in any event). Read literally, clause 8.6 might also suggest that the ground for termination can be triggered even where the Contractor itself has not been guilty of any offence, but its employee, agent or sub-contractor has been;
- The offence may relate to *any* breach of the Bribery Act.

Perhaps somewhat surprisingly, there is no equivalent termination ground in the NEC3. As a consequence, it is very common for Employers, particularly those in the public sector, to insert in the relevant NEC3 contract a Z clause containing similar wording to that of clause 8.6 of the SBC.

Chapter 23
Guarantees and Bonds

23.1 Guarantees

23.1.1 Introduction

It is common in construction contracts for the employer to require a guarantee of the contractor's obligations to be given by a third party. The third party may guarantee to carry out and complete the construction works and/or pay to the employer the damages they incur due to the contractor's breach of contract. Guarantees of the contractor's obligations will usually take the form of a parent company guarantee or a bond granted by a surety or bank. In certain cases the contractor may insist on a guarantee of the employer's payment obligations.

In the case of a bond, the third party providing the bond will usually levy a charge against the party whose performance it is guaranteeing as well as requiring a counter-indemnity for any payment it makes under the guarantee. The costs of the bond will normally be included in the contractor's tender price, so a parent company guarantee (which will not usually have any cost implications) may be more attractive to an employer, if only from a cost perspective.

The requirement for any type of guarantee depends on the circumstances of each transaction, the creditworthiness of the contracting parties, and balancing the additional cost (if any) of the guarantee against the risk of not having third party backing, while taking into account the particular risks against which the employer is seeking additional security. As mentioned, a bond comes at a price, will usually be for a fixed amount and will subsist for a limited period, typically 10% of the contract price and until either practical completion or making good of defects. On the other hand, a parent company guarantee normally has no cost and is potentially unlimited in value and duration, subject to any express contractual limitations and the statutory prescriptive period. This means that a typical parent company guarantee will cover latent defects, while a bond will not. Against that, insolvency of the contractor will in the majority of cases also mean insolvency of the parent company, resulting in a worthless guarantee. Indeed, a bond and a parent company guarantee are not mutually exclusive and often an employer will require both. Each should be considered by the employer as separate elements of the contractor's overall performance security package. In general terms, a bond will mitigate the financial consequences of contractor insolvency during the construction period, while a parent company guarantee will, assuming the

MacRoberts on Scottish Construction Contracts, Third Edition. MacRoberts.

continued solvency of the parent company, offer additional security in the event of latent defects. A practical advantage of a parent company guarantee is that the guarantor will remain liable notwithstanding any changes in ownership or structure of the contractor (i.e. regardless of whether the guarantor remains the 'parent'), so that the guarantee will provide the employer with some degree of protection should at some time in the future the parent dispose of the contractor entity to a less financially stable group, or should the contractor cease trading or its assets be dissipated, whether under a corporate reorganization or otherwise.

The obligation to provide a guarantee and/or bond will usually be specified in tender documents and in turn made a condition of the building contract (see Sections 23.1.5 and 23.2.2). Delivery of the guarantee or bond in a pre-agreed form may be a suspensive condition to the building contract coming into effect or to the first payment being made. Alternatively, the contract may specify that the employer is entitled to make a specified retention from sums otherwise due until delivery, or that non-delivery or a substantial change in the financial value of the guarantor will be deemed a material breach entitling the employer to terminate the contract.

23.1.2 Nature of guarantees

In broad terms, a contract of guarantee is an undertaking by a person to secure the performance of the obligations of a party under a contract. A guarantee may (subject to the law of prescription and to any express limitations in the guarantee itself or in the underlying contract) be unlimited as to time and amount.

A bond will, except in certain circumstances where it is construed as an on demand bond (see Section 23.1.4), be regarded as a form of guarantee comprising cautionary obligations, see *City of Glasgow District Council* v. *Excess Insurance Company Ltd* (1986). In that case, the bond was held to be a guarantee, as opposed to an indemnity, and so was subject to the five-year prescriptive period under section 6 of the Prescription and Limitation (Scotland) Act 1973.

Despite the frequent use of guarantees, few of the institutions responsible for promoting standard forms of construction contract have published standard forms of guarantee (other than for performance bonds), so it is left to the parties to devise their own wording. In such an event it is important to ensure that the document in question clearly states the intention of the parties and creates enforceable obligations. For example, a 'letter of support' may not be sufficient to create a guarantee, see *Carillion Construction* v. *Zelf Hussain and another (the Joint Liquidators of Simon Carves Limited (in Liquidation))* (2013).

23.1.3 Distinction between cautionary and principal obligations

A distinction may need to be drawn between a guarantee which is an independent obligation and one which is truly a 'cautionary obligation' and thus accessory to the principal obligation. The distinction can be important as, in the absence of express wording, variation of the principal contract may discharge a cautionary obligation.

It is a question of fact whether the obligation is one of caution or a principal obligation, though clear wording should remove any doubt. The term 'principal obligor' is often applied to a guarantor to demonstrate that its obligations are principal and not cautionary (although see *WS Tankship II B.V* referred to in section 23.2.3).

23.1.4 Cautionary obligations

If the obligation is truly one of caution, it must be given the narrowest construction which the words will reasonably bear, see *Harmer* v. *Gibb* (1911). The cautioner's liability can never exceed that of the principal debtor and on payment of the debt a cautioner is entitled to recover from the principal debtor all sums which they have paid to the creditor. They are also entitled to demand from the creditor an assignation of the debt, any security held for it and any diligence done upon it, so as to enable them to enforce their right of relief against the principal debtor.

Under section 6 of the Prescription and Limitation (Scotland) Act 1973 the prescriptive period applicable to cautionary obligations is five years from the date the obligation became enforceable. See also *City of Glasgow District Council* v. *Excess Insurance Company Ltd* (1986). In the case of latent defects under a building contract, for example, the obligation on the guarantor may not arise until many years after the guarantee was entered into.

The creditor should also have regard to any provision in the guarantee which provides for service of a demand on the guarantor.

Under section 8 of the Mercantile Law Amendment (Scotland) Act 1856, unless stated expressly to the contrary, there is no need for the creditor to pursue a remedy against the principal debtor before suing the cautioner (as was previously the position under common law).

Changes to the underlying contract without the guarantor's consent or actings by the creditor which prejudice the guarantor may in certain circumstances discharge the guarantor from liability under the cautionary obligation. In *Holme* v. *Brunskill* (1878) it was held that in the event of a variation to the underlying contract, the guarantor will be discharged unless it has agreed to the variation or the variation is self-evidently insubstantial and cannot prejudice the guarantor. In *General Steam Navigation Company* v. *Rolt* (1858) it was held that actings by the creditor which are *prima facie* prejudicial to the guarantor, even in the absence of variation to the underlying contract, will discharge the guarantor. For those reasons it is common practice for bonds and guarantees to contain an 'indulgence clause' (also known as an 'anti-discharge' or 'anti-avoidance' clause) expressly stating that, *inter alia*, no alteration in the terms of the principal contract, or in the scope or nature of the work under the principal contract, or allowance of time or indulgence granted to the contractor will release the guarantor from liability.

Both the above cases were relied upon by the guarantor in seeking to resist liability under a bond in *Aviva Insurance Ltd* v. *Hackney Empire Ltd* (2012). The guarantor argued that the contract had been varied without its consent by a side agreement under which the employer advanced payments to the contractor and that the guarantor had also been prejudiced by the fact of the advance payments as these increased the risk of default under the bond. The Court of Appeal upheld the decision of the court of first instance and rejected both arguments on the grounds that the variations to the contract were self-evidently insubstantial; that, in any event, the bond contained an 'indulgence clause' of the type mentioned above; and that the payments in issue were, in fact, additional payments not advance payments of the price and so did not prejudice the guarantor. That said, the case does illustrate the potential pitfalls of making variations to the underlying contract without the guarantor's consent, as the Court of Appeal stated that an advance payment could prejudice, and as a result discharge, a bondsman since it reduces the contractor's incentive to complete and also potentially increases the employer's loss and the bondsman's liability by reducing the amount of retention held.

In *De Montfort Insurance Co. plc* v. *Lafferty* (1997), which related to a performance bond, it was held that a guarantor was not released from its obligations as a result of novation of the employer's payment obligations under the building contract, since the novation did not release the employer from its obligations but added an additional obligant and so did not prejudice or increase the risks to the contractor or the surety. This case turned very much on the restricted nature of the novation in question and in most cases a novation replacing the original employer, without the consent of the surety, could result in the discharge of the surety (provided, of course, that the instrument is not on demand, see *Meritz Fire and Marine Insurance Co Ltd* referred to in Section 23.2.3).

The importance of including an indulgence clause and other express safeguards for the beneficiary is illustrated by *Beck Interiors* v. *Russo* (2010). In that case, Dr Russo, the principal shareholder in the employer, granted the contractor a short form of guarantee, which did not contain an indulgence clause, in respect of the payment obligations of the employer. The contractor agreed with the employer to carry out additional work and subsequently obtained an adjudicator's award against the employer who shortly thereafter became insolvent without making payment. The court held that the agreement to carry out additional work was a variation which was not insubstantial but on the facts the guarantor was considered to have consented to the variation. It was also held that without express words in the guarantee the guarantor was not bound by a determination under an adjudication between the contractor and the employer, unless by his conduct he was found to have participated in the adjudication in a personal capacity.

The test as to whether a variation in respect of the underlying obligation requires to be disclosed to the guarantor is an objective one and the beneficiary must disclose known facts to the guarantor if they reveal matters which might not naturally be expected to take place between the parties to the transaction, see *North Shore Ventures Ltd* v. *Anstead Holdings Inc and others* (2011).

However, variations to underlying obligations, even if substantial, do not in all cases require the guarantor's consent. In *National Merchant Buying Society* v. *Andrew Bellamy and another* (2013), the guarantee was not related to obligations under a specific contract but was instead a free-standing 'all monies' guarantee covering the indebtedness of the debtor company to the creditor arising out of a contemplated course of dealing. Provided that the course of dealing remained within the scope contemplated, it did not matter that the guarantor (a former director) was not made aware of any variations in such dealings. Therefore an increase in the company's credit limit without the guarantor's knowledge did not allow the guarantor to escape from liability for claims in excess of the original credit limit.

An indulgence or anti-discharge clause may not prevent the guarantor being released from liability as a consequence of a post-guarantee variation where the variation is of a kind or scope which goes beyond the parties' reasonable contemplation and so lies outside the 'purview' of the guarantee, see *IMC Raffles Offshore (Singapore) Ltd and another* v. *Schahin Holding SA* (2013).

Further limitations of an indulgence or anti-discharge clause were illustrated in *Azimut-Bennetti SpA* v. *Healey* (2010). In that case, the guarantee contained an anti-discharge clause to the effect that the liability of the guarantor was not to be discharged by reason of 'the irregularity, illegality, unenforceability or invalidity in whole or in part' of the underlying contract. The relevant claim was for liquidated damages for delay in construction and delivery of a super-yacht. The court rejected the defendant's contention that the liquidated damages clause was unenforceable as a penalty. However, it stated, obiter (i.e. as an aside), that had the liquidated damages clause been held to be unenforceable, the indulgence clause did not permit the creditor to recover the damages from the guarantor; if there was no obligation to pay the damages under the underlying contract, there could be no liability on the part of the guarantor.

23.1.5 Parent company guarantees

There is no industry style of parent company guarantee but typically the trigger for calling on a guarantee will be the default or insolvency of the contractor. The parent company or holding company may undertake to physically perform the contractor's obligations and/or pay such damages to the employer as arise from the contractor's default. Although it may be potentially unlimited in amount and in time (subject to the statutory prescriptive period), a parent company guarantee will often contain a specific limitation on the level of the guarantor's liability and the duration of the guarantee. The guarantee (being a cautionary obligation as explained in Section 23.1.4) will in any event be co-extensive with the contractor's liability under the underlying construction contract and so will be subject to any financial caps or liability periods in the construction contract. Indeed, the guarantee may confer liability limitations on the guarantor in addition to any liability limitations in the construction contract.

The SBC and the SBC/DB do not contain any provisions relating to parent company guarantees, so an employer who requires such a guarantee will require to add bespoke conditions to this effect as well as a form of guarantee.

In the case of the NEC3, Secondary Option X4 can be used to impose an obligation on the Contractor to provide a parent company guarantee. The required form of guarantee is to be set out in the Works Information.

23.2 Bonds

23.2.1 Introduction

A bond is usually provided up to a maximum sum of money, which will become payable in certain circumstances should one of the parties to a contract default. Normally, a performance bond will subsist only in respect of claims made prior to an end-date, typically either the date of practical completion or of making good defects. Thus, it will not normally, as a parent company guarantee often will, cover the cost of making good latent defects. See also Section 23.1.1.

23.2.2 Performance bonds

A performance bond is the most commonly used type of bond in the construction industry, and its main purpose is to enable the employer to secure completion of the works for which they have contracted without incurring additional costs due to the non-performance, default or insolvency of the contractor. What the grantor of the bond undertakes to perform will depend on the wording of the bond, e.g. its performance obligations may be to pay a sum equivalent to the loss or damage suffered by the employer, to pay the costs of employing another contractor to complete the works, or to pay a specified sum.

As with parent company guarantees, the SBC and the SBC/DB make no provision for performance bonds, so again bespoke drafting will be required. They do, however, contain optional clauses in relation to retention bonds and advance payment bonds. See Sections 23.2.6 and 23.2.7.

Where the NEC3 is used, Secondary Option X15 can be applied to impose on obligation on the Contractor to deliver a performance bond. The surety is to be a bank or insurer, which the Project Manager may refuse to accept if 'its commercial position is not strong enough to carry the bond'. The amount of the bond is to be stated in the Contract Data and the form set out in the Works Information (which should include the expiry date).

Performance bonds fall into two general categories, namely, on demand bonds and conditional bonds as described in the following sections. For a case considering the inter-relationship between payment made pursuant to a bond and the contractor's liability cap under the underlying contract, see *SABIC UK Petrochemicals Ltd* v. *Punj Lloyd Ltd* (2013).

23.2.3 On demand bonds

Unlike a conditional guarantee, an on demand bond is not a cautionary obligation but constitutes a primary and independent obligation not dependent on first establishing the liability of a third party, i.e. (in most cases) the contractor. An on demand bond is payable upon the creditor's demand without any requirement to prove default or the amount of damages. The bondsman is then obliged to pay up to the level of the demand, subject to any monetary limit to the bond itself and subject to the demand complying with the terms of the bond. This means, for example, that the rules described above applying to conditional guarantees in respect of variations to the underlying contract do not apply to on demand bonds.

Given the onerous liability on the part of the bondsman in the event of a purported on demand bond being called, it is not surprising that one of most common challenges to a demand is that the bond is not an on demand bond at all and so cannot be called without liability under the underlying contract first being established. In answering the question as to whether a bond is on demand or conditional, the terms of the relevant document must be construed objectively. A statement within the document that the bond is on demand is not by itself conclusive.

In *WS Tankship II B.V.* v. *The Kwangju Bank Ltd and another* (2011), it was held that the advance payment guarantee under consideration was an on demand guarantee having regard to a proper construction of its wording. No special words were necessary for the instrument to constitute an on demand guarantee, and the absence of wording typically found in an on demand guarantee, such as 'unconditionally' and 'primary obligor' was of marginal relevance.

The issue was also considered in *Wuhan Guoyu Logistics Group Co Ltd and another* v. *Emporiki Bank of Greece SA* (2012). Reversing the decision at first instance, the Court of Appeal held that where, as in this case, the instrument included wording which was indicative of it being both an on demand and a conditional guarantee, the court would not adopt an approach based on the highest number of 'pointers' to one or the other. Instead the correct approach was to consider whether the instrument satisfied the four criteria specified in *Paget's Law of Banking* (13th edition (2007), para 34.4) and, if it did, there would be a presumption that the instrument was an on demand bond. These criteria are that the instrument: (1) related to an underlying transaction between parties in different jurisdictions; (2) was issued by a bank; (3) contained an undertaking to pay 'on demand'; and (4) did not contain clauses excluding or limiting the defences available to the guarantor. Even though the relevant instrument did not meet the fourth criterion, since it met the others, the court held that it was indeed an on demand bond. This case is helpful insofar as it provides some certainty that, where the instrument meets all or most of the four criteria, there is a presumption that it will be construed as an on demand bond. However, that does not mean to say that where the instrument satisfies only one criterion, namely, an undertaking to pay 'on demand', that it will necessarily be found to be a conditional guarantee. In these circumstances, it seems we are thrown back to an objective construction of all the terms and, at the very least, this highlights the importance of clear and precise wording.

Where the formal requirements of an on demand bond are not complied with in relation to the making of a demand, for example where the creditor fails to follow a stipulation in the bond that any demand must be accompanied with copies of notices to the contractor relating to the breach, then the bondsman may be entitled to refuse to respond to the demand, see *AES-3C Maritza East 1 EOOD* v. *Credit Agricole and Another* (2011).

The creditor may also be bound to comply with any restrictions in the underlying contract in respect of making demands under the bond. In *Simon Carves Ltd* v. *Ensus UK Ltd* (2011), the court granted an interim injunction restraining the creditor from making a call on the on demand bond, notwithstanding the absence of any allegation of fraud, on the grounds that there was a *prima facie* case that under the terms of the underlying contract the bond was 'null and void'. It could be argued that is inconsistent with the decision in *Meritz Fire and Marine Insurance Co Ltd* (see below).

Subject to the above limited exceptions, a court will not normally prevent enforcement of an on demand bond in the absence of fraud, see *Edward Owen Engineering Ltd* v. *Barclays Bank International Ltd* (1978). The party challenging the demand must be able to show that the only realistic inference from the facts is that the demand was fraudulent and that the bondsman was aware of the fraud. In most cases the bondsman will be prepared to pay upon demand, without challenge, because it has obtained a counter-indemnity from the contractor.

The question also arises as to the consequences where a bond is improperly called. It is suggested that the employer should account to the contractor for the proceeds of the bond where the employer has not in fact suffered a loss in respect of the matter for which the bond was allegedly called. In *Speirsbridge Property Developments Ltd* v. *Muir Construction Ltd* (2008) it was held that a term could be implied into the contract that the duty to account for an overpayment made under a bond was owed by the employer, who had made the erroneous demand, to the contractor rather than the guarantor.

Where an instrument is truly an on demand bond, the demand must be honoured, in the absence of fraud and, if properly made, without reference to the relationship between the parties to the underlying transaction and to the terms of the underlying contract itself. Thus variations to the underlying contract and changes to the corporate identity of the debtor by means of novation of the underlying contract, which had the instrument been a conditional guarantee might have discharged the guarantor for the reasons described in Section 23.1.4, will not affect liability under an on demand bond, see *Meritz Fire and Marine Insurance Co Ltd* v. *Jan de Nul and another* (2011); and also *WS Tankship II B.V.*

23.2.4 Conditional or default bonds

Unlike an on demand bond, a conditional bond will normally provide that it can only be called upon the occurrence of certain events relative to the contractor's obligations under the principal construction contract. If there is no liability under the principal contract, there will be no liability under the bond.

A conditional bond is payable upon the creditor's demand, which will usually require to be accompanied by evidence that the condition entitling a call on the bond has been satisfied. The creditor may also be required to provide evidence that the amount claimed reflects the actual amount of damages it has suffered. In *The Royal Bank of Scotland Ltd* v. *Dinwoodie* (1987), the bondsman paid out monies pursuant to a performance bond. The bond was supported by a counter-indemnity from individual guarantors who argued that the bank was wrong to make payment under the performance bond. On the wording of the bond, which guaranteed 'the damages sustained by the employer' by the contractor's default, it was held by the Court of Session that the surety bank was obliged to be satisfied, at the very least, that damages had been sustained by the employer and also as to the quantification of those damages.

A conditional or default bond, unlike an on demand bond, is a cautionary obligation and so the principles applying to such obligations described in Section 23.1.4 will apply to this type of bond.

The triggers for calling the bond should be clearly set out in the bond. These may include, for example, the employer establishing that the contractor is in breach of contract and the extent of the damages arising from such breach; the insolvency of the contractor; or presentation of an arbiter's award or court decree or possibly the decision of an adjudicator pursuant to the 1996 Act, see *Beck Interiors* v. *Russo*.

23.2.5 The ABI model form bond

There are a number of standard forms of bond and one of the most commonly used is the model form of the Association of British Insurers (ABI). This was published in 1995 (and revised in 2004) primarily in response to the Court of Appeal decision in *Trafalgar House Construction (Regions) Ltd* v. *General Surety and Guarantee Co. Ltd* (1995), in which a form of bond then in common use (described by the Court of Appeal as 'archaic') was treated effectively as an on demand bond, despite earlier assumptions to the contrary. This decision was later overruled by the House of Lords, which followed *inter alia* the decision of *City of Glasgow District Council* v. *Excess Insurance Co. Ltd* (1986) that a performance bond in similar terms was a cautionary obligation.

The ABI bond provides that the guarantor will 'satisfy and discharge the damages sustained by the employer as established and ascertained pursuant to and in accordance with the provisions of or by reference to the contract'. Thus, the employer's entitlement is linked expressly to the contract itself. If, as with most standard forms, the contract contains a mechanism for ascertainment of loss and damages following breach by the contractor, then this mechanism must be followed before any money is payable under the bond. However, the wording of the bond does not go so far as to state that the employer has to establish the amount of his loss, if necessary by going to court or to arbitration, and it therefore leaves some uncertainty as to exactly at what stage and in what circumstances the bond can be called.

The ABI bond expressly limits liability as to time and money. The expiry date for making a call on the bond is a matter for negotiation but typically the date will be stated as the date of practical completion of the works or the issue of the certificate of completion of making good defects.

The bond contains a prohibition on assignation by the employer without the prior written consent of the guarantor and the contractor. In practice, this can lead to difficulties where there is a change in the employer, either due to novation or because a funder has stepped into the building contract on the employer's default. This particular difficulty can be overcome by inserting additional wording stating that the bond will be assignable to any successor to the employer under the contract.

The ABI bond also makes clear that the bond operates as a guarantee, i.e. it is accessory to the principal contract. Therefore, if there is no liability or limited liability under the principal contract, liability under the bond will be similarly excluded or limited. In order to avoid the potential difficulties associated with cautionary obligations referred to above, it is made clear that the guarantor shall not be discharged or released by any alteration of any of the terms, conditions and provisions of the principal contract.

One of the most likely situations in which an employer would wish to call on a bond is the insolvency of the contractor. This will of course depend upon the wording of the bond but in *Perar BV* v. *General Surety and Guarantee Co. Ltd* (1994) the Court of Appeal held that insolvency and consequent termination of the contractor's employment were not a breach of contract which could trigger the bond. Although a termination event, insolvency was not itself a breach of the contract and the right to call for payment of the bond would not arise until, for example, the contractor had failed to make payment of any sums consequently due to the employer.

The ABI bond does not provide that the contractor's insolvency would allow the employer to demand payment of the bond amount, which some might view as limiting the efficacy of this form of bond. This difficulty is often addressed by the insertion of additional wording to specify that insolvency will be treated as a breach of contract by the contractor and establishing a method of determining what level of damages is then payable to the employer.

23.2.6 Retention bonds

Retention bonds are becoming more commonplace, as an alternative to the employer making a cash retention from the contract sum, and have obvious cash-flow attractions for contractors. The bond, backed up by a bank or insurance company, will secure the level of retention until the contractual date for release. To ensure that the employer has the same level of security as if it has made a cash retention, the retention bond will normally be on demand (or at least will be drafted with that intention). The bond will also normally provide that the maximum amount of the bond will reduce at practical completion by the same level as retention would have reduced, had it been applicable. If the contractor fails to honour its obligations to remedy defects, the employer can call upon the bondsman to pay the requisite sum up to the maximum amount of the bond. Clause 4.19 of the SBC and clause 4.17 of the SBC/DB contain optional drafting for the provision of a retention bond and a form of retention bond is to be found in Part 2 of Schedule Part 6 to the SBC and the SBC/DB. The NEC3 does not make provision for a retention bond.

23.2.7 Advance payment bonds

Clause 4.8 of the SBC and clause 4.6 of the SBC/DB provide that, if so stated in the Contract Particulars, an advance payment may be made by the Employer to the Contractor which shall be reimbursed on the terms set out in the Contractor Particulars. Although this provision is not often used, it may be of value in cases where the Contractor requires to expend significant amounts of money, e.g. for pre-ordering specialist materials, at the commencement of the contract. In such circumstances, the Employer may obtain security for reimbursement of the advance payment by means of an advance payment bond, a form of which is contained in Part 1 of Schedule Part 6 to the SBC and the SBC/DB.

Similar arrangements can be made under the NEC3 where Secondary Option X14 is used. The form of bond is not provided but is to be set out in the Works Information.

Tables of Cases

The following abbreviations of Reports are used:

AC	Law Reports, Appeal Cases
All ER	All England Law Reports
BLR	Building Law Reports
CH	Law Reports, Chancery
CILL	Construction Industry Law Letter
CLD	Construction Law Digest
Con LR	Construction Law Reports
Const LJ	Construction Law Journal
CSIH	Court of Session Inner House
CSOH	Court of Session Outer House
D	Dunlop's Session Cases 1838–62
DLR	Dominion Law Reports
ECR	European Court Reports
EGLR	Estate Gazette Law Reports
EuLR	European Law Reports
EWCA	England and Wales Court of Appeal Cases
EWHC	England and Wales High Court Cases
F	Fraser's Session Cases 1898–1906
FC	Faculty Collection
FCR	Family Court Reports
GWD	Greens Weekly Digest
H&N	Hurlstone & Norman
Hudson's BC	Hudson's Building Contracts
Lloyd's Rep	Lloyd's Law Reports
M	Macpherson's Session Cases 1862–73
Macq	Macqueens Scotch Appeal Cases
Mor	Morison's Dictionary of Decisions
PD	Law Reports, Probate Division
PNLR	Professional Negligence Law Reports

MacRoberts on Scottish Construction Contracts, Third Edition. MacRoberts.

QB	Law Reports, Queens Bench
SLR	Scottish Law Reporter
S	P Shaw's Session Cases 1821 – 38
SC	Session Cases 1907 –
SCLR	Scottish Civil Law Reports
SGCA	Singapore Court of Appeal
Sh Ct Rep	Sheriff Court Reports
SLT	Scots Law Times
TCLR	Technology and Construction Law Reports
UKHL	United Kingdom, House of Lords
UKSC	United Kingdom, Supreme Court
WLR	Weekly Law Reports
W&S	Wilson & Shaw

Table of Statutes

MacRoberts on Scottish Construction Contracts, Third Edition. MacRoberts.

Table of Statutory Instruments

MacRoberts on Scottish Construction Contracts, Third Edition. MacRoberts.

Table of References

JCT Standard Form of Building Contract (2011 Edition) and the Standard Building Contract with Quantities for Use in Scotland SBC/Q/Scot (Revised November 2011)

MacRoberts on Scottish Construction Contracts, Third Edition. MacRoberts.

Table of References to the Design and Build Contract for Use in Scotland, DB/Scot 2011 (Revised November 2011)

Table of References to the NEC3 Engineering and Construction Contract (Revised September 2011)

Index

MacRoberts on Scottish Construction Contracts, Third Edition. MacRoberts.

Also of interest

The JCT Standard Building Contract 2011
David Chappell
Paperback 978 1 118 81975 3
ePub

The Expert Witness in Construction
Robert Horne and John Mullen
Hardback 978 0 470 65593 1
ePub

Cornes & Lupton's Design Liability in the Construction Industry
Fifth Edition
Sarah Lupton
Hardback 978 1 4443 3006 9
ePub

Evaluating Contract Claims
Second Edition
Peter Davison & John Mullen
978 1 4051 5920 3
ePub

Building Contract Claims
Fifth Edition
David Chappell
978 0 470 65738 6
ePub

A Practical Guide to the NEC3 Engineering and Construction Contract
Michael Rowlinson
978 1 4443 3688 7
ePub

The NEC3 Engineering and Construction Contract: A commentary
Second Edition
Brian Eggleston
978 0 632 05386 5
ePub

The FIDIC Form of Contract
Third Edition
Nael Bunni
978 1 4051 2031 9
ePub

Building Law Encyclopaedia
David Chappell, Michael Cowlin and Michael Dunn
978 1 4051 8724 4
ePub